Handbook of Research on Technologies and Systems for E-Collaboration During Global Crises

Jingyuan Zhao
University of Toronto, Canada

V. Vinoth Kumar
Jain University, India

A volume in the Advances in Social Networking and Online Communities (ASNOC) Book Series

Published in the United States of America by
IGI Global
Information Science Reference (an imprint of IGI Global)
701 E. Chocolate Avenue
Hershey PA, USA 17033
Tel: 717-533-8845
Fax: 717-533-8661
E-mail: cust@igi-global.com
Web site: http://www.igi-global.com

Library of Congress Cataloging-in-Publication Data

Names: Zhao, Jingyuan, 1968- editor. | Kumar, V. Vinoth, 1988- editor.
Title: Handbook of research on technologies and systems for e-collaboration during global crises / Jingyuan Zhao and V. Vinoth Kumar, editors.
Description: Hershey, PA : Information Science Reference, 2021. | Includes bibliographical references and index. | Summary: "This book focuses on emerging technologies and systems, strategies and solutions for e-collaboration especially highlighting the importance of technologies and systems for e-collaboration in dealing with emerging crisis such as unpredicted pandemic diseases like COVID -19"-- Provided by publisher.
Identifiers: LCCN 2021035395 (print) | LCCN 2021035396 (ebook) | ISBN 9781799896401 (hardcover) | ISBN 9781799896425 (ebook)
Subjects: LCSH: Business enterprises--Computer networks. | Business networks. | Communication--Technological innovations. | Organizational change. | COVID-19 (Disease)--Economic aspects.
Classification: LCC HD30.37 .T434 2021 (print) | LCC HD30.37 (ebook) | DDC 658/.05--dc23
LC record available at https://lccn.loc.gov/2021035395
LC ebook record available at https://lccn.loc.gov/2021035396

This book is published in the IGI Global book series Advances in Social Networking and Online Communities (ASNOC) (ISSN: 2328-1405; eISSN: 2328-1413)

British Cataloguing in Publication Data
A Cataloguing in Publication record for this book is available from the British Library.

Advances in Social Networking and Online Communities (ASNOC) Book Series

Hakikur Rahman
Ansted University Sustainability Research Institute, Malaysia

ISSN:2328-1405
EISSN:2328-1413

Mission

The advancements of internet technologies and the creation of various social networks provide a new channel of knowledge development processes that's dependent on social networking and online communities. This emerging concept of social innovation is comprised of ideas and strategies designed to improve society.

The **Advances in Social Networking and Online Communities** book series serves as a forum for scholars and practitioners to present comprehensive research on the social, cultural, organizational, and human issues related to the use of virtual communities and social networking. This series will provide an analytical approach to the holistic and newly emerging concepts of online knowledge communities and social networks.

Coverage

- Community Practices
- Learning Utilities
- Knowledge Management Practices and Future Perspectives
- Measuring and Evaluating Knowledge Assets
- Meta-data Representation and Management (e.g., Semantic-Based Coordination Mechanisms, Use of Ontologies, etc.)
- Networks and Knowledge Communication in R&D Environments
- Whole-Network Properties and Knowledge Communication
- Leveraging Knowledge Communication in Social Networks
- Knowledge Management System Architectures, Infrastructure, and Middleware
- Best Practices for Mobile Computing

IGI Global is currently accepting manuscripts for publication within this series. To submit a proposal for a volume in this series, please contact our Acquisition Editors at Acquisitions@igi-global.com or visit: http://www.igi-global.com/publish/.

The Advances in Social Networking and Online Communities (ASNOC) Book Series (ISSN 2328-1405) is published by IGI Global, 701 E. Chocolate Avenue, Hershey, PA 17033-1240, USA, www.igi-global.com. This series is composed of titles available for purchase individually; each title is edited to be contextually exclusive from any other title within the series. For pricing and ordering information please visit http://www.igi-global.com/book-series/advances-social-networking-online-communities/37168. Postmaster: Send all address changes to above address.

Titles in this Series

For a list of additional titles in this series, please visit: http://www.igi-global.com/book-series/advances-social-networking-online-communities/37168

Information Manipulation and Its Impact Across All ndustries
Maryam Ebrahimi (Independent Researcher, Germany)
Information Science Reference • © 2022 • 234pp • H/C (ISBN: 9781799882350) • US $195.00

E-Collaboration Technologies and Strategies for Competitive Advantage Amid Challenging Times
Jingyuan Zhao (University of Toronto, Canada) and Joseph Richards (California State University, Sacrameto, USA)
Information Science Reference • © 2021 • 346pp • H/C (ISBN: 9781799877646) • US $195.00

Analyzing Global Social Media Consumption
Patrick Kanyi Wamuyu (United States International University – Africa, Kenya)
Information Science Reference • © 2021 • 358pp • H/C (ISBN: 9781799847182) • US $195.00

Global Perspectives on Social Media Communications, Trade Unionism, and Transnational Advocacy
Floribert Patrick C. Endong (University of Calabar, Nigeria)
Information Science Reference • © 2020 • 300pp • H/C (ISBN: 9781799831389) • US $195.00

Electronic Hive Minds on Social Media Emerging Research and Opportunities
Shalin Hai-Jew (Kansas State University, USA)
Information Science Reference • © 2019 • 358pp • H/C (ISBN: 9781522593690) • US $205.00

Hidden Link Prediction in Stochastic Social Networks
Babita Pandey (Lovely Professional University, India) and Aditya Khamparia (Lovely Professional University, India)
Information Science Reference • © 2019 • 281pp • H/C (ISBN: 9781522590965) • US $195.00

Cognitive Social Mining Applications in Data Analytics and Forensics
Anandakumar Haldorai (Sri Eshwar College of Engineering, India) and Arulmurugan Ramu (Presidency University, India)
Information Science Reference • © 2019 • 326pp • H/C (ISBN: 9781522575221) • US $195.00

Modern Perspectives on Virtual Communications and Social Networking
Jyotsana Thakur (Amity University, India)
Information Science Reference • © 2019 • 273pp • H/C (ISBN: 9781522557159) • US $175.00

701 East Chocolate Avenue, Hershey, PA 17033, USA
Tel: 717-533-8845 x100 • Fax: 717-533-8661
E-Mail: cust@igi-global.com • www.igi-global.com

List of Contributors

Table of Contents

Detailed Table of Contents

Section 1
COVID-19: Smart Healthcare Applications Using E-Collaboration During a Global Crisis

Section 1 introduces smart healthcare systems as well as their innovations and applications that are responsible for making complete health monitoring together disease diagnosis using e-collaboration tools. Section 1 is organized into 10 chapters.

Acute lymphocytic leukemia (ALL) is a variety of malignant somatic cell cancer that influences children and teenagers. The goal of the study is to create a system that can detect cancer from blood corpuscle images mechanically. This method employs a convolutional network that takes images of blood corpuscles and determines whether or not the cell is cancer infected. The appearance of cancer in blood corpuscle images is frequently ambiguous, overlaps with other diagnosis, and can be mistaken for a variety of benign abnormalities. Machine-assisted cancer identification from blood corpuscle images at the level of skilled medical staff would be extremely beneficial in clinical settings and also in the delivery of healthcare to populations with limited access to diagnostic imaging specialists. Here, the authors proposed a convolutional neural network (CNN)-based methodology to distinguish between outdated as well as irregular somatic cell photos. With the dataset and 1188 somatic cell images, the proposed methodology achieves an accuracy of up to 96.6%.

The spinal cord and heart in the body are major organs. Diagnosis of diseases in these organs is very complex using MRI and CT images. The conventional methods like post segmentation, pre-image processing, and text feature extraction mechanisms cannot handle accurate diagnosis. Therefore, advanced techniques are needed. In this work, pixel-based convolution neural networks with centered deep learning processes are proposed to cross over the problems. The projected PCNN has four pixel-based convolution neural networks. Here disease objects are identified through grading framework. The entire mechanism is working based on sequential part of PCNN segmentation process. The spinal cord and heart image MRI-based diagnosis process is very difficult with conventional methods. But the proposed method provides accurate results and outperforms the standard methodology performance measures in accuracy, precision, and F1score.

In our busy lives, most of us hardly have time to read books. This habit of reading is slowly diminishing because of people's busy lives. The situation is significantly difficult for persons who are visually challenged or have lost their vision. As a result, the authors provide a method based on the sense of sound that is better and more accurate than the sense of touch for visually impaired people. This chapter discusses an effective method for condensing books into important keywords in order to avoid having to read the entire text each time. This work employs a variety of APIs and modules, including Gensim, Text Ranking Algorithm, and other functions for translating summary text to speech, allowing the system to assist even the blind.

Rama Krishna Garigipati, Koneru Lakshmaiah Education Foundation, India
Kasula Raghu, Mahatma Gandhi Institute of Technology, India
K. Saikumar, Koneru Lakshmaiah Education Foundation, India

This chapter proposes that employee attrition is the major circumstance faced in many organizations. Usually, organizations face this attrition when there is pressing need of employees due to mass retirements or while expanding the organization. Generally, any organization faces higher attrition rate for employment when they have more employment opportunities in market or recession time. Due to the demand for software goods across all industries, the software industry once suffered a significant attrition rate from employers due to large openings globally in the software business. The purpose of this research is to look at how objective elements influence employee attrition in order to figure out what factors influence a worker's decision to leave a company and to be able to predict whether a particular employee will leave the company using machine learning algorithms.

Swapna B., Dr. M. G. R. Educational and Research Institute, India
S. Manivannan, Dr. M. G. R. Educational and Research Institute, India
M. Kamalahasan, Dr. M. G. R. Educational and Research Institute, India

A sensor centers on using detectors beneath the surface of the soil. The applications require the sending of sensors beneath the ground surface. Henceforth, the sensors turn out to be a piece of the detected condition and may convey more exact detecting. Sensors like NPK (nitrogen, phosphorus, and potassium), soil moisture, and humidity are underground and impart through soil. Most of the applications for sensors are shrewd farming, natural observing of the soil, etc. In this chapter, moisture substance, NPK level of the soil in land is estimated utilizing the sensors, which send it to the centralized server through internet of things for checking. The authors introduce propelled channel models to portray the underground remote channel to consider the qualities of the expansion of electromagnetic waves in the soil. From this detection of soil, one can increase crop production as per the wealth and nutrient levels of soil.

Section 2
E-Collaboration Techniques: Exploration of Current and Future Implications

Section 2 improves the understanding of technology-supported collaboration in order to achieve individual and organizational success with the adoption, use, and implementation of e-collaboration in a pandemic and post-pandemic world. Section 2 is organized into 16 chapters.

This chapter focuses on uses and gratifications of social media use among college students in the United Arab Emirates and in Kuwait for three social media platforms: Twitter, Instagram, and Snapchat. Mixed methodologies are duly applied (quantitative and qualitative) to explore various use and gratifications factors, as well as other social factors among a youth that contributes to the adoption of these social network sits (SNSs). Moreover, several statistical tests were performed to analyze collected data. A few research articles are published about new and social media platform use in the region; however, comparative studies were rarely noticed regarding this subject. The survey includes (N=190) samples between Kuwaiti and Emirati students. Conclusively, the study reveals that the main use and gratification reason for using the abovementioned social media platforms amongst college youth is entertainment, while the main social reason is identification.

In a smart system, a software-defined network (SDN) is frequently used to monitor and manage the communication organisation. Large-scale data analysis for SDN-based bright networks is gaining popularity. It's a potential technique to deal with a large amount of data created in an SDN-based shrewd lattice using AI advancements. Nonetheless, the disclosure of personal security information must be considered. Client power conduct examination, for example, may result in the disclosure of personal security information due to information bunching. Clustering is an approach for displaying models' observations, data items, or feature vectors in groups. Batching addresses has been catered to in various interesting circumstances and by masters in distinct requests; it gleams far-reaching attractiveness and assistance as one of the ways in exploratory data examination and moreover increases the genuine assessment of data. In this chapter, the authors conduct a study of packing and its various types and examine the computation. Finally, they use it to create an outline model.

Big data analytics as well as data mining play vital roles in extracting the hidden statistics. Customary advances for investigation and extraction of hidden information from data may not exert efficiently for big data because of its complex, elevated volume nature. Data clustering is a data mining technique that exacts the useful data from the data by grouping data into clusters. In big data as the data is complex and of very large volume, individual clustering techniques may not consider all the samples, which may lead to inaccurate results. To overcome this inaccuracy, the proposed method is the combination of dynamic

attack, which enables security threats to amplify and spread quickly in the cloud; injecting malicious code; and finally information deletion. MapReduce and dynamic decomposition-based distributed algorithm with the help of Hadoop and JavaBeans in the live forensic method is used to find solution to the problem. MapReduce is a software framework and live forensics is the method attempting to discover, control, and eliminate threats in a live system environment. This chapter uses Hadoop's cloud simulation techniques that can give a live result.

Education is an essential factor in the development of a nation. We should make it appropriate according to the changing scenario of the country. Learning is an opportunity to reflect upon the social, economic, cultural, and moral issues faced by human beings. Nowadays most of the institutions are working only for a degree, students and teachers are running after attaining or providing degrees, and not towards knowledge and wisdom. A good teacher can bring the entire world to the classroom. The overall scenario of higher education in India does not match the global quality standards. Education has to develop appropriately according to the time and changing scenarios of the world. It contributes to national development through specialized knowledge and skills. So, higher education has to come out of the static environment and become more dynamic and more futuristic. The solution to all the problems is providing quality education, and teachers are the main ingredients in giving quality education.

The emergence of embedded and connected smart devices, systems, and technologies has given rise to the concept of smart cities in modern metropolises. They've made it possible to connect "anything" to the internet. As a result, the internet of vehicles (IoV) will play a critical role in newly constructed smart cities in the approaching internet of things age. The IoV has the ability to efficiently tackle a variety of traffic and road safety issues in order to prevent fatal crashes. Furthermore, ensuring quick, secure transmission and accurate recording of data is a particular problem in the IoV, particularly in vehicle-to-vehicle (V2V) and vehicle-to-infrastructure (V2I) connections. Furthermore, the authors qualitatively examined the suggested overall system performance and resiliency against popular security assaults. The proposed method solves the primary issues of vehicle-to-x (V2X) communications, such as centralization, lack of privacy, and security, according to computational experiments.

Instead of sending human beings into volcanoes, drone-bot is used to measure the live lava temperature, and it alerts the ground station to protect people near the surroundings. The thermocouple is used as a temperature sensor. It can measure a wide range of higher temperatures, and it can be interfaced with the TTGo T-Call development board to process and send the temperature data to the ground station through GSM as short message service (SMS). Also the ESP-32 CAM is interfaced with that development board to capture the snapshot of the mountain if the temperature is high and the same snap is shared to the ground station through Wi-Fi. The GPS module is also interfaced with the development board to know the location of the volcano.

The process of incorporating robotic technology and autonomous vehicles are increasing in all applications where for all real-time application developments time and energy can be saved for every single movement transfer as compared to human classifications. Thus, considering the advantage of autonomous process without any presence of an individual, the supply chain management can be designed using robotic technology. The robotic technology provides an informal route where all goods can be transported to different places within a short span of time, and any false identification in transfer of goods can also be easily identified. To drive the autonomous vehicle towards correct location, a precise protocol is chosen, which is termed common industrial protocol (CIP) where proper solutions can be achieved for all control applications using time synchronization model. Further, the data monitoring process is trailed using an online contrivance which is termed as internet router (IR) where short distance can be identified using corresponding addressing scheme.

Preface

Today more and more organizations are utilizing new technologies to become more efficient and employees frequently collaborate by using e-tools at work. E-collaboration (Electronic collaboration) is defined as teamwork that is facilitated by the use of electronic technologies to accomplish a common task (Kock & D'Arcy, 2002; Knock, 2008; Linnes, 2020). E-collaboration is conducted without face-to-face interaction among individuals or members of virtual teams engaged in common task using information and communication technologies. Taking North America as example, 93.9% of the population is connected to the internet, with the world average at 65.6%, which makes it possible to collaborate at a distance (Internet World Stats, 2021).

Technologies and systems for e-collaboration refer to tools and platforms designed to better facilitate group work remotely and enable comprehensive distance collaboration for product development, manufacturing, and marketing. Because of their popularity, most major vendors are competing to enter this market or increase their market share in this fast-growing field (Bidgoli, 2012). Collaboration technologies and systems are increasingly transforming work and organizations, bringing changes in both the work processes and practices of individual employees and organizations and the spaces where work takes place (OECD, 2016; Flecker, 2010; Schubert & Williams, 2022). To provide support for collaborative work, organizations are combining multiple systems, tools and applications for the different forms of collaborative activities. In parallel to these changing work arrangements, the range of tools and systems that are available to support collaborative work has grown rapidly. These trends are also reflected in global market forecasts for collaboration technologies. In 2021, the global collaboration technologies market is forecast at $US4.5 billion, representing an increase of 17.1% from 2020 levels, with upward growth predicted to continue into 2022 to $US5.1 billion, a further increase of 14% from 2021 levels (Gartner, 2021; Schubert & Williams, 2022).

COVID-19 (Novel Coronavirus), the most recent pandemic that is ravaging in many parts of the world with untold human loss and suffering has resulted in an unprecedented global economic upheaval. Compared with the financial crisis of 2008, the size and rapidity of the coronavirus crisis are bigger and have a broader global impact (Cassidy, 2020). The COVID-19 pandemic impacts almost every individual, business, government and institution in a unique way, a crisis such as this tends to expose the weakest parts and links in an economy and a society and wreaks havoc on them. The pandemic has demonstrated that it is not only an issue of health care, but also a huge challenge for the global economy, business and society. COVID-19 worsens several preexisting conditions in the global economy and poses greater challenges to the more vulnerable parts of the global economy (Song & Zhou, 2020). This crisis affects and transforms global and regional economic relationships, geopolitical constellations and alliances, business strategies and competitive positions, national politics and priorities, technological paradigms and forms of communication (Karabag, 2019, 2020; Hafiz, Oei, Ring & Shnitser, 2020).

The proliferation of the COVID-19 has created a global, regional, national, political, societal, economic and commercial crisis. The crisis could be characterized as a disruptive period of instability, uncertainty, and danger. But the crisis can also be perceived as a period of accelerated diffusion of digital technologies and micro-level initiatives, and a consideration of established resource-intensive forms of communication and globalized supply and outsourcing chains (Karabag, 2020; Mitchell, 2021). Obviously, the COVID-19 pandemic has put a major emphasis on the essential impact that technologies and systems for e-collaboration play in an organization's overall strategy implementation.

The *Handbook of Research on Technologies and Systems for E-Collaboration During Global Crises* aims to explore e-collaboration technologies and systems affordances from virtual collaboration and remote work during global crises. The purpose of this exploration is to improve the understanding of technology-supported collaboration in order to achieve individual and organizational success with the adoption, use and implementation of e-collaboration in a pandemic and post-pandemic world. This book identifies topics of e-collaboration success as well as challenges related to organizational transitions during COVID-19, provides insight into the complexities of e-collaboration in these areas while also making recommendations for the post-pandemic future. This research makes a contribution through the analysis of a unique set of data elaborating on participant experiences during a global pandemic as well as through the exploration of future implications.

This handbook focuses on technologies and systems, tools and platforms, strategies and solutions for e-collaboration. This book assesses the importance of technologies and systems for e-collaboration in dealing with global crises. Covering topics such as deep learning processes, machine vision, profit-sharing models, Internet of Things, Big Data, social networks, security challenges, it is an essential resource for computer scientists, public officials, engineers, students and professors of higher education, healthcare administration, programmers, researchers, and academicians.

ORGANIZATION OF THE BOOK

This book has been divided into two sections:

Section 1, "COVID-19: Smart Healthcare Applications Using E-Collaboration During Global Crisis," introduces smart health care systems, as well as their innovations and applications that are responsible for making complete health monitoring together disease diagnosis using e-collaboration tools. Section 1 is organized into 10 chapters. A synopsis of each chapter is given below.

Chapter 1 (Deep-CNN Model for Acute Lymphocytic Leukemia [ALL] Classification Using Microscopic Blood Images: Global Research) proposes a Convolutional Neural Network (CNN) mainly based methodology to distinguish between outdated as well as irregular somatic cell photos. With the dataset and 1188 somatic cell images, the proposed methodology achieves an accuracy of up to 96.6 percent.

Chapter 2 (A PCCN-Based Centered Deep Learning Process for Segmentation of Spine and Heart: Image Deep Learning) claims that the projected PCNN has four pixel-based convolution neural networks, here disease objects are identified through grading framework. The proposed method providing accurate results and outperforms the methodology. At final calculating the performance measures such as accuracy, precision and F1score.

Chapter 3 (Multilingual Novel Summarizer for Visually Challenged Peoples) discusses an effective method for condensing books into important keywords in order to avoid having to read the entire text each time. This work employs a variety of APIs and modules, including Gensim, Text Ranking Algorithm, and other functions for translating summary text to speech, allowing the system to assist even the blind.

Chapter 4 (An India's Remote Medical Monitoring System Using Big Data and MapReduce Hadoop Technologies: Big Data With Healthcare) discusses the role of Big Data Analytics and Hadoop, as well as their effect on providing healthcare services to all at the lowest possible cost.

Chapter 5 (A Secure and Effective Image Retrieval Based on Robust Features) discusses the most typical approaches, which are Content-Based Image Retrieval systems. Content primarily based picture retrieval may be one in all the image retrieval techniques that uses user visual options of an image like color, form, and texture. The objective is to retrieve the set of pictures quickly and economically supported color and texture options.

Chapter 6 (Computer Vision for Weed Identification in Corn Plant Using Modified Support Vector Machine) presents three procedures such as segmentation, feature extraction and classification, for weed plant identification in detail. The proposed modified SVM was compared with CNN using performance analyzes and produce high accuracy of 98.56% compared to existing systems. The farmers are expected to adopt these technologies to overcome the agricultural problems.

Chapter 7 (AI-Based Motorized Appearance Acknowledgement Scheme for Attendance Marking System) proposes that with the help of deep learning and datasets scheme senses the position of the creature look and recognizes which appearance goes to which ID formerly spot the attendance in the datasheet. Then exportation the presence as excel sheet then excluding it to exact position. All apprehend resemblance and datasets are protected to the attendants.

Chapter 8 (Analysts and Detection of Concealed Weapon Using IR Fusion With MMW Support Imaging Technology) presents a study for concealed weapon detection using Passive Millimeter Wave Imaging Sensors combining with Image Processing and Convolutional Neural Networks. This eliminates the ambiguity of using Millimeter Wave Imaging alone. The proposed system will perform the fusion of Passive MMW images with corresponding IR images followed by YOLO and VGG Net detection models.

Chapter 9 (Detection and Identification of an Employee Attrition Using Machine Learning Algorithm: Machine Learning) discusses how objective elements influence employee attrition in order to figure out what factors influence a worker's decision to leave a company and to be able to predict whether a particular employee will leave the company using Machine Learning Algorithms.

Chapter 10 (IoT-Based Design and Execution of Soil Nutrients Monitoring) introduces propelled channel models to portray the underground remote channel to consider the qualities of the expansion of Electromagnetic waves in the soil. From this detection of soil, authors increase crop production as per the wealth and nutrient level of soil.

Section 2, "E-Collaboration Techniques Exploration of Current and Future Implications," provides to improve the understanding of technology-supported collaboration in order to achieve individual and organizational success with the adoption, use and implementation of e-collaboration in a pandemic and post-pandemic world. Section 2 is organized into 16 chapters. A synopsis of each chapter is given below.

Chapter 11 (Efficient Data Verification Systems for Privacy Network) presents and compares various verification strategies based on the crypt arithmetic methodology for various set-valued data. It primarily checks for privacy risks in the sharing of details and information between the publisher, admin, and customers, and automatically stops the person who is attempting to inject the vulnerability code using the technique, and all of this information is kept in the database.

Chapter 12 (Profit Sharing Models for Social Media in Big Data Commercialized Crises) explores the users' motivation for using social network media and their opinions on the issues related to data collected, privacy and profit sharing. This study helps understand the common views of users of social media platforms on the collection and sharing of their data.

Chapter 13 (Study of Social Media Indulgence Among College Students in UAE and Kuwait: Case Study) focuses on uses and gratifications of social media use among college students in the United Arab Emirates and in Kuwait for three social media platforms: Twitter, Instagram and Snapchat. The study reveals that the main use and gratification reason for using the abovementioned social media platforms amongst college youth is entertainment, while the main social reason is identification

Chapter 14 (Efficient Data Clustering Techniques for Software-Defined Network Centre) claims that batching addresses numerous has been catered to in various interesting circumstances and by masters in distinct requests; it gleams far-reaching attractiveness and assistance as one of the ways in exploratory data examination and moreover increases the genuine assessment of data. This chapter proposes to conduct a study of packing and its various types, as well as examine the computation. Finally, we use it to create an outline model.

Chapter 15 (Hybrid Clustering Technique to Cluster Big Data in Hadoop Ecosystem: Big Data Application) presents that in Big Data as the data is complex and of very large volume, individual clustering techniques may not consider all the samples it may leads to inaccurate results. To overcome this inaccuracy this proposed method is the combination of dynamic k-means and hierarchical clustering algorithms. This proposed method can be called as hybrid method. Being hybrid method will overcome few drawbacks like static k value.

Chapter 16 (Planning a Three-Year Research Based on the Community of Inquiry Theory: An Approach to Monitor the Learning of English as a Second Language in the EU Academic Environments) aims to take advantage of five annual English as a Second Language (ESL) courses, in two faculties of the "Sapienza" University of Rome, during a three-year period. The research plan takes shape within the Community of Inquiry Framework (CoI) and this theory encompasses the conceptual boundaries.

Chapter 17 (Mining Perspectives for News Credibility: The Road to Trust Social Networks) introduces a background that highlighted the related aspects, the relation between these domains, and the news credibility. The study also presents the recent research in these fields with demonstrating the roles of these techniques for the required study target. The study could support as the foundation of future text mining studies on social networks data.

Chapter 18 (Design and Develop a Decision-Making Assistance Model for Agriculture Product Price Prediction: Deep Learning) develops a decision-assistance model for agricultural product price forecasting. Farming decisions may be made using this method, which takes into consideration elements like yearly rainfall, WPI, and so on. A year's worth of forecasts are available from the technology. The system employs a machine learning regression approach known as decision tree regression.

Chapter 19 (Prediction and Prevention of Malicious URL Using ML and LR Techniques for Network Security: Machine Learning) detects the malicious URLs by using machine learning algorithm SVM. One of the essential features that an apparatus acquire is to permit the benevolent URLs that are demanded by the customer and avoid the malicious URLs before attainment the user. Blacklisting is one of the basic and trivial mechanisms in detecting malicious URLs.

Chapter 20 (Security Challenges in Internet of Things) analyses several challenges for Secure in IoT, Security Concerns in IoT, and which Safeguard the IoT assets viz devices and data in contradiction of hacking and stealing. At last, we will scrutiny the challenges were integrated in WSN and IOT for ecosystem observing.

Chapter 21 (A Forensic Way to Find Solution for Security Challenges in Cloud Server Through MapReduce Technique) claims that cloud computing is a large and distributed platform repository of user information. But it also extensively serves the security threats in the research aspect. This chapter attempts to find the solution to the security challenges through the MapReduce technique in a forensic way.

Chapter 22 (The Interaction Between Technologies, Techniques, and People in Higher Education Through Participatory Learning) suggests that higher education has to come out of the static environment and become more dynamic and more futuristic. The solution to all the problems is by providing quality education, and teachers are the main ingredients in giving quality education.

Chapter 23 (Blockchain-Based Incentive Announcement Network for Communications of Smart Vehicles) proposes that the IoV has the ability to efficiently tackle a variety of traffic and road safety issues in order to prevent fatal crashes. The authors qualitatively examined the suggested overall system performance and resiliency against popular security assaults. The proposed method solves the primary issues of Vehicle-to-X (V2X) communications.

Chapter 24 (The Categorization of Development Boards to Implement the Embedded Systems and Internet of Things With Cloud Database Using Volcano Monitoring Drones) puts forward that the thermocouple is used as a temperature sensor, and the ESP-32 CAM is interfaced with that development board to capture the snapshot of the mountain if the temperature is high and the same snap is shared to the ground station through Wi-Fi. The GPS module is also interfaced with the development board to know the location of the volcano happening mountain.

Chapter 25 (Autonomous Robotic Technology and Conveyance for Supply Chain Management Using 5G Standards) proposes that to drive the autonomous vehicle towards correct location a precise protocol is chosen which is termed as Common Industrial Protocol (CIP) where proper solutions can be achieved for all control applications using time synchronization model. Further, data monitoring process is trailed using an online contrivance which is termed as Internet Router (IR) where short distance can be identified using corresponding addressing scheme.

Chapter 26 (E-Collaboration for Management Information Systems Using Deep Learning Techniques) presents a thorough foundation for an e-collaboration platform that was established during the successful implementation of an e-collaboration solution at the Management Information Systems. The solution makes use of cutting-edge web portal technology and a digital asset management system to create a uniform, centralized platform for system users to collaborate, communicate, and exchange information.

Jingyuan Zhao
University of Toronto, Canada

V. Vinoth Kumar
Jain University, India
2022 January

REFERENCES

Bidgoli, H. (2012). E-collaboration: New productivity tool for the twenty-first century and beyond. *Business Strategy Series, 13*(4), 147–153. doi:10.1108/17515631211246212

Cassidy, J. (2020). An Economic-History Lesson for Dealing with the Coronavirus. *New Yorker (New York, N.Y.).* https://www.newyorker.com/news/our-columnists/an-economic-history-lesson-for-dealing-with-the-coronavirus

Flecker, J. (2010). *Space, Place and Global Digital Work.* Palgrave Macmillan.

Gartner. (2021). *Gartner Forecasts Worldwide Social Software and Collaboration Market to Grow 17% in 2021: Press Release.* https://www.gartner.com/en/newsroom/press-releases/2021-03-23-gartner-forecasts-worldwide-social-software-and-collaboration-market-to-grow-17-percent-in-2021

Hafiz, H., Oei, S. Y., Ring, D. M., & Shnitser, N. (2020). *Regulating in Pandemic: Evaluating Economic and Financial Policy Responses to the Coronavirus Crisis.* Boston College Law School Legal Studies Research Paper, 527.

Internet World Stats. (2021). *Internet Usage Statistics: The Internet Big Picture World Internet Users and 2021 Population Stats.* Retrieved from: https://www.internetworldstats.com/stats.htm

Karabag, S. F. (2019). Factors impacting firm failure and technological development: A study of three emerging economy firms. *Journal of Business Research, 98*, 462–474. doi:10.1016/j.jbusres.2018.03.008

Karabag, S. F. (2020). An Unprecedented Global Crisis! The Global, Regional, National, Political, Economic and Commercial Impact of the Coronavirus Pandemic. *Journal of Applied Economics and Business Research, 10*(1), 1–6.

Knock, N. (2008). A Basic Definition of E-Collaboration and its Underlying Concepts. In N. Kock (Ed.), *Encyclopedia of E-Collaboration* (pp. 48–53). IGI Global.

Kock, N., & D'Arcy, J. (2002). Resolving the e-collaboration paradox: The competing influences of media naturalness and compensatory adaption. *Information Management and Consulting, 17*(4), 72–78.

Linnes, C. (2020). Embracing the Challenges and Opportunities of Change Through Electronic Collaboration. *International Journal of Information Communication Technologies and Human Development, 12*(4), 37–58. doi:10.4018/IJICTHD.20201001.oa1

Mitchell, A. (2021). Collaboration technology affordances from virtual collaboration in the time of COVID-19 and post-pandemic strategies. *Information Technology & People.*

OECD. (2016). *New Forms of Work in the Digital Economy. OECD Digital Economy Papers, No 260.* OECD Publishing.

Schubert, P., & Williams, S. P. (2022). Enterprise Collaboration Platforms: An Empirical Study of Technology Support for Collaborative Work. *Procedia Computer Science, 196*, 305–313. doi:10.1016/j.procs.2021.12.018

Song, L., & Zhou, Y. (2020). The COVID-19 Pandemic and Its Impact on the Global Economy: What Does It Take to Turn Crisis into Opportunity? *China & World Economy, 28*(4), 1–25.

Section 1

COVID-19: Smart Healthcare Applications Using E-Collaboration During a Global Crisis

Section 1 introduces smart healthcare systems as well as their innovations and applications that are responsible for making complete health monitoring together disease diagnosis using e-collaboration tools. Section 1 is organized into 10 chapters.

Chapter 1
Deep-CNN Model for Acute Lymphocytic Leukemia (ALL) Classification Using Microscopic Blood Images:
Global Research

Prasanna Ranjith Christodoss
https://orcid.org/0000-0003-4778-7915
University of Technology and Applied Sciences, Shinas, Oman

Rajesh Natarajan
https://orcid.org/0000-0003-1255-9621
University of Applied Science and Technology, Shinas, Oman

ABSTRACT

Acute lymphocytic leukemia (ALL) is a variety of malignant somatic cell cancer that influences children and teenagers. The goal of the study is to create a system that can detect cancer from blood corpuscle images mechanically. This method employs a convolutional network that takes images of blood corpuscles and determines whether or not the cell is cancer infected. The appearance of cancer in blood corpuscle images is frequently ambiguous, overlaps with other diagnosis, and can be mistaken for a variety of benign abnormalities. Machine-assisted cancer identification from blood corpuscle images at the level of skilled medical staff would be extremely beneficial in clinical settings and also in the delivery of healthcare to populations with limited access to diagnostic imaging specialists. Here, the authors proposed a convolutional neural network (CNN)-based methodology to distinguish between outdated as well as irregular somatic cell photos. With the dataset and 1188 somatic cell images, the proposed methodology achieves an accuracy of up to 96.6%.

DOI: 10.4018/978-1-7998-9640-1.ch001

INTRODUCTION

Blood is made up of plasma and three different types of cells: white blood cells, red blood cells, and platelets, each of which has a specific function. Chemical elements are transported from the lungs to the body's tissues by red blood cells, and vice versa. The body's ability to fight diseases and infections is aided by white blood cells. Platelets aid in clotting and bleeding control (A. Sindhu & S. Meera., 2015). Malignant neoplastic illness is a type of blood cancer in which the number of white blood cells grows rapidly and the cells become immature, interfering with other blood cells, including red blood cells and platelets (Chatap, N., & Shibu, S., 2014). The white blood corpuscle to white blood corpuscle magnitude ratio in our body is 1000:1. It means that there is one white blood cell for every thousand red blood cells. There are two types of white blood cells that can develop into leukaemia:

- Lymphoid cells
- Myeloid cells

White blood corpuscle or lymphocytic leukaemia is caused by tumour cells, while myelogenous or chronic leukaemia is produced by myeloid cells (Athira Krishnan, S. K. (2014). Leucaemia is divided into two categories: acute and chronic, based on the rate at which the cells multiply. In leukaemia, the aberrant blood cells are sometimes immature blasts (young cells) that don't work correctly (S., D. Mohapatra et al., 2010). These cells are rapidly multiplying. If leukaemia isn't treated right away, it will swiftly worsen. In leukaemia, young blood cells are produced, but mature purposeful cells are also produced. Leukemia causes blasts to grow slowly (Ritter, N & Cooper, J., 2007). It takes the disease longer to progress. The four most common types of leukaemia are:

- Acute lymphoblastic leukaemia (ALL)
- Acute myelogenous leukaemia (AML)
- Chronic lymphocytic leukaemia (CLL)
- Chronic myelogenous leukaemia (CML

Leukemia is a potentially lethal illness that puts many sufferers' lives at jeopardy. Its rate of remission will be significantly improved if it is detected early (Dhiman, G et al., 2021). Using transfer learning, this project presents two machine-driven classification models supported by blood microscopic images to detect leukaemia, as opposed to traditional approaches that have numerous drawbacks. Blood microscopic images are pre-processed in the first model, and options are retrieved by a pre-trained deep convolutional neural network termed AlexNet, which makes classifications in accordance with a number of well-known classifiers. When pre-processing the pictures in the second model, AlexNet is fine-tuned for each feature extraction and classification. Experiments were carried out on a dataset of 2820 images, indicating that the second model outperforms the main due to its 100% classification accuracy (Zhao J et al., 2016).

Deep learning has recently made improvements in a variety of disciplines, including laptop vision, speech processing, and visual perception. Convolutional neural networks (CNNs) and deep neural networks (DNNs) are utilised to develop computer-aided diagnostic systems.

Deep neural network analysis and training, on the other hand, are time-consuming and difficult processes. As a result, rather than developing a deep neural network from the ground up, we prefer to employ the concept of transfer learning, in which a deep network that has found success in resolving one drawback is tweaked to overcome another.

Scope of the Work

At the level of prodigious active medical personnel, we design a technology that identifies cancer from blood corpuscle images. This technology will improve the availability of medical imaging expertise and enhance access to it in areas of the world where qualified medical experts are few.

Methodology

The magnifier staining method's exposure determines the quality of the microscopic image. The usage of exposure settings that are too high or too low may cause detection issues. Image sweetening strategies are a collection of approaches that aim to change an image into a shape that is better suited to human or computer interpretation.

CNN techniques

1. From stained blood smear images, presented a segmentation algorithm that uses a color-based cluster to extract the nucleus region and living substance area. SVM classifiers are used in conjunction with appropriate alternatives to produce satisfying results.
2. White blood cells (WBCs) are automatically detected from peripheral blood images and classified into five types: white corpuscle, basophil, neutrophil, monocyte, and WBC. SVM used a graininess characteristic to classify white corpuscle and white corpuscle from various WBCs at first. Different 3 types of WBCs are then identified using a convolutional neural network to extract options, which are subsequently used by random forest to classify those WBCs.

The images used in this research were taken from the Kaggle dataset, which is a publicly available dataset. This data was separated into two categories. There were a total of 4961 coaching photos, with 2483 photos from healthy people and 2478 photos from patients with blood cancer. We tend to test the model using a total of 1240 images, 620 from each category. These images were 320*240 pixels in size.

CNN classification

For the feature extraction and classification of blood samples, we used CNN. A CNN is a multi-layered neural network with a unique design for detecting advanced knowledge options. CNNs are used in image identification, robot vision, picture text recognition, and self-driving automobiles (Sundar, P. P et al., 2020).

The CNN is made up of layers of neurons and is designed to recognise patterns in two dimensions. Convolutional, pooling, and completely linked layers are the three types of layers used by CNN. Except for the input layer, our network has eleven layers. During the processing of an RGB colour image, the input layer processes each colour channel separately. Convolution layer is the first of the convolution network's layers. A pair of convolution layers is used to apply sixteen 3*3 filters on a picture within the layer. The image is processed by the opposing 2 layer, which uses thirty two 3*3 filters. As a result, the final two layers of convolution apply a total of sixty-four 3*3 filters on a picture. The ReLU activation function is used in the nonlinear transformation sublayer. The Georgia home boy pooling sublayer applies a 2*2 filter to the image, resulting in a [*fr1] size reduction (Litjens G et al., 2016). The convolution network retrieves sixty-four possibilities at this moment, each represented by a 32*32 array for each colour channel. The flatten layer is the ninth layer. The flatten layer converts a multidimensional array into a one-dimensional array by simply concatenating the array's items together. This flatten layer produces a 4800-by-4800-by-4800-by-4800- by-4800-by-4800-by-4800-by-4800-by-4800-by-4800 The fully linked ANN with the ReLU activation function, which maps 4800 input values to 64 output values, is the ninth layer. The dropout layer is the tenth layer. To address the problem of overfitting, 50% of the input values flowing into the layer are set to zero. The eleventh and final layer is a fully connected ANN that maps 64 input values to two class labels using the sigmoid activation function (Basheer, S et al., 2020).

First, we use the data in the training set to train a convolutional network to discover acceptable filter weights in the three convolutional sublayers, as well as weights that provide the lowest error in the two fully connected layers. The validation error and cross-entropy loss are then calculated using the data from the validation set to evaluate the convolution network. We repeat the convolution network training in this manner until we have completed 10 epochs. Finally, we assess the convolution network's performance using data from the test set. Neural networks are employed in the automatic detection of cancer in blood samples. Neural networks were chosen because of its well-known approach as a quick classification tool. One of the most crucial aspects in constructing an accurate process model using CNNs is the training and validation process (Song Y et al., 2014). The training and validation datasets are divided into two parts: a training features set that is used to train the CNN model, and a testing features set that is used to check the accuracy of the learned model using the feed-forward back propagation network. Connection weights were always updated in the training section until they achieved the set iteration number or a suitable error (Kumar, V.V et al., 2021). In the automatic identification of cancer in blood samples, neural networks are used. Because of its well-known methodology as an effective classifier for many real-world applications, neural networks were chosen as a classification tool. One of the most crucial aspects in constructing an accurate process model using CNNs is the training and validation process (S. Mohapatra et al., 2012).

RELATED WORKS

An image analysis methodology is used in this work for automatic detection, pre-processing (smoothing, enhancement, segmentation), feature extraction (morphological and calorimetric), and then identification and classification of specific cells, particularly cancer cells from normal cells. Automatic Otsu's Thresholding for blood cell segmentation was proposed in this project, along with image enhancement and arithmetic for WBC segmentation. To distinguish blast cells from normal lymphocyte cells, the K-NN classifier was used. This research describes a study that used K-NN to classify blasts in acute leukaemia into two major forms: ALL and AML. From blood pictures, 12 key features representing size, colour, and form were retrieved. In order to determine optimal parameters to apply in the approach of classifying the blasts, the k values and distance metric of k-NN were examined. This research discusses preliminary research of using microscopic blood sample photographs to detect leukaemia kinds. It will explore changes in texture, geometry, colour, and statistical analysis using features found in microscopic images. Changes in these characteristics will be sent into the classifier (Kouser, R.R et al., 2018).

The gradient magnitude, Thresholding, morphological operations, and watershed transform are all used in this research to perform cell segmentation. The proposed method was tested on 50 photos, and the findings demonstrated that the system was capable of producing qualitatively satisfactory segmentation results. To increase image quality, a global contrast stretching and segmentation based on HIS colour space is applied in this research. For both ALL and AML images, an image enhancement approach is employed to isolate the nucleus region in the WBC image sample using the same threshold value (Shalini A et al., 2018). They offered a new public dataset of blood samples for this project, which was specially created for the evaluation and comparison of segmentation and classification algorithms. The categorization of the cells is given for each image in the dataset, as well as a specific set of figures of merits to compare the performances of different algorithms objectively. The number of blood cells counted will be used to compute the blood cell ratio for leukaemia detection.

PROPOSED SYSTEM

Medical history and physical examination: A record of current symptoms as well as previous difficulties. A person's family's medical history might also aid in the diagnosis of leukaemia.

- **Complete blood count (CBC):** Blood is drawn and the quantity of RBCs, WBCs, and platelets is measured under a microscope.
- **Bone marrow aspiration:** Bone marrow is extracted from the bone using a needle. The extracted sample is examined under a microscope for aberrant cells.
- **Cytogenetic examination:** A cytogenetic examination uses blood or bone marrow to help identify individual chromosomes. It reveals chromosomal abnormalities that aid in the diagnosis and classification of cancer. Results can take up to three weeks to appear on the market.
- **Immunohistochemistry:** In this technique, a blood sample of cells is treated with specific antibodies. The colour change may be noticed under the magnifier. It aids in determining the types of cells that are gifted.

The main goal of transfer learning with extremely deep CNNs is to adapt a pre-trained deep network that was previously fitted to a large dataset like ImageNet (approximately 1.2 million images with another 50,000 images for validation and a hundred,000 images for testing, on a thousand different categories) to solve a specific image classification problem. Because the network has previously learned relevant picture alternatives from a generic coaching dataset, it may be used to target a specific image type in order to solve a classification problem.

We chose a well-liked and dependable CNN design called VGGNet, which has sixteen conv (convolutional) and three FC (completely connected) layers. The number of channels (width of the conv layers) is quite small, ranging from 64 in the first layer to 512 in the final layer, increasing by a factor of two with each max-pooling operation. The mounted size of the input layer is 224 224 pixels. A stride is superimposed to keep abstraction resolution, as every image is tried and true a stack of conv layers. Over a given window, five max-pooling layers conduct pooling, with stride following some but not all conv layers. A stack of conv layers with different depths in different architectures is followed by three FC layers with 4096 channels in the first two, and classification in the third. This layer only has two channels in our situation (one for every class). A soft-max layer could be the last layer. The non-linearity of all buried layers has been rectified. The weighted binary cross-entropy loss is the parameter to be optimised for each image X of study type T of coaching knowledge.

Tree-Based Classifier for Feature Importance

1. Tree-based methods are popular for classification because of their high level of accuracy and ease of use, as well as their durability. Furthermore, they provide two direct methods for selecting selections. Each individual node in a call tree, as is well known, might be a condition on a single characteristic, dividing the set of knowledge into two parts.
2. As a result, similar responses should be found in the same collection. Impurity is defined as the ability to choose the finest geographically available condition. During the coaching of a tree, one will calculate how much each feature reduces the tree's weighted impurity. As a result, the impurity decrease will be averaged per feature, and the options will be ranked according to the impurity decrease. There is one drawback to this method: when a dataset contains two or more similar alternatives, there is no preference for one over the other, and any of those options might be used as the predictor. However, as soon as one of them is picked, the value of the others is immediately diminished, because whatever impurity that they might remove has already been removed by the major trait. This was resolved by eliminating all associated options from the first step.

In a given blood sample, the present approach may detect many kinds of leukaemia. With the use of blood samples, it is an effective and automated method for identifying cancer cells. The present approach can differentiate between three forms of leukaemia. The patient can be diagnosed at a lower cost and with more accuracy.

Figure 1. General architecture

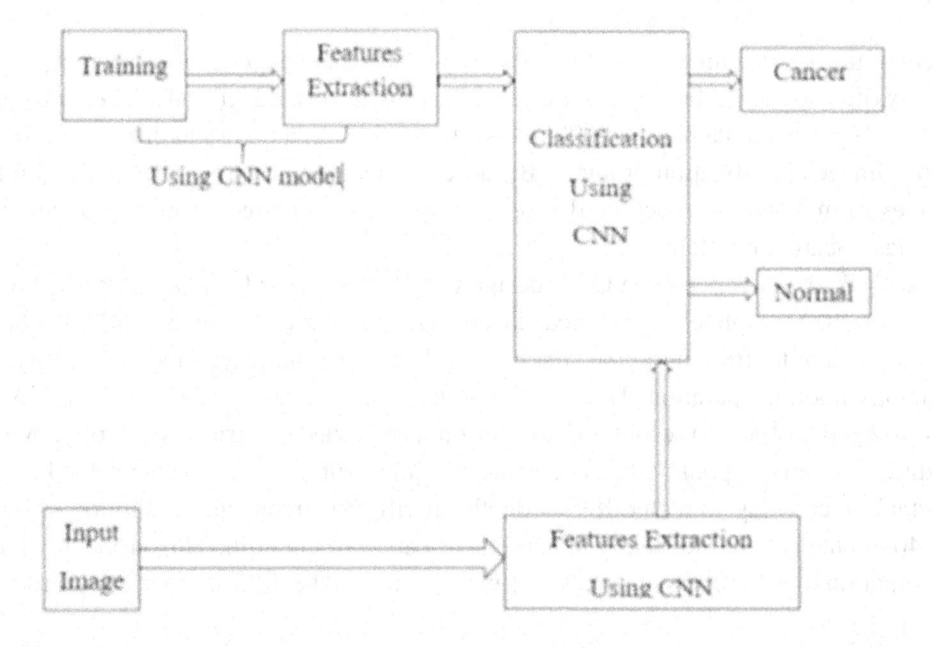

Figure 1 shows the proposed cancer diagnosis model architecture diagram first the input image is sent to training model and then features are extracted using CNN and then classification of image is done by using CNN and after classification we can determine whether it is cancer or normal.

Figure 2. Implementation Procedure

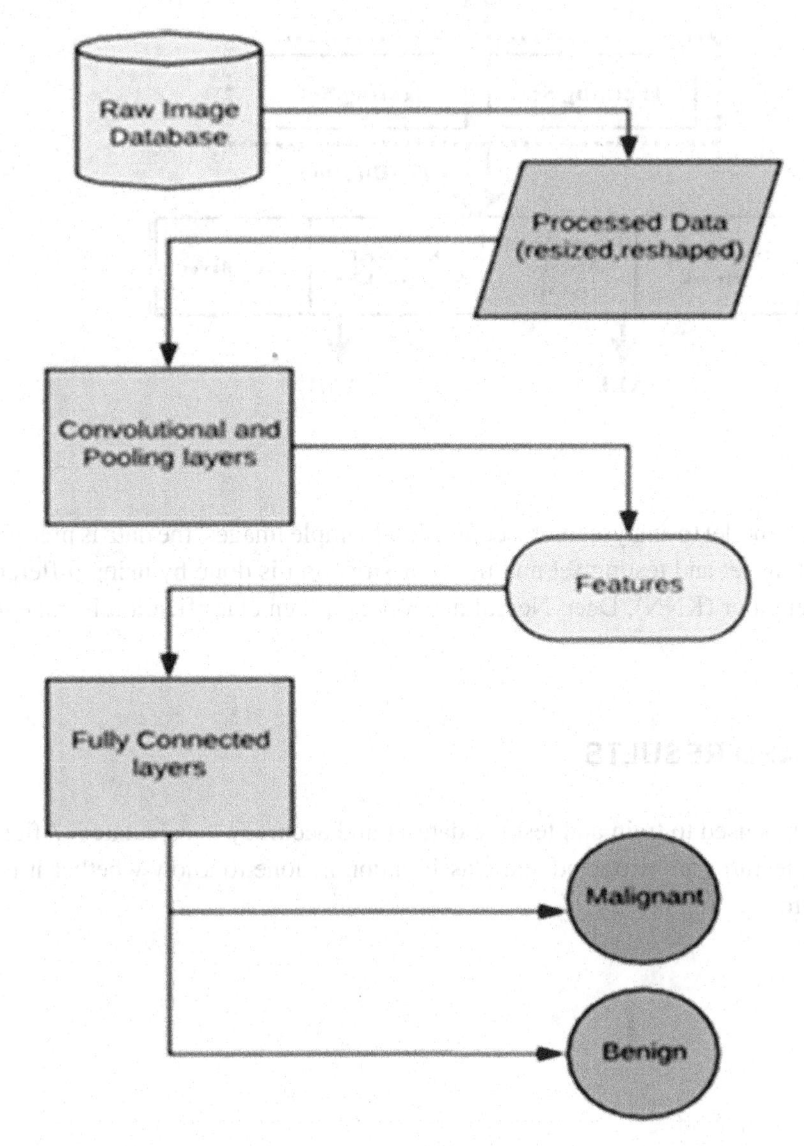

Figure 2 shows the implementation procedure of our proposed model. first the input blood image is send for detection and then sent for de-noising, after de-noising is completed it then moves to the segmentation and after segmenting the features are extracted and then classification of the image is done. the input image is processed and goes through 3 stages de-noising, segmentation and feature extraction . After these 3 stages the output image is classifies as cancer or normal.

Figure 3. ML model for proposed blood cancer detection using microscopic images

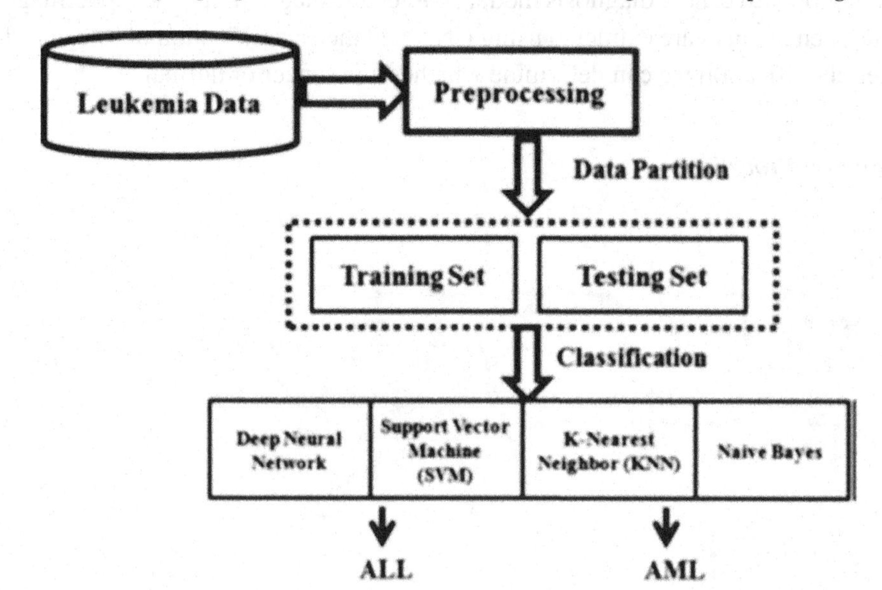

Figure 3 depicts the ML model to analyse microscopic blood sample images. the data is pre-processed and partitioned into training set and testing set and then classification is done by using different algorithms like K-Nearest Neighbor (KNN), Deep Neural network and then classification is done to know whether it is cancer or not.

IMPLEMENTATION AND RESULTS

In implementation python is used to train and test the dataset and accuracy is calculated. After testing the blood sample image, features are extracted and classification is done to know whether it is cancer or normal shown in Figure 4.

Deep-CNN Model

Figure 4. Accuracy level trained blood sample images

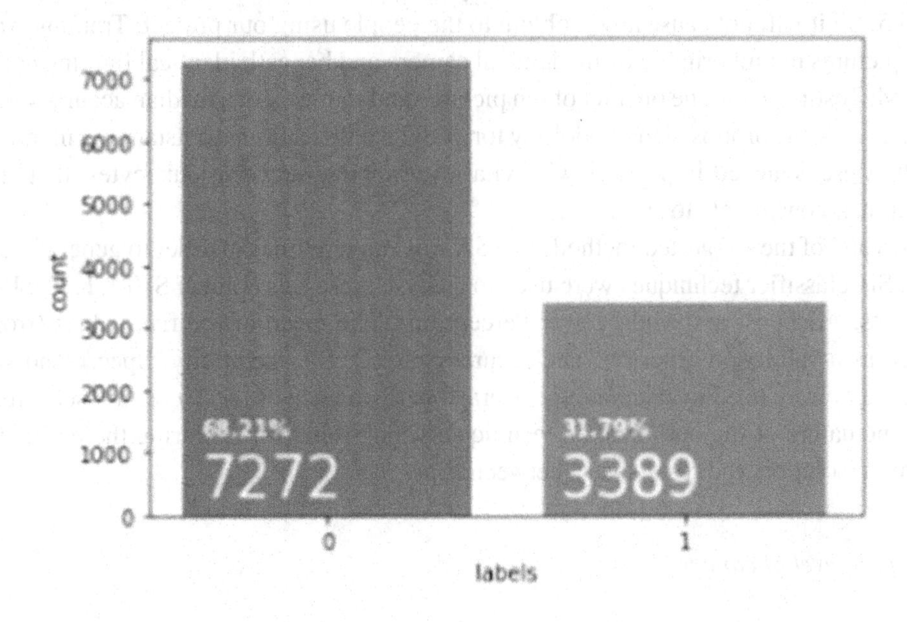

Figure 5. Microscopic blood sample images

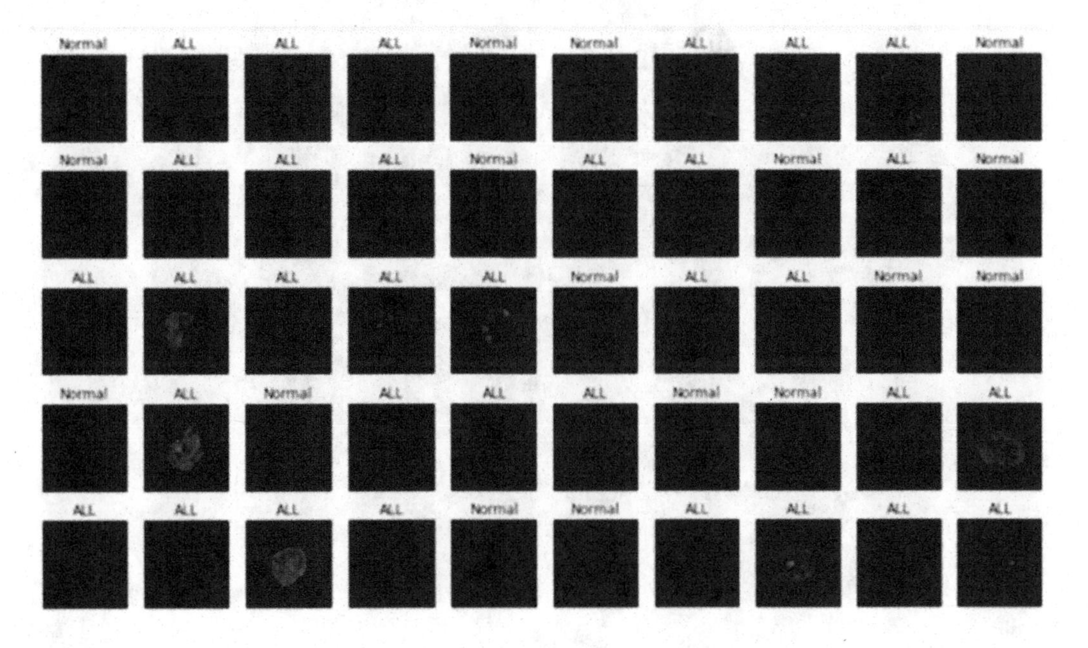

The testing is done for finding out that there are any errors or mistakes after completing the project shown in Figure 5. So it will not cause any problem to the people using our project. Training was done on a set of forty pictures noninheritable from identical camera and beneath identical lighting conditions employing a SVM. Testing was done on a set of ten pictures and that we got a median accuracy of eighty six. The performance of the proposed methodology for WBC identification is outstanding in most cases. The worst results were achieved in pictures with vital overlapping between leukocytes, that area unit even tough for human consultants to spot.

To test the efficacy of the suggested method, SESSA was run ten times in order to generate ten separate feature sets. Six classifier techniques were used to analyse these sets (Linear SVM, KNN, Decision Trees, Naive Bayes, Adaboost and Multi-Layer Perceptron). The mean of the five values (from each fold) was utilised as a validation criterion. The accuracy (Acc.), F1, specificity (Spec.), and sensitivity (Sens.) parameters were used to calculate the average performance of the 10 feature sets created by SESSA. Due to the nature of the optimization technique, which relies on exploring the problem space to find the optimum solution, the results vary between runs.

Figure 6. Prediction level of cancer

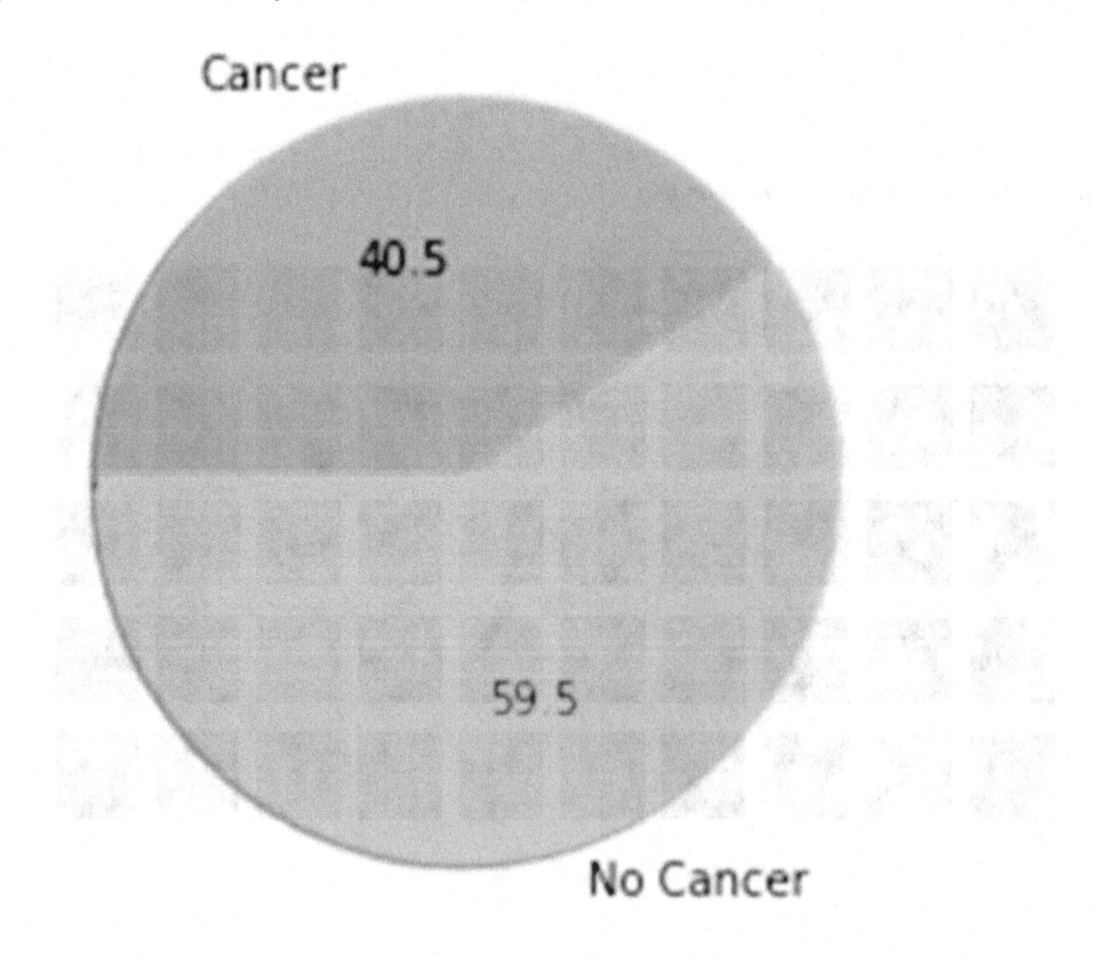

We are using only the training dataset folder for our training and validation too. We are using 80% of the training dataset to train and 20% of it to validate. If we are using everything in that dataset it will take time to train and validate. finally we testing for the accuracy of the model, ours is around 83% to 87%. We are using inceptionV3 model to compute the dataset and it uses CNN - Convolution Neural Network to compute it because these are images of cells. In that training dataset we have a class of hem - normal hemoglobin and ALL - Acute Lymphocytic Leukemia. So these are 2 classes we take the first 80% to train our model and next 20% to test it out. After that model out will be saved and we use it for future purpose. finally we can test our validator models accuracy by running as much as test cases or passes or we call it as Epochs for this on it runs upto 100 epochs. The plot indicates red is cells with Leukimia and Blue without cancer cellos Below that random 25 cells are taken as examples with ALL and Normal The plot is from training set not validated. This model summary output indicates the layers of CNN model done to the training set CNN uses many hidden layers depending upon the type of data. Here it is 11 layers. The shape which is mentioned are the tensors which is like a multi dimensional array for each layers it will get reduced for computation purpose. So among the validation data we got the results shown in Figure 6.

Efficiency of the Proposed System

We offer a method for automatically identifying and classifying leukocytes using microscopic images, as well as an automated procedure to support the popularity of ALL. Our findings show that the proposed methodology is capable of quickly determining the WBCs present in a photograph and accurately classifying leukoblasts. The next phase in this project could include further development of the identification,

thresholding, and segmentation sections. These improvements are required to improve the accuracy of calculating WBCs and segmentation, as good segmentation results in a large number of robust form extraction choices, which is critical in this type of problem. Furthermore, it will be important to examine and analyse the use of the most recent options, which will be critical for this type of research. The best level of accuracy can therefore be achieved by selecting the most discriminatory options. Additional improvement of the proposed methodology could affect the separation of nearby leukocytes, which is critical to account for all leukocytes in the image.

Comparison of Existing and Proposed System

Many old computer-aided systems used image processing and machine learning techniques that include pre-processing, segmentation, feature extraction, and classification, among other things. Every step, however, is dependent on the success of the one before it. For example, classification success is dependent on the success of the previous feature extraction, which is dependent on the performance of the previous segmentation. As a result, high classification accuracy necessitates the success of all processes, each of which is complex and problem specific. Deep learning has recently made improvements in a variety of domains, including computer vision, speech processing, and sight. Convolutional neural networks (CNNs) and deep neural networks (DNNs) will be utilised to create computer-aided diagnostic systems. Deep neural network analysis and training, on the other hand, are time-consuming and difficult processes. As a result, rather of creating a deep neural network from the ground up, we prefer to employ the concept of transfer learning, in which a deep network that has been successful in determining one explicit drawback is tweaked to uncover another drawback shown in Figure 7.

Advantages of the Proposed System

- The therapy of blood cancer will be more effective if it is detected early.
- This research proposed two categorization methods for leukaemia free and leukaemia-affected blood microscopic images.
- Transfer learning is used in each model. The discriminant options are extracted using a pre-trained CNN called AlexNet, and alternative well-known classifiers including DT, LD, SVM, and K-NN are employed for classification in the first model.
- The ubiquity of the SVM classifier has been proven by experiments. AlexNet is used in the second model for feature extraction and classification.
- Experiments for this model show that it outperforms the primary model on a variety of performance metrics.
- Instead of simply labelling photos as leukaemia-free or leukaemia affected, a future study could identify between the many types of blood cancer.

Figure 7. Accuracy and loss of proposed system

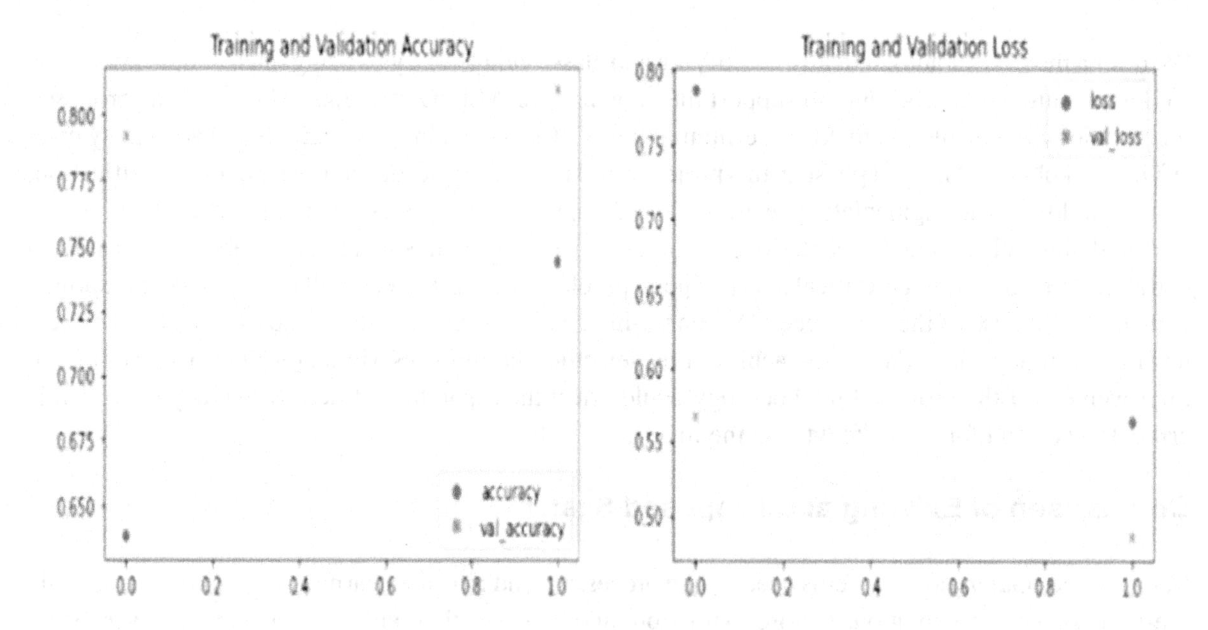

CONCLUSION

The proposed system is the ability to reliably observe cancer using the iterations collected, the loss and accuracy graph, and hence the confusion matrix. We've run twenty iterations and can clearly see that the loss is decreasing with each cycle. The term "loss" refers to how accurate the model is in terms of proportion. As a result, we needed to lower the loss rate, and as a result, our model has consistently decreased the loss rate from start to finish, with each iteration bringing us closer to the minimum. Following that, we usually execute a loss and accuracy curve to get the best results from our model. These learning curves (loss and accuracy curves) depict our model's performance on coaching and validation as a function of a number of coaching iterations. We have a loss curve that is dropping with each iteration, indicating that loss is being minimised and the most effective solution is being achieved. On the other side, we've created an accuracy curve that gets better with each iteration, implying that our model is better at learning, recovering, convalescing, recouping, and recovering.

REFERENCES

Athira Krishnan, S. K. (2014). A survey on image segmentation and feature extraction methods for acute myelogenous leukemia detection in blood microscopic images. *International Journal of Computer Science and Information Technologies*, *5*(6), 7877-7879.

Basheer, S., Anbarasi, M., Sakshi, D. G., & Kumar, V. V. (2020). Efficient text summarization method for blind people using text mining techniques. *International Journal of Speech Technology*, *23*(4), 713–725.

Chatap, N., & Shibu, S. (2014). Analysis of blood samples for counting leukemia cells using Support vector machine and nearest neighbour. *IOSR Journal of Computer Engineering, 16*(5), 79-87.

Dhiman, G., Kumar, V. V., Kaur, A., & Sharma, A. (2021). DON: Deep Learning and Optimization-Based Framework for Detection of Novel Coronavirus Disease Using X-ray Images. *Interdisciplinary Sciences, Computational Life Sciences*, 1–13.

Kouser, R. R., Manikandan, T., & Kumar, V. V. (2018). Heart disease prediction system using artificial neural network, radial basis function and case based reasoning. *Journal of Computational and Theoretical Nanoscience*, *15*(9-10), 2810–2817.

Kumar, V. V., Raghunath, K. M., Muthukumaran, V., Joseph, R. B., Beschi, I. S., & Uday, A. K. (2021). Aspect based sentiment analysis and smart classification in uncertain feedback pool. *International Journal of System Assurance Engineering and Management*, 1-11.

Litjens, G., Sánchez, C. I., Timofeeva, N., Hermsen, M., Nagtegaal, I., Kovacs, I., ... Van Der Laak, J. (2016). Deep learning as a tool for increased accuracy and efficiency of histopathological diagnosis. *Scientific Reports*, *6*(1), 1–11.

Mohapatra, S., Patra, D., & Satpathi, S. (2010, December). Image analysis of blood microscopic images for acute leukemia detection. In *2010 International Conference on Industrial Electronics, Control and Robotics* (pp. 215-219). IEEE. 10.1109/IECR.2010.5720171

Mohapatra, S., Patra, D., & Satpathy, S. (2012). Leukemia diagnosis using color-based clustering and unsupervised blood microscopic image segmentation. *International Journal of Computer Information Systems and Industrial Management Applications*, 477–485.

Ritter, N., & Cooper, J. (2007). Segmentation and border identification of cells in images of peripheral blood smear slides. *30th Australasian Computer Science Conference Conference in Research and Practice in Information Technology*, 161-169.

Shalini, A., & Jayasuruthi, L., & VinothKumar, V. (2018). Voice recognition robot control using android device. *Journal of Computational and Theoretical Nanoscience*, *15*(6-7), 2197–2201.

Sindhu, A., & Meera, S. (2015). A survey on detecting brain tumorinmri images using image processing techniques. *International Journal of Innovative Research in Computer and Communication Engineering*, *3*(1), 16.

Song, Y., Zhang, L., Chen, S., Ni, D., Li, B., Zhou, Y., . . . Wang, T. (2014, August). A deep learning based framework for accurate segmentation of cervical cytoplasm and nuclei. In *2014 36th Annual International Conference of the IEEE Engineering in Medicine and Biology Society* (pp. 2903-2906). IEEE.

Sundar, P. P., Ranjith, D., Karthikeyan, T., Kumar, V. V., & Jeyakumar, B. (2020). Low power area efficient adaptive FIR filter for hearing aids using distributed arithmetic architecture. *International Journal of Speech Technology*, *23*(2), 287–296.

Zhao, J., Zhang, M., Zhou, Z., Chu, J., & Cao, F. (2016). Convolutional neural networks are used to detect and classify leukocytes automatically. *Medical & Biological Engineering & Computing*, *55*(8), 287–1301.

Chapter 2
A PCCN-Based Centered Deep Learning Process for Segmentation of Spine and Heart:
Image Deep Learning

K. Uday Kiran
Koneru Lakshmaiah Education Foundation, India

Gowtham Mamidisetti
Bhoj Reddy Engineering College for Women, India

Chandra shaker Pittala
MLR Institute of Technology, India

V. Vijay
Institute of Aeronautical Engineering, India

Rajeev Ratna Vallabhuni
Independent Researcher, USA

ABSTRACT

The spinal cord and heart in the body are major organs. Diagnosis of diseases in these organs is very complex using MRI and CT images. The conventional methods like post segmentation, pre-image processing, and text feature extraction mechanisms cannot handle accurate diagnosis. Therefore, advanced techniques are needed. In this work, pixel-based convolution neural networks with centered deep learning processes are proposed to cross over the problems. The projected PCNN has four pixel-based convolution neural networks. Here disease objects are identified through grading framework. The entire mechanism is working based on sequential part of PCNN segmentation process. The spinal cord and heart image MRI-based diagnosis process is very difficult with conventional methods. But the proposed method provides accurate results and outperforms the standard methodology performance measures in accuracy, precision, and F1score.

DOI: 10.4018/978-1-7998-9640-1.ch002

INTRODUCTION

The rapid growth in deep learning technology it can provide extreme results in image processing and any application specific feature extraction. Especially MRI medical image features are extracted through lower, front, up and down view disease diagnosis is very critical. These problems are overcome by proposed PCNN deep learning technology. The LBP and HBP mechanisms are very useful and creating pixel-based repair in processing time. It is usually very complex through old conventional methods. The nature of MRI image features extraction and disease finding is very simple through proposed PCNN process. The behaviors of abnormal conditions in MRI, heart and spine treatment are very difficult to track the exact damaged area. This can be very easy through proposed PCNN deep learning mechanism. The multi scale and slices segmentation process is extracting the pixels, where disease is located (Ahammad, S. H., 2020). The SVM model was trained using 1,040 pictures from 26 pregnant women and evaluated with 800 photographs from a different group of 24% women. The suggested approach has a successfulness of 97.12% on the training dataset and 94.2% on the test set. In 45 of the instances, the trained support vector machine model correctly identified the appropriate needles injection location (intervertebral area) using 46 off-line recorded Scans (Ahammad, S. H.2019).

The continuous monitorization of MRI medical image process providing an accurate observation result. But this functionality is very difficult by using machines and humans. Therefore, a deep learning mechanism is called to promote above complex work to make easy. In this impartial addition to marketing to mind when you think of any picture classification or image segmentation. Maybe something more along the lines of: provide a 512x512 picture as an input and receive a 10x1 classifier result. In the middle, there are convolution and max-pooling layers, which gradually reduce picture resolution yet preserving essential characteristics. Rather than using a traditional neural net like the one described above, Google Research and Brain Team developed SpineNet at CVPR-2020. This network, which comprises of scale-permuted intermediate features and bridge, may be a viable solution for a variety of extracting features purposes such as object recognition and picture classifications (Ahammad, S. K., 2018) .

In this discussion the calculations of spinal cord extraction very difficult in shape or appearance point of view. The adjusted conventional models can't suggest by researchers and scientist due to performance and accuracy point of view. The past mechanisms and their applications are majorly depending on old programming languages, those are not that much robust. The programming methods like python and spring java providing accurate image training mechanism and easily getting disease diagnosis area as well as location(Ahammad, S. H., 2019)..

The Long Short-Term Memory Network (LSTMN) is an enhanced RNN (sequential network) that enables information to be stored indefinitely. It can deal bwith the vanishing gradient issue that RNN has. For permanent memory, an RNN, also known as CNN model, is employed. Let's suppose you recall the prior scene when viewing a video or you know what occurred in the previous chapter while reading a book. RNNs operate in a similar way; they remember past information and utilize it to update the financial input. Because of the diminishing gradient, RNNs are unable to recall long-term connections. Long-term dependence issues are expressly avoided using LSTMs (Vijaykumar, G.2017). In this case, the encoding features extracted (which may also be used for picture classification) and the decoding recovers those features to give bounding box localization. Encoders and decoders are often linked using residual connections (ResNet) to preserve essential information by combining features from various resolutions. This encoder is a scale-decrease network, which has also been utilized in YOLOv3, Faster-RCNN, and other models, and is often referred to as a model's backbone. The rationale for this level architectural design is explained by LeCun et al: "High resolution may be required to identify the existence of a feature, but its precise location may not need the same level of precision." This article proposes a level networks with the goal of retaining spatial information as the model grows deeper and facilitating multi-scale face detection.

Another of the network's present faults is that it was created with the assistance of NAS (Network Assistance system). We're still unsure how we'd go about creating another permuted network on our own since the rationale behind the various links is still a good research (Inthiyaz, S. & Prasad, M. V. D., 2019). Although they demonstrate in the article that it improves in object recognition and picture training institutes. More tests with other systems are needed to establish how well the SpineNet-49 functions in other situations (Kumar, M. S. & Inthiyaz, S.2019).

TECHNIQUES AND DATASETS

In this section Dataset and PCNN technique is discussed, using all these raw image pixels as specific characteristics is the easiest method to generate characteristics from a picture. Considering the same scenario with our picture (the number '8') - the image's dimensions are 28 x 28. Feature extraction is a method for identifying significant data characteristics or properties. Pattern recognition and finding similar themes within a huge piece of text are two examples of this approach. Spam-detection software is an example of feature extraction that we can all relate to. Feature extraction is used in artificial intelligence, pattern recognition, and image processing to create derived values (features) that are meant to be useful and non-redundant, allowing for faster teaching and generalization and, in some instances, superior human performance. Extracting features aids in the reduction of unnecessary data in a data collection. Finally, reducing the data allows the computer to construct the models with less effort, as well as speed up the training and generalization stages in the data mining process. (Ahammad, S. H., 2019) (Siva Kumar, M.2019)

Machine learning (ML) and intelligent systems (IS) are quickly growing rapidly in healthcare (Ahammad, S. H., 2020) particularly in the area of medical imaging, and are expected to radically alter clinical practice (Ahammad, S. H.,2019), in the next years (Myla, S.2019). ML is a more efficient of analysis and processing that utilizes a mathematical model combined with training data to learn how to make predictions, whereas AI refers to the wider application of machines that perform tasks that are characteristic of human intelligence, such as extrapolate findings from computation or induction. ML learns variables from instances rather than directly calculating outcomes from a set of preset rules and would have the ability to perform better at a job like identifying and distinguishing trends in information by being subjected to more examples(Kumar V, 2021). The most sophisticated machine learning (ML) methods, also known as deep learning (DL), are particularly well-suited for this task (Fig. 1). Because it relied heavily on complicated strategy, such as multifaceted contrast processes, as well as the need for reliable and accurate segmenting the market and quantifying of biological markers predicated on procured data to help guide diagnosis and therapy governance, cardiac MRI is a research area that gives itself more to ML (Karthikeyan T, 2019).

Figure 1. Representation for the network of multi-scale segmentation

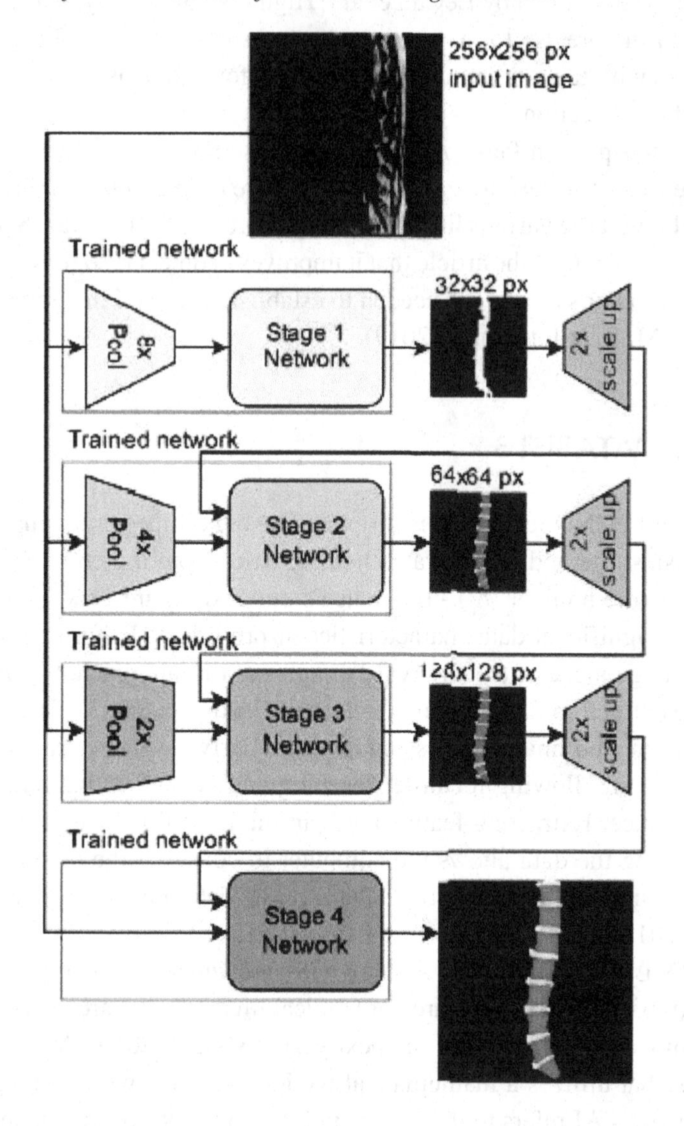

System Diagram

A multidimension based pixel convolution (CNN) has been suggested for the dependency of the system in the neural networks of NN to be utilized in each system of pixels for the region to be located. Further, every definite system holds the alternate images for the randomness in the course of principal system to be at the stage of depicters for 1/8th amount of cost in the adaption of the image in the strategy of the picture in the available dataset, this relates end route for measures for the clouded part towards the aim disintegrating various tissues (Raj Kumar, A.2019) & (Gattim, N. K.2019). Calculations to be done for replicating the estimated quantity of the state of art of system in the division of computation associated with the system of the segments for an image in the midst of estimations of 32x32px. Exactly when the image is scaled back tremendous applicable features on image indenture with estimation, empowering toward neural network-based frameworks for interpreting immense image features deprived of getting ready pixels further. The principal division for the resulted stage to scale up for closing an element having configuration 2.1 (approximately 64x64pixellinked to the framework) commitment not long before the accompanying stage. The second point divides the image at 1/4thextent just as the model continues (Ahammad, S. H., 2019) & (Srinivasa Reddy, K.2019). The second point adds its respect the commitment for the third stage (half level), later then permits the division yield in coordination to enhance the development for final limited version of range estimate (Narayana, V. V.2019) & (Kumar, S.2019).

Mathematical Analysis and Preparing

Important features or qualities for completing a specific function are retrieved from pictures using a predefined set of features in a typical ML model. Image extraction noise characteristics, smoothness, and movement are examples of characteristics for cardiac outlines in the image below. After the designing phase is completed, ML algorithms must be trained on real-world data. The parameters of the feature set model are learnt during this training period. The parameters of the model govern the real predictions produced. A model is any function of the features utilised for prediction. The algorithm may then be used it to generate predictions for information that was not observed during the training period. Classifier, in which discrete labels such as the presence or absence of illness are identified, or regress, in which observed data such as T1 are calculated, are both possible with ML models. Because the algorithms train from instances, having a suitably big dataset with realistic variation accessible for performance is essential. It is critical to keep the data sets used in model creation and fine-tuning distinct from the test data used to assess the model's performance when evaluating the precision. And during learning phase, additional data (typically referred to as the verification dataset) is utilized to assist in determining the best architecture for the ML model. This data is being used to improve input variables and prevent overfitting. (Nagageetha, M.2017).

$$\mathcal{L}(W, C, C^*) = \overline{W \circ \|C_{i,j} - C^*_{i,j}\|} \quad (1)$$

$$W_1 * \sum_{i,j \in c_1} e_{i,j} \approx W_2 * \sum_{i,j \in c_2} e_{i,j} \quad (2)$$

The W1 and W2 qualities having the weighted surfaces of the local grid in the conditional feature of the arrangement at the frontal region of C1 pixels, it may be at the dialogue side of the circles having the real condition of the define for the components to ne contrasted to the foundation of C2 pixels in the vertebrate. The grouping of the single species can be merged with the normal concern of the pixel per image of the goal reached at the perseverance for the support needed in the category of the weights for the preparation of image in the weighted path of heaviness exclusive for the deduction of the procedure. Setting up the involved framework starts through the essential stage in addition to continues in a repetitive solicitation, as showed up in fig: 2. to look at the multi/scale framework making progress toward various plans, each filled arrangement attaining the feature of daintiness at the point of individual age etc (Kumar, D. S.2020) & (Soumya, N.2019).

Design for Stage-Network

The best discriminant information for a particular CNN finding is one of the most important stages in developing ML systems. This has proved to be very difficult. DL is a branch of machine learning that can help with this problem. Unlike traditional machine learning techniques, Ductility may learn neural network, eliminating the requirement for discriminative characteristics to be handcrafted. In the case of estimating the position of myocardial contours, DL techniques learn the picture characteristics that are most important for reducing the contours' placement. (Saikumar, K., & Rajesh, V. 2020).

Figure 2. U-Net CNN process

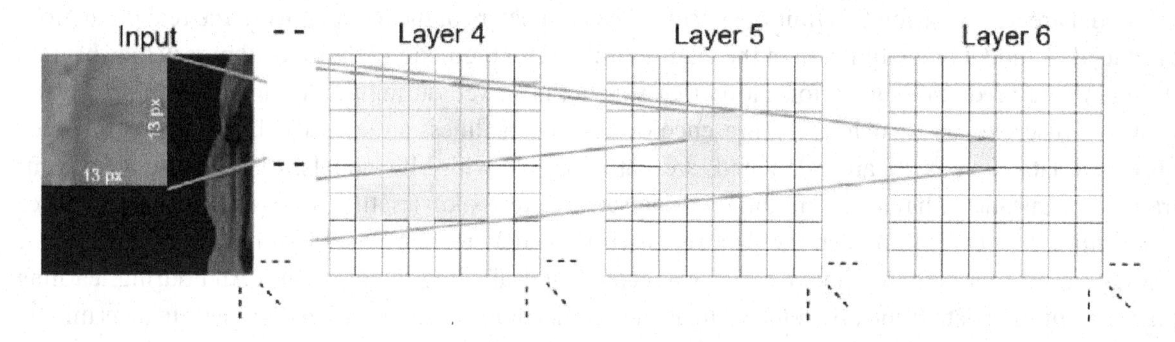

Four symbiotic breakthroughs have helped fuel latest DL achievements: 1) the accessibility of large tons of extra digital image data for training; 2) the capacity of methodologies to learn necessary details directly from images deprived of the need for handcrafted features; 3) reduced strong cpu (GPU) equipment; and 4) open - source software library resources and working examples. These advancements have resulted in the creation of multilayered neural networks, which are referred to as "deep" in DL. The convolutional neural network (CNN), a kind of deep learning network, is mostly used for image processing.

A typical CNN system is collected of many layers, each of which has its own design (Fig. 3). Convolutional layer uses a series of filters to apply to a picture in order to generate scattering characteristics for the following layer. The goal is to figure out what the best values for the filters (also known as weights) are so the next layers may produce characteristics that are most relevant to the job. Down sampling spatial data using max - pooling (e.g., max pooling or regular pooling) makes characteristics more appropriate for the job. A fully linked layer, in which each component is linked to all other nodes in the layer, is used for multi - class nets. To restore the picture dimensions to the input image size, segmented systems often employ down sampling processes. Skipping levels are often used to recover fine imaging ways to enhance gradient propagation during training by allowing small details to propagate from one layer to another. Ultimately, a non-linear process is performed by a SoftMax layer, which rescales the mechanisms to give each pixel class a non-negative likelihood. In the hidden layers, this guarantees that all outputs add up to 1. Many billions of parameters are often used in deep CNN solutions. Even though the knowledge included in the features generated from the convolutions in the intermediate layers is relevant to the job, it may be hard to understand how the network makes predictions or why it fails. For medical image analysis, however, DL is the most often used ML structure. In the six years leading up to 2017, more than 300 DL articles were submitted to the field of medicine image analysis, including CMR, according to a recent study, with quantities increasing rapidly (Devaraju, V. K.2019).

Approachable field

Four interdependent advances have helped to latest DL dataset achievements: 1) the accessibility of big pixels of extra digital image data for training; 2) the capacity of methodologies to learn necessary details directly from images without the need for handcrafted features; 3) reduced strong (GPU) equipment; and 4) open - source software library resources and working examples. These advancements have resulted in the creation of multilayered neural networks, which are referred to as "deep" in DL. The convolutional neural network (CNN), a kind of deep learning network, is mostly used for image processing (K, Saikumar & V, Rajesh.2020). A typical CNN system is composed of many layers, each of which has its own design (Fig. 3). Convolutional layer uses a series of filters to apply to a picture in order to generate scattering characteristics for the following layer. The goal is to figure out what the best values for the filters (also known as weights) are so the next layers may produce characteristics that are most relevant to the job. Down sampling spatial data using max - pooling (e.g., max pooling or average pooling) makes characteristics more appropriate for the job. A fully related layer, in which each component is linked to all additional nodes in the layer, is used for multi - class nets. To restore the picture dimensions to the input image size, segmented systems often employ downsampling processes. Skipping levels are often used to recover fine imaging ways to enhance gradient propagation during training by allowing small details to propagate from one layer to another. Ultimately, a non-linear process is performed by a SoftMax layer, which rescales the mechanisms to give each pixel class a non-negative likelihood. In the hidden layers, this guarantees that all outputs add up to 1. Many billions of parameters are often used in deep CNN solutions. Even though the knowledge included in the features generated from the difficulties in the middle layers is relevant to the job, it may be hard to understand how the network types calculations or why it fails. For medical image analysis, however, DL is the most often used ML structure. In the six years leading up to 2017, more than 300 DL articles were submitted to the field of medicine image analysis, including CMR, according to a recent study, with quantities increasing rapidly (Saikumar, K., Rajesh, V. 2020).

Examining with U-net

Generally, the assessment of the overall system is made by different layers, it with an asserting scientific design strategy developed in CT/MRI tomography field, the U- net model. Table 1 clearly explains about the training data of the DL results in the past measurements of the prepared system that as set for the testing system. Both the plate and vertebral division of body has been subjected to the outcome appeared in below figure. 3

Figure 3. Column based spinal cord U-Net segmentation

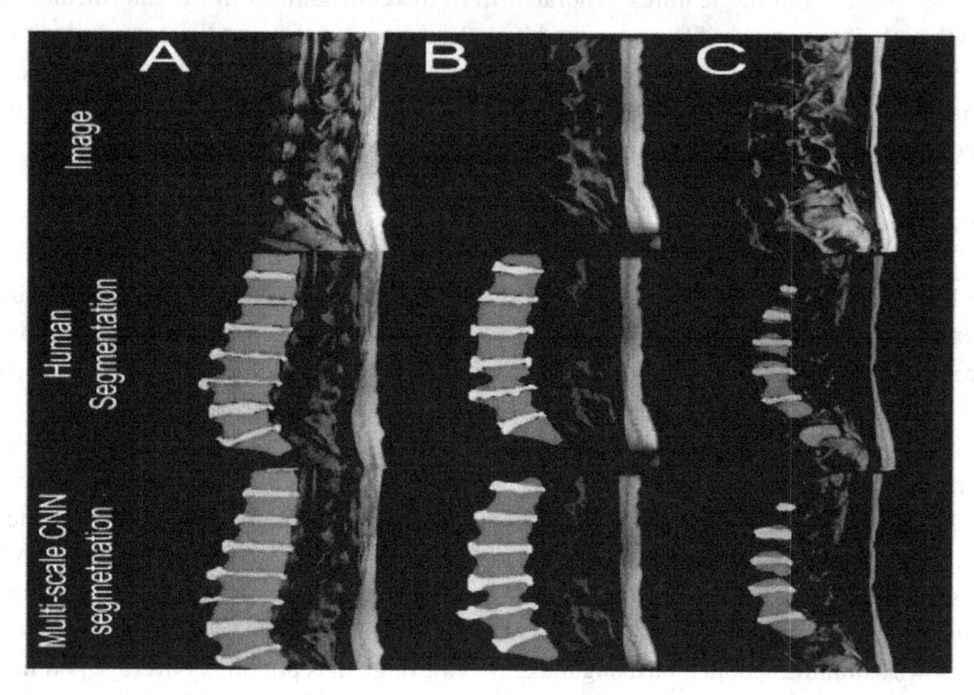

Figures 3 and 4 demonstrate the dissimilar result for the U-net has been stranded for the bolster made in the multiscale n be empowered with the rate of dependency of the spine system life span in the cross-sectional way to the automated machines for the growth of the internal.

Figure 5 establishes the DSI result on test data for the picture through the region of span of materials at each stage of system. The highest situation is made for initial ages and reliability is reached for the proposed system than the U-net.

Figure 4. Based on the partial visualizing disk and vertebrae the network performance outperforms

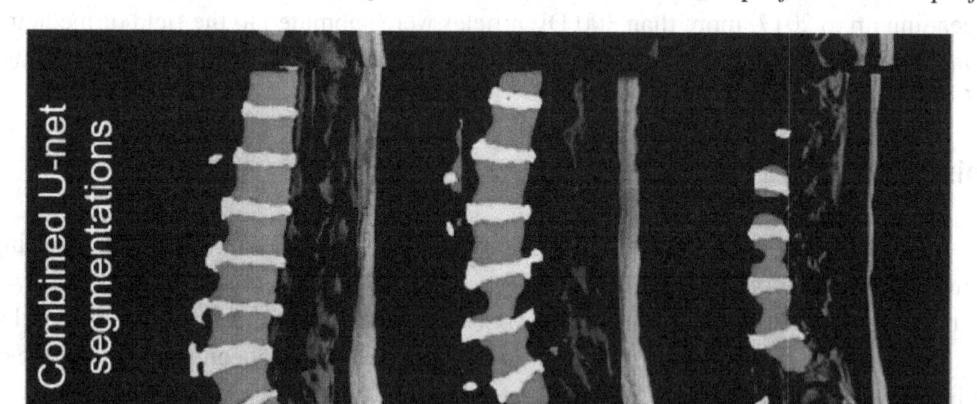

Figure 5. Various transforms for the fusion methods and edge association

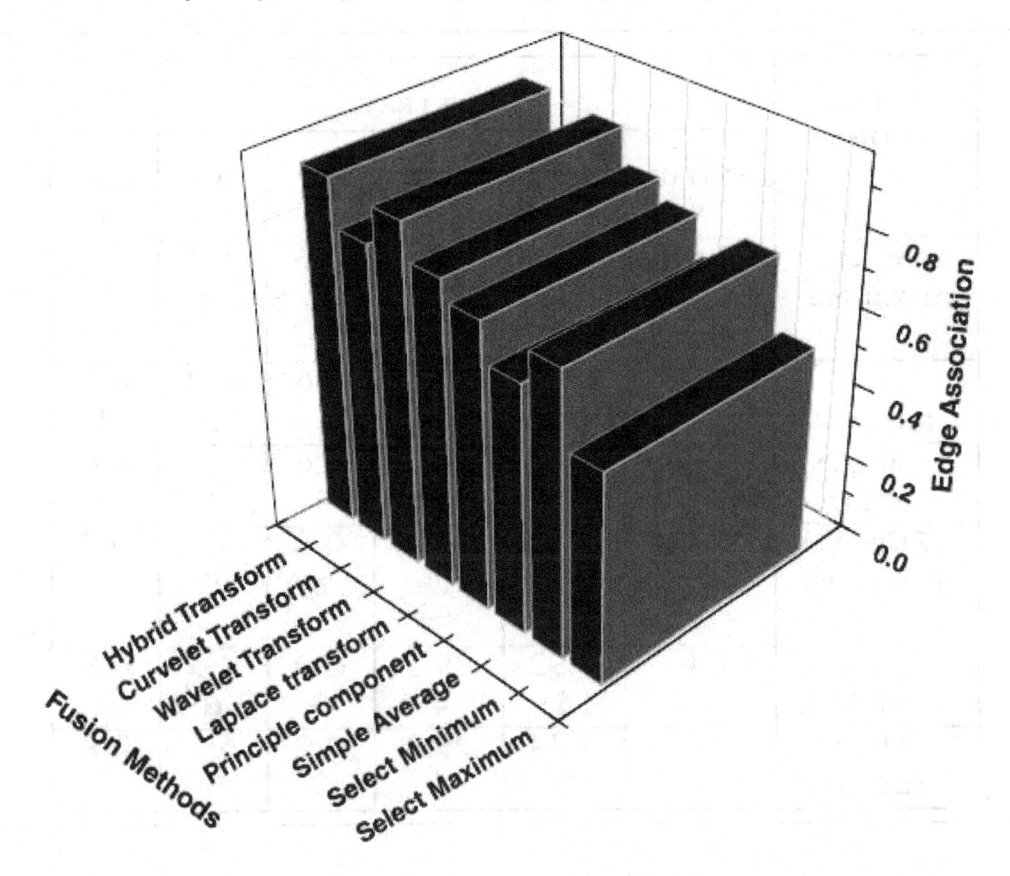

Table 1. Statistical results of various techniques

FUSION METHODS	METRICS					
	ENTROPY	RMSE	PSNR	CC	MI	QAB/F
Select maximum	6.63	4.248	29.56	0.61	5.23	0.55
Select minimum	2.89	15.23	23.25	0.67	6.92	0.74
Simple average	4.23	13.23	27.32	0.67	4.81	0.65
Principle component	6.34	3.421	36.12	0.88	5.89	0.79
Laplace transform	7.45	3.921	39.24	0.90	4.56	0.80
Hybrid transform	7.77	3.442	29.33	0.92	6.23	0.89

CONCLUSION

In this research work an MRI spinal cord and heart images disease diagnosis is estimating using pixel based convolutional neural network mechanism. It is a multi-level scalling technology to extracting damaged pixels in the input image. The training process contain 2lakh MRI images and testing process has around 1000 images from various hospitals. After the training process estimating the accuracy and confusion matrix calculations using following mechanism measuring presentation measures such as exactness, sensitivity, F1 score and recall. The U-net based PCNN technology had been improving disease detection functionality compared to conventional models. Due to parallel channel mechanism and U-Net technology had been helping achievement of better accuracy score i.e. 99.5% and sensitivity 98.73%. The spinal cord columns and heart vessels information can be attained from segmentation process. The pixel-based segmentation can help the object and background variation as well as providing better testing accuracy.

REFERENCES

Ahammad, S. H., Rajesh, V., Hanumatsai, N., Venumadhav, A., Sasank, N. S. S., Gupta, K. B., & Inithiyaz, S. (2019). MRI image training and finding acute spine injury with the help of hemorrhagic and non hemorrhagic rope wounds method. *IJPHRD*, *10*(7), 404. doi:10.5958/0976-5506.2019.01603.6

Ahammad, S. H., Rajesh, V., Neetha, A., Sai Jeesmitha, B., & Srikanth, A. (2019). Automatic segmentation of spinal cord diffusion MR images for disease location finding. *Indonesian Journal of Electrical Engineering and Computer Science*, *15*(3), 1313–1321. doi:10.11591/ijeecs.v15.i3.pp1313-1321

Ahammad, S. H., Rajesh, V., & Rahman, M. Z. U. (2019). Fast and accurate feature extraction-based segmentation framework for spinal cord injury severity classification. *IEEE Access: Practical Innovations, Open Solutions*, *7*, 46092–46103. doi:10.1109/ACCESS.2019.2909583

Ahammad, S. H., Rajesh, V., Rahman, M. Z. U., & Lay-Ekuakille, A. (2020). A hybrid CNN-based segmentation and boosting classifier for real time sensor spinal cord injury data. *IEEE Sensors Journal*, *20*(17), 10092–10101. doi:10.1109/JSEN.2020.2992879

Ahammad, S. H., Rajesh, V., Venkatesh, K. N., Nagaraju, P., Rao, P. R., & Inthiyaz, S. (n.d.). Liver segmentation using abdominal CT scanning to detect liver disease area. *International Journal of Emerging Trends in Engineering Research*, *7*(11), 664-669.

Ahammad, S. K., & Rajesh, V. (2018). Image processing based segmentation techniques for spinal cord in MRI. *Indian Journal of Public Health Research & Development*, *9*(6), 317. doi:10.5958/0976-5506.2018.00571.5

Devaraju, V. K., & Rao, S. S. (n.d.). A Real and Accurate Vegetable Seeds Classification Using Image Analysis and Fuzzy Technique. *Turkish Journal of Physiotherapy and Rehabilitation*, *32*, 2.

Gattim, N. K., Pallerla, S. R., & Bojja, P. (2019). Plant Leaf Disease Detection Using SVM Technique. *International Journal of Emerging Trends in Engineering Research*, *7*(11), 634–637. doi:10.30534/ijeter/2019/367112019

Inthiyaz, S., Prasad, M. V. D., Lakshmi, R. U. S., Sai, N. B. S., Kumar, P. P., & Ahammad, S. H. (2019). Agriculture Based Plant Leaf Health Assessment Tool: A Deep Learning Perspective. *International Journal of Emerging Trends in Engineering Research*, *7*(11), 690–694. doi:10.30534/ijeter/2019/457112019

Karthikeyan, T., Sekaran, K., Ranjith, D., Vinoth Kumar, V., & Balajee, J. M. (2019). Personalized Content Extraction and Text Classification Using Effective Web Scraping Techniques. *International Journal of Web Portals*, *11*(2), 41–52. doi:10.4018/IJWP.2019070103

Kumar, D. S., Kumar, C. S., Ragamayi, S., Kumar, P. S., Saikumar, K., & Ahammad, S. H. (2020). A test architecture design for SoCs using atam method. *Iranian Journal of Electrical and Computer Engineering*, *10*(1), 719.

Kumar, M. S., Inthiyaz, S., Vamsi, C. K., Ahammad, S. H., Sai Lakshmi, K., Venu Gopal, P., & Bala Raghavendra, A. (2019). Power optimization using dual sram circuit. *International Journal of Innovative Technology and Exploring Engineering*, *8*(8), 1032–1036.

Kumar, V. V., Ramamoorthy, S., Kumar, V. D., Prabu, M., & Balajee, J. M. (2021). Design and Evaluation of Wi-Fi Offloading Mechanism in Heterogeneous Networks. *International Journal of e-Collaboration*, *17*(1), 60–70. doi:10.4018/IJeC.2021010104

Myla, S., Marella, S. T., Goud, A. S., Ahammad, S. H., Kumar, G. N. S., & Inthiyaz, S. (n.d.). *Design Decision Taking System For Student Career Selection For Accurate Academic System*. Academic Press.

Myla, S., Marella, S. T., Goud, A. S., Ahammad, S. H., Kumar, G. N. S., & Inthiyaz, S. (n.d.). *Design Decision Taking System For Student Career Selection For Accurate Academic System*. Academic Press.

Nagageetha, M., Mamilla, S. K., & Hasane Ahammad, S. (2017). Performance analysis of feedback based error control coding algorithm for video transmission on wireless multimedia networks. *Journal of Advanced Research in Dynamical and Control Systems*, *9*, 626–660.

Narayana, V. V., Ahammad, S. H., Chandu, B. V., Rupesh, G., Naidu, G. A., & Gopal, G. P. (2019). Estimation of Quality and Intelligibility of a Speech Signal with varying forms of Additive Noise. *International Journal of Emerging Trends in Engineering Research*, *7*(11), 430–433. doi:10.30534/ijeter/2019/057112019

Raj Kumar, A., Kumar, G. N. S., Chithanoori, J. K., Mallik, K. S. K., Srinivas, P., & Ahammad, S. H. (2019). Design and Analysis of a Heavy Vehicle Chassis by using E-Glass Epoxy & S-2 Glass Materials. *International Journal of Recent Technology and Engineering*, *7*(6), 903–905.

Saikumar, K., & Rajesh, V. (2020). Diagnosis of coronary blockage of artery using MRI/CTA images through adaptive random forest optimization. *Journal of Critical Reviews*, *7*(14), 591–600.

Saikumar, K., & Rajesh, V. (2020). Coronary blockage of artery for Heart diagnosis with DT Artificial Intelligence Algorithm. *International Journal of Research in Pharmaceutical Sciences.*, *11*(1), 471–479. doi:10.26452/ijrps.v11i1.1844

Saikumar, K., & Rajesh, V. (2020). A novel implementation heart diagnosis system based on random forest machine learning technique. *International Journal of Pharmaceutical Research.*, *12*, 3904–3916.

Siva Kumar, M., Inthiyaz, S., Venkata Krishna, P., Jyothsna Ravali, C., Veenamadhuri, J., Hanuman Reddy, Y., & Hasane Ahammad, S. (2019). Implementation of most appropriate leakage power techniques in vlsi circuits using nand and nor gates. *International Journal of Innovative Technology and Exploring Engineering*, *8*(7), 797–801.

Srinivasa Reddy, K., Suneela, B., Inthiyaz, S., Kumar, G. N. S., & Mallikarjuna Reddy, A. (2019). Texture filtration module under stabilization via random forest optimization methodology. *International Journal of Advanced Trends in Computer Science and Engineering*, *8*(3), 458–469. doi:10.30534/ijatcse/2019/20832019

Vijaykumar, G., Gantala, A., Gade, M. S. L., Anjaneyulu, P., & Ahammad, S. H. (2017). Microcontroller based heartbeat monitoring and display on PC. *Journal of Advanced Research in Dynamical and Control Systems*, *9*(4), 250–260.

Chapter 3
Multilingual Novel Summarizer for Visually Challenged Peoples

Amalraj Irudayasamy
University of Technology and Applied Sciences, Nizwa, Oman

Prasanna Ranjith Christotodoss
https://orcid.org/0000-0003-4778-7915
University of Technology and Applied Sciences, Shinas, Oman

Rajesh Natarajan
https://orcid.org/0000-0003-1255-9621
University of Technology and Applied Sciences, Shinas, Oman

ABSTRACT

In our busy lives, most of us hardly have time to read books. This habit of reading is slowly diminishing because of people's busy lives. The situation is significantly difficult for persons who are visually challenged or have lost their vision. As a result, the authors provide a method based on the sense of sound that is better and more accurate than the sense of touch for visually impaired people. This chapter discusses an effective method for condensing books into important keywords in order to avoid having to read the entire text each time. This work employs a variety of APIs and modules, including Gensim, Text Ranking Algorithm, and other functions for translating summary text to speech, allowing the system to assist even the blind.

INTRODUCTION

Giving machines the power to think has always been a far-fetched ambition for humans since the dawn of humanity. People with normal vision may be able to benefit from information received through their ears. In any case, according to WHO data, 285 million individuals worldwide will be designated as visibly impaired by October 2013:Visually impaired are 39 million men, and low vision is 246. Unfortunately, a large number of visually impaired and blind peoples are unable to read conventional news stories in the same way that normal people can, and must instead rely on their fingerprints or special clothing to

DOI: 10.4018/978-1-7998-9640-1.ch003

read braille, which adds greatly to their weight. Braille is an extraordinary frame of loads of images made of small rectangular braille cells containing small significant knocks called elevated dabs used by the visually impaired and outwardly impaired. People with visual impairments encounter several challenges around the world as they try to establish themselves in a modern, complicated, and competitive world dominated by able-bodied folks. Disabled people are frequently excluded from social activities and are not treated equally to their able-bodied peers.Text summarization is the process of distilling from a source (or source) the most important information to create a shortened version for a specific user (or user) and task (or task).

Humans are generally good at this function because we can comprehend the essence of a written message, extract outstanding aspects, and characterise the documents using human language. However, in today's environment, where there is a plethora of data and a lack of manpower and time to understand it, automatic text summarization methods are essential. Automatic Text Analysis can be beneficial for a variety of reasons. There are numerous reasons why Automatic Text Analysis is useful:

1. Summaries reduce the time it takes to read.
2. Summaries support the selection process when reviewing papers.
3. Automatic summarising improves indexing effectiveness.
4. Automatic algorithms of the summary are less biased than those of human summarizers.
5. Contextual summaries are useful as they provide contextual data in question-answering systems.
6. Using automated or semi-automatic synthesis systems, commercial abstract services can increase the number of text documents they can process.

The approach we present in this paper is beneficial for people with visual impairment as well as for people who have tight schedules are unable to read long novels. Using the PDFminer module that runs through the PDF document, it acts as an interface to retrieve the content, whatever it may be, in textual form.Upon receiving the message, the material is now condensed into a few relevant and self-explanatory keywords by applying the Message rank algorithm, thus saving the effort to go through the entire article. In the next step, this summarized content is converted into a voice that can be perceived directly by people with visual impairment. Thus, with no help from anyone, the content can be delivered to them accurately (Aone, C et al., 1998).

Based on the survey by the World Health organization in 2010, the total population in India is 1181.4 million out of which people who suffer from blindness, low vision, and visual impairment are 152.238 Million. Impaired vision may have negative effects on learning and social interaction, according to Dr. Bjorn. People who are visually impaired cannot be recovered with the help of glasses. This condition affected the reading process's duration and tired the ears. A tool for reading the article is needed to help improve the quality of life for people with low vision (Praveen Sundar, P. V et al., 2020). The level of vision impairment can vary with low vision in each person. Therefore another sensory feature was used by a system built in this work to receive information from a message.The device is designed especially for low vision people. Therefore, they can use this device easily without having to ask others for help, and they can use this tool for educational and intellect skills (Barzilay, R., & Lee, L. 2003).

LITERATURE SURVEY

Most of the previous methods of summarization are extraction-based, which ranks and extracts existing sentences directly in a document set to summarize them. Typical approaches include centroid-based (Barzilay et al., 2003), NeATS, supervised learning-based methods, graph-based rankings, integrated linear programming and submodular function programming and in addition, a cross-language document summary has been investigated, but the task is focused on how to select good content quality from translated sentences Baxendale, P. B. (1958). We can see that all current description programs are introduced for normal people, but not for the blind or visually impaired people (Brandow, R et al., 1995). For the blind and visually challenged, document review must take place and should be discussed. It has taken a long time to let the blind and visually impaired surf information, as convenient as it is for ordinary people. The long-term goal is to do this by creating unique instruments. Following Braille's popularity, a variety of braille display devices have been developed for braille reading. The majority of research in this area, on the other hand, concentrated on how to make internet data from blind individuals more usable Edmundson, H. P. (1969).

Consumer demographics, requirements, trends, and behaviour are frequently conducted or compiled by organisations that supply to print-disabled individuals (Erkan, G., et al 2004). They are likely to be up to date on the latest vision aids and technology and are already educating customers or assisting with adaptive device issues. Because many blind individuals use both a library and these agencies' services, there are significant prospects for collaboration in terms of enabling adaptive technologies and understanding consumer demands. Instead than duplicating services that are sometimes rather costly to begin with, libraries could take advantage of these relationships (Freitas, D., & Kouroupetroglou, G. 2008). It compares the difference between the methods based on the degree and the methods based on the centroid and concludes that the approaches based on the degree are better. In this method, the main drawback is that the technique used is insensitive to the data noise. Shows and proves that the part of the text that is present in the content and that repeats a lot has a high likelihood of being present on the content summarized or are the better terms to be present on the content summarized. This paper (Gillick, D., & Favre, B. 2009). has coined a term weight on each content (which is the frequency count in the content of that particular text), which, if kept in the summarized content, gives good results. Clients of blind libraries must build solid policies and procedures to ensure that they may acquire information and leisure reading materials in the formats of their choosing and choose the most efficient distribution methods. These policies should spell out the terms of service for returning borrowed materials to the library, including when and how they should be returned (Li, C., Qian, X., & Liu, Y. 2013).

Another unsupervised technique was explored in to paraphrase using Multi-Sequence Alignment, in which they addressed the problem of sentence-level text-to-text paraphrasing, which is even more difficult than word-level paraphrasing (Kumar V V et al, 2021).They present an approach that applies the alignment of multiple sequences to sentences collected from the unnoted corpora. The system is made to learn a set of paraphrasing patterns, and then it determines automatically how to apply these patterns to paraphrasing sentences. The researchers have developed a system that correctly extracts paraphrases by using machine learning principles, thus outperforming the baseline systems. Another unsupervised approach to word sense disambiguation has been proposed which excels in quality. The researchers also developed a system that uses the principles of machine learning to correctly extract paraphrases, thereby outperforming the baseline systems. The algorithm eliminates the need for any kind of valuable hand-tagged training data as it uses powerful human language properties, namely one sense of collocation and one sense per discourse, to achieve the outcome. Manual summarizing requires a great deal of human effort and time. Therefore, automated text analysis is added, which saves the time of the legal expert (Lin, C. Y., & Hovy, E. 2003). The overview activity includes defining linguistic positions interpreting a legal decision document's sentences. The search task is to identify related past cases as per the legal query in question. We have introduced a hybrid system for these two tasks, which is a combination of various techniques. Keywords or essential phrase matching techniques and case-based techniques are the strategies used in our hybrid system (Lin, H., & Bilmes, J. 2010).

A survey conducted on various techniques of text summary that have occurred in the recent past. This paper has put particular emphasis on summarizing the legal text as it is one of the key areas in the legal domain. We begin with a general introduction to the text summary, briefly touch on recent progress in the single and multi-document summary, and then dive into the extraction-based summary of legal text. We discuss various data sets and metrics used to summarize and compare the performance of different approaches, first generally and then focused on legal text. It offers an approach to the use of incoming and outgoing citation data for the review of documents.Our work aims to automatically generate catchphrases for legal case reports, using the text of cited cases and cases citing the current situation as well as the full text.We propose methods for using catchphrases and quoted/citing case sentences to extract catchphrases from the target case text. We created a corpus of cases, slogans, and quotations, and performed a ROUGE-based evaluation that shows our citation-based methods' superiority over full-text-only methods (Karthikeyan T et al., 2019).

A sequence of tests can be represented using a wide range of machine learning methods to determine the rhetorical status of sentences. Experimental results were presented for rhetorical classification, showing the contribution of a sentence to the overall argumentative structure of legal decisions using four learning algorithms from the 5 Weka kit (C4.5, naïve Bayes, Winnow and SVMs). Results are also documented in a generic classification system as well as in a sequence labeling framework using maximum entropy models. The SVM classifier and the maximum entropy sequence tagger produce the most promising results. Statistical description methods were extended to the legal domain by the SUM campaign. We explain some experiments with the House of Lord's judgments, Where we performed a small sample collection of automated linguistic annotations to examine the connection between linguistic features and arguments (Muthukumaran V et al., 2021) . They used state-of-the-art NLP techniques to perform linguistic annotation using XML-based tools and a mixture of rule-based and quantitative methods. Here we focus on the predictive potential of the tense and aspect characteristics of a classifier.

Similar research has been proposed and tested on methods of text analysis. In the summary process, there are two main methods, namely extraction, and abstraction. Extraction involves concatenating extracts taken from the corpus into a summary, while abstraction involves generating new sentences from the corpus data.Extractive summary techniques can be further categorized into two groups: supervised techniques based on pre-existing summary pairs of documents, and unsupervised techniques based on properties and heuristics extracted from the text. Supervised extractive definition techniques consider the sentence level analysis method as a question category of two classes. Several unsupervised methods were developed to summarize the study by using different characteristics and phrase relationships. Early text analysis work uses different features such as word frequency, phrase position, phrase cue phrases, phrase length and upper case letter etc.Today, corpus-based approaches play an important role in text summarization. It is possible to learn from a corpus of documents and their associated summaries by using machine learning algorithms. The main advantage is that corpus-based methods are easy to implement. Much of the summary work done so far has not been extended to summary usage, instead of describing the purpose of the method. The main weakness of extractive summarization is its low readability. Since readability is an essential component of text-com-understanding, in extractive summarization there is no ordering of continuity and sentence.Extractions must be ordered correctly to achieve a succinct description (Maithili, K et al., 2018).

METHODOLOGY

PDF to Text using PDFMiner

PDFMiner is a tool in which data is extracted from PDF files. Similar to other PDF-related software, it focuses entirely on the processing and review of text data PDFMiner allows you to accurately identify the text of a document and other details such as fonts or lines. It includes a PDF converter that can convert PDF files to other types of text (e.g. HTML).

Figure 1. The workflow of the proposed system

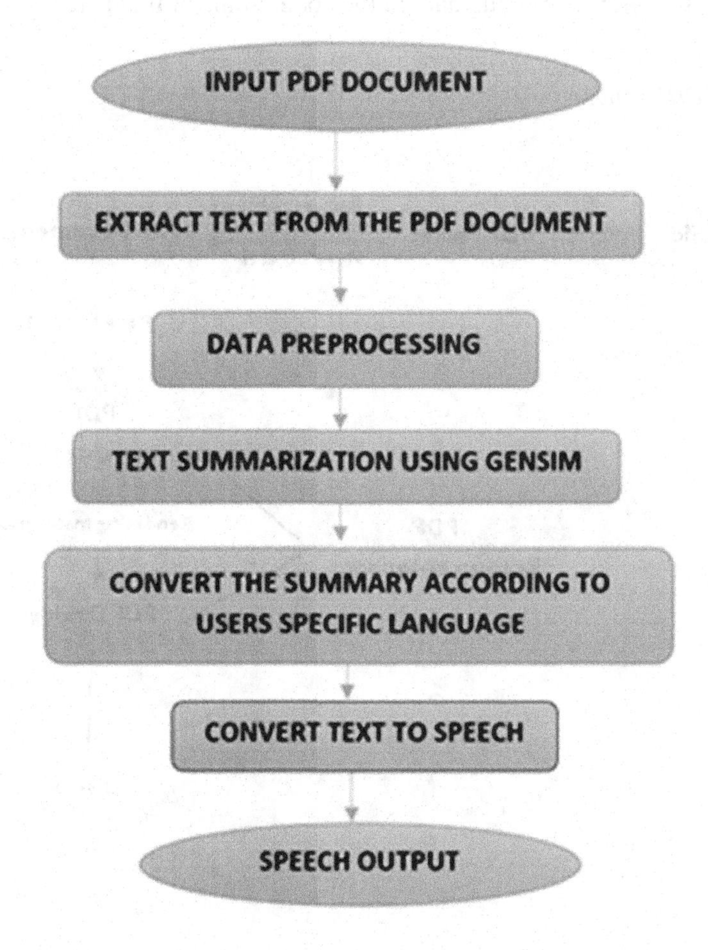

PDFMiner allows you to precisely locate the text of a document, as well as other information such as fonts or lines.This provides a PDF translator that can convert PDF files to other text (e.g. HTML) formats. It has an extensible PDF parser that can be used for purposes of non-text analysis shown in Figure 1.

Data Preprocessing

Data pre-processing is a data mining technique for transforming raw data into a usable format shown in Figure 2.
Steps involved in data pre-processing:

Data Cleaning

The findings contain a number of trivial and missing sections. In order to control this fraction, results cleansing is performed. This involves dealing with missing software, noisy data, and so forth.

Missing data

This situation arises when there is missing data in the code. Some of them are:

Figure 2. Working of PDFminer

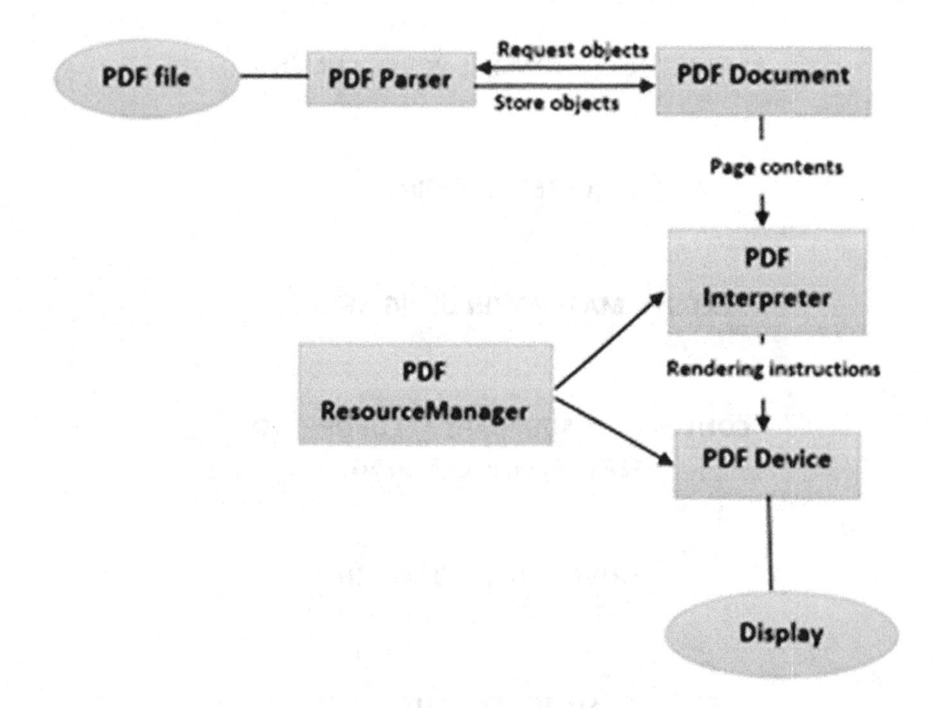

Remove the tuples

This method should only be used if the dataset is huge and there are many missing values within a tuple.

Fill the Missing Values

This task can be completed in a variety of ways. You can manually fill in the missing values using the mean attribute or the most likely value.

Noisy data

Noisy data is worthless information that computers cannot interpreter. It can be caused by faulty data collecting, data entry problems, and other factors. It can be dealt with in the following way:

Binning method

This method works on sorted data in order to smooth it out. The whole data is divided into sections of equal size and different methods are then used to complete the task. Separately, the portion is dealt with. To complete the task, all information in a section can be replaced by the mean or limit values.

Regression

Data can be easily changed to a regression function here. The regression used may be linear (with a single variable) or hierarchical (with multiple independent variables).

Clustering

In a cluster, similar data are clustered in this method. It is not possible to observe the outliers or collapse outside the clusters.

Data transformation

This move is taken in order to turn the data into suitable forms for the mining process. It includes the following possibilities:

Normalization

It is performed in a specified range to scale data values (-1.0 to 1.0 or 0.0 to 1.0).

Attribute Selection

New attributes are generated to assist the mining process from the given set of attributes in this strategy.

Discretization

In replace raw values of numeric attributes with interval or logical classes, this is done.

Concept Hierarchy Generation

Here attributes are translated from level to higher level in the hierarchy. You can convert the "city" attribute to "country," for example.

Data Reduction

Since data mining is a technique used to manage large quantities of data. In such cases, when dealing with a large volume of data, analysis became more complicated. To get rid of this, we use the data reduction technique. This is intended to increase processing capacity and reduce data storage and research costs. The various steps to data reduction are:

Data Cube Aggregation

Aggregation operation is applied to data for the construction of the data cube.

Attribute Subset Selection

The highly relevant attributes must be used; the rest can be discarded. For the choice of attributes, the attribute with a p-value greater than the degree of significance can be discarded. It is possible to use the attribute with a p-value greater than the significance level.

Numerosity Reduction

This requires the data model to be stored instead of data as a whole, like Regression Models.

Dimensionality Reduction

It reduces the size of information by encoding mechanisms. It can be without loss or without loss. This reduction is called lossless reduction if original data can be restored after retrieval from compressed data, otherwise, it is called loss reduction. The two effective wavelet transforms and methods for reducing dimensionality are PCA (Principal Component Analysis).The text extracted from the pdf document includes uninformative characters to be deleted. Such figures increase the efficiency of our models by having excessive ratios of the count. The method below uses a series of regex search and replace function as well as a list-understanding to substitute these characters with a blank space. The functions below have been used for this approach and the resulting text contains only alphanumeric characters shown in Figure 3.

Figure 3. Steps involved in pre-processing

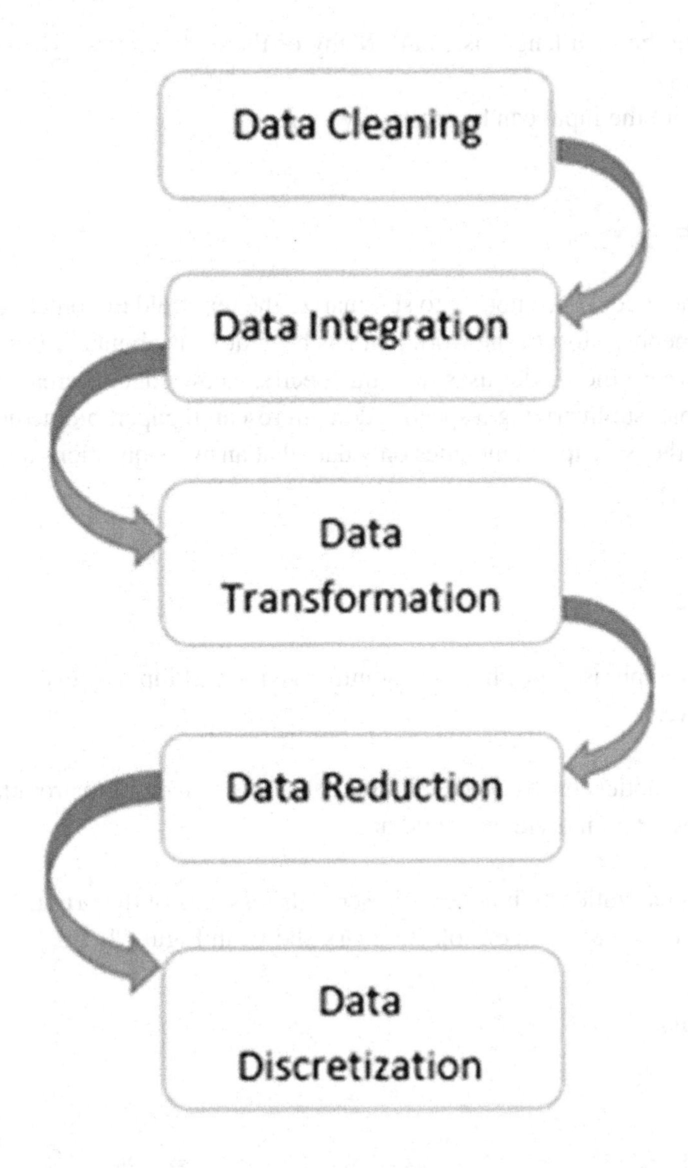

Text Summarizer

Methods of text summary can be classified into various types.

Based on Input Type

1. The single file where the data length is small. Many of the early review systems dealt with the single report overview.
2. Multi-document, where the input can be arbitrarily long

Based on the Purpose

1. Traditional, where the model does not try to summarize the text field or content and considers all inputs to be homogeneous. Most of the work that has been done is about a traditional synthesis.
2. A specific domain where the model uses domain-specific knowledge to create a more accurate summary. For example, summarizing a specific domain research paper, biomedical papers, etc.
3. Query-based, where the description includes only data that answers questions about the input text in the natural language.

Based on Output Type

1. Extractive, where essential phrases are chosen to summarize the text input. Most of today's overview methods are extractive.

2. Abstract, where the model constructs the phrases and phrases in order to provide a more concise explanation of what an individual would do.

We used an extractive summarization technique to preserve the essence of the original novel. The ranking algorithm for text is used to create an eBook summary shown in Figure 4.

Figure 4. Text rank algorithm

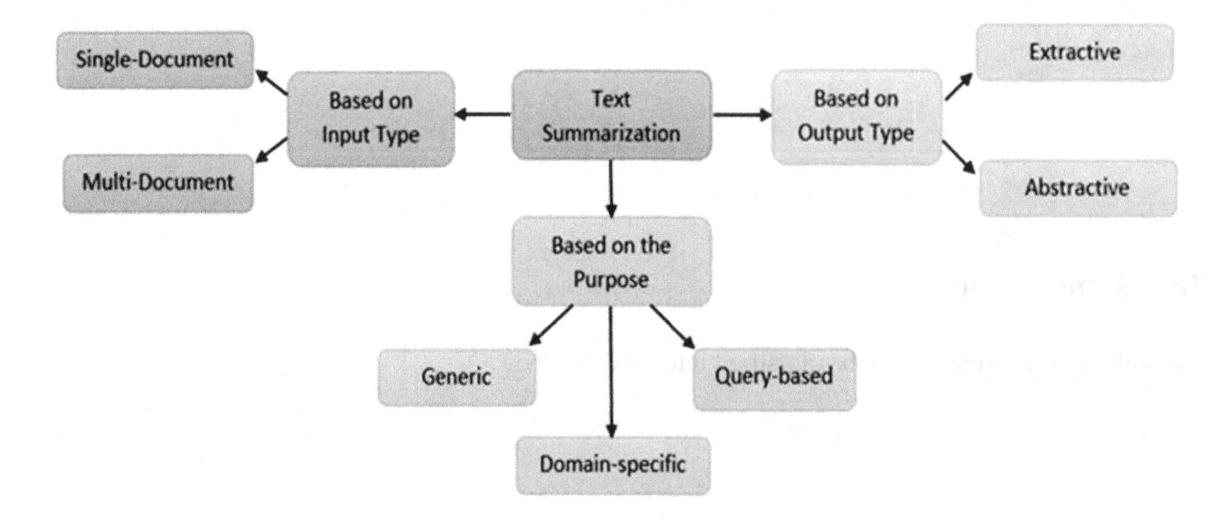

Identify Relevant Keywords

The text rank algorithm builds a word network to find relevant keywords. This network is built by looking at the words that follow each other. When they follow each other, a relationship is formed between two words, the link becomes a higher weight if these two words appear more often in the text next to each other which is shown in Figure 5.

Figure 5. Text summarization methods

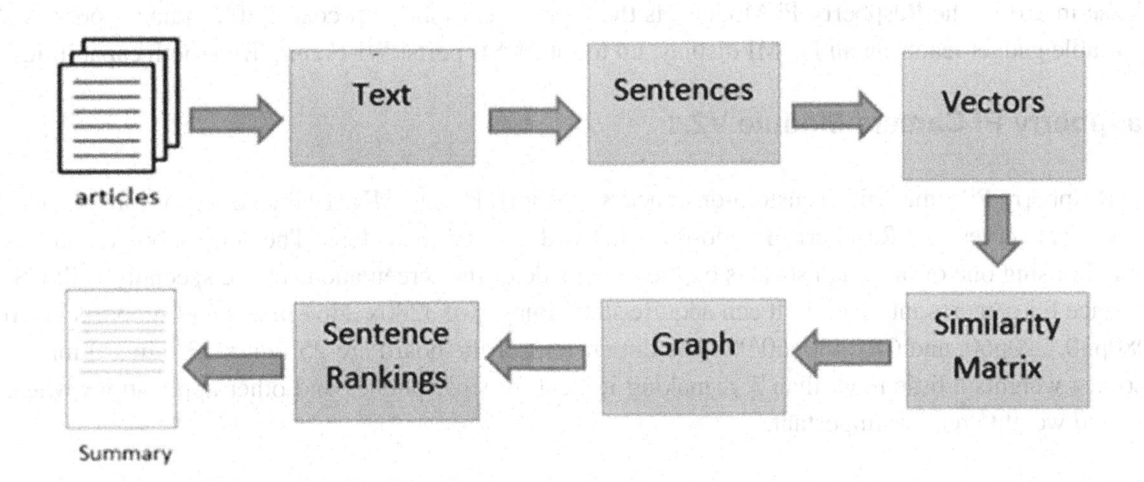

The Pagerank algorithm is applied to get the meaning of each word on top of the resulting network. Of all these terms, the top 1/3 is retained and considered relevant. After that, a able of keywords is created by combining the relevant words if they appear in the text to match each other.
Working of the TextRank algorithm under Gensim:

1. The input is delivered as pre-processed text.
2. Use sentences as the vertices of a graph.
3. The network has edges at the vertices that represent the similarity between the two texts.
4. On this weighted graph, run the PageRank algorithm.
5. Select the vertices with the greatest scores and add them to the summary.
6. The number of vertices to be chosen is determined by the ratio or word count.

The word limit for the summarization can also be specified in case the user wishes to read a smaller summary. However, it is not required to determine the word limit. The summarizer can automatically create an outline according to the necessity of each sentence if the word limit is not specified.

Once this text is summarized it can be converted to whichever language the user wishes to read the summary.

HARDWARE AND SOFTWARE

Raspberry Pi 3

The Raspberry Pi is a small computer that fits in the palm of your hand. It runs on Linux as well as a few other low-power operating systems. It was created by the Raspberry Pi Foundation. Raspberry Pi's official operating system is Raspbian. Other operating systems exist, but they are mostly intended for specific purposes. It features a single-board computer with a 1.2GHz Broadcom BCM2837 64bit Quad Core Processor and 1 GB of RAM. The Raspberry Pi has gone through multiple versions since its first release in 2012. The Raspberry Pi Model 3 is the most recent model. It costs 3200 Indian rupees. It's a portable gadget featuring an HDMI display, up to four USB ports, Wi-Fi, and Bluetooth capabilities.

Raspberry Pi Camera Module V2.1

The Raspberry Pi comes with a custom-designed 8-megapixel Sony IMX219 image sensor camera module V2 that serves as a Raspberry Pi add-on panel with a fixed focus lens. The add-on board attaches to the Pi using one of the small sockets on the upper side of the screen and uses the specially built CSI interface for cameras interfacing. It can acquire static images of 3280x 2464 pixels and supports video 1080p30, 720p60, and 640x480p60/90. The dimensions of the board are 25 mm x 23 mm x 9 mm. It also just weights a little more than 3 g, making it ideal for smartphones and other applications where size and weight aren't as important.

Python IDLE

Python is a high-level programming language established by Guido van Rossum that is mostly used for general-purpose programming. This language includes elements that make it possible to develop simple programmes on both a small and large scale. Python supports several programming paradigms, including object-oriented, functional, and imperative programming, and has a dynamic sorting system and automated memory management. It has a massive and diverse generic library. Python is a popularc programming language with a wide range of applications. It is simple to use and allows us to put our ideas into reality using the Raspberry Pi. Python syntax is simple, with a focus on readability and the use of Standard English terms. It begin with the IDLE prompt. The quickest way to learn Python is to use IDLE, a Python Integrated Development Environment.

System Specification and Design

Based on the following constraints, the device is designed:

1. The reading distance range is 15-30 cm.
2. The length of the character is at least 8 pt.
3. Total reading content size can vary.
4. The maximum text row tilt is 5 degrees from the Vertical.
5. Character type includes the types Roman, Egyptian or Sans Serif.

Because it is fitted over the board's enclosure using two L-clamps, the module is designed to be used to carry the pi cam module without any physical equipment or stand-like structure. The lens of the Pi camera has been changed in order to get a clear image of the writing. The distance between the cam module and the script is 15 to 30 cm, which is the minimum distance required for a human eye to read a script.

SYSTEM ARCHITECTURE

Figure 6. Architecture of the device

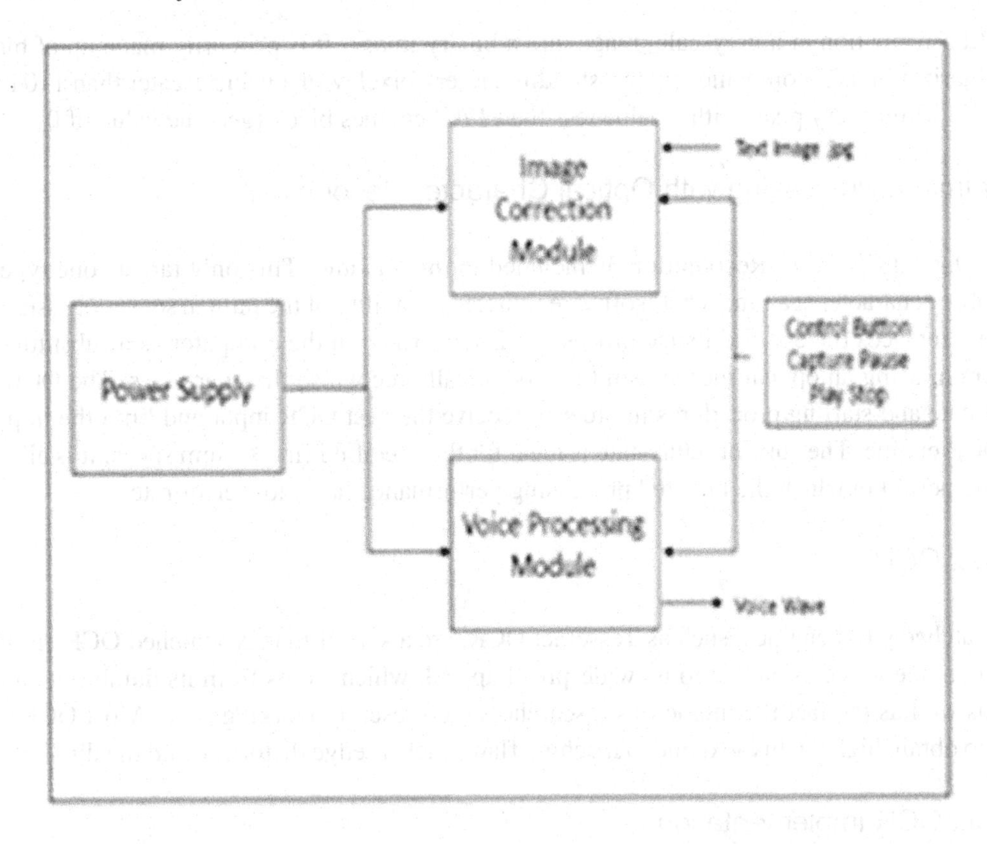

The module is designed to be used to carry the pi cam module in such a way that there is no physical equipment or stand-like structure as it is placed over the board's enclosure using two L-clamps. Each of these modules is elaborated in the consequent sections shown in Figure 6.

Image Correction Module

GrayScaling

It is the conversion of a digital or pixel image to a grayscaled image. A pixel's value is defined as a single sample because it only contains information about intensity. These are made entirely of grey tones, ranging in intensity from the weakest (black) to the strongest (white), with values ranging from 0 to 255.

Binarization

It is the transformation of a greyscale image into a binary image. It is primarily made up of black and white. Binarization is accompanied by thresholding. Every pixel with a value greater than 170 becomes white (255) while every pixel with a value less than 170 becomes black (gets the value of 0).

Unit for İmage Processing with Optical Character Recognition

OCR, or Optical Character Recognition, is included in this module. This only targets one typewritten text, glyph, or character at a time. OCR software imitates the ability of the human senses of vision, where the camera replaces the eye and image processing is performed in the computer as a substitute for the human brain, using an optical mechanism to automatically recognise the characters. The OCR engine required state and starting procedures in order to receive the best OCR input and limit the impairment of this OCR engine. The current setup state is ideal for the specified first system specs. It is also a short processing period in which the targeted processing performance has a low error rate.

Tesseract OCR

Matrix-matched OCR engines, such as Tesseract OCR, are a sort of matrix-matched OCR engine. The Tesseract engine was chosen due to its widespread appeal, which stems from its durability and extensibility, as well as the fact that numerous researchers are presently working on it. Most OCR systems struggle to obtain high quality text due to machine flaws such as edge distortion and the dim light effect.

Tesseract OCR ımplementation

The camera's input image has a resolution of 8 MP (pixels per inch) and a width of 215 PPI (pixels per inch). The minimal character size that can be read by the Tesseract OCR engine is 20 pixels uppercase characters. With an 8pt font size, Tesseract OCR's accuracy would suffer.

The application processes and converts the input image to text format. The image is captured by the user using the interrupt function on a GPIO pin attached to the touch key

TTS Correction and Voice Module

For operating systems such as Linux and Windows, ESpeak is a lightweight open-source speech synthesizer for English and other languages.ESpeak uses a method of "formant synthesis." It enables the supply of many languages in a small size. The voice is simple and at high speeds, it can be used.

TTS (Text-to-Speech) is a device capable of translating text input into speech. In theory, Text-to-Speech consists of two subsystems:

Text to Phoneme Converter

The text to phoneme converter converts the text input of a sentence in a specific language into a set of codes that typically indicate the tone, duration, and pitch of phoneme codes. The language has an impact on this portion.

Phoneme to Speech Converter

Phoneme to Speech Converter accepts coded input as well as the pitch and length of phonemes provided by the previous section.

DESIGN IMPLEMENTATION

Figure 7. The system architecture of the proposed system

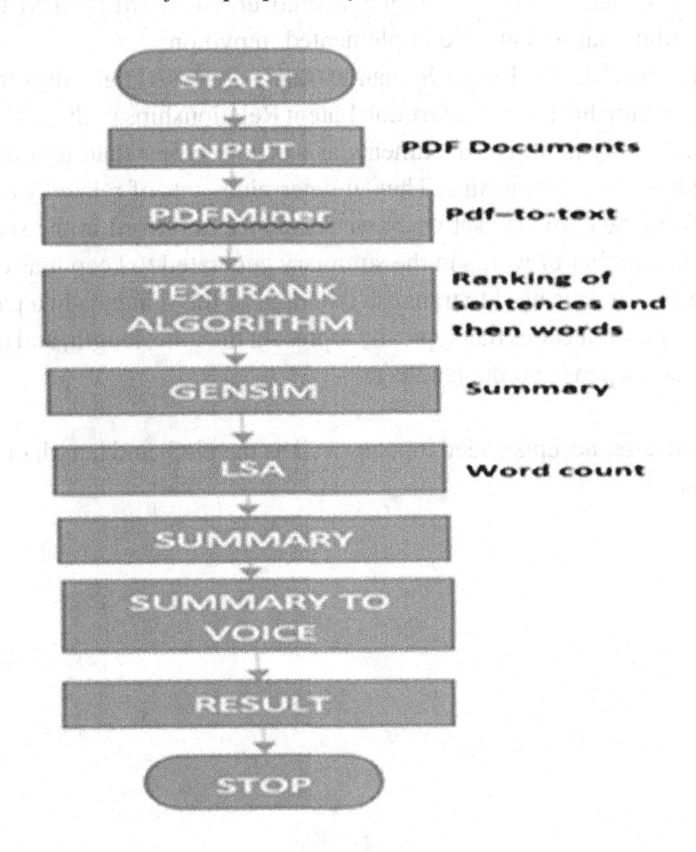

1. The text is then extracted using pdfminer from the pdf document shown in Figure 7. This feature uses the python library pdf-miner to extract all characters from a pdf document except the photos (although this may change me to accommodate this). The role simply takes in the home folder the name of the pdf document, extracts from it all the characters, and outputs the extracted texts as a python string list. PDFMiner is a PDF file data extraction tool. In contrast to other PDF-related resources, this focuses entirely on gathering and analyzing text informationPDFMiner allows you to accurately locate a document's text as well as other details such as fonts or rows. This provides a PDF translator capable of converting PDF files to other text formats (e.g. HTML).
2. One library that we have used is the Gensim Python Library, which is also an open-source library used for Natural Language Processing (NLP) with a subject modeling specification.It can also perform similarity detection and retrieval (IR) and document indexing when provided with large corpora. The target users are the NLP and the IR community.
3. The task of summarization of the given piece of text is a classic one and has been studied from different perspectives from time to time. We followed the Text Ranking Algorithm for this. The initial step is to pick a subset of words from the text so that the summary determined by it is as close to the original text as possible. The subset, named the summary, should be logical and understandable. This does not mean that the system determines the most common words only, but the most relevant words available. The task is not about picking the most common words or entities. We will use a Naive way to perform neural network training. Moreover, GENSIM is an NLP (Natural Language Processing) algorithm that will also be implemented in python.
4. The nextalgorithm used is the Latent Semantic Analysis (LSA)algorithm in python. LAS us a mathematical algorithm that tries to determine Latent Relationships within a collection of the given documents, thus, looking at all the documents as a whole rather than looking at each document separately to identify the relationships. Thus, it determines sets of related words and includes the relevant results from the complete set when we search for any word in the set. We have used this module to limit the number of words in the summary generated to keep it as compact as possible.
5. Text to phoneme converter is used to translate the input of the sentence in a particular language in the form of text into a set of codes that typically represent the tone, length, and pitch of the phoneme codes. This segment depends on the language.

Phoneme to Speech Converter accepts coded input as well as the pitch and length of phonemes provided by the previous section.

RESULTS AND DISCUSSION

Figure 8. The first page of an ebook

THE DEPRESSED PERSON

An ebook is shown in Figure 8. It is given as an input to the application to convert to a summary. The first step is to extract all the text out of the pdf document which is done with the help of the library PDFMiner. The testing was carried out with the following requirements using the Raspberry Pi platform:

1. SBU Raspberry Pi 3 900 MHz Quad Code ARM

2. Cortex-A7
3. Raspberry Pi 8MP Camera Board Module
4. Bootable SanDisk Ultra 16GB micro SD Card

From the experimental results, the following constraints are known to the image processing unit. We are the maximum size taken from Magazine for the input image. Any image input using the letter fonts of the block will work fine. The size of the minimum font is 8 rows.

We checked each module to see each step's effectiveness. The results of an accuracy check and an average time after entering the image processing unit are reported in Table 1.

Table 1. Results of accuracy testing

Distance in cm	Total words	Errors	Processing Time in seconds	Percentage of Error
30	254	5	54	2.36
15	78	1	23	1.28
18	156	1	38	0.64
25	234	2	47	0.85
26	312	4	54	0.64
20	213	5	52	1.27
24	213	6	45	2.81

From the results of this experiment, it is shown that, depending on the number of input words processed with an average error, the average time of image processing is about one minute or less. This is because the additional condition and state give the OCR system better feedback.

CONCLUSION

The way people in the world are approaching to make the world a better place to live in is through optimization and automation. The model we present in this concept-oriented paper can prove to be useful to the blind people because of the results we obtained. The model summarizes the content presented to it, based on neural network algorithms and that summarized content is finally converted to speech. This brings a major change in the lives of blind people a lot, which is a step of success for us. In the end, we were able to extract the basic summary of any piece of text we presented to it. This summary was tested for correctness based on the number of keywords that were obtained as the output and their importance in the text, and we were able to achieve good efficiency and accuracy, thus, bringing the proposed system to the expected conclusion.To visually impaired people, the text-to-speech system is introduced that can transform the text picture input into audio. The device's output is good enough, reaching a readability sensitivity of less than 2 percent, with the total processing time of different paper and font sizes of less than one minute.For good lighting, the average error rate of the image processing unit is higher. This is a portable device and does not require internet connection, so people with low vision, visual impairment can use it independently. This device also has a user interface that makes it easy for people to interact.

REFERENCES

Aone, C., Okurowski, M. E., & Gorlinsky, J. (1998, August). Trainable, scalable summarization using robust NLP and machine learning. In *Proceedings of the 17th international conference on Computational linguistics-Volume 1* (pp. 62-66). Association for Computational Linguistics.

Barzilay, R., & Lee, L. (2003, May). Learning to paraphrase: an unsupervised approach using multiple-sequence alignment. In *Proceedings of the 2003 Conference of the North American Chapter of the Association for Computational Linguistics on Human Language Technology-Volume 1* (pp. 16-23). Association for Computational Linguistics. 10.3115/1073445.1073448

Baxendale, P. B. (1958). Machine-made index for technical literature—An experiment. *IBM Journal of Research and Development*, *2*(4), 354–361. doi:10.1147/rd.24.0354

Brandow, R., Mitze, K., & Rau, L. F. (1995). Automatic condensation of electronic publications by sentence selection. *Information Processing & Management*, *31*(5), 675–685. doi:10.1016/0306-4573(95)00052-I

Edmundson, H. P. (1969). New methods in automatic extracting. *Journal of the Association for Computing Machinery*, *16*(2), 264–285. doi:10.1145/321510.321519

Erkan, G., & Radev, D. R. (2004). Lexpagerank: Prestige in multi-document text summarization. In *Proceedings of the 2004 Conference on Empirical Methods in Natural Language Processing* (pp. 365-371). Academic Press.

Freitas, D., & Kouroupetroglou, G. (2008). Speech technologies for blind and low vision persons. *Technology and Disability*, *20*(2), 135–156. doi:10.3233/TAD-2008-20208

Gillick, D., & Favre, B. (2009, June). A scalable global model for summarization. In *Proceedings of the Workshop on Integer Linear Programming for Natural Langauge Processing* (pp. 10-18). Association for Computational Linguistics. 10.3115/1611638.1611640

Karthikeyan, T., Sekaran, K., Ranjith, D., Vinoth Kumar, V., & Balajee, J. M. (2019). Personalized Content Extraction and Text Classification Using Effective Web Scraping Techniques. *International Journal of Web Portals*, *11*(2), 41–52. doi:10.4018/IJWP.2019070103

Kumar, V. V., Ramamoorthy, S., Kumar, V. D., Prabu, M., & Balajee, J. M. (2021). Design and Evaluation of Wi-Fi Offloading Mechanism in Heterogeneous Networks. *International Journal of e-Collaboration*, *17*(1), 60–70. doi:10.4018/IJeC.2021010104

Li, C., Qian, X., & Liu, Y. (2013). Using supervised bigram-based ILP for extractive summarization. In *Proceedings of the 51st Annual Meeting of the Association for Computational Linguistics (*Volume 1*: Long Papers)* (pp. 1004-1013). Academic Press.

Lin, C. Y., & Hovy, E. (2003). Automatic evaluation of summaries using n-gram co-occurrence statistics. In *Proceedings of the 2003 Human Language Technology Conference of the North American Chapter of the Association for Computational Linguistics* (pp. 150-157). 10.3115/1073445.1073465

Lin, H., & Bilmes, J. (2010, June). Multi-document summarization via budgeted maximization of submodular functions. In *Human Language Technologies: The 2010 Annual Conference of the North American Chapter of the Association for Computational Linguistics* (pp. 912-920). Academic Press.

Maithili, K., Vinothkumar, V., & Latha, P. (2018). Analyzing the Security Mechanisms to Prevent Unauthorized Access in Cloud and Network Security. *Journal of Computational and Theoretical Nanoscience*, *15*(6), 2059–2063. doi:10.1166/jctn.2018.7407

Muthukumaran, V., Kumar, V. V., Joseph, R. B., Munirathanam, M., & Jeyakumar, B. (2021). Improving Network Security Based on Trust-Aware Routing Protocols Using Long Short-Term Memory-Queuing Segment-Routing Algorithms. *International Journal of Information Technology Project Management*, *12*(4), 47–60. doi:10.4018/IJITPM.2021100105

Praveen Sundar, P. V., Ranjith, D., Karthikeyan, T., Vinoth Kumar, V., & Jeyakumar, B. (2020). Low power area efficient adaptive FIR filter for hearing aids using distributed arithmetic architecture. *International Journal of Speech Technology*, *23*(2), 287–296. doi:10.100710772-020-09686-y

Chapter 4
India's Remote Medical Monitoring System Using Big Data and MapReduce Hadoop Technologies:
Big Data With Healthcare

Koppula Srinivas Rao
MLR Institute of Technology, India

S. Saravanan
B. V. Raju Institute of Technology, India

Kasula Raghu
Mahatma Gandhi Institute of Technology, India

V. Rajesh
Koneru Lakshmaiah Educational Foundation, India

Pattem Sampath Kumar
https://orcid.org/0000-0003-2565-1590
Malla Reddy Institute of Engineering Technology, India

ABSTRACT

The data analytics and Hadoop applications are the most prominent elements in big data analytics to analyze the large volumes of data. The developing countries mainly concentrate on medical, economic, and emerging issues. This chapter focuses on the importance of big data management and Hadoop, as well as their influence on delivering medical services to everyone at the lowest feasible costs.

DOI: 10.4018/978-1-7998-9640-1.ch004

INTRODUCTION

Big Data is a relatively recent development in the world of information systems that has emerged as a result of the rapid growth of data over the past decade. Big Data remains a term utilized towards describe datasets that exceed the capabilities of conventional data collection then storage systems. A modern form of data analytics known as Big Data Analytics has arisen as a result of the need to handle and analyses those massive datasets Feng, C.(2020). It entails processing large amounts of data of various forms in order to uncover secret blueprints, unidentified associations, and other valuable data Shilo, S.(2020). Many companies are rapidly turning to Big Data analytics to obtain deeper visibility into their operations, improve revenue and profitability, and gain a strategic edge over competitors Siegel, J. (2012).

The Apache Hadoop Framework's basic programming models make it easier to spread the analysis of large data sets across groups of systems Au-Yong-Oliveira (2021). The architecture remains designed towards scale from a single server to thousands of devices, each with native computing then storage capabilities Dash, S.,(2019). At the software layer, the library remains structured in such a way that errors can be detected & dealt with. As a result, the architecture is capable of providing continuous operation on top of a set of applications that are vulnerable to failure Senthilkumar,S.(2018). Hadoop's bright future draws a wide range of businesses and organizations to use it for both science and manufacturing Harris, T. (2010).

HEALTHCARE IN INDIA

India ranks second in the world in terms of population. India's health-care system is being overburdened by the country's growing population. Among a wide number of individuals, economic scarcity leads to a weak approach to health care. GDP per capita are the most important indices of human growth. Longevity is linked to income and education and affects the status of one's wellbeing. The health-care sector's vulnerability can have a detrimental impact on longevity. India's Human Development Index (HDI) is low (115th) among world countries. The main reasons for India's high disease burden are a lack of access to preventative and therapeutic health facilities White, T. (2010).

"Growth in national income by itself is not enough, if the gains do not manifest themselves in the form of more food, greater access to health, and education," said Amartya Kumar Sen, an Indian economist & Nobel laureate Zulkernine, (2013).

In a lecture, Dr. MC Misra, the director of the All India Institute of Medical Sciences (AIIMS), said, "Advances in medical technology and modern medicines are indeed a blessing, but to operate in India, they must be value for money." Also free services in public hospitals are out of reach for the majority of people."

However, according to World Bank data, 99% of India's population cannot afford to pay for these facilities. As per the survey, out-of-pocket medical costs push 39 million individuals into poverty per year, with households dedicating about 5.8% of their income to medical care Yang Song. Alatorre, G. (2013).

In India, less than 10% of the population has signed up for health insurance. Hospitalization costs alone account for 58% of a typical Indian's gross annual spending. According to World Bank data, over 40% of people borrow heavily or sell properties towards cover hospitalization costs, pushing 39 million people into poverty per year.

Figure 1. Healthcare spending by the public and private sectors

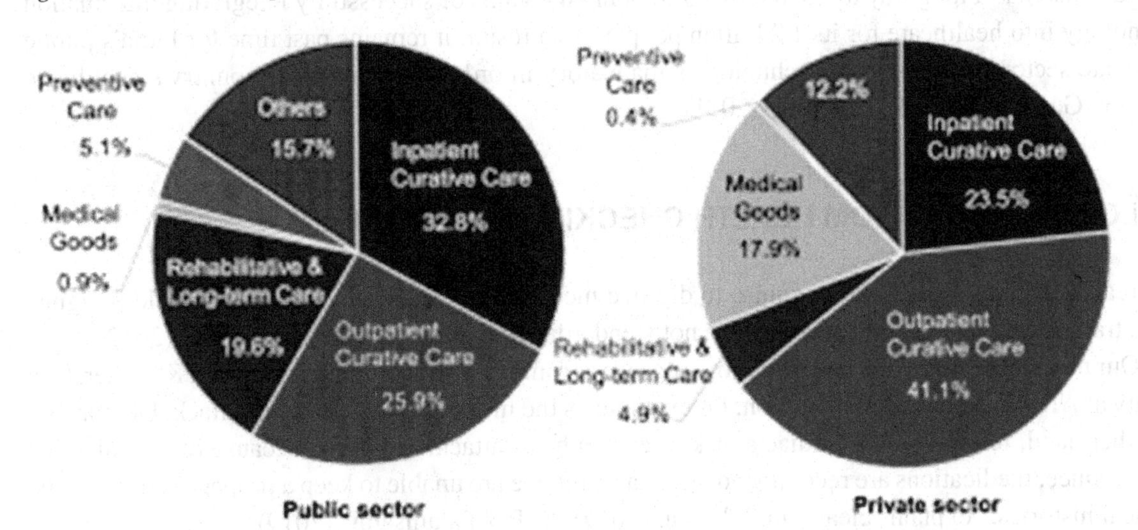

Table 1. Expenditures on healthcare by countries

Country	Total % of GDP spent on Healthcare	Private Expenditure %	Per capita spent on	
			Health care (USD)	Healthcare (USD) by Government
India	4.01	70.08	133	40
USA	17.09	46.09	8363	4439
UK	9.06	16.01	3481	2921
South Africa	8.09	55.09	936	414
China	5.01	46.04	380	205
Brazil	9.00	53.01	1029	486
Pakistan	2.02	61.05	60	25
Nigeria	5.01	62.01	123	48
Russia	5.01	37.09	999	622

Statistics show that healthcare is available and accessible to the average person in society, especially in developed countries like India, Mukherjee, (2012), Deepak Kumar, B. (2011).

Not only does availability apply to the accessibility of hospitals and medical staff, but it also relates to the accessibility of essential and trustworthy data on the screen [15-16] show in table 1. Not only does accessibility provide the ability to attend clinics, but it also includes awareness of the patient, prescriptions, and other relevant material, Weiyi Shang. Zhen Ming Jiang. Hemmati, (2013), Hao, Chen. Ying, Qiao. (2011).

India also has a long way to go in terms of reaping the gains of successfully integrating information technology into healthcare for its 1.2 billion people. As a result, it remains past time for India's public & private sectors towards make technologies mandatory in order to improve the country's healthcare situation Garcia, T. (2013), Li Xiang. (2011).

INFLOWING DATA FROM HEALTH CHECKING PROCEDURES

The reality of the government's promise to digitize medical information is not as one would imagine. Also, traditional medical terminologies are not standardized in the world, Xu Zhengqiao. (2012).

"Our medical terminology and names of drugs vary from place to location," said Sushil K Meher, tech faculty at AIIMS. Myocardial infarction, for example, is the medical term for a heart attack. Doctors, on the other hand, refer to it as a cardiac attack or even a heart attack. Similarly, because of the salts that they produce, medications are recommended. As a result, we are unable to keep a proper list of patients' medical histories," explains clearly in, S.Prabu, (2019), JS P.N.Palanisamy, (2019).

Figure 2. Inflow of healthcare data

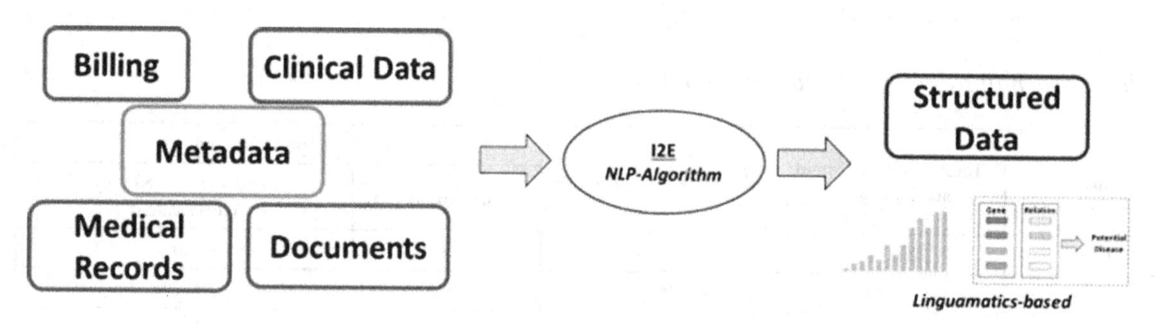

A health institution's index can be seen as medical history. Medical reports reveal details on a Healthcare Center's beginnings and development, as well as retrospective and future predictive analysis, the nature of cases admitted to the hospital, and so on. For the benefit of any practitioner in the healthcare industry, medical records must be properly & methodically gathered, conserved, & secluded. Health Records are more than just a source of medical & science information & inputs for the government's preparation & budgeting for the country's health-care system. The need of the hour is for consistency in the storage of medical records under different Acts show in fig 2.

"We cannot assess the health situation in India without data," said at a medical informatics meeting, AIIMS programming faculty. The Indian government's health ministry recently announced that medical records in public hospitals in Delhi would be digitized, enabling every hospital in the country to access patient information.

According to a study by Grant Thornton India, the Indian health check system and equipment industry remains projected to rise towards about US$ 5.8 billion by 2014, & US$ 7.8 billion by 2016, rising at a CAGR of 15.5 percent. This exponential growth generates a massive volume of data that is beyond human comprehension.

THE RELATIONSHIP BETWEEN BIG DATA AND BIG DATA ANALYTICS AND HEALTHCARE

Over time, various devices attached to the patient generate poorly organized data, which is collected by health care management systems. And since these are massive, complicated systems, powerful algorithms to process the raw data are needed, as well as a lot of computing power. Big data refers to data collected by a variety of sensors, such as medical, traffic, and social data. Amount, velocity, and meaning are three features of big data show in fig 3.

Figure 3. Big Data – 4 "Vs"

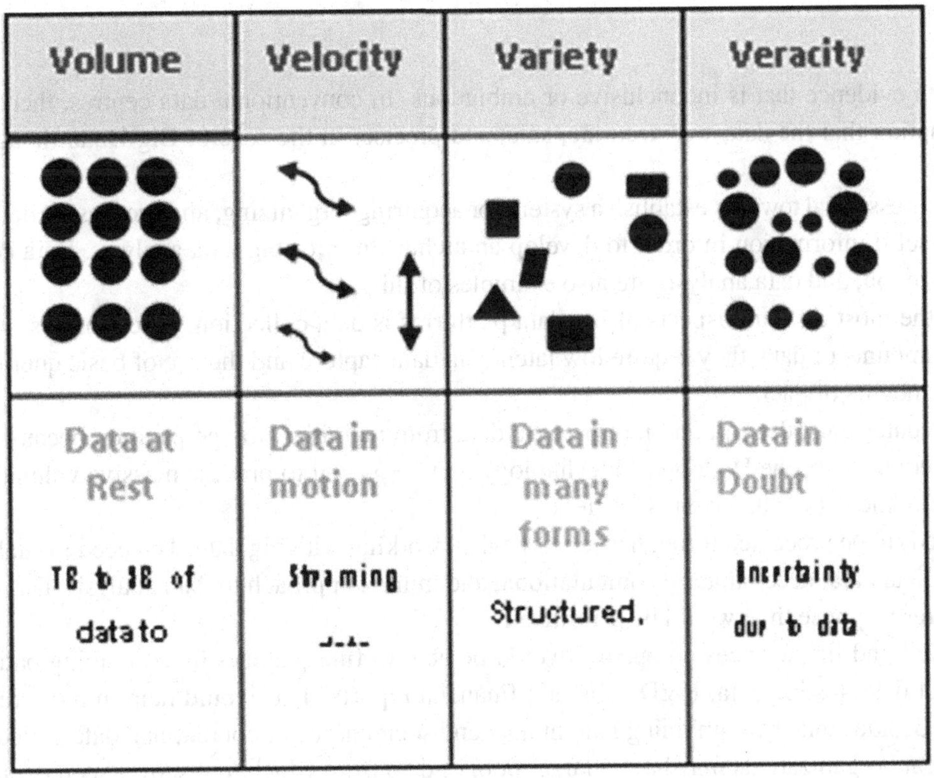

Volume	Velocity	Variety	Veracity
Data at Rest	Data in motion	Data in many forms	Data in Doubt
TB to 3B of data to	Straming	Structured,	Incertainty due to data

Features of Big Data

Volume

As compared to conventional data, the volume of data produced by different medical devices is greater.

Velocity

The annual data storage capability of an entire healthcare system is significantly less than the volume of data streamed by medical networks.

Diversity

Traditional data formats are less flexible when it comes to various types of sensor data, implementation, and other variables. On the other side, nontraditional info, such as surgical instruments, is easily adaptable to transformation.

Veracity

It deals with evidence that is inconclusive or ambiguous. In conventional data centres, there was still the presumption that the data was accurate, safe, and precise. In the case of Big Data, though, this is not the case.

It remains essential towards establish a system for acquiring, organizing, and processing data in order to extract useful information in order to develop an architecture for big data analysis. Data collection, data organization, and data analysis are also examples of this.

One of the most difficult aspects of big data platforms is data collection. Since these systems deal with large amounts of data, they require low latency in data capture and the use of basic queries to process large amounts of data.

The computer must decrypt and interpret the data from its initial storage position because the data is on a larger size. Apache Hadoop is a technology that helps you to process massive volumes of data while keeping the data in its original clusters.

Data needs to be processed through a network when working with big data. The need to analyze data, such as medical data, necessitates a computational and mining approach to data analysis. Data delivery with a shorter response time would be prioritized.

Healthcare and life sciences companies would be able to find patterns by combining patient data, opioid effect data, testing data, R&D data, and financial reports. This would help in more responsive healthcare. Besides that, by combining patient data and social network content in a data collection system, healthcare organizations may have a larger pool of data from which to discover secret connections show in fig 4.

The hopeful expectation remains that the healthcare industry would be able towards gather data from either source, organize it, and investigate it to discover solutions that will help patients.

1. Cost savings
2. Time savings
3. New science progress and improved results
4. Correct decision-making is aided by better diagnosis.

Figure 4. Analytics in Big Data

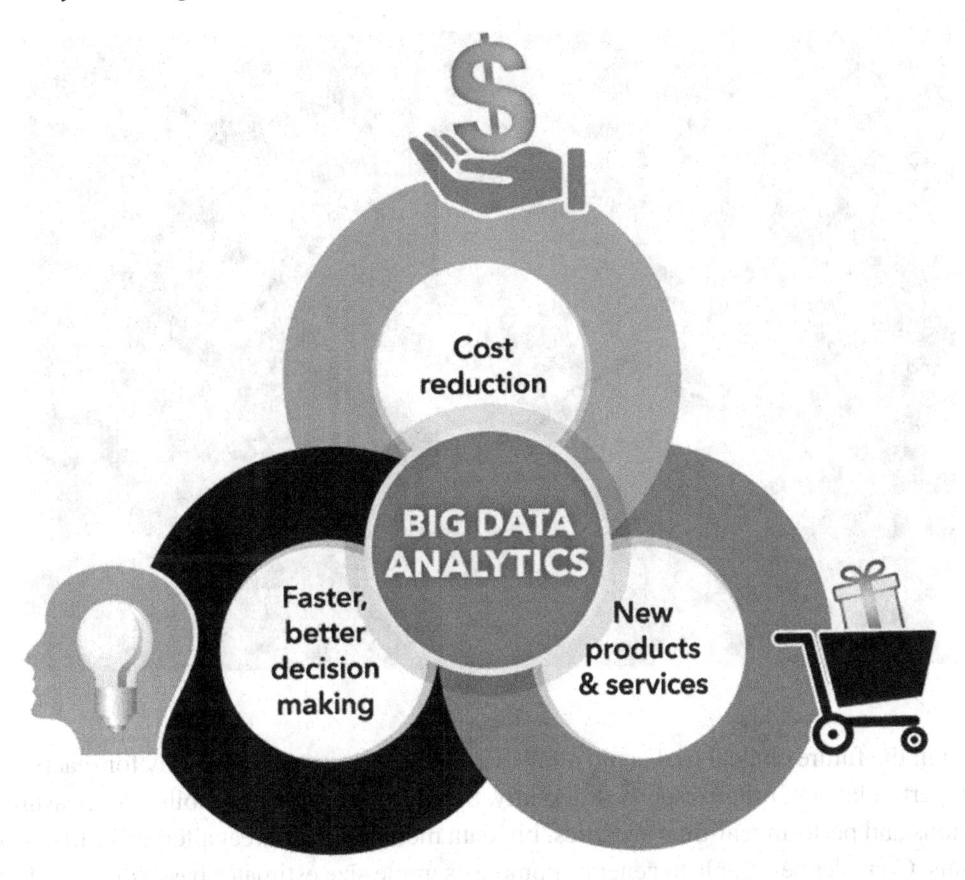

A compromise that could satisfy four main criteria

1. In terms of scalability and dependability,
2. Storage is an important aspect of every company.
3. Infrastructure for processing
4. Capabilities of search engines for retrieving articles of high availability (HA)
5. With HA, you can retrieve statistics from a scalable real-time shop.

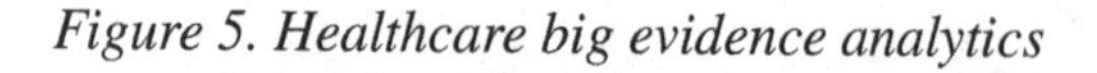

Figure 5. Healthcare big evidence analytics

Big data in the future clinical trial could allow enrollees to be watched not only for reaction but also to see how particular subgroups respond differently. Since they improve the ability to measure population variations and perform real-time analytics, big data methods are a great alternative to conventional clinical trials. Consider being able to generate complex sample size estimates based on new clinical trial results. The opportunity to refine clinical trial design, allowing for shorter and more effective experiments, is the secret to big data's appeal show in fig 5.

In our first big data in healthcare instance, we'll look at a basic issue that every shift supervisor faces: how often employees should I put on staff at any given time? You run the danger of incurring excessive labour expenses if you hire too several people. When there are too few employees, customer service suffers, which may remain deadly aimed at patients in that sector.

At least at a few Parisian institutions, big data is assisting in the solution of this issue. According to an Intel white paper, four hospitals in the Assistance Publique Hôpitaux de Paris have been utilizing data as of a number of sources towards estimate how many patients would be at each facility on a monthly and ongoing basis.

One of the major data sets is a decade's worth of hospital admissions records, which were crunched utilizing "time series analysis" methods by data scientists. The researchers were able to see significant trends in admittance rates as a result of their studies. They could then utilise machine learning to identify the best system gets for predicting admittance patterns in the coming.

To summarise the results of all of this work, the data science team created a web-based user interface that predicts patient loads & assists in allocation of resources management via the use of online data visualisation, with the aim of enhancing overall clinical outcomes.

In medicine, it is the most commonly used type of big data. Every patient has a permanent copy of their records, which include demographics, medical history, allergies, and laboratory test results, among other things. Records remain exchanged via secure information systems then remain accessible towards both public & private sector suppliers. Every record is made up of a single editable file, which means physicians may make changes over time without having to deal with paperwork or the risk of data duplication. EHRs may too send out alerts & reminders when a patient needs a fresh lab test, as well as monitor prescriptions towards see whether they've been followed. Despite the fact that electronic health records are a wonderful concept, many nations are still struggling to completely adopt them. According towards this HITECH study, the US has made a significant jump forward, by 94 percent of hospitals using EHRs, while the EU remains still lagging behind. However, it is hoped that an aggressive regulation proposed by the European Commission would alter this.

In the United States, Kaiser Permanente is leading the way, and it may serve as a model for the EU. They've completed the implementation called Health Connect, which exchanges data across all of their locations & simplifies the usage of EHRs. "The integrated system has improved outcomes in cardiovascular disease and generated an estimated $1 billion in spending from fewer doctor's visits and lab tests," according to a McKinsey study on big data medicine.

Other applications of data analytics in healthcare have one thing in common: real-time alerts. Clinical Decision Support (CDS) software analyses health information in real time in hospitals, giving guidance to doctors while they take prescription choices. Doctors, on the other hand, prefer that people avoid hospitals in order to avoid expensive in-house therapies. One of the hottest business intelligence buzzwords for 2019 is analytics, which has the potential to become a new strategy. Wearables will continuously collect and transmit patient health data to the cloud. Additionally, this data will be linked to a database on the health of the general public, allowing clinicians to analyze the data in a socioeconomic context and change delivery methods accordingly. Institutions and care managers will collect this massive data stream using sophisticated data and respond quickly if the findings are disturbing. If a patient's blood pressure increases dangerously, for example, the device will send a real-time alert to the doctor, who will contact the patient and administer pressure-lowering therapies. Asthmapolis, for example, has begun to integrate inhalers with GPS-enabled monitors to detect asthma patterns in both individuals and large groups. This data is being merged with FDA data in order to develop more effective lung treatment options.

Because of greater insights into people's motives, the use of big data in healthcare enables for strategic planning. Care managers may examine the results of check-ups among persons from various demographic groups to see what variables discourage people from seeking medical help.

The City of Orlando created heat maps for a variety of problems, including population growth and chronic illnesses, using Google Maps & free public health data. Educators then matched this information to the availability of medical care in the most hot regions. They were able to evaluate their delivery strategy then add additional care units towards the most problematic regions as a result of the information they gained. The Cancer Mars mission initiative is another fascinating example of data in healthcare. Donald Trump devised this plan before the conclusion of his second administration, with the aim of making ten years' worth of progress toward finding a cure for cancer in half the time. Large amounts of data about cancer patients' treatment plans and recovery rates can be used by cancer specialists to find patterns and therapies that have the best success rates in the real world. For example, researchers may examine tumor samples in biobanks linked to patient treatment data. This information might be used by scientists to investigate how individual mutations and cancer proteins interact with various medicines, as well as to spot patterns that could lead to better patient care.

This information may too lead to unanticipated advantages, such as discovering that the depression Awkward spaces can help treat some kinds of lung cancer.

Nevertheless, patient datasets from other organisations, including as hospitals, colleges, and NGOs, must be connected in order to make these types of insights more accessible. Researchers may then, for instance, access patient biopsy results from different establishments. Hereditarily sequencing malignant growth tissue tests from a clinical preliminary participants & providing these data accessible towards the larger important in the pathogenesis is one of the possible large data use case in oncology.

HADOOP IN HEALTH CARE DATA

This disparity in economics has drawn a lot of interest, and Hadoop will become the focal point for most large-scale data processing and analysis operations and analyses, either integrating or originating from it.

The industry has grown by one to two orders of magnitude as increasingly powerful SQL technologies have been coupled to Hadoop technology, adding the whole SQL-based environment to the world of Hadoop. Hadoop is no longer just the realm of experts.

Figure 6. Map reduction in Hadoop

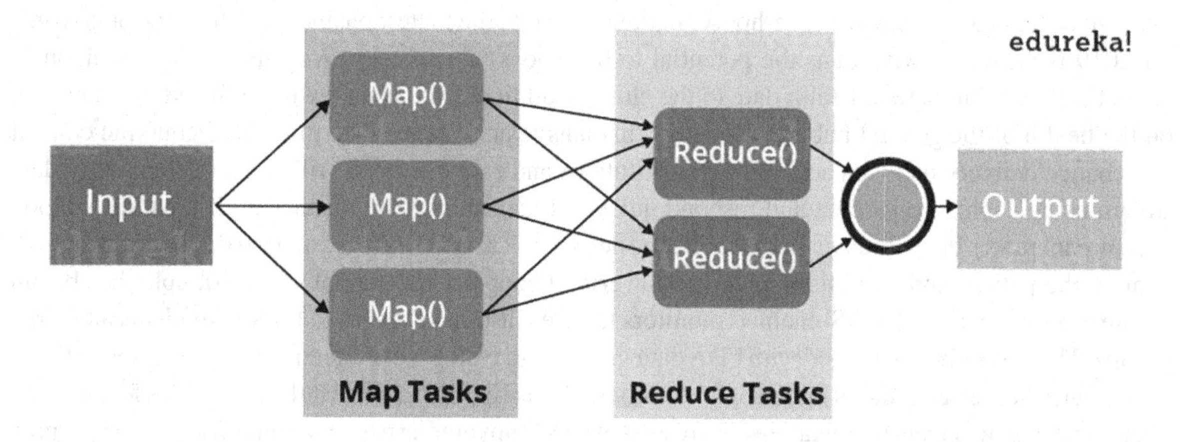

Another stumbling block is relational technologies that operate on their own. When it comes to unstructured/multi-structured/loosely structured files, the genie is out of the bottle. We can now analyze more and diverse types of data with Hadoop than we can with relational catalogs show in fig 6.

Hadoop can allow dispersed parallel computing on massive volumes of data through inexpensive and high-level servers that can be scalable to infinity it is used to store and process data. Hadoop does not consider any data to be too large. Every day, a huge amount of data is generated and introduced in the exponentially expanding medical community. Because of Hadoop's performance and efficacy, data that were once valuable for research are now worthless.

Hadoop may use a distributed file system to store data through several servers in a cluster. Hadoop hides the position of data being viewed in the cluster, enabling end users to refer to files as though they were on a local computer.

Figure 7. Hortonworks application using MapReduce

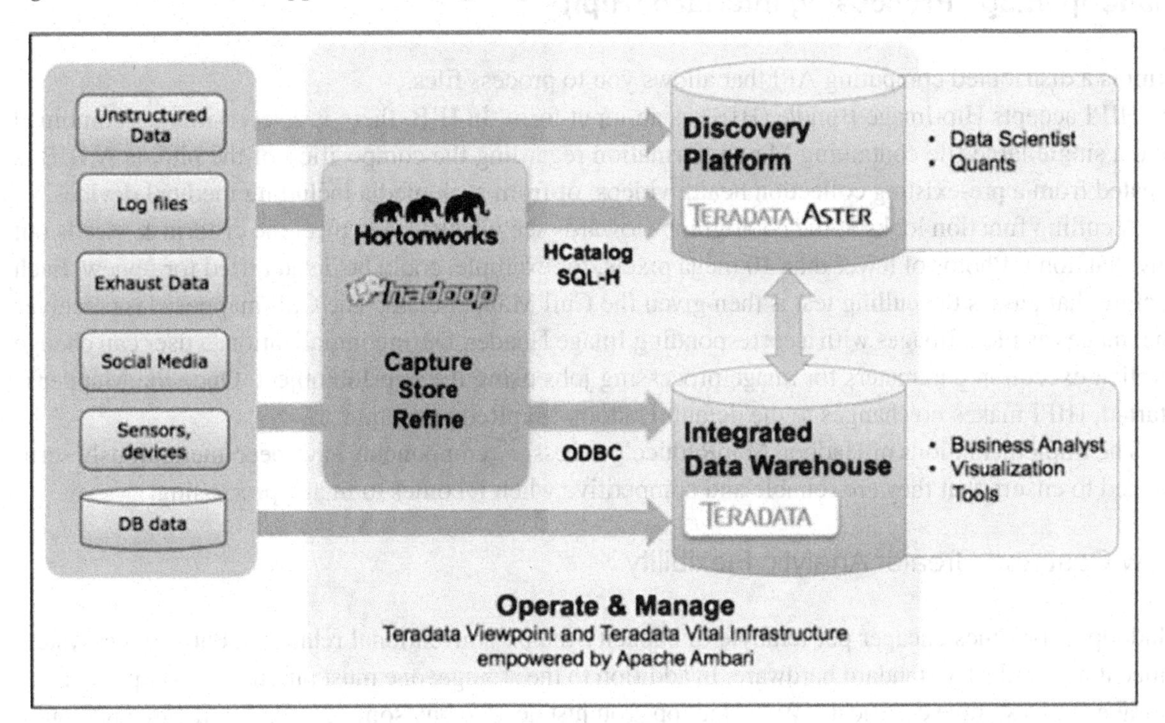

Through processing and interpreting Big Data, the healthcare industry may make important predictions. In the other hand, since 80% of medical data is "unstructured," it must be processed in order to do accurate data mining & investigation. Hadoop remains a fundamental proposal aimed at processing Big Data with the aim of making it usable for analysis. Because of its simplicity, completeness, and convergence, Hortonworks Data Platform (HDP) may be used to build applications for healthcare including Hadoop and Big Data. The diagram below depicts Hadoop's effort to chart and minimize tasks show in fig 7.

Hadoop aims to implement Map & Reduce operations on the systems where the data being stored remains located while doing Map Reduce function, eliminating the need aimed at data towards remain copied amid systems. The program demonstrates that MapReduce activities are more effective where only one large file is used as input rather than a large number of small files. Since the tiny files are spread over several computers, copying them to the MapReduce device requires substantial overheating. However, because large files are stored on a single system, this is not the case. The program claims that tiny file overhead slows the runtime by ten to one hundred times. When the data being analyzed is localized to the devices doing the operation, it is clear that the MapReduce architecture performs well.

The healthcare sector faces many obstacles in handling data in order to provide excellent treatment to all parties involved. Medical image recognition is one of the topics covered. Hadoop offers a solution for analyzing a growing number of medical images from different sources and extracting the information needed to make an accurate diagnosis. The Hadoop Image Processing Interface (HIPI) describes the steps involved in image processing.

Hadoop Image Processing Interface (Hipi)

Hipi is a distributed computing API that allows you to process files.

HIPI accepts HipiImage Bundle (HIB) as an input form. In HIB, there is a set of images combined into a single large file containing Meta information regarding the composition of the photos A HIB is created from a pre-existing collection health videos, or from such media including medical devices.

A culling function looks at the photographs towards see whether they meet the criteria & weeds out any that don't. Photos of fewer than 10 mega pixels, for example, could be disqualified for review. Each picture that passes the culling test is then given the Cull Mapper class. The Cull mapper class receives the images as Float Images with a corresponding Image Header. During initialization, a user can change explicit execution parameters for image processing jobs using the HipiJob object. Once the Mapper is started, HIPI makes no changes to the default Hadoop MapReduce output.

The implementations of Hadoop MapReduce's necessary components have been meticulously scrutinized to ensure that they are reliable and competitive when it comes to image processing.

Low Cost and Greater Analytic Flexibility

Hadoop is 10 times cheaper per terabyte of capacity than a conventional relational data center system since it uses industry standard hardware. In addition to the storage, one must purchase a computer, SAN storage, and a storage certificate. With Hadoop, you just need to buy some popular hardware and you're ready to go.

It also helps consumers to run analyses on the cumulative compute and storage, allowing them to get more bang for their buck. The previous strategies we had in place didn't even come close to meeting the need. Although, the advantage of storing information in a Hadoop-type approach is outweighs the value of storing data in a database if costs are equal."

WORK APPROACHES FOR THE FUTURE

Even though human life is on the line, healthcare necessitates the most thorough investigation practicable in order to collect reliable results for tests that cannot be tampered with. Despite the reality that our Big Data Analytics & Hadoop for healthcare analysis assist us in reducing costs and improving treatment for the country's last and least guy, the challenges mentioned below must be overcome in order to achieve the best results.

- If big data comes in an unstructured medium including text, picture, audio, or video, it may be difficult for non-technical medical professionals to understand and use it.
- The second hurdle is collecting the most critical data in real-time, such as big surgeries, & communicating it towards the appropriate people aimed at further interpretation.
- A third roadblock remains data collection, as well as understanding and analyzing large amounts of data using the minimal computing resources available.

The healthcare industry needs real-time data interaction, while Hadoop will batch process the use cases. Transactions cannot be handled in Hadoop because it does not support indexing and does not explicitly follow the ACID paradigm.

Image processing is critical for the future since it allows images to be linked towards other forms of data for mining. As a result, medical image processing research takes the lead too may make a significant contribution. The development of advanced Cloud storage technologies could be able to solve the data storage issue show in fig 8.

Figure 8. HIPI

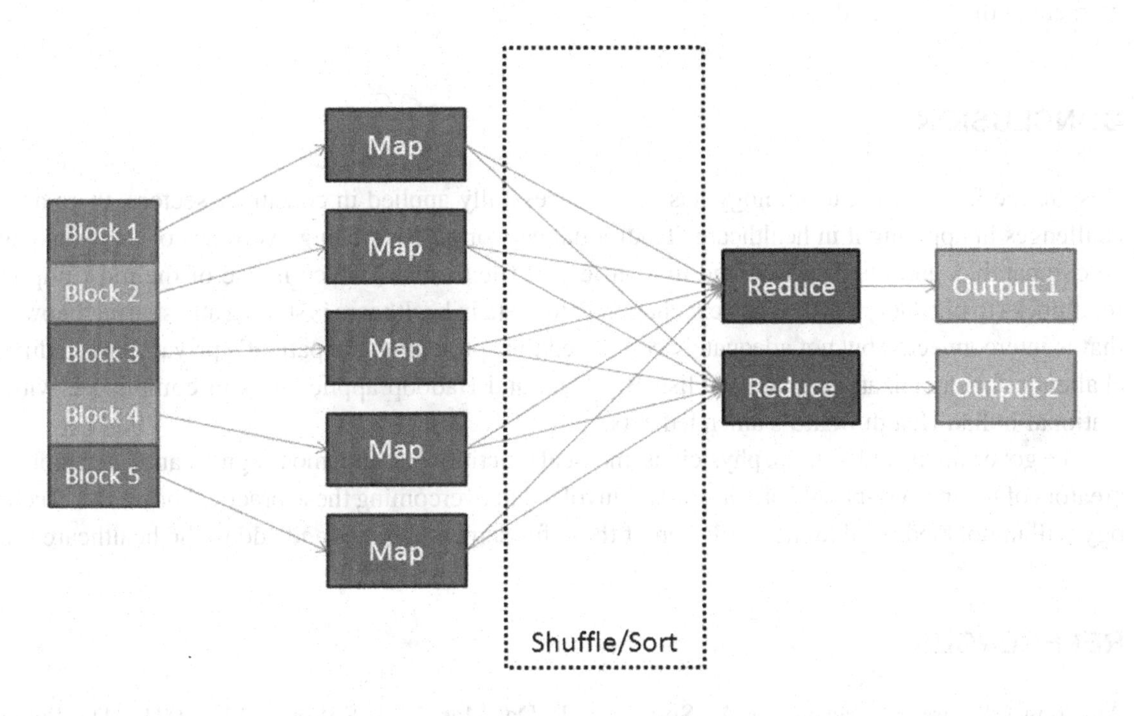

In health care, big data analytics refers to massive amounts of data generated using numerical technology to collect patient records then aid in the administration of hospital performance, which would otherwise be too large & complicated for traditional technologies.

The utilization of enormous information examination in medical care offers an assortment of positive, even lifesaving, repercussions. Generally, large style information alludes to enormous measures of information created by the digitalization of everything, which is consolidated and dissected by particular innovations. When applied to medical care, it will make utilize of accurate health data from a community (or an organization) to help prevent epidemics, treat illnesses, cut costs, and so on.

As people have lived longer, treatment methods have altered, and the majority of these changes have been driven by data. Doctors strive towards learn as much as they can about their patients as early as possible in their lives so that they can recognize warning signs of serious illness as soon as they occur. It is significantly easier and less expensive to deal with an early diagnosis. When it comes to pharmaceutical data analytics and key performance indicators, prevention remains better than cure, then being able towards develop a thorough picture of a customer will allow assurance to give a customized package. This remains the industry's attempt to address the information silos that a patient's information has: bits and pieces remain acquired & stored in hospitals, clinics, surgeries, and other sites, by the inability towards connect efficiently.

Collecting large amounts of data for medical purposes, on the other hand, has been expensive and tedious for quite a long time. With the present consistently improving innovation, it's simpler to gather such data as well as to order extensive medical services reports and transform them into urgent bits of knowledge that might be utilized to improve treatment. This is what is the issue here: utilizing information driven bits of knowledge to not just anticipate and address an issue before it's past the point of no return yet in addition to survey systems and medicines quicker, monitor stock, and support healthcare workers in their own wellness.

CONCLUSION

Despite the fact that big technology has been successfully applied in consumer sectors, there are still challenges in applying it in healthcare. Traditional networks provide huge volumes of data that cannot be comparable, and the data is stored in a variety of file formats, which is one of the most important roadblocks to big data growth. The next challenge for data in healthcare is storing and sharing knowledge that is interconnected but not adequately connected thus protecting the patient's privacy. It is a difficult challenge for organizations that establish Big Data and Hadoop applications in compliance with the National Indian Health Board's amended acts.

The government and its laws, physicians, medical practitioners, and, most significantly, technological creators of technology-based software are all involved in overcoming these practical obstacles. Technology will undoubtedly aid in the resolution of these fundamental issues and add to the healthcare sector.

REFERENCES

Au-Yong-Oliveira, M., Pesqueira, A., Sousa, M. J., Dal Mas, F., & Soliman, M. (2021). The Potential of Big Data Research in HealthCare for Medical Doctors' Learning. *Journal of Medical Systems*, *45*(1), 1–14. doi:10.100710916-020-01691-7 PMID:33409620

Dash, S., Shakyawar, S. K., Sharma, M., & Kaushik, S. (2019). Big data in healthcare: Management, analysis and future prospects. *Journal of Big Data*, *6*(1), 1–25. doi:10.118640537-019-0217-0

Deepak Kumar, B. (2011). Evaluation of the Medical Records System in an Upcoming Teaching Hospital—A Project for Improvisation. *Journal of Medical Systems*.

Feng, C., Adnan, M., Ahmad, A., Ullah, A., & Khan, H. U. (2020). Towards energy-efficient framework for IoT big data healthcare solutions. *Scientific Programming*, *2020*, 2020. doi:10.1155/2020/7063681

Garcia, T., & Wang, T. (2013). Analysis of Big Data Technologies and Method - Query Large Web Public RDF Datasets on Amazon Cloud Using Hadoop and Open Source Parsers. *Semantic Computing (ICSC), IEEE Seventh International Conference.* http://architects.dzone.com/

Hao, C., & Ying, Q. (2011). Research of Cloud Computing Based on the Hadoop Platform. *Computational and Information Sciences (ICCIS), 2011 International Conference.*

Harris, T. (2010). *Cloud Computing- An Overview, Whitepaper.* Torry Harris Business Solutions.

Mukherjee, A., Datta, J., Jorapur, R., Singhvi, R., Haloi, S., & Akram, W. (2012). Shared disk big data analytics with Apache Hadoop. *High Performance Computing (HiPC), 19th International Conference.*

Palanisamy, J. S. P. N., & Malmurugan, N. (2019). FPGA implementation of deep learning approach for efficient human gait action recognition system. International Journal of Innovations in Scientific and Engineering Research, 6(11).

Prabu, S., & Kalaivani, M. (2019). An intelligent power adaptive model using machine learning techniques for wsn based smart health care devices. International Journal of Innovations in Scientific and Engineering Research, 6(12).

Senthilkumar, S. A., Rai, B. K., Meshram, A. A., Gunasekaran, A., & Chandrakumarmangalam, S. (2018). Big data in healthcare management: A review of literature. *American Journal of Theoretical and Applied Business*, *4*(2), 57–69. doi:10.11648/j.ajtab.20180402.14

Shang, W., Zhen, M. J., Hemmati, H., Adams, B., Hassan, A.E., & Martin, P. (2013). Assisting developers of Big Data Analytics Applications when deploying on Hadoop clouds. *Software Engineering (ICSE), 35th International Conference.*

Shang, W., Zhen, M. J., Hemmati, H., Adams, B., Hassan, A.E., & Martin, P. (2013). Assisting developers of Big Data Analytics Applications when deploying on Hadoop clouds. *Software Engineering (ICSE), 35th International Conference.* hipi.cs.virginia.edu/ 2014/

Shilo, S., Rossman, H., & Segal, E. (2020). Axes of a revolution: Challenges and promises of big data in healthcare. *Nature Medicine*, *26*(1), 29–38. doi:10.103841591-019-0727-5 PMID:31932803

Siegel, J., & Perdue, J. (2012). Cloud Services Measures for Global Use: The Service Measurement Index (SMI). *SRII Global Conference (SRII).* 10.1109/SRII.2012.51

Song, Y., Alatorre, G., Mandagere, N., & Singh, A. (2013). Storage Mining: Where IT Management Meets Big Data Analytics. *Big Data (BigData Congress), IEEE International Congress.*

White, T. (2010). *Hadoop: the Definitive Guide* (2nd ed.). O'Reilly Media.

Xiang, L. (2011). *Analysis on architecture of cloud computing based on Hadoop.* Computer Era.

Xu, Z., & Zhao, D. (2012). Research on Clustering Algorithm for Massive Data Based on Hadoop Platform. *Computer Science & Service System (CSSS), 2012 International.*

Zulkernine, F., Martin, P., Ying, Z., Bauer, M., Gwadry-Sridhar, F., & Aboulnaga, A. (2013). Towards Cloud-Based Analytics-as-a-Service (CLAaaS) for Big Data Analytics in the Cloud. *Big Data (BigData Congress), IEEE International Congress.*

Chapter 5
A Secure and Effective Image Retrieval Based on Robust Features

Swapna B.
https://orcid.org/0000-0002-7186-2842
Dr. M. G. R. Educational and Research Institute, India

Arulmozhi P.
Karpagam College of Engineering, India

Kamalahasan M.
Dr. M. G. R. Educational and Research Institute, India

Anuradha V.
Dr. M. G. R. Educational and Research Institute, India

Meenaakumari M.
Dr. M. G. R. Educational and Research Institute, India

Hemasundari H.
Dr. M. G. R. Educational and Research Institute, India

Aathilakshmi T.
Dr. M. G. R. Educational and Research Institute, India

ABSTRACT

The most typical approaches are content-based image retrieval systems. Content-based picture retrieval may be the only one in all the image retrieval techniques that uses user visual options of an image like color, form, and texture. The objective is to retrieve the set of pictures quickly and economically by supported color and texture options. Color is the foremost authoritative and utilized visual option that is invariant to image dimension and adjustment. Color car correlogram includes the special correlation and figures the mean color of all components of intensity about a distance k-th of a pixel of intensity the picture. Next, the feel feature may be a powerful region-based descriptor to provide a life of attributes like smoothness, coarseness, and regularity. Block distinction probabilities and block variation of native correlation features are analysed to speed up the retrieval method. BDIP may be a block-based approach to extract color and intensity features and live native brightness variation from the photographs.

DOI: 10.4018/978-1-7998-9640-1.ch005

INTRODUCTION

With the fast proliferation of the web and therefore the worldwide-web, the number of digital image information accessible to users has full-grown tremendously. Image databases are getting larger and additional widespread, and there is a growing want for effective and economical image retrieval (IR) systems. Most IR systems adopt the subsequent ballroom dancing method to look picture information.

1. Classification for every picture in database a column vector obtaining sure fundamental parts of the picture measured and hold on in article information.
2. Looking out addressed a question image, its feature vector measured, related to the feature vectors within the feature info, and pictures most almost like the question picture area unit came to the user (Chandan Singh and Kanwal Preet Kaur, 2016).

Advances in information storage and image acquisition technologies have enabled the creation of enormous image datasets. Supported that, it is necessary to develop acceptable info systems with efficiency manage these collections. BVLC employed to extract form and texture smoothness within the image. This approach outperforms the quick and economical retrieval technique on COREL and UKBENCH DATASET. From 10,000 pictures of the COREL dataset, thirty pictures are willy-nilly elite and processed. From 10,200 pictures of the UKBENCH dataset, twenty-six pictures are willy-nilly elite and processed. About question image, similar pictures are retrieved supported various options and for performance analysis, accuracy is calculated for every option and compared severally.

LITERATURE SURVEY

Feature Extraction Methods

The extraction method describes a quick and sturdy color categorization technique; specifically motorcar color correlation supported a color correlogram for obtaining and categorization low-level options of pictures (Jing Huang et al., 2011; Meenakshi Sharma and Anjali Batra 2014). It will scale back the processing circumstances of the color correlogram method from O (m2d) to O (md). In addition, it consumes less interval than the opposite algorithms.

An economical CBIR methodology supported the mixture of multiresolution intensity and surface options where described (Nirmala and Subramani 2013). The intensity and texture options square measure selected in the multiresolution rippling field including combined. The dimension about the consolidated feature vector is decided to some extent wherever the retrieval accuracy becomes saturated (Young Deok Chun et al., 2008; Jacob et al., 2011).

Accuracy attained by using retires away techniques. An evolving topic below the image process is content-primarily based image retrieval was described (Ritendra Datta et al., 2008). They propose a changed combined method that removes low-level image options square measure intensity, depth, form, and character. A feature detection and outline system that uses in feature extraction ways (Anxo Conde and Jorge Dominguez 2018; Swapna et al., 2019). Feature detection is that the method wherever mechanically extract options of a picture, in such a fashion that we tend to square measure ready to observe AN object supported its options in numerous pictures.

Similarity Activity Ways

A distinct agglomeration methodology to photos taken by multiple travellers (Swapna and Kamalahasan. 2020). It overcomes the main issue of good management for cluster photos for an individual cluster. They projected many supervised and unattended agglomeration ways for cluster photos. Supervised ways to acquire true clusters from a user and build different photos into bespoken clusters.

Similarity measures for pictures was described (Tao and Grosky, 1999). They perform the image quality analysis and live the image quality. Microphone judge and compare the various algorithms to unravel the matter by comparing two fuzzy sets. A similarity live could be a fuzzy binary relation and min-transitive. The similarity measures for pictures (Thenkalvi and Murugavalli, 2014).

They perform the image quality analysis and live the image quality. Microphone judge and compare the various algorithms to unravel the matter by comparing two fuzzy sets. A similarity live could be a fuzzy binary relation and min-transitive phase was measured in the field programmable gate array (Young Deok Chun et al., 2003).

The similarity live for the image databases (Tungkasthan et al., 2009). He designed to check numerous similarity measures for application to image databases. They live on crisp logic similarly to formal logic. Crisp logic measures square measure supported a category of distance functions. It perform many analyses like similarity matrix for every live, sensitivity analysis, and transition analysis determined garbage filling system by using internet of things (Van der Weken et al., 2002).

IMAGE RETRIEVAL SYSTEM

An image retrieval method affords the user by some method to obtain, survey, and regain pictures expeditiously from databases. These databases area unit utilized in a spread of fields together with fingerprint classification, variety data system, digital museum, crime hindrance, medical imaging, ancient archives (Dinakaran and Vijayarajan, 2013). Image retrieval supported content aims at looking out similar pictures through the analysis of image content thus; image representations and similarity live have become important for such a task. The sector of content-based image retrieval consists of loads of various classification structures, schemes, and ways aiding the connected retrieval tasks. The goal of the classification structures is to require associate degree accessible dataset and manufacture a curt and easier to handle index, which may be wont to rummage around for similar content (Gayathri S et al., 2020).

Figure 1. Image retrieval system

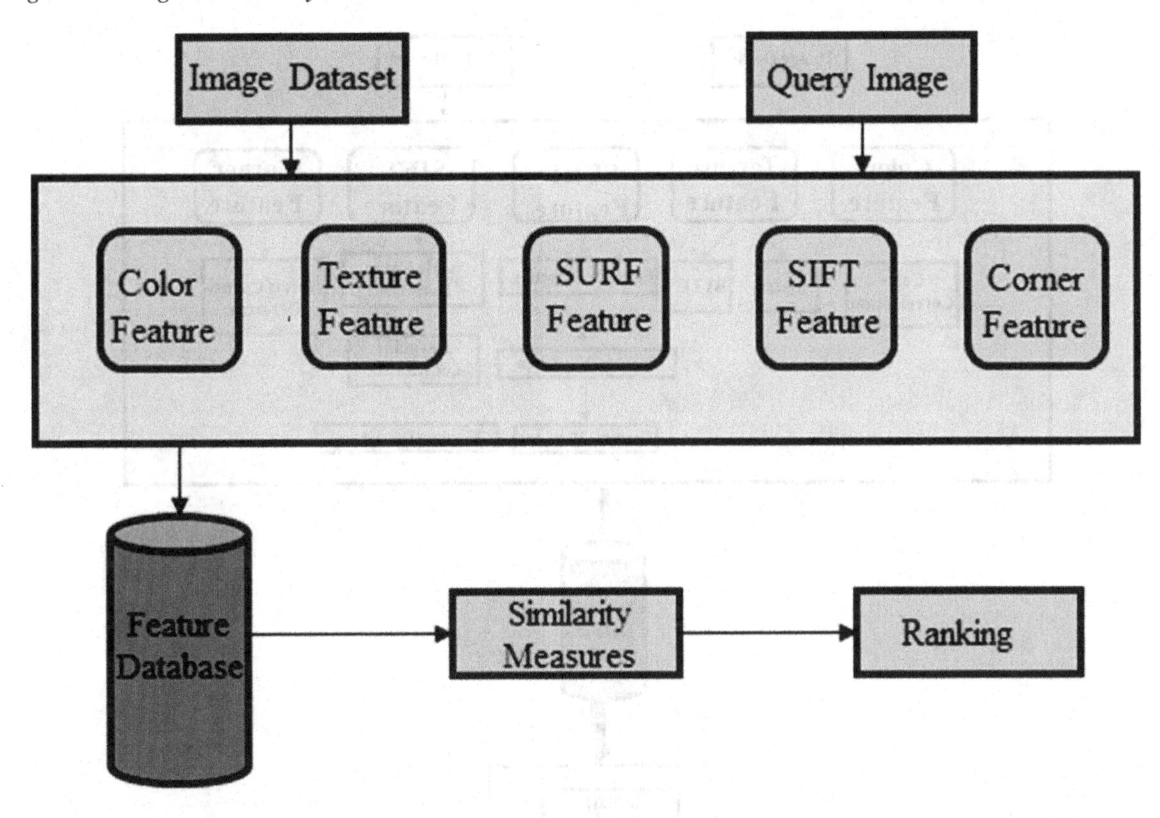

In the pre-processing analysis, the question image gets from a picture dataset. For each image dataset and question image color and texture, options extracted and the feature vector of dataset pictures is holding on within the feature info. The image feature vector compared with the feature info to seek out the space matrix similarity measures. Then the highest relevant pictures are retrieved supported question image victimization sorting method. The general design Diagram as shown in Figure 1 and block diagram as shown in Figure 2.

Figure 2. Block diagram of image retrieval system

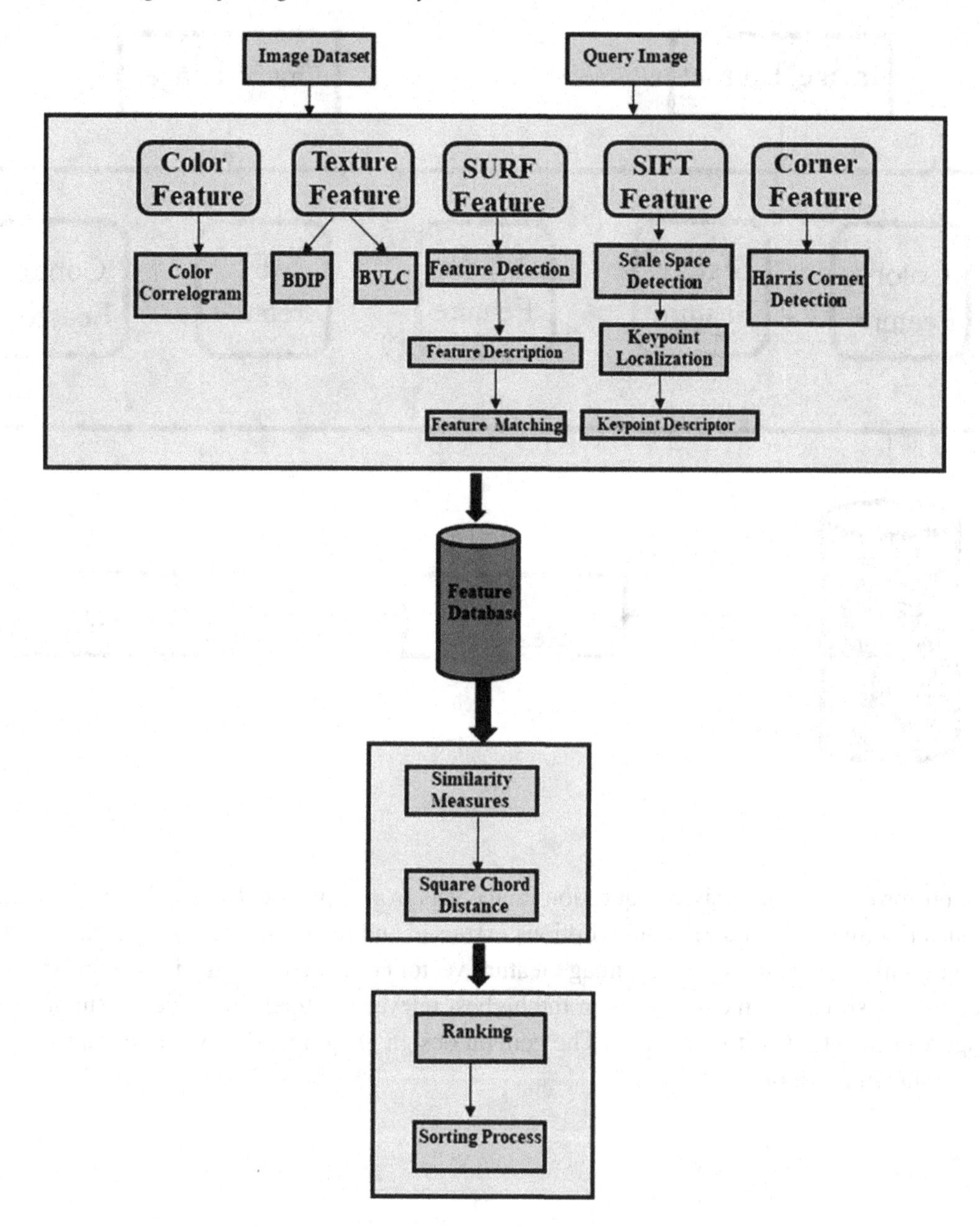

Feature Extraction

Features square measure the data obtained of pictures in words of mathematical conditions that square measure tough to grasp also compare by humans. Assume the picture as information the data derived from the info understood as options. Usually, options are removed from a picture square measure of a rather more under dimension than that first picture. The reduction in spatiality decreases the expenses of the process of a bunch of pictures. Primarily their square measure 2 varieties of options square measure extracted from the photographs supported the appliance. They are native and international options.

Options square measure typically related to as descriptors additionally. International descriptors usually employed in image retrieval, target detection, and classification, whereas the native descriptors used for target identification/description. There is an outsized distinction within disclosure and classification. Disclosure is determining the existence of an object wherever Recognition is determining the identity of an object. International options describe the picture as a full to conclude the complete object wherever because the native options describe the image patches of a picture. International options embrace shape descriptions, form descriptors, and texture options and native options represent the feel in a picture patch. Usually, for low-level utilization like object disclosure and analysis, international options square measure practiced and for higher-level relationships like seeing, native options square measure used. A combination of world and native options increases the efficiency of the popularity with the facet result of process expenses.

Color Correlogram

The color correlogram shows however, the abstraction similarity of combinations of colors changes with distance. It describes the worldwide distribution of native abstraction correlations of colors. In addition, additionally, it defines the color level pairs at a distance d. In image retrieval, the processing time is proportional to the length of the feature vector employed in image info categorization. The 2 vital edges of color correlogram are lightened the process price by decreases of greyscale within the pictures. Division of the image into sets of comparable levels generalizes the image content that improves the retrieval results. The color correlogram is not a picture partitioning methodology neither a bar graph learning methodology. In contrast to strictly native properties, like pel position, gradient direction, or strictly international properties, like color distribution, correlograms take under consideration the native color abstraction correlation similarly because of the international distribution of this abstraction correlation. whereas any theme that's supported strictly native properties is probably going to be sensitive to massive look changes, correlogram square measure a lot of stable to those changes; whereas any theme that's supported strictly international properties is liable to false positive matches, correlogram encourage be sufficient for content-based image retrieval from outsized picture info. AN economical algorithmic program that is AN enlargement of the correlogram method for color categorization. First, the scale of the specific feature vectors. This does the retrieval method potential wherever correct and quick retrieval outcomes square measure needed. This issue is kind of material once the image databases square measure considerably massive. Frequently, the users do not like long latent periods albeit such a combination of options might yield higher retrieval efficiency. Second, merging options from completely various modalities is not a simple task. AN motorcar color correlation expresses the way to cipher the mean color of all pels of color Cj at a range k-th from a pixel of color C, in the picture. There square measure 2 major issues related to combining completely different options.

Algorithm

Color correlogram offers data regarding the options of colors. It includes spatial color correlations that describes the worldwide pattern of the native spatial relationship of colors and does extremely straight forward to work out victimization.

The pseudocode of color correlogram explained in the Algorithm 3.1.

Input: RGB image.
Output: Color Correlogram background vectors.

Algorithm 3.1 Auto Color Correlogram algorithm

1: For every K distance
2: For every X position
3: For every Y position
4: Ci current pixel
5: while (Ci Get neighbours pixel of Ci at distance K)
6: For every color Cm
7: If (Cm =Ci and Ci 6=Ci)
8: count Color++ color R[Cm] = color R[Cm]+color R Ci
190: color G[Cm] = color G[Cm]+color G Ci
11: color B[Cm] = color B[Cm]+color B Ci
12: mean Color R=sum (color R[Cm])
13: mean Color R=sum (color R[Cm])
14: mean Color R=sum (color R[Cm])

Implementation

One will select the pictures from the specialization of the image into sets of comparable levels generalizes the image content, which improves the retrieval results. the pictures in COREL and United Kingdom of Great Britain and Northern Ireland BENCH dataset are chosen {randomly|indiscriminately|haphazardly|willy-nilly|arbitrarily|atrandom|every that way} and may perform the feature extraction which is referred in Figure 3 and have vector table is shown in Figure 4

Figure 3. Dataset pictures

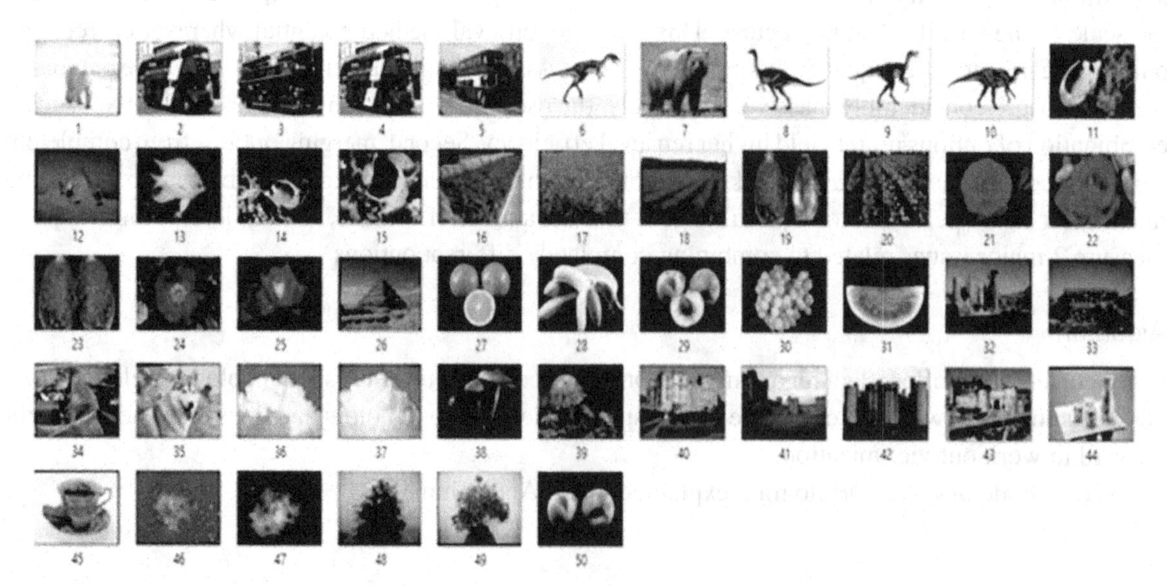

The datasets contain 10,000 pictures in every of the dataset. Any pictures taken and Color Correlogram feature vectors extracted for each info picture and question image.

In automotive vehicle Color Correlogram, the color image is representing in unit matrix and finding the sixty-four measure color in numerous in-norm distance vectors (64n ×1).

Figure 4. Feature vector tabulation

img_5	img_6	img_7	img_8	img_9	img_10	img_11	img_12	img_13	img_14	img_15	img_16	img_17	img_18	img_19	img_20
0.61198	0.19844	0.37303	0.30252	0.3052	0.18729	0.43993	0.26273	0.37115	0.20708	0.26193	0.33529	0.59519	0.43155	0.49081	0.57235
0.28654	0.03019	0.04024	0.14003	0.09975	0.03213	0.24098	0.57215	0.23837	0.0868	0.33256	0.10972	0.14729	0.03794	0.18466	0.11617
0.41655	0.12218	0.05622	0.16019	0.15249	0.07877	0.19631	0.31369	0.46129	0.16516	0.45505	0.2115	0.09584	0.12157	0.19109	0.08217
0.28146	0.60485	0.44272	0.11275	0.05081	0.32088	0.40246	0.15658	0.13733	0.28909	0.39665	0.34326	0.19808	0.27733	0.08376	0.11438
0.17646	0.06792	0.09302	0.10758	0.01465	0.0813	0.25924	0.56703	0.50342	0.13477	0.24684	0.0979	0.18839	0.34276	0.705	0.20783
0.10339	0.10707	0.1197	0.49961	0.06107	0.26928	0.12654	0.55763	0.17005	0.08025	0.29767	0.19545	0.07598	0.09787	0.78939	0.04328
0.15156	0.07561	0.19962	0.15025	0.24981	0.06417	0.36466	0.67661	0.28499	0.19786	0.38411	0.08806	0.21815	0.13578	0.13788	0.16392
0.25928	0.4801	0.12978	0.09664	0.35008	0.22469	0.12912	0.25113	0.42703	0.19872	0.59242	0.12835	0.08483	0.65641	0.07395	0.15747
0.31206	0.38307	0.07174	0.09256	0.31932	0.35412	0.35631	0.28983	0.41845	0.10217	0.27719	0.17607	0.09389	0.13374	0.04718	0.11573
0.46878	0.15111	0.47232	0.14041	0.21899	0.18991	0.5287	0.55191	0.31169	0.51034	0.25077	0.11839	0.09489	0.24098	0.19638	0.09684
0.53058	0.04973	0.11494	0.32185	0.23002	0.0703	0.2114	0.42315	0.41542	0.12888	0.26328	0.08543	0.14899	0.58799	0.32917	0.17934
0.30822	0.24649	0.15343	0.22376	0.14276	0.15913	0.31124	0.66374	0.46519	0.16655	0.25987	0.58782	0.1553	0.23834	0.17111	0.34551
0.09689	0.12713	0.14459	0.49928	0.13428	0.45509	0.2204	0.50283	0.15965	0.16915	0.21339	0.20734	0.11072	0.15147	0.70924	0.56225
0.12917	0.14936	0.20753	0.33271	0.30052	0.16537	0.14661	0.58335	0.48719	0.02768	0.55502	0.13771	0.29009	0.13569	0.09683	0.119
0.27851	0.31096	0.14704	0.01379	0.32821	0.12485	0.20257	0.39822	0.6303	0.31886	0.20602	0.34398	0.14427	0.14286	0.08907	0.07789
0.37337	0.04154	0.10387	0.21508	0.03663	0.0597	0.27195	0.78476	0.33566	0.11724	0.41967	0.4771	0.2237	0.6508	0.67528	0.07305
0.18474	0.07312	0.24732	0.17143	0.21451	0.10537	0.2551	0.24285	0.17882	0.2154	0.31776	0.10566	0.11072	0.10182	0.07134	0.1492
0.29453	0.01081	0.04278	0.14699	0.03564	0.14347	0.39647	0.64348	0.2304	0.12005	0.45243	0.10403	0.22989	0.13448	0.07455	0.14111
0.38907	0.03125	0.15661	0.10201	0.15385	0.14767	0.47276	0.27101	0.49726	0.39057	0.22351	0.80609	0.10445	0.2304	0.1595	0.1961
0.17417	0.28571	0.06331	0.13622	0.07824	0.26904	0.11128	0.68396	0.13605	0.80727	0.43165	0.22889	0.08398	0.36807	0.05591	0.05569
0.20238	0.44434	0.09332	0.79771	0.08012	0.13401	0.36422	0.80177	0.77329	0.1388	0.20715	0.32207	0.19598	0.31787	0.10342	0.15573
0.56841	0.06236	0.62852	0.12483	0.05902	0.03622	0.20747	0.2946	0.51211	0.09033	0.23374	0.08606	0.08993	0.38724	0.33322	0.24823
0.7261	0.07029	0.25372	0.2749	0.02268	0.01187	0.27748	0.08902	0.1287	0.0877	0.24116	0.27709	0.13232	0.11607	0.11496	0.15303
0.24677	0.09524	0.19158	0.11964	0.75431	0.05452	0.50298	0.4767	0.44495	0.12109	0.3	0.07641	0.08544	0.11596	0.06362	0.10892
0.18981	0.12874	0.09238	0.08084	0.23166	0.15274	0.2311	0.42266	0.20581	0.07561	0.2363	0.28045	0.13463	0.19973	0.0924	0.13245
0.1287	0.15107	0.2179	0.08602	0.35384	0.02721	0.10077	0.33755	0.51205	0.24937	0.2071	0.41552	0.09962	0.36552	0.10828	0.26067

Block Distinction of Inverse Chances

The distinction of inverse chances (DIP) is Associate in the Nursing director as removing configuration options that accommodate valleys and edges directed to native intensities. In the DIP, the magnitude relation of a pel power in a picture window to the addition of all pel intensities in an exceeding window considered as a chance. So, the title DIP indicates that the distinction among the inverse of the chance for the canter pel in an exceeding window which for the pel of most intensity within the window BDIP, that is one in all the planned texture options, could be a block-based version of the DIP. It is outlined because of the distinction among the amount of pels in an exceedingly block and the magnitude relation of add of pixel intensities within the block to the utmost within the block and its calculated victimization.

BDIP expeditiously live native brightness variation several. BDIP derived from 2 × 2 non-overlapped windows. It is a block-based mostly process approach to select color and intensity (Brightness) features of the pictures. It measures native brightness variation. It will extract twenty-two non-overlapped windows that divide from a picture I(x, y). It represents the sides and bounds of the image regions.

Algorithm

The block distinction of inverse chances options is in algorithm 3.2

Input: RGB image.

Output: Texture feature point.

Algorithm 3.2 Block Difference algorithm

1: Read the input image.
2: Get the texture image BDIP(x,y).
3: Prepare a couple of 2-D texture images BDIP(x,y) within two 1-D arrays.
4: Make the two 1-D sorted arrays be denoted by A(x).
5: Partition the array A(x) within a couple of classes with practicing the middle of A(x) being that threshold.
6: Copy the method as an individual of the elements to achieve four classes.
7: Additional copied to partition individual of the four classes into two classes.
8: Define the valleys and edges for all the classes.
9: Fine the Mean and approved variation of each class.

Implementation

The images in COREL and UK BENCH dataset are choosed randomly. The input image and their BDIP texture feature are extracted is presented in Figure 5.

Figure 5. BDIP Input picture and their texture output

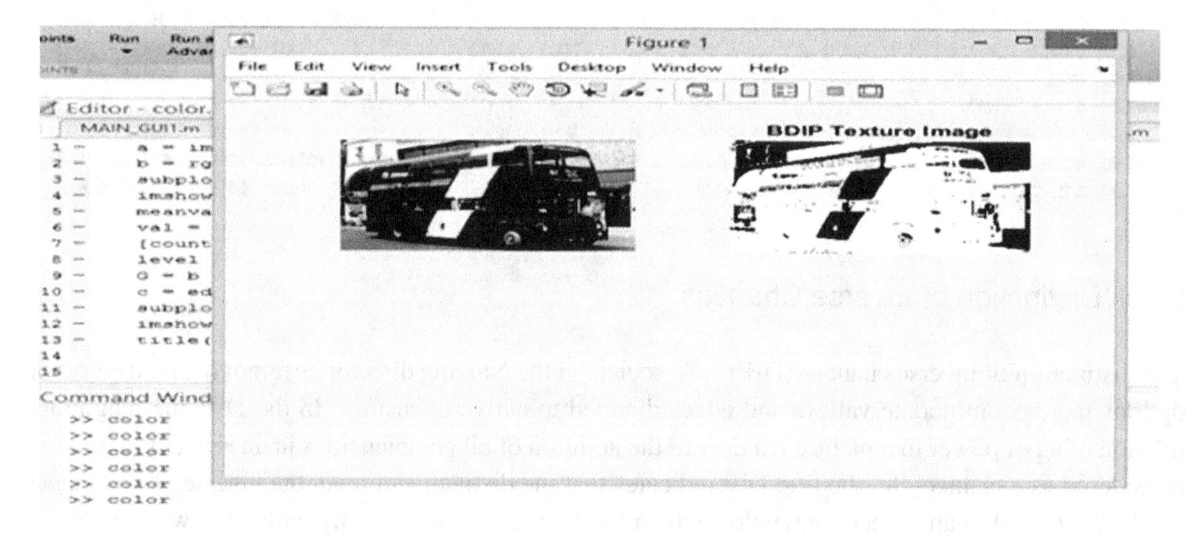

Block Variation of Social Correlation Coefficients

Block-based local correlation coefficients (BVLC) are associated to include surface smoothness. It is defining since the change, or the contrast within the highest and least, of local correlation coefficients according to four familiarizations. Several local correlation coefficients are represent as local covariance normalized by local variance.

Algorithm

The block variation of local correlation coefficients features explain in the Algorithm 3.3 and obtained as follows,

Input: RGB image.

Output: Texture feature points.

Algorithm 3.3 Block Difference algorithm

1: Read the input image.
2: Get some texture image BVLC(x,y).
3: Prepare these couple 2-D texture pictures into 1-D arrays.
4: Find the Local correlation coefficients for all the classes.
5: Find the Mean and approved variation of a specific class.

This is also block-based access to obtain shape and surface smoothness. It means the difference within the maximum and minimum of local correlation coefficients according to the six orientations. It measures the local texture smoothness. In the image, the degree of roughness is high then the value of BVLC is large. It represents the surface during the entire concept region wherever intensity variation is low.

Implementation

The images in COREL and UK BENCH datasets chosen randomly. The input image and their BVLC texture feature are extracted are presented in Figure 6 and Figure 7 each.

Figure 6. Input image

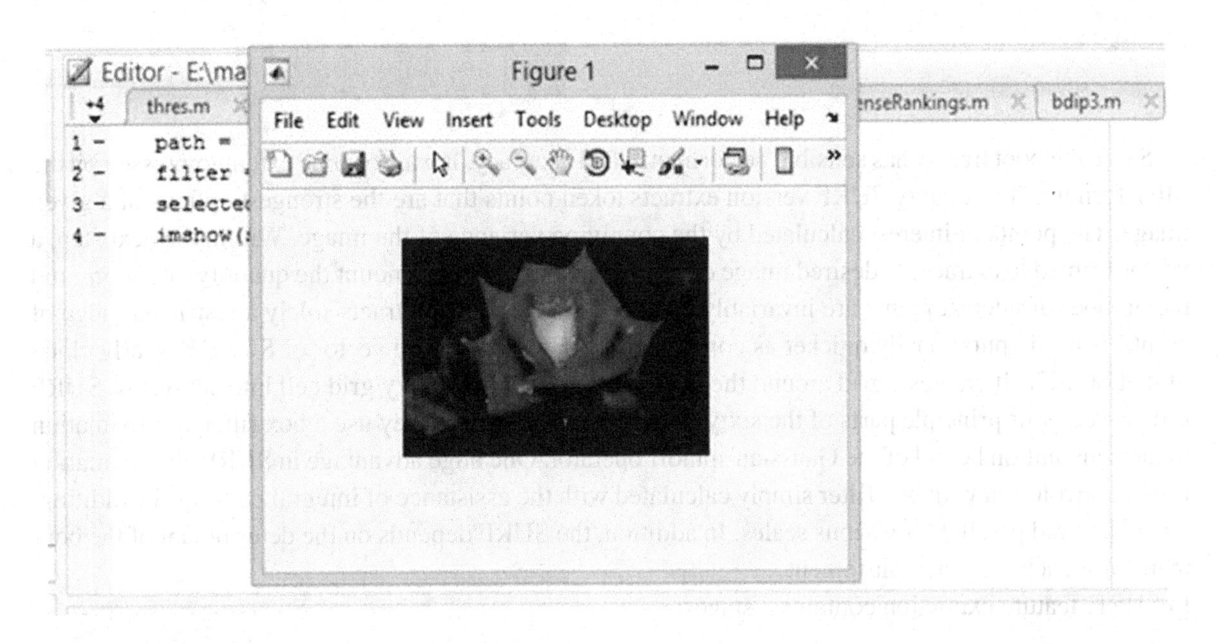

Speed-Up Robust Features

SURF (Speed up Robust Features) algorithm, does base on multi-scale area theory, including this feature indicator based on a Hessian pattern.

Figure 7. BVLC texture output

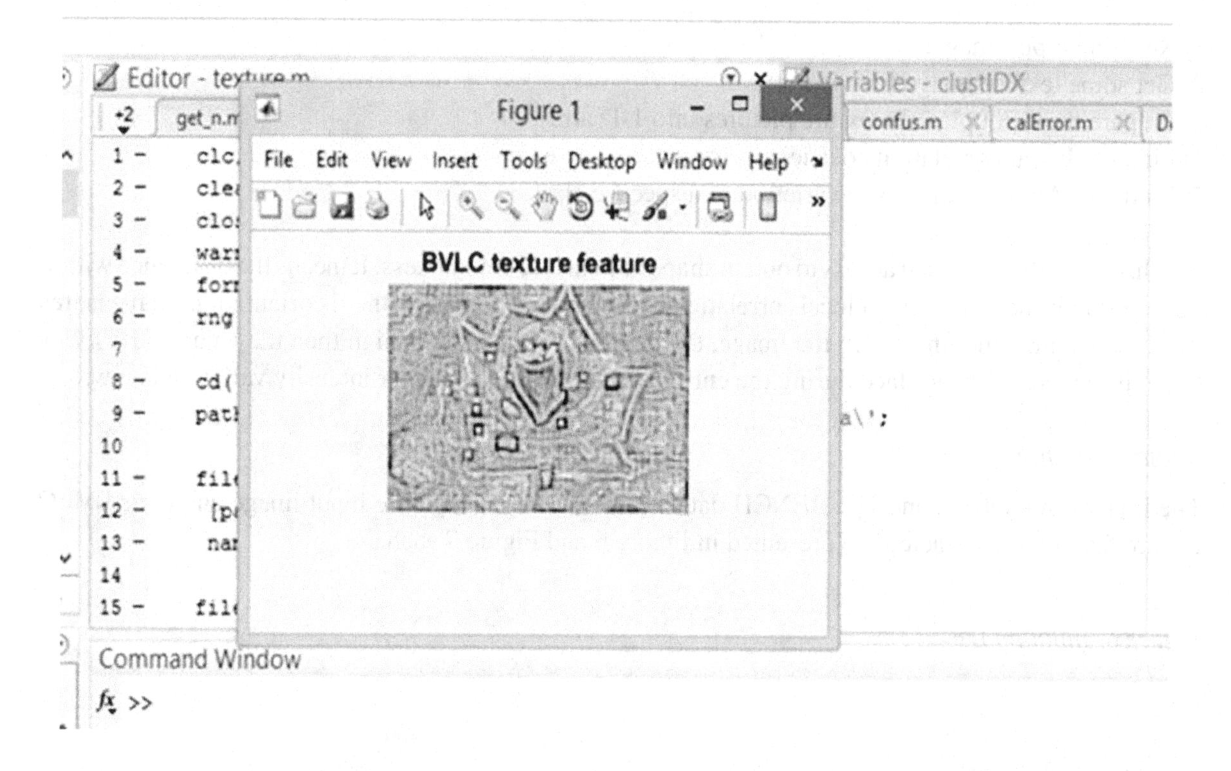

Since the boot matrix has sensible performance and accuracy. It was to some extent impressed by the SIFT formula. The quality SURF version extracts token points that are the strongest options of a given image. The points of interest calculated by the conniving variance of the image. Within the next step, a vector formed to extract the desired image options. Thus, variety the amount the quantity of options and the number of interest points are invariably the same. Since SURF extracts solely a restricted range of points, it is computationally quicker as compared to SIFT. The feature vector of SURF is nearly clone of that of SIFT. It creates a grid around the key points and divides every grid cell into sub-grids. SURF use 36-vector of principle parts of the sixty-four vectors for speed. They use a box filter approximation to the convolution kernel of the Gaussian spinoff operator. One huge advantage in SURF approximation is that convolution with box filter simply calculated with the assistance of integral pictures. In addition, it may be tired parallel for various scales. In addition, the SURF depends on the determinant of the boot matrix for each scale and placement.
In SURF, feature extraction contains 3 stages:

- Feature Disclosure: Automatically recognize attractive characteristics, interest points that need to be prepared robustly. This corresponding characteristic detected regardless of position.
- Feature Classification: Any interest point should have a single account that gives not depend upon the scale and rotation of the features.
- Feature Matching: For a provided input image, determine which objects it contains, and possibly a transformation of the object, based on predetermined interest points.

SURF algorithm

The SURF algorithm based on identical principles as SIFT. The algorithm has three main parts: interest point exposure, social community information, and matching. It explained in Algorithm 3.4 and 3.5 respectively.
Algorithm 3.4 Selection of features

1: input: o, i, DoH(u)
2: output: list Key Points
3: I←– Image (I)
4: DoH(u)←– Determinant of Hessian (U, L)
5: list Key Points←– list Key Points + Key Points (o, i, DoH(u))
6: return list Key Points

Algorithm 3.5 Construction of SURF Descriptor

1: input: Input image I, key point X: (x, y, L)
2: output: SURF Descriptor, orientation of key point and sign of Laplacian
3: function build Descriptor (I, x, y, L)
4: θ ←Orientation (u, x, y, L)
5: for i: =1 to 4 do
6: for j: =1 to 4 do
7: for u: =-9.5 to 9.5 step5 do
8: for v: =-9.5 to 9.5 step5 do
9: (x,y) =s(u,v)
10: (dx (u;v);dy(u;v))
11: end for
12: end for
13: end for
14: end for

Implementation

One can choose the images from the COREL and UKBENCH dataset randomly and can perform the feature extraction, which refers in the Figure 8. In Speeded-up Robust Features, the total local feature points are 256. Among that, maximum 10 points chosen to cluster the closest points to number of classes.

Figure 8. SURF feature output

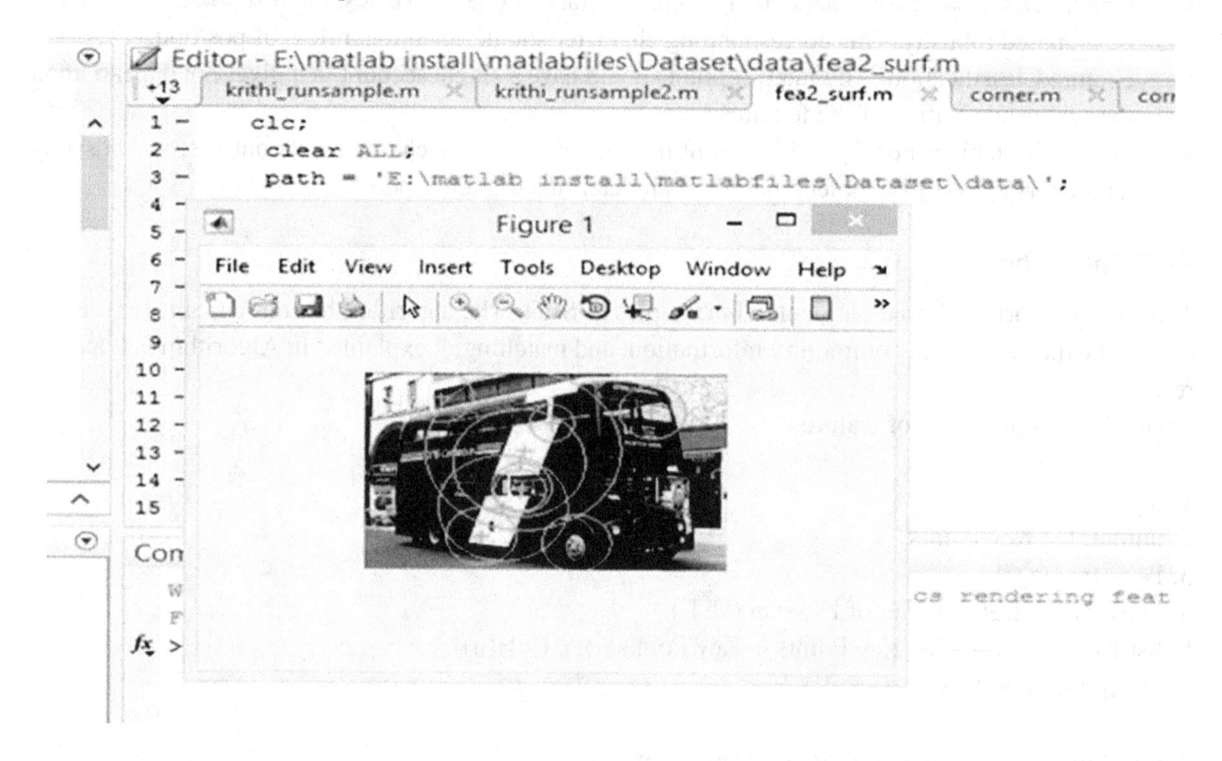

Harris Corner Detection

Harris Corner Detection could be a corner exposure operative that is ordinarily utilize in pc image algorithms to extract corners and infer options of a picture. Harris corner detector considers the differential from the corner score. With relation to direction right, rather than victimization shifting patches for every forty-five-degree angle, and has tested to be additionally correct in identifying between edges and corners. Since it is changed and utilized in several algorithms to pre-process pictures for ensuant applications. A corner also described in any area that square measure 2 aggressive and completely separate turn directions. Associate in nursing interest design could be a purpose in a picture that incorporates a well-defined position and maybe robustly detected.

Corner Algorithm

Its key points with respect to its target region mainly define the corner detection algorithm as follows.
Algorithm 3.6 Corner Detection Descriptor

1: input: Input C= set of corners resulting from Harris corner detection.
2: input: Input SF=sf1, sf2, sft, is the set of (t) scales considered.
3: input: Input TP=target product to be detected.
4: output: Output R=image windows
5: PRE-SELECTION (c, sf, tp) R=θ
6: list Key points← list Key points + Key points(c, sf (i+1),tp (i))

7: return list Key Points =0

Implementation

The corner feature points are extracted based on the selected regions and is shown in the Figure 9.

Figure 9. Harris Corner Output

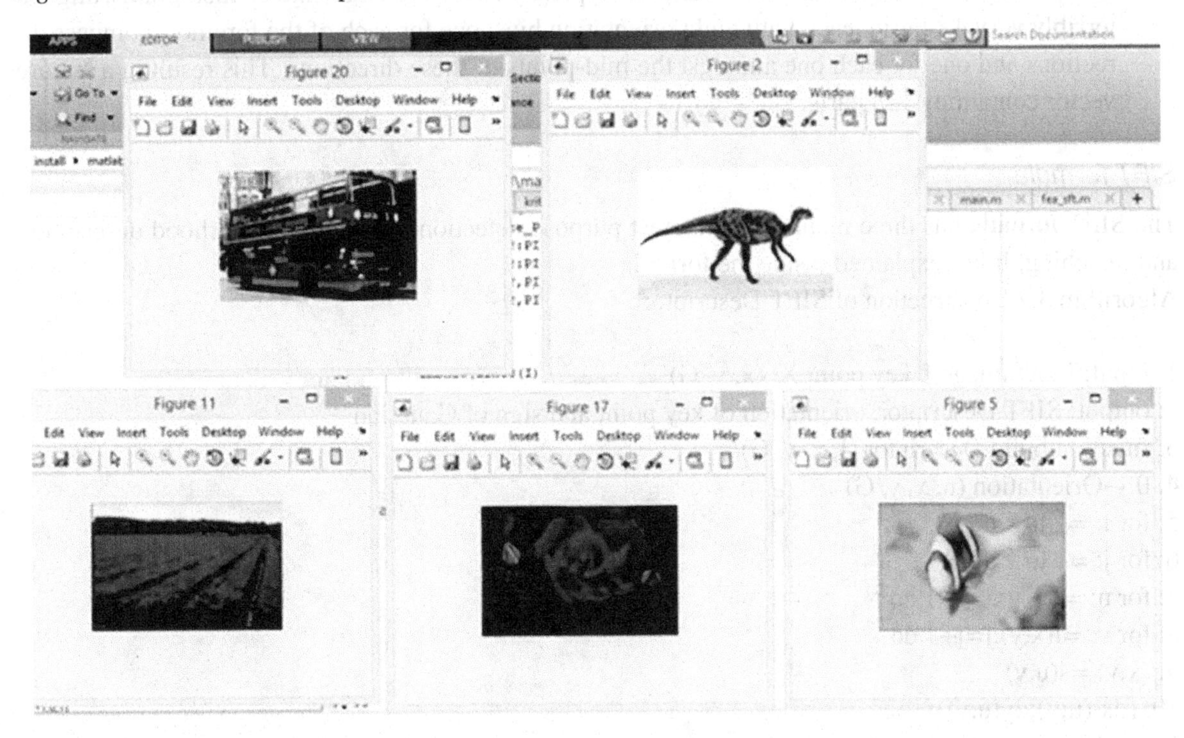

Scale-Invariant Feature Remodel

For any object, their unit of measurement many choices, attention-grabbing points on the article, which will extracted to supply a feature description of the article. This description used once attempting to seek out the article in an exceeding image containing many various objects. There unit of measurement many considerations once extracting these choices and therefore the thanks to record them. SIFT image choices provides a cluster of choices of an object that are not laid low with many of the complications experienced in alternative routes, like object scaling and rotation. Whereas permitting an object to be recognized terribly very larger image, SIFT image choices collectively afford objects in multiple footage of constant location, taken from utterly completely different positions at intervals the setting, to be recognized. SIFT choices are resilient to the results of noise inside the image. The SIFT approach, for image feature generation, takes an image and transforms t into large assortment of native feature vectors. Each of these feature vectors is invariant to any scaling, rotation, or translation of the image. This approach shares many choices with vegetative cell responses in primate vision. To assist the extraction of these choices the SIFT formula applies four stages filtering approach:

1. Scale-Space Extrema Detection: This stage of the filtering attempts to identify those locations and scales that unit of measurement identifiable from utterly completely different views of constant object. This is expeditiously achieves using a "scale-space" Perform. Extra it has shown at a lower place cheap assumption it ought to be supported the scientist perform.
2. Key point Localization: This stage tries to eliminate extra points from a listing of key points by finding those that have an occasional distinction or poorly localized on an edge.
3. Key point Descriptor: Key points descriptors typically uses a bunch of sixteen histograms, aligned terribly very 4×4 grid, each with eight orientation bins, one for each of the foremost compass directions and one for each one amongst the mid-points of these directions. This result in a feature vector containing 128 parts.

SIFT formula

The SIFT formula has three main parts: interest purpose detection, native neighbourhood description and matching. It has explained inside the formula
Algorithm 3.7 Construction of SIFT Descriptor

1: input: Input image I, key point X: (x, y, G)
2: output: SIFT Descriptor, orientation of key point and sign of Gaussian
3: function build Descriptor (I, x, y, G)
4: θ ←Orientation (u, x, y, G)
5: for i: =1 to x do
6: for j: =1 to y do
7: for u: =I(x,y),i=i+1 do
8: for v: =I(x,y),j=j+1 do
9: (x,y) =s(u,v)
10: (dx (u;v);dy(u;v))
11: end for
12: end for
13: end for
14: end for

Implementation

The SIFT matching feature points are extracted from the input image and it is shown in the Figure 10.

Figure 10. SIFT feature output

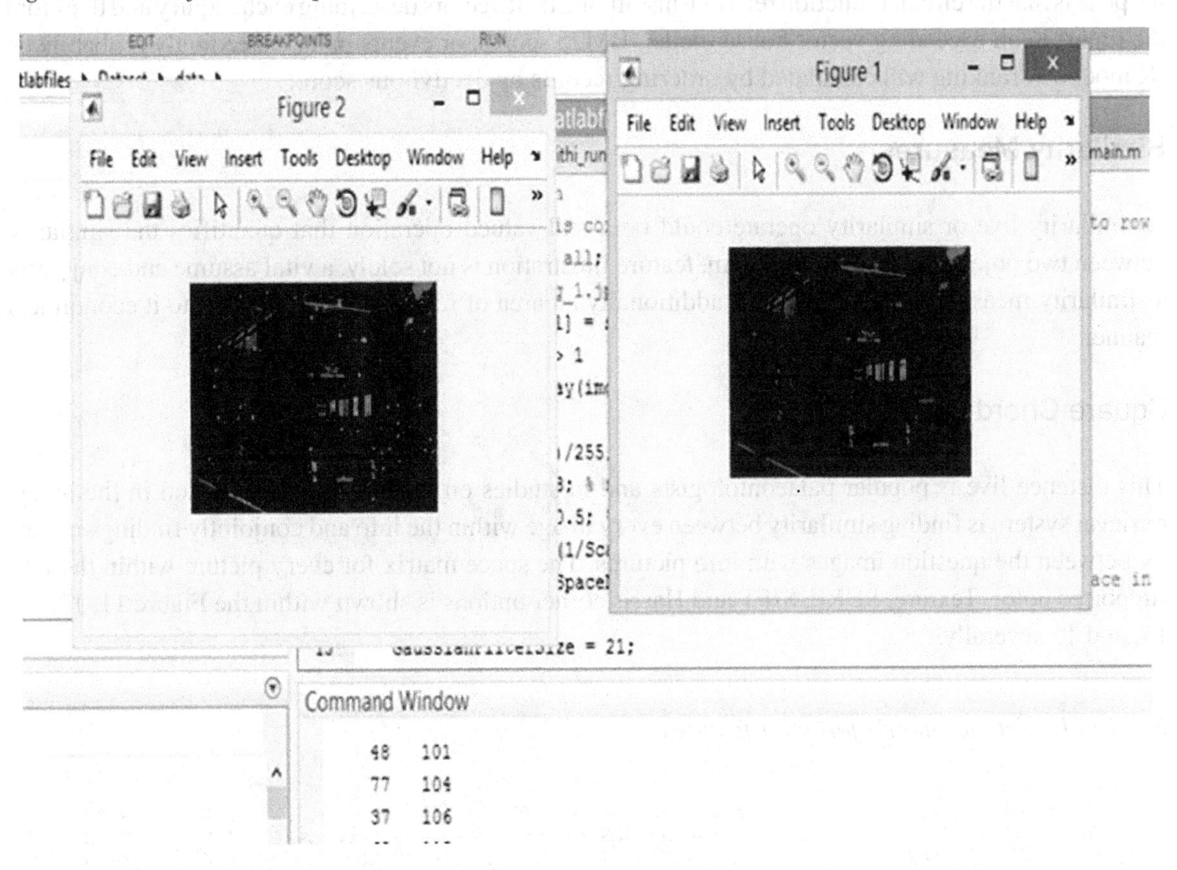

RANKING AND SIMILARITY MEASURES

Ranking

A ranking could be a relationship between a collection of things specified, for any 2 things, the primary is either 'ranked higher than', 'ranked lower than' or 'ranked equal to' the second. Supported the sorting method, the highest relevant pictures are retrieved. The retrieval performance mistreatment color and texture options. If the elements of knowledge the info the information} instance vectors are all iconstant physical units then it's doable that the straightforward square the chord distance metric is sufficient to with success cluster similar data instances. The space between 2 points is the length of the line connecting them this is often the alleged square chord distance. The ranking is that the method of sorting the highest relevant pictures with user interactions. It involves active sample choice during which the user labels the pictures as relevant or digressive. Ranking of question effects does one in every one of the elemental issues in data retrieval (IR), the systematic/engineering training following research engines. Given Query|a question |a question q and a set D of records that even the question, the matter is to order, this is, kind, the records in D in step with any model so the most excellent results seem early within the event list presented the user. Classically, grading standards phrased in terms of connection of records concerning associate degree data would like expressed within the question. The ranking is commonly decreased to the number of numeric scores on question/report pairs; a baseline score operate for this purpose is that the circular function relationship among tfidf vectors describing each inquiry and therefore the report in an exceeding vector house model, BM25 scores, or events in an exceedingly probabilistic IR model. A ranking will calculated by ordering records by declivitous score.

Similarity Measures

A similarity live or similarity operate could be a real-valued operation that quantifies the similarity between two objects. In the CBIR system, feature illustration is not solely, a vital assume and conjointly a similarity measure and illustration is additionally an area of retrieval system to create it economical manner.

Square Chord Distance

This distance live is popular palaeontologists and in studies on spore. The next method in the image retrieval system is finding similarity between every image within the info and conjointly finding similarity between the question images with info pictures. The space matrix for every picture within the info supported color, Texture, SURF, SIFT and Harris corner options is shown within the Figure 11, 12, 13, 14, and 15 severally.

Figure 11. Distance matrix for color features

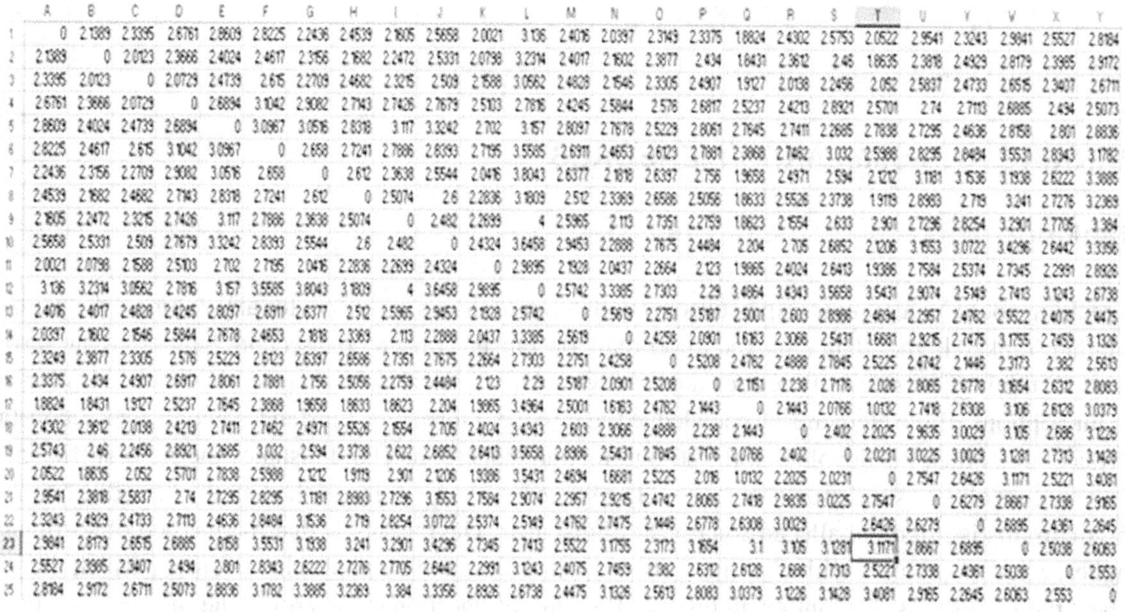

	A	B	C	D	E	F	G	H	I	J	K	L	M
1	0	2.1389	2.3395	2.6761	2.8609	2.8225	2.2436	2.4539	2.1805	2.5658	2.0021	3.136	2.4016
2	2.1389	0	2.0123	2.3666	2.4024	2.4617	2.3156	2.1682	2.2472	2.5331	2.0798	3.2314	2.4017
3	2.3395	2.0123	0	2.0729	2.4739	2.615	2.2709	2.4682	2.3215	2.509	2.1588	3.0562	2.4828
4	2.6761	2.3666	2.0729	0	2.6894	3.1042	2.9082	2.7143	2.7426	2.7679	2.5103	2.7816	2.4245
5	2.8609	2.4024	2.4739	2.6894	0	3.0967	3.0516	2.8318	3.117	3.3242	2.702	3.157	2.8097
6	2.8225	2.4617	2.615	3.1042	3.0967	0	2.658	2.7241	2.7886	2.8393	2.7195	3.5585	2.6911
7	2.2436	2.3156	2.2709	2.9082	3.0516	2.658	0	2.612	2.3638	2.5544	2.0416	3.8043	2.6377
8	2.4539	2.1682	2.4682	2.7143	2.8318	2.7241	2.612	0	2.5074	2.6	2.2836	3.1809	2.512
9	2.1805	2.2472	2.3215	2.7426	3.117	2.7886	2.3638	2.5074	0	2.482	2.2699	4	2.5965
10	2.5658	2.5331	2.509	2.7679	3.3242	2.8393	2.5544	2.6	2.482	0	2.4324	3.6458	2.9453
11	2.0021	2.0798	2.1588	2.5103	2.702	2.7195	2.0416	2.2836	2.2699	2.4324	0	2.9895	2.1928
12	3.136	3.2314	3.0562	2.7816	3.157	3.5585	3.8043	3.1809	4	3.6458	2.9895	0	2.5742
13	2.4016	2.4017	2.4828	2.4245	2.8097	2.6911	2.6377	2.512	2.5965	2.9453	2.1928	2.5742	0
14	2.0397	2.1602	2.1546	2.5844	2.7678	2.4653	2.1818	2.3369	2.113	2.2888	2.0437	3.3385	2.5619
15	2.3249	2.3877	2.3305	2.576	2.5229	2.6123	2.6397	2.6586	2.7351	2.7675	2.2664	2.7303	2.2751
16	2.3375	2.434	2.4907	2.6917	2.8061	2.7881	2.756	2.5056	2.2759	2.4494	2.123	2.29	2.5187
17	1.8824	1.8431	1.9127	2.5237	2.7645	2.3868	1.9658	1.8633	1.8623	2.204	1.9865	3.4964	2.5001
18	2.4302	2.3612	2.0138	2.4213	2.7411	2.7462	2.4971	2.5526	2.1554	2.705	2.4024	3.4343	2.603
19	2.5743	2.46	2.2456	2.8921	2.2685	3.032	2.594	2.3738	2.622	2.6852	2.6413	3.5658	2.8986
20	2.0522	1.8635	2.052	2.5701	2.7838	2.5988	2.1212	1.9119	2.901	2.1206	1.9386	3.5431	2.4694
21	2.9541	2.3818	2.5837	2.74	2.7295	2.8295	3.1181	2.8983	2.7296	3.1553	2.7584	2.9074	2.2957
22	2.3243	2.4929	2.4733	2.7113	2.4636	2.8484	3.1536	2.719	2.8254	3.0722	2.5374	2.5149	2.4762
23	2.9841	2.8179	2.6515	2.6885	2.8158	3.5531	3.1938	3.241	3.2901	3.4296	2.7345	2.7413	2.5522
24	2.5527	2.3965	2.3407	2.494	2.801	2.8343	2.6222	2.7276	2.7705	2.6442	2.2991	3.1243	2.4075
25	2.8184	2.9172	2.6711	2.5073	2.8836	3.1782	3.3885	3.2369	3.384	3.3356	2.8926	2.6738	2.4475

	N	O	P	Q	R	S	T	U	V	W	X	Y
1	2.0397	2.3149	2.3375	1.8824	2.4302	2.5753	2.0522	2.9541	2.3243	2.9841	2.5527	2.8184
2	2.1602	2.3877	2.434	1.8431	2.3612	2.46	1.8635	2.3818	2.4929	2.8179	2.3965	2.9172
3	2.1546	2.3305	2.4907	1.9127	2.0138	2.2456	2.052	2.5837	2.4733	2.6515	2.3407	2.6711
4	2.5844	2.576	2.6817	2.5237	2.4213	2.8921	2.5701	2.74	2.7113	2.6885	2.494	2.5073
5	2.7678	2.5229	2.8061	2.7645	2.7411	2.2685	2.7838	2.7295	2.4636	2.8158	2.801	2.8836
6	2.4653	2.6123	2.7881	2.3868	2.7462	3.032	2.5988	2.8295	2.8484	3.5531	2.8343	3.1782
7	2.1818	2.6397	2.756	1.9658	2.4971	2.594	2.1212	3.1181	3.1536	3.1938	2.6222	3.3885
8	2.3369	2.6586	2.5056	1.8633	2.5526	2.3738	1.9119	2.8983	2.719	3.241	2.7276	3.2369
9	2.113	2.7351	2.2759	1.8623	2.1554	2.633	2.901	2.7296	2.8254	3.2901	2.7705	3.384
10	2.2888	2.7675	2.4494	2.204	2.705	2.6852	2.1206	3.1553	3.0722	3.4296	2.6442	3.3356
11	2.0437	2.2664	2.123	1.9865	2.4024	2.6413	1.9386	2.7584	2.5374	2.7345	2.2991	2.8926
12	3.3385	2.7303	2.29	3.4964	3.4343	3.5658	3.5431	2.9074	2.5149	2.7413	3.1243	2.6738
13	2.5619	2.2751	2.5187	2.5001	2.603	2.8986	2.4694	2.2957	2.4762	2.5522	2.4075	2.4475
14	0	2.4258	2.0901	1.6163	2.3066	2.5431	1.6681	2.9215	2.7475	3.1755	2.7459	3.1326
15	2.4258	0	2.5208	2.4762	2.4888	2.7845	2.5225	2.4742	2.1446	2.3173	2.382	2.5613
16	2.0901	2.5208	0	2.1151	2.238	2.7176	2.026	2.8065	2.6778	3.1654	2.6312	2.8083
17	1.6163	2.4762	2.1443	0	2.1443	2.0766	1.0132	2.7418	2.6308	3.106	2.6128	3.0379
18	2.3066	2.4888	2.238	2.1443	0	2.402	2.2025	2.9635	3.0029	3.105	2.686	3.1226
19	2.5431	2.7845	2.7176	2.0766	2.402	0	2.0231	3.0225	3.0029	3.1281	2.7313	3.1428
20	1.6681	2.5225	2.016	1.0132	2.2025	2.0231	0	2.7547	2.6426	3.1171	2.5221	3.4081
21	2.9215	2.4742	2.8065	2.7418	2.9635	3.0225	2.7547	0	2.6279	2.8667	2.7338	2.9165
22	2.7475	2.1446	2.6778	2.6308	3.0029		2.6426	2.6279	0	2.6895	2.4361	2.2645
23	3.1755	2.3173	3.1654	3.1	3.105	3.1281	3.1171	2.8667	2.6895	0	2.5038	2.6063
24	2.7459	2.382	2.6312	2.6128	2.686	2.7313	2.5221	2.7338	2.4361	2.5038	0	2.553
25	3.1326	2.5613	2.8083	3.0379	3.1226	3.1428	3.4081	2.9165	2.2645	2.6063	2.553	0

Figure 12. Distance matrix for texture features

0	0.0501	0.0648	0.0215	0.0142	0.0237	0.0908	0.0152	0.0092	0.0212	0.065	0.0093	0.0559	0.057	0.0678	0.009	0.0288	0.0058	0.0234	0.0254	0.0626	0.041	0.0532	0.0675	0.0749
0.0501	0	0.0147	0.0286	0.0643	0.0738	0.0809	0.0653	0.0593	0.0713	0.0149	0.0534	0.0058	0.0069	0.0077	0.0591	0.0213	0.0559	0.0267	0.0217	0.0247	0.0125	0.0091	0.0174	0.0248
0.0648	0.0147	0	0.0433	0.079	0.0885	0.0956	0.08	0.074	0.086	2.00E-04	0.0681	0.0089	0.0078	0.003	0.0738	0.036	0.0706	0.0414	0.0364	0.0394	0.0022	0.0238	0.0027	0.0101
0.0215	0.0286	0.0433	0	0.0357	0.0452	0.0523	0.00367	0.0907	0.0427	0.0435	0.0248	0.0344	0.0355	0.0463	0.0305	0.0073	0.0273	0.0019	0.0069	0.0039	0.0411	0.0195	0.046	0.0534
0.0142	0.0643	0.079	0.0357	0	0.0095	0.0166	0.001	0.005	0.007	0.0792	0.0109	0.0701	0.0712	0.082	0.0052	0.043	0.0084	0.0376	0.0426	0.0396	0.0768	0.0552	0.0817	0.0891
0.0237	0.0738	0.0885	0.0452	0.0095	0	0.0071	0.0085	0.0145	0.0025	0.0887	0.0204	0.0796	0.0807	0.0915	0.0147	0.0525	0.0179	0.0471	0.0521	0.0491	0.0863	0.0647	0.0912	0.0986
0.0908	0.0809	0.095	0.0523	0.0166	0.0071	0	0.0156	0.0216	0.0096	0.0958	0.0275	0.0867	0.0878	0.0986	0.0218	0.0596	0.025	0.0542	0.0592	0.0562	0.0934	0.0718	0.0983	0.1057
0.0152	0.0653	0.08	0.0967	0.001	0.0085	0.0156	0	0.006	0.006	0.0802	0.0119	0.0711	0.0722	0.083	0.0062	0.044	0.0094	0.0386	0.0436	0.0406	0.0778	0.0562	0.0827	0.0901
0.0092	0.0593	0.074	0.0907	0.005	0.0145	0.0216	0.006	0	0.012	0.0742	0.0059	0.0651	0.0662	0.077	2.00E-04	0.038	0.0034	0.0326	0.0376	0.0346	0.0718	5.02E-02	0.0767	0.0841
0.0212	0.0713	0.086	0.0427	0.007	0.0025	0.0096	0.006	0.012	0	0.0862	0.0179	0.0771	0.0782	0.089	0.0122	0.05	0.0154	0.0446	0.0496	0.0466	0.0838	0.0622	0.0887	0.0961
0.065	0.0149	2.00E-04	0.0435	0.0792	0.0887	0.0958	0.0802	0.0742	0.0862	0	0.0532	0.0683	0.0091	0.008	0.0028	0.074	0.0362	0.0708	0.0366	0.0396	0.0024	0.024	0.0025	0.0099
0.0093	0.0534	0.0681	0.0248	0.0109	0.0204	0.0275	0.0119	0.0059	0.0179	0.0532	0	0.0592	0.0603	0.0711	0.0057	0.0321	0.0025	0.0267	0.0317	0.0287	0.0659	0.0443	0.0708	0.0182
0.0559	0.0053	0.0089	0.0344	0.0701	0.0796	0.0867	0.0711	0.0651	0.0771	0.0683	0.0592	0	0.0011	0.0119	0.0649	0.0271	0.0617	0.0325	0.0275	0.0305	0.0067	0.0149	0.0116	0.019
0.057	0.0069	0.0078	0.0355	0.0712	0.0807	0.0878	0.0722	0.0662	0.0782	0.0091	0.0603	0.011	0	0.0108	0.066	0.0282	0.0628	0.0336	0.0286	0.0316	0.0056	0.016	0.0105	0.0179
0.0678	0.0177	0.003	0.0463	0.082	0.0915	0.0986	0.083	0.077	0.089	0.008	0.0711	0.0119	0.0108	0	0.0768	0.039	0.0736	0.0444	0.0394	0.0424	0.0052	0.0268	3.00E-04	0.0071
0.09	0.0591	0.0738	0.0305	0.0052	0.0147	0.0218	0.0062	2.00E-04	0.0122	0.0028	0.0057	0.0649	0.066	0.0768	0	0.0378	0.0032	0.0324	0.0374	0.0344	0.0716	0.05	0.0765	0.0839
0.0288	0.0213	0.036	0.0073	0.043	0.0525	0.0596	0.044	0.038	0.05	0.074	0.0321	0.0271	0.0282	0.039	0.0378	0	0.0346	0.0054	4.00E-04	0.0034	0.0338	0.0122	0.0387	0.0461
0.0058	0.0259	0.0706	0.0273	0.0084	0.0179	0.025	0.0094	0.0034	0.0154	0.0362	0.0025	0.0617	0.0628	0.0736	0.0032	0.0346	0	0.0292	0.0342	0.0312	0.0684	0.0468	0.0733	0.0807
0.0234	0.0267	0.0414	0.0019	0.0376	0.0471	0.0542	0.0386	0.0326	0.0446	0.0708	0.0267	0.0325	0.0336	0.0444	0.0324	0.0054	0.0292	0	0.005	0.002	0.0392	0.0176	0.0441	0.0515
0.0254	0.0217	0.0364	0.0069	0.0426	0.0521	0.0592	0.0436	0.0376	0.0496	0.0366	0.0317	0.0275	0.0286	0.0394	0.0374	4.00E-04	0.0342	0.005	0	0.003	0.0342	0.0126	0.0391	0.0465
0.0626	0.0247	0.0394	0.0039	0.0396	0.0491	0.0562	0.0406	0.0346	0.0466	0.0396	0.0287	0.0305	0.0316	0.0424	0.0344	0.0034	0.0312	0.002	0.003	0	0.0372	0.0156	0.0421	0.0495
0.041	0.0125	0.0022	0.0411	0.0768	0.0863	0.0934	0.0778	0.0718	0.0838	0.0024	0.0659	0.0067	0.0056	0.0052	0.0716	0.0338	0.0684	0.0392	0.0342	0.0372	0	0.0216	0.0049	0.0123
0.0532	0.0091	0.0238	0.0195	0.0552	0.0647	0.0718	0.0562	0.0502	0.0622	0.024	0.0443	0.0149	0.016	0.0268	0.05	1.22E-02	0.0468	0.0176	0.0126	0.0156	0.0216	0	0.0265	0.0339
0.0675	0.0174	0.0027	0.046	0.0817	0.0912	0.0983	0.0827	0.0767	0.0887	0.0025	0.0708	0.0116	0.0105	3.00E-04	0.0765	0.0387	0.0733	0.0441	0.0391	0.0421	0.0049	0.0265	0	0.0074
0.0749	0.0248	0.0101	0.0534	0.0891	0.0986	0.01057	0.0901	0.0841	0.0961	0.0099	0.0782	0.019	0.0179	0.0071	0.0839	0.0461	0.0807	0.0515	0.0465	0.0495	0.0123	0.0339	0.0074	0

Figure 13. Distance matrix for SURF features

SURF Distance Matrix

A	B	C	D	E	F	G	H	I	J	K	L	M	N	O	P	Q	R	S	T	U	V	W	X	Y
0	0.3799	0.3122	0.3398	0.1946	0.2803	0.2435	0.1878	0.3569	0.2985	0.2426	0.247	0.2674	0.3293	0.261	0.3281	0.1373	0.1953	0.4096	0.2872	0.3272	0.2963	0.3816	0.2156	0.2006
0.3799	0	0.2732	0.3291	0.2419	0.3685	0.2573	0.3106	0.2194	0.3334	0.2418	0.3773	0.3531	0.2458	0.1752	0.217	0.3726	0.1733	0.25	0.2205	0.2425	0.2853	0.3147	0.2465	0.2457
0.3122	0.2732	0	0.2396	0.2248	0.1137	0.2797	0.2026	0.2158	0.2488	0.2625	0.2163	0.1645	0.2223	0.217	0.2551	0.2203	0.1895	0.2276	0.1776	0.167	0.0644	0.2486	0.1507	0.197
0.3398	0.3291	0.2396	0	0.2142	0.2183	0.2567	0.1833	0.2918	0.2027	0.1991	0.1816	0.1617	0.2041	0.2979	0.205	0.2629	0.1653	0.1792	0.145	0.1639	0.1989	0.1492	0.1109	0.2132
0.1946	0.2419	0.2248	0.2142	0	0.1657	0.3084	0.1993	0.2785	0.2749	0.2763	0.2684	0.157	0.2064	0.2775	0.2395	0.2251	0.2096	0.2956	0.1944	0.2017	0.2128	0.2978	0.1919	0.2306
0.2803	0.3685	0.1137	0.2183	0.1657	0	0.2757	0.1771	0.2604	0.3282	0.3445	0.1761	0.1433	0.1352	0.3185	0.2412	0.1554	0.2185	0.2419	0.1715	0.2216	0.1034	0.2882	0.1408	0.2847
0.2435	0.2573	0.2797	0.2567	0.3084	0.2757	0	0.2276	0.2897	0.1681	0.1726	0.1333	0.3623	0.2207	0.2101	0.2679	0.1545	0.1762	0.1682	0.2242	0.2821	0.1963	0.2763	0.1867	0.2339
0.1878	0.3106	0.2026	0.1833	0.1993	0.1771	0.2276	0	0.1781	0.2971	0.2591	0.1943	0.316	0.3183	0.2615	0.1767	0.1669	0.3037	0.1633	0.2398	0.2405	0.2156	0.1818	0.1714	0.1654
0.3569	0.2194	0.2158	0.2918	0.2785	0.2604	0.2897	0.1781	0	0.3794	0.2716	0.3149	0.2006	0.287	0.2182	0.283	0.314	0.2558	0.1993	0.1823	0.2179	0.2757	0.1851	0.2523	0.2444
0.2985	0.3334	0.2488	0.2027	0.2749	0.3282	0.1681	0.2971	0.3794	0	0.2524	0.1328	0.3272	0.2733	0.2348	0.2373	0.241	0.2212	0.2165	0.2581	0.1817	0.2128	0.2912	0.203	0.181
0.2426	0.2418	0.2625	0.1991	0.2763	0.3445	0.1726	0.2591	0.2716	0.2524	0	0.2254	0.3364	0.3033	0.105	0.3101	0.2081	0.228	0.1775	0.1859	0.2733	0.2037	0.2236	0.1642	0.135
0.247	0.3773	0.2163	0.1816	0.2684	0.1761	0.1333	0.1943	0.3149	0.1328	0.2254	0	0.2834	0.2788	0.2905	0.2818	0.1377	0.2159	0.2154	0.1743	0.214	0.1794	0.2293	0.1049	0.1961
0.2674	0.3531	0.1645	0.1617	0.157	0.1433	0.3623	0.316	0.2006	0.3272	0.3364	0.2834	0	0.1954	0.3075	0.2174	0.3099	0.1984	0.2519	0.1963	0.1308	0.2041	0.2704	0.1771	0.2349
0.3293	0.2458	0.2223	0.2041	0.2064	0.1352	0.2207	0.3183	0.287	0.2733	0.3033	0.2788	0.1954	0	0.2432	0.1811	0.2576	0.1765	0.1665	0.1847	0.2427	0.1232	0.3416	0.1657	0.3324
0.261	0.1752	0.217	0.2979	0.2775	0.3185	0.2101	0.2615	0.2182	0.2348	0.105	0.2905	0.3075	0.2432	0	0.2688	0.1893	0.2171	0.1587	0.2564	0.194	0.2002	0.294	0.1803	0.1143
0.3281	0.217	0.2551	0.205	0.2395	0.2412	0.2679	0.1767	0.283	0.2373	0.3101	0.2818	0.2174	0.1811	0.2688	0	0.3389	0.1614	0.2336	0.1536	0.1757	0.2609	0.2311	0.237	0.2254
0.1373	0.3726	0.2203	0.2629	0.2251	0.1554	0.1545	0.1669	0.314	0.241	0.2081	0.1377	0.3099	0.2576	0.1893	0.3389	0	0.1928	0.2476	0.2626	0.2764	0.1719	0.2718	0.1299	0.19
0.1953	0.1733	0.1895	0.1653	0.2096	0.2185	0.1762	0.3037	0.2558	0.2212	0.228	0.2159	0.1984	0.1765	0.2171	0.1614	0.1928	0	0.2578	0.1965	0.153	0.1992	0.2571	0.131	0.1579
0.4096	0.25	0.2276	0.1792	0.2956	0.2419	0.1682	0.1633	0.1993	0.2165	0.1775	0.2154	0.2519	0.1665	0.1587	0.2336	0.2476	0.2578	0	0.2304	0.2338	0.1481	0.1562	0.1876	0.2377
0.2872	0.2205	0.1776	0.145	0.1944	0.1715	0.2242	0.2398	0.1823	0.2581	0.1859	0.1743	0.1963	0.1847	0.2256	0.1536	0.2626	0.1965	0.2304	0	0.2034	0.1493	0.179	0.1078	0.2149
0.3272	0.2425	0.167	0.1639	0.2017	0.2216	0.2821	0.2405	0.2179	0.1817	0.2733	0.214	0.1308	0.2427	0.194	0.1757	0.2764	0.153	0.2338	0.2034	0	0.2362	0.2448	0.1812	0.1292
0.2963	0.2853	0.0644	0.1989	0.2128	0.1034	0.1963	0.2156	0.2757	0.2128	0.2037	0.1794	0.2041	0.1232	0.2002	0.2609	0.1719	0.1992	0.1481	0.1493	0.2362	0	0.2525	0.1126	0.2153
0.3816	0.3147	0.2486	0.1492	0.2978	0.2882	0.2763	0.1818	0.1851	0.2912	0.2236	0.2293	0.2704	0.3416	0.294	0.2311	0.2718	0.2571	0.1562	0.179	0.2448	0.2525	0	0.2156	0.2555
0.2156	0.2465	0.1507	0.1109	0.1919	0.1408	0.1867	0.1714	0.2523	0.203	0.1642	0.1049	0.1771	0.1657	0.1803	0.237	0.1299	0.131	0.1876	0.1078	0.1812	0.1126	0.2156	0	0.1772
0.2006	0.2457	0.197	0.2132	0.2306	0.2847	0.2339	0.1654	0.2444	0.181	0.135	0.1961	0.2349	0.3324	0.1143	0.2254	0.19	0.1579	0.2377	0.2149	0.1292	0.2153	0.2555	0.1772	0

Figure 14. Distance matrix for SIFT features

SIFT Distance Matrix

2.1389	2.3395	2.6761	2.8609	0.2803	0.2435	0.1878	0.3569	0.2985	0.065	0.0033	0.0559	0.057	0.0678	2.3375	1.8824	2.4302	2.5753	2.0522
0	2.0123	2.3666	2.4024	0.3685	0.2573	0.3106	0.2194	0.3334	0.0149	0.0534	0.0058	0.0069	0.0077	2.434	1.8431	2.3612	2.46	1.8635
2.0123	0	2.0729	2.4739	0.1137	0.2797	0.2026	0.2158	0.2488	2.00E-04	0.0681	0.0089	0.0078	0.003	2.4907	1.9127	2.0138	2.2456	2.052
2.3666	2.0729	0	2.6894	0.2183	0.2567	0.1833	0.2918	0.2027	0.0435	0.0248	0.0344	0.0355	0.0463	2.6817	2.5237	2.4213	2.8921	2.5701
2.4024	2.4739	2.6894	0	0.1657	0.3084	0.1993	0.2785	0.2749	0.0792	0.0109	0.0701	0.0712	0.082	2.8061	2.7645	2.7411	2.2685	2.7838
2.4617	2.615	3.1042	3.0967	0	0.2757	0.1771	0.2604	0.3282	0.0887	0.0204	0.0796	0.0807	0.0915	2.7881	2.3868	2.7462	3.032	2.5988
2.3156	2.2709	2.9082	3.0516	0.2757	0	0.2276	0.2897	0.1681	0.0958	0.0275	0.0867	0.0878	0.0986	2.756	1.9658	2.4971	2.594	2.1212
2.1682	2.4682	2.7143	2.8318	0.1771	0.2276	0	0.1781	0.2971	0.0802	0.0119	0.0711	0.0722	0.083	2.5056	1.8633	2.5526	2.3738	1.9119
2.2472	2.3215	2.7426	3.117	0.2604	0.2897	0.1781	0	0.3794	0.0742	0.0059	0.0651	0.0662	0.077	2.2759	1.8623	2.1554	2.633	2.901
2.5331	2.509	2.7679	3.3242	0.3282	0.1681	0.2971	0.3794	0	0.0862	0.0179	0.0771	0.0782	0.089	2.4484	2.204	2.705	2.6852	2.1206
2.0798	2.1588	2.5103	2.702	0.3445	0.1726	0.2591	0.2716	0.2524	0	0.0532	0.0683	0.0091	0.008	2.123	1.9865	2.4024	2.6413	1.9386
3.2314	3.0562	2.7816	3.157	0.1761	0.1333	0.1943	0.3149	0.1328	0.0532	0	0.0592	0.0603	0.0711	2.29	3.4864	3.4343	3.5658	3.5431
2.4017	2.4828	2.4245	2.8097	0.1433	0.3623	0.316	0.2006	0.3272	0.0683	0.0592	0	0.0011	0.0119	2.5187	2.5001	2.603	2.8986	2.4694
2.1602	2.1546	2.5844	2.7678	0.1352	0.2207	0.3183	0.287	0.2733	0.0091	0.0603	0.011	0	0.0108	2.0901	1.6163	2.3066	2.5431	1.6681
2.3877	2.3305	2.576	2.5229	0.3185	0.2101	0.2615	0.2182	0.2348	0.008	0.0711	0.0119	0.0108	0	2.5208	2.4762	2.4888	2.7845	2.5225
2.434	2.4907	2.6917	2.8061	0.2412	0.2679	0.1767	0.283	0.2373	0.0028	0.0057	0.0649	0.066	0.0768	0	2.1151	2.238	2.7176	2.026
1.8431	1.9127	2.5237	2.7645	0.1554	0.1545	0.1669	0.314	0.241	0.074	0.0321	0.0271	0.0282	0.039	2.1443	0	2.1443	2.0766	1.0132
2.3612	2.0138	2.4213	2.7411	0.2185	0.1762	0.3037	0.2558	0.2212	0.0362	0.0025	0.0617	0.0628	0.0736	2.238	2.1443	0	2.402	2.2025
2.46	2.2456	2.8921	2.2685	0.2419	0.1682	0.1633	0.1993	0.2165	0.0708	0.0267	0.0325	0.0336	0.0444	2.7176	2.0766	2.402	0	2.0231
1.8635	2.052	2.5701	2.7838	0.1715	0.2242	0.2398	0.1823	0.2581	0.0366	0.0317	0.0275	0.0286	0.0394	2.016	1.0132	2.2025	2.0231	0
2.3818	2.5837	2.74	2.7295	0.2216	0.2821	0.2405	0.2179	0.1817	0.0396	0.0287	0.0305	0.0316	0.0424					

Figure 15. Distance matrix for corner features

A	B	C	D	E	F	G	H	I	J	K	L	M	N	O	P	Q	R	S	T	U
0	2.1389	2.3395	2.6761	2.8609	0.2803	0.2435	0.0152	0.0092	0.0212	0.065	0.0033	0.0559	0.057	0.0678	0.009	0.0288	0.0058	0.0234	0.0254	0.3272
2.1389	0	2.0123	2.3666	2.4024	0.3685	0.2573	0.0653	0.0593	0.0713	0.0149	0.0534	0.0058	0.0069	0.0077	0.0591	0.0213	0.0559	0.0267	0.0217	0.2425
2.3395	2.0123	0	2.0729	2.4739	0.1137	0.2797	0.08	0.074	0.086	2.00E-04	0.0681	0.0089	0.0078	0.003	0.0738	0.036	0.0706	0.0414	0.0364	0.167
2.6761	2.3666	2.0729	0	2.6894	0.2183	0.2567	0.00367	0.0307	0.0427	0.0435	0.0248	0.0344	0.0355	0.0463	0.0305	0.0073	0.0273	0.0019	0.0069	0.1639
2.8609	2.4024	2.4739	2.6894	0	0.1657	0.3084	0.001	0.005	0.007	0.0792	0.0109	0.0701	0.0712	0.082	0.0052	0.043	0.0084	0.0376	0.0426	0.2017
2.8225	2.4617	2.615	3.1042	3.0967	0	0.2757	0.0085	0.0145	0.0025	0.0887	0.0204	0.0796	0.0807	0.0915	0.0147	0.0525	0.0179	0.0471	0.0521	0.2216
2.2436	2.3156	2.2709	2.9082	3.0516	0.2757	0	0.0156	0.0216	0.0096	0.0958	0.0275	0.0867	0.0878	0.0986	0.0218	0.0596	0.025	0.0542	0.0592	0.2821
2.4539	2.1682	2.4682	2.7143	2.8318	0.1771	0.2276	0	0.006	0.006	0.0802	0.0119	0.0711	0.0722	0.083	0.0062	0.044	0.0094	0.0386	0.0436	0.2405
2.1605	2.2472	2.3215	2.7426	3.117	0.2604	0.2897	0.006	0	0.012	0.0742	0.0059	0.0651	0.0662	0.077	2.00E-04	0.038	0.0034	0.0326	0.0376	0.2179
2.5658	2.5331	2.509	2.7679	3.3242	0.3282	0.1681	0.006	0.012	0	0.0862	0.0179	0.0771	0.0782	0.089	0.0122	0.05	0.0154	0.0446	0.0496	0.1817
2.0021	2.0798	2.1588	2.5103	2.702	0.3445	0.1726	0.0802	0.0742	0.0862	0	0.0532	0.0683	0.0091	0.008	0.0028	0.074	0.0362	0.0708	0.0366	0.2733
3.136	3.2314	3.0562	2.7816	3.157	0.1761	0.1333	0.0119	0.0059	0.0179	0.0532	0	0.0592	0.0603	0.0711	0.0057	0.0321	0.0025	0.0267	0.0317	0.214
2.4016	2.4017	2.4828	2.4245	2.8097	0.1433	0.3623	0.0711	0.0651	0.0771	0.0683	0.0592	0	0.0011	0.0119	0.0649	0.0271	0.0617	0.0325	0.0275	0.1308
2.0397	2.1602	2.1546	2.5844	2.7678	0.1352	0.2207	0.0722	0.0662	0.0782	0.0091	0.0603	0.011	0	0.0108	0.066	0.0282	0.0628	0.0336	0.0286	0.2427
2.3249	2.3877	2.3305	2.576	2.5229	0.3185	0.2101	0.083	0.077	0.089	0.008	0.0711	0.0119	0.0108	0	0.0768	0.039	0.0736	0.0444	0.0394	0.194
2.3375	2.434	2.4907	2.6917	2.8061	0.2412	0.2679	0.0062	2.00E-04	0.0122	0.0028	0.0057	0.0649	0.066	0.0768	0	0.0378	0.0032	0.0324	0.0374	0.1757
1.8824	1.8431	1.9127	2.5237	2.7645	0.1554	0.1545	0.044	0.038	0.05	0.074	0.0321	0.0271	0.0282	0.039	0.0378	0	0.0346	0.0054	4.00E-04	0.2764
2.4302	2.3612	2.0138	2.4213	2.7411	0.2185	0.1762	0.0094	0.0034	0.0154	0.0362	0.0025	0.0617	0.0628	0.0736	0.0032	0.0346	0	0.0292	0.0342	0.153
2.5743	2.46	2.2456	2.8921	2.2685	0.2419	0.1682	0.0386	0.0326	0.0446	0.0708	0.0267	0.0325	0.0336	0.0444	0.0324	0.0054	0.0292	0	0.005	0.2338
2.0522	1.8635	2.052	2.5701	2.7838	0.1715	0.2242	0.0436	0.0376	0.0496	0.0366	0.0317	0.0275	0.0286	0.0394	0.0374	4.00E-04	0.0342	0.005	0	0.2034
2.9541	2.3818	2.5837	2.74	2.7295	0.2216	0.2821	0.0406	0.0346	0.0466	0.0396	0.0287	0.0305	0.0316	0.0424	0.0344	0.0034	0.0312	0.002	0.003	0

CONCLUSION AND FUTURE WORK

To conclude, this work presents an image retrieval system based on global features, which tested using COREL and UKBENCH datasets. Firstly, the datasets images chosen randomly and the feature vector extracted. This achieved by implementing a color correlogram feature vector for color features and the solid variation of inverse chances, block a variety of local exchange coefficients feature vector for texture features are extracted from an image. Next, similarity measures are determined based on the distance metrics values between the two images.

The distance metric is calculating how far the similar points are present. Based on that the similar points are classify into a separate group. Now, the ranking process employed, which used to determine the most relevant images retrieved based on the query image. Fix the minimum threshold value to the distance matrix and below the threshold, value images eliminated. The 25 images are randomly selected, processed and stored in the database. Now, retrieval process performed. Hence, the efficient retrieval system based on global features created with the most relevant images. In this work, it proved that the color correlogram and BDIP, BVLC features makes the retrieval system more efficient with high accuracy.

However, there are some issues is to be addressed. The main issue is in color features is color correlogram will give a result based on spatial information, because of that when the object is spatially changed it can't be analyzed correctly. Further, this work employed in combination of global and local features to give a better performance in more effective manner.

REFERENCES

Chun; Seo; Kim. (2003). Image retrieval using bdip and bvlc moments. *IEEE Transactions on Circuits and Systems for Video Technology*, *13*(9), 951–957. doi:10.1109/TCSVT.2003.816507

Chun, Y. D., Kim, N. C., & Jang, I. H. (2008). Content-based image retrieval using multiresolution color and texture features. *IEEE Transactions on Multimedia*, *10*(6), 1073–1084. doi:10.1109/TMM.2008.2001357

Conde & Dominguez. (2018). Scaling the chord and hellinger distances in the range [0, 1]: An option to consider. *Journal of Asia-Pacific Biodiversity*, 76–83.

Datta, Joshi, Li, & Wang. (2008). Image retrieval: Ideas, influences and trends of the new age. *ACM Computing Surveys*, *40*, 5.

Dinakaran, M., & Vijayarajan, V. (2013). Feature based image retrieval using fused sifts and surf features. *Research Gate*, *10*, 2500–2506.

Gayathri, S., Swapna, B., Kamalahasan, M., & Balavinoth, S. (2020). Retire away Essential Accuracy for Darkness Discovery and Elimination. *Test Engineering and Management*, *83*, 2411–2417.

Huang, Ravi Kumar, Mandar, Zhu, & Ramin. (1997). Image indexing using color correlograms. *Computer Vision and Pattern Recognition, Proceedings. IEEE Computer Society Conference*, 762-768.

Jacob, Toft, & Pedersen. (2011). Study group surf: Feature detection description. *Research Gate*, *4*, 52-56.

Nirmala, K., & Subramani, K. (2013). Content based image retrieval system using auto color correlogram. *Jisuanji Yingyong*, *6*, 67–73.

Sharma & Batra. (2014). Analysis of distance measures in content based image retrieval. *Global Journal of Computer Science and Technology*, 28-32.

Singh, C., & Preet Kaur, K. (2016). A fast and efficient image retrieval system based on color and texture features. *Journal of Visual Communication and Image Representation*, *41*, 225–238. doi:10.1016/j.jvcir.2016.10.002

Swapna, B., & Kamalahasan, M. (2020). Phase Measurement Analysis in Field Programmable Gate Array. *Test Engineering and Management*, *82*, 14225–14230.

Swapna, B., Kamalahsan, M., Sowmiya, S., Konda, S., & SaiZignasa, T. (2019). Design of smart garbage landfill monitoring system using Internet of Things. *IOP Conference Series. Materials Science and Engineering*, *561*(1), 012084. doi:10.1088/1757-899X/561/1/012084

Tao, Y., & Grosky, W. I. (1999). Spatial color indexing: a novel approach for content-based image retrieval. *Proceedings IEEE International Conference on Multimedia Computing and Systems*, *1*, 530-535. 10.1109/MMCS.1999.779257

Thenkalvi, B., & Murugavalli, S. (2014). Image retrieval using certain block-based difference of inverse probability and certain block based variation of local correlation coefficients integrated with wavelet moments. *Journal of Computational Science*, *10*(8), 1497–1507. doi:10.3844/jcssp.2014.1497.1507

Tungkasthan, A., Intarasema, S., & Premchaiswadi, W. (2009). Spatial color indexing using acc algorithm. *7th International Conference on ICT and Knowledge Engineering*, 113-117.

Van der Weken, D., Nachtegael, M., & Kerre, E. E. (2002). An overview of similarity measures for images. *IEEE International Conference on Acoustics, Speech and Signal Processing*, *2*, 3317-3320.

Chapter 6
Computer Vision for Weed Identification in Corn Plants Using Modified Support Vector Machine

Archana K. S.
Vels Institute of Science, Technology, and Advanced Studies, India

Arul Stephen C.
Vels Institute of Science, Technology, and Advanced Studies, India

Sivakumar B.
SRM Institute of Science and Technology, India

Vijayalakshmi A.
Vels Institute of Science, Technology, and Advanced Studies, India

Siva Prasad Reddy K.V
JNTUA College of Engineering Pulivendula(JUTUACEP), India

Ebenezer Abishek B.
VelTech Multitech Engineering College, India

ABSTRACT

Weed plants are unwanted plants growing in between host plants. There are more than 8000 weed species in the agriculture field. This is the global issue that leads to loss in both the quality and quantity of the product. So, attention has to be taken to avoid these losses and save manpower. In this chapter, the three procedures, segmentation,feature extraction, and classification, for weed plant identification are presented in detail. To separate the region of interest, threshold segmentation method was applied. Then the important features, shape, and textures were analysed with the help of GLCM method, which are discussed in this review. Finally, in the image classification method, modified support vector machine was used to separate the weed and host plants. Finally, this modified SVM was compared with CNN using performance analyses and produced high accuracy of 98.56% compared to existing systems. Hence, the farmers are expected to adopt these technologies to overcome the agricultural problems.

DOI: 10.4018/978-1-7998-9640-1.ch006

INTRODUCTION

Agriculture is very important in Indian economy(Jones et al., 2017). In recent years, due to climate change effects, diseases, pests, human error the agriculture faces many problems. So, the most challenging task of agriculture is increasing both quality and quantity of the product from these critical issues (Sanjeevi et al., 2020).Agriculture is the backbone of the Indian economy. For more than half of India's population, it is a source of income(Shuping & Eloff, 2017). The cultivation from agriculture is important to offer the source of livelihood for any employees in farming, international trading, national revenue, raw material, saving source, foreign exchange resources, economic development and significance of transport. So there is a need to develop the agriculture because of less production in food products and reduction of cultivation and also to develop new techniques (Fritz et al., 2018). However, people from different places in India die due to food scarcity because farmers from different places highly depend on Indian. Agriculture controlling weeds that grow among plantation crops is one of the most important difficulties in agribusiness.

Accurate identification and precision treatment are needed for both types of plants such as weeds and crops(Thornton et al., 2017). But both these treatment and identifications are subjected to predict the error in affect crops field. Hence, research in agriculture is focused to increase quality and quantity of the of the product. Manual diagnosis is challenging task for agriculture area. So automatic detection is needed for human error and man power saving. In recent years, the agriculture and farming systems has become a worldwide development with well-growing technology (Jones et al., 2017)(Ragul Krishna et al., 2020). With the help of image processing techniques automatic system is introduced in agriculture area.

Weeds are the most challenges in agriculture field because these weeds present anywhere in the crops field. As a result of this weeds the crop yields get more loss(Alexander et al., 2017). Farmers are currently manually removing weeds wherever possible, or spraying weed killers/herbicides all over the field to keep them under control(Donatelli et al., 2017). This method is ineffective since chemicals are sprayed on plantation crops as well, damaging the environment and causing human health concerns. A sophisticated weed management method should be used to avoid these outcomes. The system must be capable of detecting weeds in the field and notifying farmers to their precise positions. So those pesticides are only applied in certain areas. It focuses on decreasing the usage of pesticides that impair plant growth and cause major health problems in humans. So, attention has to be taken to identify the weeds present in the crop field. Hence, the yield loss occurs due to unneeded weed plants, plant disease, nutrition deficiency and quality of yield. The farmers are benefited according to the well growing technology such as image processing and communication systems(Hugar, 2016)(Liakos et al., 2018).

The amount of well growing technology with computer vision applications are quality of yields, disease identification, monitoring irrigation and water stress management(Gao et al., 2018)(Jayanthi et al., 2019). Hence, the weeds are identified and classified by using different machine learning techniques(Qin et al., 2016). Initially to show the weeds different features were extracted and categorized into different classes including color, shape and texture. Shape, colour, and size characteristics distinguish weed recognised from photos utilising image processing techniques. Different weeds and crop species are classified based on these features. Images of the plantation rows are taken at regular intervals using image processing methods (Contreras-Medina et al., 2012). Herbicides are sprayed by robots if the weed is recognised. Probabilistic neural networks may also identify weeds. The automatic threshold for weed segments is based on Otsu's technique. Weed is detected using angular cross-section intensities, and support vector machines (SVM) can be utilised for high accuracy. Weed identification and classification are critical in the agricultural industry from a technical and economic standpoint (Wu et al., 2021). Identification and extraction of weeds boost production on a wide scale, resulting in increased revenue. Early weed identification can help to protect intended plants from being harmed. As a result, weed detection and categorization play a significant role in agriculture.

RELATED WORK

The weed detection and classification done with the following four methods

1. Image Acquisition: The images are taken in the direct field.
2. Image Segmentation: Based on the segmentation techniques weeds are detected.
3. Feature Extraction: After segmentation, the relevant features are identified for further classification
4. Image Classification: At last, the weeds are classified based on the classification techniques(Alam et al., 2020).

(Brinkhoff et al., 2020) proposed a method to locate and classify the perennial crops by using remote sensing method in New South Wales, Australia. To classify the crops, they used with supervised SVM classification. The author improved the accuracy through this image analysis techniques using SVM classifier. Finally, whereas the accuracy was 90.9% to achieve the result. (Ahmed et al., 2012) proposed an algorithm to identify weed plants in chilly farm. The main goal of this research is the author proposed SVM algorithm to classify the weed plants from chilly plant. The author collected five different weeds from chilly farm in Bangladesh. Then the weed plants are segmented by using global thresholding-based algorithm. Finally, SVM algorithm used to classify the weed plant from chilly plant.

(Desai, 2015) The Author proposed that the weed classification and identification play a major role in faming industries and has been considered as the most important technical and economic significance in the farming industry. weeds are being extracted using images by image processing technique which is described using shape, color, and size features (Tang et al., 2017). These features mentioned here are used to classify weeds which are similar and also crop species which are similar. There are several techniques to differentiate between weed and crop these techniques are SVM, CNN, DA and methods like otsu method, 2G-R-B. The author worked on these different techniques to analyze the weed and crop and to detect the weed present in each and every crop by using the technique image processing. The weed can be visualized by using the machine vision technique. This technique uses a unique image processing technique. The weeds which are present on the agricultural land are detected by using different properties such as shape, size, spectral reflection, texture features. In this document they have used size feature has an important technique for weed detection. To get the clear image of the weed an excessive green algorithm was developed to remove the soil and other unwanted things from the image. to eliminate the noises from the image we use image enhancement techniques, then for extracting each component from the image, sized based on area-based features like area and perimeter. The label algorithm technique used to calculate selecting suitable threshold value for crop segmentation has been done to detect.

(Satish et al., 2016) The author concluded that controlling weed was very essential and a critical operation which could affect the crop yield. This document is proposed with 2 important methods to differentiate between crop and weed which consist crop row detection in images from agriculture field. This crop row detection method includes 3 major processes which are – image filtering, image segmentation using the ostu's method and also crop row detection technique, further classification between the weed and crop is carried by using the technique box plotting. The proposed technique did not work against the lighting due to the environmental condition.(Nathalia et al., 2016)The author proposed an application for detecting the unnecessary weed in crop from one area with extra agricultural impact using computer vision. To get the region of attention which was developed by image processing using processed neural network. The author proposed three methods like image acquisition, segmentation and ANN. they enhanced in the method by introducing herbicides, exacting case of this application, image processing. These were the important aspect since identification of regions of interest, light intensity, and it was major challenge.

METHODOLOGY

This methodology is for identifying and classifying the weed plant in the farm. It involves several tasks such as image acquisition, segmentation, feature extraction and classification. Figure 1 displays the general structure of weed plant disease identification and classification.

Figure 1. Architecture of weed detection

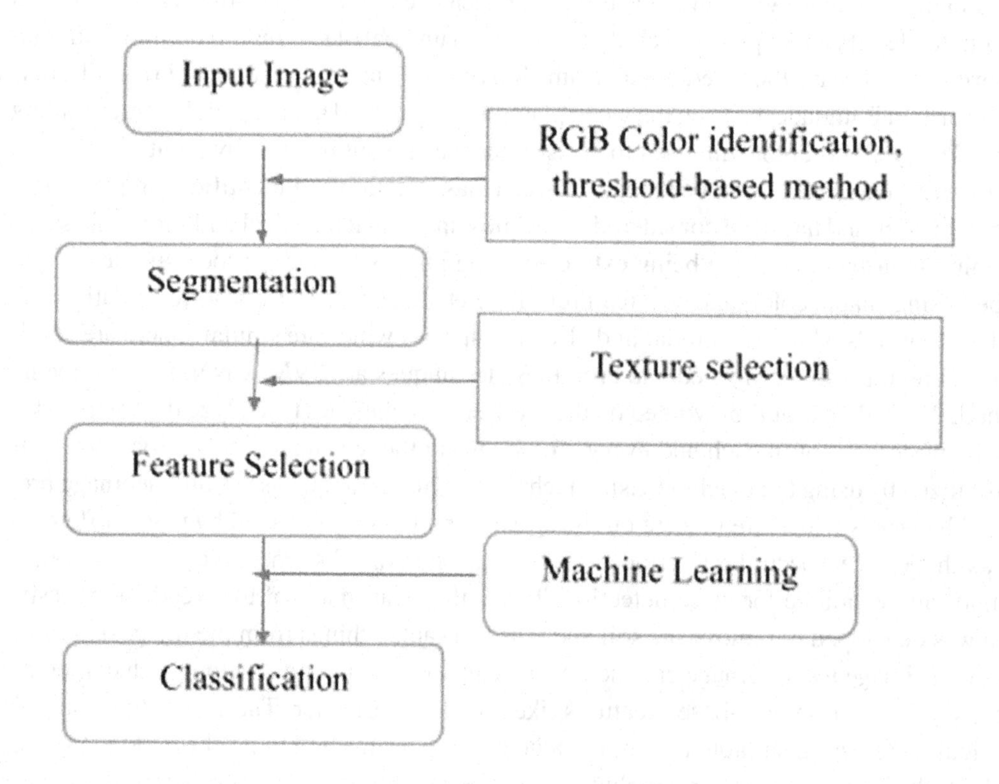

IMAGE ACQUISITION

Developing a quality database, the images are very important for developing algorithm based on dataset. To acquire the plant leaf, the image acquisition is the initial process. The higher solution plant leaf images are considered for this process. Here the input images are taken from .jpeg format which is available in the dataset folder. Hence, there are varieties of physical device to acquire the image such as cameras, x-ray devices, radar, electron microscopes, ultrasound and used for various purpose. In this research work initially, the RGB color image of the leaves are taken from using the digital camera or from farmer's web portal and taken this input for further processing. The acquisition process completely depends on hardware device. The image can be accrued through three ways such as single sensor, strips and sensor arrays. Hence, the combination of electrical energy and the sensor used to take the scene being imaged. Hence, the input images are acquired through sensors camera mounted on drones. So, with the help of this camera the images are taken directly from the agriculture field as shown in the Figure 2 . The important features of the camera are Sony 6000D with high resolution lens at 210 mm focal length at a speed of 1.4 m/s and 7 m above the ground. Hence this camera used to collect the image automatically(Wang et al., 2019). Based on the weather condition, climate condition high resolution digital camera was used to collect picture (Pulido-Rojas et al., 2016).

Figure 2. Input image of corn plant

IMAGE SEGMENTATION

The captured images were transferred to computer for processing using machine leaning. Since, the images were captured under different lightning condition, due to that noises are addressed which is to be problematic during classification(Hamuda et al., 2018). Hence, the noises are removed using the adaptive median filter (Archana & Sahayadhas, 2018). The main objective of image segmentation is separate the weed plant from all plants using some various HSV methods. Here, the weeds are identified by two approaches. The approaches here is one is inter-line and another one is intra-line approach. Generally, the plants such as sugar beet, paddy, maize are planted in line with clear spacing. But the weed plants are rowed outside of the line from the host crops(Reddy & Basha, 2017). Hence those weeds are identified easily from the host crops (Reddy & Basha, 2017) Figure 3 described the histogram and edge detection method to addressed the segmentation and to point out the region of interest from the background image by using the following formula.

$$I_{bin}(x,y)=\begin{cases}0, IMedian(x,y)<t\\1, IMedian(x,y)\geq t\end{cases}$$

Figure 3. HSV image and Histogram image of corn plant

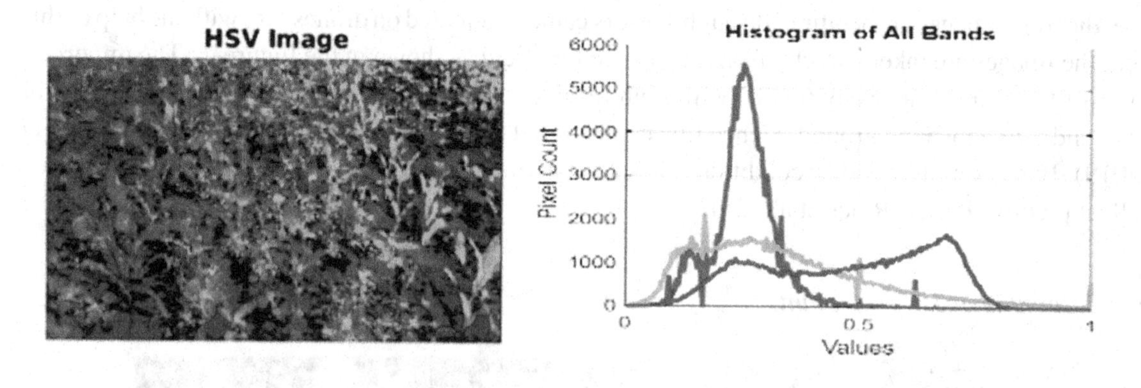

FEATURE EXTRACTION AND CLASSIFICATION

Feature extraction method aims to extract meaningful information from the segmented object and to represent the attributes for further processing. On account of that, the independent features were identified through color, shape and texture. In general, feature extraction means to extract the information from the existing data to help the classification. Hence, the new information is extracted to find the unique features by using the algorithm of color, texture and shape. The unique features are identified through feature extraction using the proposed algorithm based on color, shape and texture. Finally, the selected unique features were extracted for the final classification to differentiate the disease. In this work, after segmenting the preprocessed image, the process of feature extraction is carried out to extract the information. Generally, there are several features from the input image such as the important color, texture and shape features. That features are identified for the final classification. The first step of feature extraction starts with the color features are extracted by using a Novel intensity based color feature. And so there are 28 relevant color features extracted from the segmented image for further processing. Finally, to find the weed plants and host plants, the important features are involved to distinguish between two plants.

The additional features such as shape and textures are selected and extracted from the input image to classify the weed and host plants(Asad & Bais, 2019).One of the most important features are texture analysis. The major characteristics of the textures such as structural features, model-based features, statistical features, and transform-based features were analyzes. Based on the region of interest the important information of textures is extracted from the image. Some other features such as area of interest from the shape features are analyzed. After extracting important features, all the features were combined to select the important feature to find the weed plants based on the trained models (Arif et al., n.d., 2011). The independent features were identified through color, shape and texture. In general, feature extraction means to extract the information from the existing data to helpful for classification. The new information's are extracted to find the unique features by using the algorithm of color, texture and shape. The unique features are identified through feature extraction using the proposed algorithm based on color, shape and texture. Finally, the selected unique features were extracted for final classification to differentiate the weed plant from normal plant (Lu et al., 2017). The following are the most common methods used to select the unique features from image segmentation.

- Color based feature extraction
- Shape based feature extraction
- Texture based feature extraction

In image classification one of the most important visual feature extractions is color based method. The fundamental method to represent the color contents in is color histogram through some important color model. Hence, the most common color model used to represent the color features are RGB (Red, Green and Blue) and HSV (hue, saturation, value). Then, the feature extraction is the basis of measuring the similarities between the shapes by their features. The shape of the features can be identified through two main characteristics one is region character and other one is characteristics border. Accordingly, the shape of the severity can be identified through the mathematical model of simple geometric features such as area, perimeter and boundary of the image. To select the unique features from the shape the important features was identified from area and diameter. Next, the Texture feature helps to calculate the information about the selected region of an image(Huang et al., n.d., 2016).The importance of texture feature is the most useful and unique features are used for final classification method(Wei et al., 2018). There are different ways to represent the texture feature particularly GLCM are used to identify the unique features. In texture feature the huge amount of information is retrieved from the image content through GLCM(Chaki et al., 2015). Hence, to estimate the image properties one of the most well-known texture analysis methods is Gray level co-occurrence matrix (GLCM). It gives the information of grey level intensities as G.

Finally, classification methods are used to distinguish between weed and host plants. Machine Learning approaches are used to automatically analyze to make decision. Finally support vector machine used to separate the host and weed plants(Basavarajeshshwari & Madhavanavar, 2017). In computer vision, the image classification makes an important task to extract and classify the information in various applications (Le et al., 2013). Through the input data, the classification method makes a decision to analyze the image based on the trained data and tested data for the conclusion (Halder et al., 2018). Hence, the classification is separated into two types:

- Supervised classification and
- Unsupervised classification

The supervised data is obtained from the classification method to train the data model from the given dataset for classification and for the final performance the test samples are collected from the unsupervised classification. Finally, this comparison is based on various parameters like accuracy, precision and recall. Generally, this method is based on the hypothesis of one or more features and these features were organized into different categories. Therefore, these different categories belong to several distinct and exclusive classes from various features (Lai & Deng, 2018). The protocols of each class are followed by supervised classification and unsupervised classification.

RESULT AND DISCUSSION

Here, the main motivation of SVM is separate the dataset in best way with the help of margin, hyperplane and given dataset to detect the weed plants (Brinkhoff et al., 2020). The following steps are involved to maximize marginal hyper plane:

Step 1: First, hyper plane is chosen to separate two classes.

Step 2: Net, fix the margin to separate nearest data points and hyper plane to classify the given data sets.

Step 3: According to the SVM selection, the class is segregated accurately with the help of selected and at the same time the margin level also to maximize based on the classes along with new additional features.

Hence, the SVM uses the techniques called kernel tick which used to find the accurate classifier(Jan & Koo, 2018). The main function of kernel is the given input dataset is transformed to required form along with various types function such as: linear, polynomial, and radial basis function. Based on the practical applications it gives high performance. For weed detection, two different types of datasets were used such as paddy and corn. The dataset such as paddy and corn were taken in direct field. Finally, these mixed datasets were utilized for experiment. The last procedure is to separate the different plant varieties such as crop and weed with proper classifiers. Hence, the algorithm used to classify the weed and plant is conventional neural network and the modified support vector machine(Tang et al., 2017). To estimate the proposed methodology, with the help of train data and the test data the accuracy level was identified.

The result of the testing was arranged in confusion matrix with true positive, true negative, false positive and false negative. Here, true positive identifies the weed plants; true negative identifies the crops; false positive identifies the incorrect weeds and false negative identify the incorrect crops are described in Figure 4, (Islam et al., 2021). Training data is a common concept in machine learning. Training data is also called as AI training data, training set, training dataset, or learning set is the information used to train an algorithm. The training data includes both input data and the corresponding expected output(Oppenheim & Shani, 2017).Training data can be used for various machine learning algorithms, such as sentiment analysis, Natural language processing (NLP) and chat bot training(Lu et al., 2017).Train data is an important part of evaluating data mining models. The most of the data are used for training data by using similar data's for training it can be minimized the effects of data discrepancies and better understand the characteristics of the model. In image processing Training is used for classifying pixels in order to segment different objects. The classifier is trained on this set of pixels, and the remainders of the pixels are attributed to one of the classes by the classifier.

Figure 4. Threshold image of corn plant

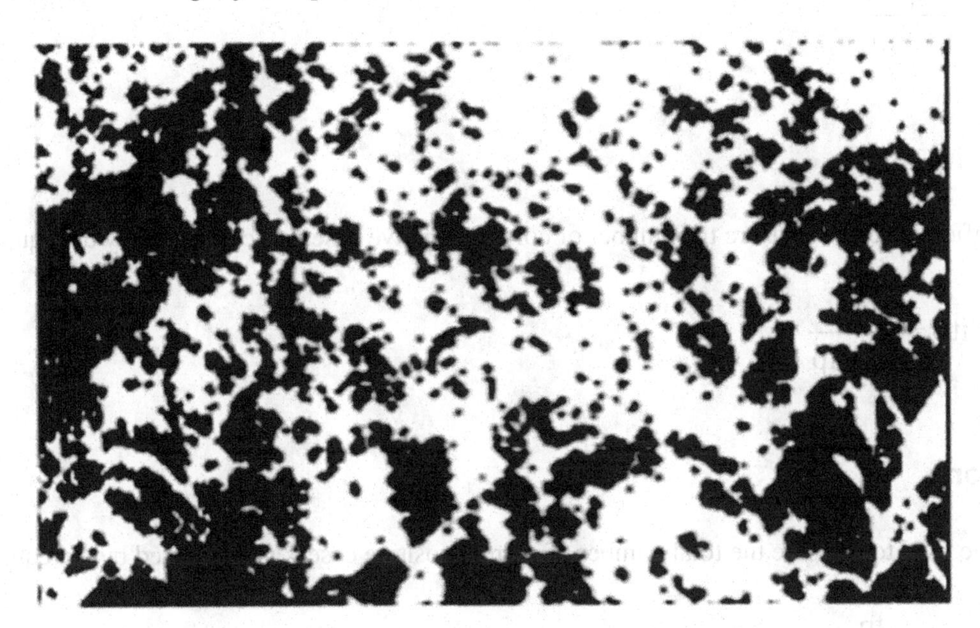

During image classification, the performance metrics are the important aspect to evaluate the data models(Zhu et al., 2016)(Arif et al., n.d.). In machine learning, the technique which used to measure and evaluates the data is performance metrics. Generally, it used for refining the parameters and to select the appropriate model based on the techniques. The most common performance metrics are Accuracy, Sensitivity, Specificity, Precision (Jiao & Du, 2016). The performance metrics is systematically reviewed the error report for the classification problem. Based on the confusion matrix of unseen data the evaluation metrics are used to measure the quality of the data in trained classifier with total correctness based on the best future performance(Espinoza et al., 2016).The predictive analytics of error rate represents the confusion matrix such as true positives, false positives, false negatives, and true negatives respectively. Generally, the confusion matrix takes two rows and two columns to measure the correct speculation. Hence, in order to analyze the performance accuracy level was measured using this following formula.

Accuracy

The accuracy measures the total number of accurate predictions of true positive and the true negative by the total number of the dataset, as calculated by using the equation:

$$\text{Accuracy} = \frac{\text{tp} + \text{tn}}{\text{tp} + \text{tn} + \text{fp} + \text{fn}} \tag{1}$$

Recall

The recall used to calculate the number of true positive cases, as calculated by the equation:

$$\text{Recall} = \frac{\text{tp}}{\text{tp} + \text{fn}} \quad (2)$$

Specificity

The specificity used to measure the number of correct negative cases, as calculated by the equation

$$\text{Specificity} = \frac{\text{tn}}{\text{tn} + \text{fp}} \quad (3)$$

Precision

The metric used to measure the total number of correct positive cases, as calculated by the equation:

$$\text{Precision} = \frac{\text{tp}}{\text{tn} + \text{fp}} \quad (4)$$

TP = The weed plants are detected correctly.

- TN = The crop and bare land is detected correctly.
- FP = The weed plants are detected incorrectly.
- FN = The crop and bare land is detected incorrectly.

Based on this formula, the classifier is compared with SVM and CNN algorithm to classify the weed plants from the normal plants. The performance metrics such as accuracy, precision, recall, specificity was calculated to identify the weed plants (Sharpe et al., 2020),

Figure 5. Confusion matrix of the crops and weeds performance analyzer for classifier

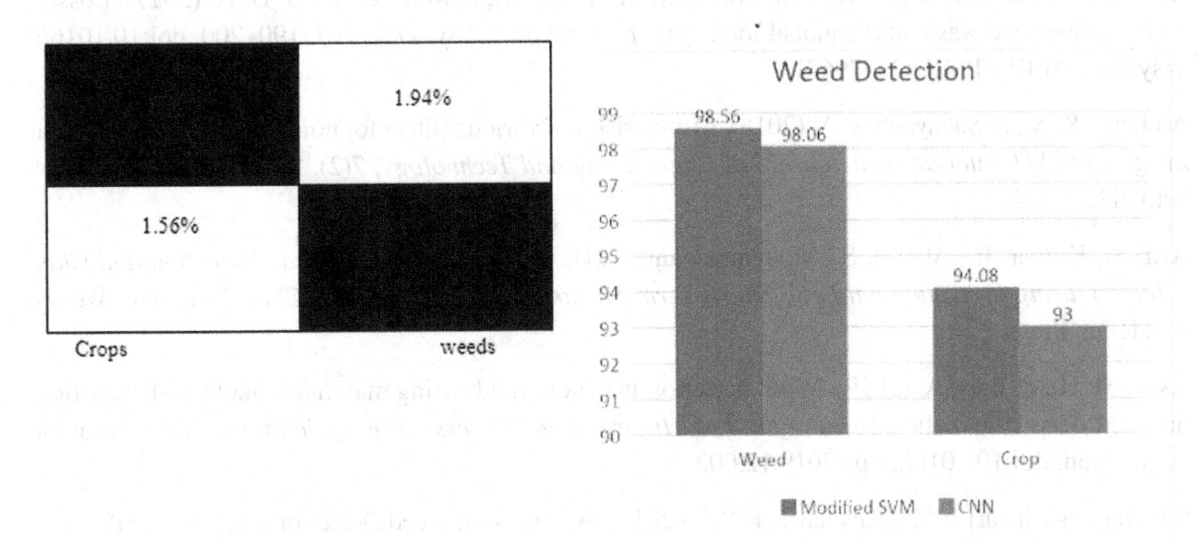

This modified SVM classifies weed and crop effectively. Figure 6 described the final method of modified support vector machine and it shows the better accuracy compared to conventional neural network.

CONCLUSION

This paper presents the automatic detection of weed plants in image processing. The algorithm designed to automatically identify the weed plants in crop field. Here, the image processing technique is used to detect the weed plants through machine vision. In this paper, the dataset of 150 image was collected and done various processing such as segmentation, feature extraction and classification. Weed and the crops are detected through various methods and it is classified by using the methods namely modified SVM and the CNN. Finally, this proposed modified SVM was compared with CNN using performance analyzes and produce high accuracy of 98.56% compared to existing systems.

REFERENCES

Ahmed, F., Al-Mamun, H. A., Bari, A. S. M. H., Hossain, E., & Kwan, P. (2012). Classification of crops and weeds from digital images: A support vector machine approach. *Crop Protection (Guildford, Surrey), 40*, 98–104. doi:10.1016/j.cropro.2012.04.024

Alam, M., Alam, M. S., Roman, M., Tufail, M., Khan, M. U., & Khan, M. T. (2020). Real-Time Machine-Learning Based Crop/Weed Detection and Classification for Variable-Rate Spraying in Precision Agriculture. *2020 7th International Conference on Electrical and Electronics Engineering, ICEEE 2020*, 273–280. 10.1109/ICEEE49618.2020.9102505

Alexander, P., Brown, C., Arneth, A., Finnigan, J., Moran, D., & Rounsevell, M. D. A. (2017). Losses, inefficiencies and waste in the global food system. *Agricultural Systems*, *153*, 190–200. doi:10.1016/j.agsy.2017.01.014 PMID:28579671

Archana, K. S., & Sahayadhas, A. (2018). Comparison of various filters for noise removal in paddy leaf images. *IACSIT International Journal of Engineering and Technology*, *7*(2), 372–374. doi:10.14419/ijet.v7i2.21.12444

Arif, S., Kumar, R., Abbasi, S., Mohammadani, K. H., & Dev, K. (n.d.). *Weeds Detection and Classification using Convolutional Long-Short- Term Memory 2 Methodology for Classification of Weeds*. Academic Press.

Asad, M. H., & Bais, A. (2019). Weed detection in canola fields using maximum likelihood classification and deep convolutional neural network. *Information Processing in Agriculture*. Advance online publication. doi:10.1016/j.inpa.2019.12.002

Basavarajeshshwari & Madhavanavar, P. S. P. (2017). A Survey on Weed Detection using Image Processing. *International Journal of Engineering Research & Technology, 5*(6), 1–3.

Brinkhoff, J., Vardanega, J., & Robson, A. J. (2020). Land cover classification of nine perennial crops using sentinel-1 and -2 data. *Remote Sensing*, *12*(1), 1–26. doi:10.3390/rs12010096

Chaki, J., Parekh, R., & Bhattacharya, S. (2015). Plant leaf recognition using texture and shape features with neural classifiers. *Pattern Recognition Letters*, *58*, 61–68. Advance online publication. doi:10.1016/j.patrec.2015.02.010

Contreras-Medina, L. M., Osornio-Rios, R. A., Torres-Pacheco, I., Romero-Troncoso, R. de J., Guevara-González, R. G., & Millan-Almaraz, J. R. (2012). Smart sensor for real-time quantification of common symptoms present in unhealthy plants. *Sensors (Basel)*, *12*(1), 784–805. doi:10.3390120100784 PMID:22368496

Desai, R. (2015). Removal of weeds using Image Processing: A Techni- cal Review. *International Journal of Advancements in Computing Technology*, *4*(1), 27–31.

Donatelli, M., Magarey, R. D., Bregaglio, S., Willocquet, L., Whish, J. P. M., & Savary, S. (2017). Modelling the impacts of pests and diseases on agricultural systems. *Agricultural Systems*, *155*, 213–224. doi:10.1016/j.agsy.2017.01.019 PMID:28701814

Espinoza, K., Valera, D. L., Torres, J. A., López, A., & Molina-Aiz, F. D. (2016). Combination of image processing and artificial neural networks as a novel approach for the identification of Bemisia tabaci and Frankliniella occidentalis on sticky traps in greenhouse agriculture. *Computers and Electronics in Agriculture*, *127*, 495–505. doi:10.1016/j.compag.2016.07.008

Fritz, S., See, L., Bayas, J. C. L., Waldner, F., Jacques, D., Becker-Reshef, I., Whitcraft, A., Baruth, B., Bonifacio, R., Crutchfield, J., Rembold, F., Rojas, O., Schucknecht, A., Van der Velde, M., Verdin, J., Wu, B., Yan, N., You, L., Gilliams, S., ... McCallum, I. (2017, December). (2018). A comparison of global agricultural monitoring systems and current gaps. *Agricultural Systems*. Advance online publication. doi:10.1016/j.agsy.2018.05.010

Gao, J., Nuyttens, D., Lootens, P., He, Y., & Pieters, J. G. (2018). Recognising weeds in a maize crop using a random forest machine-learning algorithm and near-infrared snapshot mosaic hyperspectral imagery. *Biosystems Engineering*, *170*, 39–50. doi:10.1016/j.biosystemseng.2018.03.006

Halder, M., Sarkar, A., & Bahar, H. (2018). Plant Disease Detection By Image Processing: A Literature Review. *SDRP Journal of Food Science & Technology*, *3*(6), 534–538. doi:10.25177/JFST.3.6.6

Hamuda, E., Mc Ginley, B., Glavin, M., & Jones, E. (2018). Improved image processing-based crop detection using Kalman filtering and the Hungarian algorithm. *Computers and Electronics in Agriculture*, *148*(February), 37–44. doi:10.1016/j.compag.2018.02.027

Huang, S., Zhong, S., & Chen, K. (n.d.). *A Novel Method of Stone Surface Texture Image Recognition*. Academic Press.

Hugar, S. M. (2016). *A new approach for weed detection in agriculture using image processing techniques*. Academic Press.

Islam, N., Rashid, M. M., Wibowo, S., Xu, C. Y., Morshed, A., Wasimi, S. A., Moore, S., & Rahman, S. M. (2021). Early weed detection using image processing and machine learning techniques in an australian chilli farm. *Agriculture (Switzerland)*, *11*(5), 387. Advance online publication. doi:10.3390/agriculture11050387

Jan, S. U., & Koo, I. (2018). A Novel Feature Selection Scheme and a Diversified-Input SVM-Based Classifier for Sensor Fault Classification. *Journal of Sensors*, *2018*, 1–21. doi:10.1155/2018/7467418

Jayanthi, G., Archana, K. S., & Saritha, A. (2019). Analysis of automatic rice disease classification using image processing techniques. *International Journal of Engineering and Advanced Technology*, *8*(3).

Jiao, Y., & Du, P. (2016). Performance measures in evaluating machine learning based bioinformatics predictors for classifications. *Quantitative Biology*, *4*(4), 320–330. doi:10.100740484-016-0081-2

Jones, J. W., Antle, J. M., Basso, B., Boote, K. J., Conant, R. T., Foster, I., Godfray, H. C. J., Herrero, M., Howitt, R. E., Janssen, S., Keating, B. A., Munoz-Carpena, R., Porter, C. H., Rosenzweig, C., & Wheeler, T. R. (2017). Brief history of agricultural systems modeling. *Agricultural Systems*, *155*, 240–254. doi:10.1016/j.agsy.2016.05.014 PMID:28701816

Lai, Z., & Deng, H. (2018). Medical Image Classification Based on Deep Features Extracted by Deep Model and Statistic Feature Fusion with Multilayer Perceptron. *Computational Intelligence and Neuroscience*, *2018*, 1–13. doi:10.1155/2018/2061516 PMID:30298088

Le, T. H., Tran, H. S., & Nguyen, T. T. (2013). Applying Multi Support Vector Machine for Flower. *Institute for Computer Sciences, Social Informatics and Telecommunications Engineering*, 268–281. doi:10.1007/978-3-642-36642-0_27

Liakos, K. G., Busato, P., Moshou, D., Pearson, S., & Bochtis, D. (2018). Machine learning in agriculture: A review. *Sensors (Switzerland)*, *18*(8), 1–29. doi:10.339018082674 PMID:30110960

Lu, Y., Yi, S., Zeng, N., Liu, Y., & Zhang, Y. (2017). Identification of rice diseases using deep convolutional neural networks. *Neurocomputing*, *267*, 378–384. Advance online publication. doi:10.1016/j.neucom.2017.06.023

Nathalia, B., Panqueba, S., Arturo, C., & Medina, C. (2016). A computer vision application to detect unwanted weed in early stage crops 3 Problem Solution 2 Problem Formulation. *Wseas Transactions on Computer Research, 4*, 41–45.

Oppenheim, D., & Shani, G. (2017). Potato Disease Classification Using Convolution Neural Networks. *Advances in Animal Biosciences, 8*(2), 244–249. Advance online publication. doi:10.1017/S2040470017001376

Pulido-Rojas, C. A., Molina-Villa, M. A., & Solaque-Guzmán, L. E. (2016). Machine vision system for weed detection using image filtering in vegetables crops. *Revista de la Facultad de Ingeniería, 2016*(80), 124–130. doi:10.17533/udea.redin.n80a13

Qin, F., Liu, D., Sun, B., Ruan, L., Ma, Z., & Wang, H. (2016). *Identification of Alfalfa Leaf Diseases Using Image Recognition Technology*. doi:10.1371/journal.pone.0168274

Ragul Krishna, P., Ahamed, J., & Sivakumar, B. (2020). Green house monitoring using raspberry Pi and cloud. *International Journal of Advanced Science and Technology, 29*(6), 2813–2819.

Reddy, R. A., & Basha, M. (2017). Image Processing For Weed Detection. *International Journal of Engineering Technology. Management and Applied Sciences, 5*(4), 485–489.

Sanjeevi, P., Prasanna, S., Siva Kumar, B., Gunasekaran, G., Alagiri, I., & Vijay Anand, R. (2020). Precision agriculture and farming using Internet of Things based on wireless sensor network. *Transactions on Emerging Telecommunications Technologies, 31*(March), 1–14. doi:10.1002/ett.3978

Satish, A. N., Pandey, S., Jain, R., Sayeed, M. A., & Shashikala, G. (2016). Detection of Weeds in a Crop Row Using Image Processing. *Imperial Journal of Interdiciplinary Research, 2*(8), 1108–1111.

Sharpe, S. M., Schumann, A. W., Yu, J., & Boyd, N. S. (2020). Vegetation detection and discrimination within vegetable plasticulture row-middles using a convolutional neural network. *Precision Agriculture, 21*(2), 264–277. doi:10.100711119-019-09666-6

Shuping, D. S., & Eloff, J. (2017). The use of plants to protect plants and food against fungal pathogens: A review. *African Journal of Traditional, Complementary, and Alternative Medicines, 14*(4), 120–127. doi:10.21010/ajtcam.v14i4.14 PMID:28638874

Tang, J. L., Wang, D., Zhang, Z. G., He, L. J., Xin, J., & Xu, Y. (2017). Weed identification based on K-means feature learning combined with convolutional neural network. *Computers and Electronics in Agriculture, 135*, 63–70. doi:10.1016/j.compag.2017.01.001

Thornton, P. K., Schuetz, T., Förch, W., Cramer, L., Abreu, D., Vermeulen, S., & Campbell, B. M. (2017). Responding to global change: A theory of change approach to making agricultural research for development outcome-based. *Agricultural Systems, 152*, 145–153. doi:10.1016/j.agsy.2017.01.005

Wang, A., Zhang, W., & Wei, X. (2019). A review on weed detection using ground-based machine vision and image processing techniques. *Computers and Electronics in Agriculture, 158*(November), 226–240. doi:10.1016/j.compag.2019.02.005

Wei, L., Gan, Q., & Ji, T. (2018). Skin Disease Recognition Method Based on Image Color and Texture Features. *Computational and Mathematical Methods in Medicine*, *2018*, 1–10. doi:10.1155/2018/8145713 PMID:30224935

Wu, Z., Chen, Y., Zhao, B., Kang, X., & Ding, Y. (2021). Review of weed detection methods based on computer vision. *Sensors (Basel)*, *21*(11), 1–23. doi:10.339021113647 PMID:34073867

Zhu, H., Cen, H., Zhang, C., & He, Y. (2016). *Early Detection and Classification of Tobacco Leaves Inoculated with Tobacco Mosaic Virus Based on Hyperspectral Imaging Technique*. Advance online publication. doi:10.13031/aim.20162460422

Chapter 7
AI-Based Motorized Appearance Acknowledgement Scheme for an Attendance Marking System

Swapna B.
https://orcid.org/0000-0002-7186-2842
Dr. M. G. R. Educational and Research Institute, India

M. Kamalahasan
Dr. M. G. R. Educational and Research Institute, India

S. Gayathri
https://orcid.org/0000-0002-1676-6284
Karpagam College of Engineering, India

S. Srinidhi
https://orcid.org/0000-0001-8187-5818
Dr. M. G. R. Educational and Research Institute, India

H. Hemasundari
Dr. M. G. R. Educational and Research Institute, India

S. Sowmiya
Dr. M. G. R. Educational and Research Institute, India

S. Shavan Kumar
Dr. M. G. R. Educational and Research Institute, India

ABSTRACT

Staff attendance exists as the greatest historical overwhelming chore in each institution. Existing presence scheme is typically grounded in RFID, IRIS, impression, and uniform notepad. Those schemes necessitate corporeal communication. One and all must wait until the preceding worker goes through the queue. The authors develop an appearance acknowledgement-based staff presence scheme by means of AI. With the help of deep learning and datasets, the scheme senses the position and recognizes which appearance goes to which ID and marks attendance in the datasheet. Then it is exported as an Excel sheet. All resemblance and datasets are protected.

DOI: 10.4018/978-1-7998-9640-1.ch007

INTRODUCTION

Attendance marking is a crucial task in an educational institute and in a workplace. Many organizations are using some automatic systems such as punch cards, RFID cards, and fingerprint scanner and to some extend face recognition. The most traditional method of marking attendance is through roll calls. But this roll call method is now used only in schools and some colleges, since most of the colleges have moved towards digital methods such as swipe or punch cards to get the job done. Both the methods got some drawbacks. Roll call method is time-consuming and prone to human errors. Mistakes such as marking the present person absent and vice versa can occur in this method.

Digital methods such as swipe and punch cards can overcome the drawbacks in the previous mentioned method, but it also gives rise to new issues. Swipe cards are less time consuming but it doesn't check for the presence of the person while swiping the card, which may lead to another person marking proxy attendance of the absent person.Also, another concern is that if a person loses his or her swipe card, then he or she may have to wait sometime before getting a new card.

The other method used for attendance marking is by RFID. In this method the person has to register the attendance using RFID on the card reader. The drawback of this method is that it can lead to fraudulent entry since available persons can put entry using the ID of unavailable persons. So an absent person can be marked as present. Using biometric can help to overcome the issues existing in previous techniques. Biometric-based attendance system is used in a number of places. Biometric systems rely on a person's unique physical features in order to identify them. Fingerprint, palm print, iris, retina, face is some of the biometrics that a system used to uniquely identify a person.

Using Face recognition reduces error to a great extent, the unique identifier in a face recognition-based system is a person's face, therefore the risk of losing that unique identifier is next to zero. No human interaction is involved and is less time-consuming (Borkar & Kuwelkar, 2017, pp. 249 – 255.)

Artificial Intelligence is additionally machine insight which is meant by machines, in software engineering it is characterized as clever operators which sees the earth by taking activities and accomplishing the objectives. This paper is under the area of Artificial Intelligence and it is moderately connected with security segment (Adam Santoro et al., 2021, pp. 1-24)

Artificial intelligence is wide ranging branch of computer science concerned with building smart machines capable of performing tasks that typically require human intelligence. AI is an interdisciplinary science with multiple approaches, but advancement in Machine learning and Deep learning. AI is a broad field and it is concerned with getting computers to do tasks that require human intelligence. However, there are many tasks like complex arithmetic which computers can do easily. Conversely, there are many tasks that people will do without thinking like recognizing a face which are extremely complex to automate. AI is concerned with these complex tasks which seem to be sophisticated reasoning process and knowledge.

AI is a field that overlaps with computer science rather than being a strict subfield. Different areas of AI are more closely related to psychology, philosophy, logic, linguistics and even neurophysiology. Making us to develop with an AI based application that extraordinarily distinguishes an individual by analyzing designs dependent on facial surfaces. It is a discerning assignment where our objectives are to automate the entire work. Ordinary activities are progressively asking to deal with by machine, as an option of pencil and paper or up close and personal improvement in electronic exchanges which result in extraordinary interest for quick and definite client recognizable proof and verification.

LITERATURE REVIEW

For new things to include, we need a framework which gauges participation naturally by catching picture and highlighting participation via looking in database. Understudies participation in the study hall is imperative obligation and whenever taken hand worked squanders a great deal of moment (Navesh Sallawar et al., 2017, pp. 2156-2159). There is heaps of self-deciding strategies close by for this point for example biometric participation. Every one of these techniques drains time since understudies need to make a line to contact their thumb in the checking gadget. Primary aim of this task is to make entire work computerized. It portrays the efficient calculation that without human mediation denotes the participation without human intervening. These attendances are recorded by passing the info remotely utilizing a camera joined before the talk room that is constantly catching pictures of understudies, identifies the appearances in pictures and contrast the recognized countenances and the database, by following the systems and denoting the participation. The paper gauge the connected work in the field of participation framework at that point portrays the framework auxiliary structure, programming calculation and results (Venkata Kalyan Polamarasetty & Muralidhar Reddy Reddem, 2018, pp. 4606-4610).

Modern approach used in face recognition attendance system not only is compatible with image but also has cctv footages. This method is effective in the usage of police and other law enforcements agencies. The implementation of face recognition follows image acquisition, processing, classification and decision making. The main role is to capture and convert the image, match it with the attendance (Deepanshu Chaudhary et al., 2020, pp. 485-487). The smartphone-based face attendance system is affordable and practically implementable in universities and schools. This method follows a novel system architecture where it recognizes a group of faces at one shot. To learn facial features, we can perform both horizontal and vertical flips, where each face is embedded creating one for neutral image and one for horizontal flip. The students can use the mobile application to analyses their daily attendance (Bhat et al., 2020).

Images captured using HDR cameras are encoded which are classified under local binary pattern histogram that results in matching images. It records the attendance percentage classifying the subject code and class of the students. This system improves the images with directed gradient descriptive histogram that characterizes the facial pictures utilizing a straightforward information vector (Damale & Pathak, 2018).

AIis the recent technology which was compared with Internet of things for several applications such as agriculture, smart cities, farming, traffic control and energy meter (Swapna et al., 2019). But there is lot of trouble in accumulating the iris of the students and therefore a fast application of face recognition with minimum illumination can be used (Martin Heller, 2020, pp. 1-10).

The technology of biometric recognition system for personal identification of irises, voice prints, fingerprints and signatures formats in a passive method. This face recognition is regarded as a kind of human computer interfaces (HCIs). The image captured from a PTZ camera gives an automatic real-time multiple faces recognition through laser range finder system. The proposed face localization techniques develop feature-based approaches. In case there are a lot of images in the database, two new algorithm methods come in order which gives a grey-level template matching (Fahn & Wang, 2011).

In this system, they have proposed hybrid face recognition by combining two faces recognition technique using PCA which helps in dimensional-reduction, used often in large datasets. LDA is also used for pre-processing step for machine learning applications. This system is implemented on embedded system based on Raspberry pi 3 boards (Borkar & Kuwelkar, 2017, pp. 249 – 255.)

Many organizations, companies and institutions take periodic attendance of the employees using RFID method, Biometric Fingerprint method and Registers. More time is required for calculating attendance by using these methods. RFID (Radio Frequency Identification) is a method where in there are electromagnetic fields which are used to automatically detect and track the tags attached to persons (Lim et al., 2009).

Biometric fingerprint identification system makes use of fingerprint as a unique identity (Quan & Gang, 2012, pp. 1-6). This is the most efficient method among all. But recognition of an individual fingerprint from a set of enrolled fingerprints is a difficult process. The fingerprint system may not always reveal any detail about the original fingerprint. This may have been proved to be false as many algorithms reveal that a fingerprint can be reconstructed with minute templates. By using a real-time OpenCV system attendance of the students are automated. This saves a lot of manual time and errors. This prototype's algorithm of the dataset images stores its attendance results in my SQL database. Authentication process takes place using the human face recognition technique along with bio-metric verification also. This paper focuses on PC based inbuilt webcam system that recognizes the images and implements LBP classifier which in turn creates the attendance record. All the attendance is recorded in the training folder. RFID can invade the privacy and security of human beings in the industrial environment. RFID strategies in turn effect's the software that allows each person to be analyzed by the primary database. This environment can be easily studied by hackers. If RFID reader and receiver are not properly resonated, then less read rate occurs (Suyash et al., 2020, pp. 34-40).

Attendance of individuals plays an important part in educational organizations to calculate the overall performances. To eliminate the tedious traditional methods, biometric attendance system based on iris recognition serves a great solution. It characterizes uniqueness, immovability and time variance of iris. Images are captured in high resolution and are detected, calculated and stored in MATLAB program. Iris Recognition is another type of implementation where the iris of people is scanned, stored and then retrieved for the matching and attendance is done without fail in the server (Khatun et al., 2015).

The proposed system MTCNN is a mobile application. It has facial recognition and generates attendance sheet, shares the report through mail to the respective department and staff members. Then the detected faces are cropped and resized and stored as a testing dataset. The FaceNet used in this system adds an embedded layer extraction feature that gives high accuracy (Anitha et al., 2020, pp. 189-195).

MATERIALS AND METHODS

It is a biometric unified PC innovation being utilized in various applications which separate human faces in advanced pictures. It additionally directs to the physiological procedure by which people find and arrange to faces in a visual scene. It is utilized as human PC interface and database the board. A face Detector needs to tell whether a picture of arbitrary size contains a human face and if so where it is. Face location as shown in Figure 1 can be performed dependent on a few prompts: skin shading, facial /head shape, facial appearance or a blend of these parameters. Most face identification calculations are appearance based without utilizing different prompts. An information picture is filtered at all potential areas by a sub window. Face recognition is comprised as ordering the example in the sub opening either as a face or a non-face.

It is likewise an innovation which is effective of recognizing or checking an individual from the computerized picture. Critically it is a Biometric Artificial Intelligence based application that exceptionally distinguishes an individual by investigating design dependent on individual's literary highlights, size, and separating tourist spots and so on. There are various strategies in which facial acknowledgment framework works however for the most part individuals work utilizing a chose facial element and looking at them from the info given. It resembles external observation by which human cerebrum comprehends and deciphers the face. Well known acknowledgment calculations incorporate PCA, LDA and Multi direct subspace getting the hang of utilizing tensor portrayal. It is anyway being noticed that the current face ID procedures are not 100% competent yet. Common efficiencies go between 40 to 60 percent. By this no manual contact in the interest of the client is required. It is just biometric that enable us to perform latent recognizable proof in a one to numerous situations.

It includes lessening the measure of assets. Highlight extraction begins from an underlying set estimated from information given which infers building highlights for enlightening, encouraging, learning and furthermore for better understandings. It interfaces with dimensionality decrease. This comes when the information is enormous to be handled it is suspected to get changed in diminished arrangements of highlights. The picked highlight for this contains the appropriate data from the info information, with the goal that the ideal errand can be performed by utilizing this decreased portrayal rather than complete introductory information. Fundamentally in this we have things to focus; as our highlights are anything but difficult to ascertain however, we have to remove the blank areas, at that point we create a vital picture which best suits for include extraction making the procedure quick. So, includes are extricated from sub windows of the given picture. The eventual fate of this could lead us to various information. In any case, the face recognition step, patches are removed from pictures.

The face discovery is a significant advance of the human face acknowledgment and global positioning framework. For empowering the framework to remove facial highlights all the more productively and impeccably, we should limit the scope of face recognition first, and the presentation of the face location strategy can't be excessively low. To distinguish human faces rapidly and precisely, we exploit an AdaBoost (Adaptive Boosting) calculation to assemble a solid classifier with basic square shape highlights associated with an indispensable picture (Fahn & Wang, 2011).

Thus, regardless of what the size of a square shape include we use, the execution time is consistently steady. Some square shape highlights are taken care of to the solid classifier that can recognize positive and negative pictures. Performance of patches have a few drawbacks like utilizing this patch straightforwardly for face acknowledgment like right off the bat, each fix by and large contains more than 1000 pixels, which are considerable to assemble an overwhelming acknowledgment framework. Also, face patches might be taken from various arrangements, not at all like face looks, enlightenments, yet may experience the ill effects of stopping and disorder. To illuminate these downsides, highlight extractions are performed so to data pressing, measurement decrease, notability extraction, and cleaning (Fahn & Wang, 2011).

Henceforward, the fix is altered into a vector of fixed measurement or a lot of guardian focuses with their comparing areas. As the given example by applying methods ordinarily we get the chance to remove the highlights and afterward choice technique with extraordinary grouping. In certain papers, highlight extraction is either remembered for face location or face acknowledgment.

Despite the fact that the AdaBoost calculation invests a great deal of energy in preparing the solid classifier, the face location result accomplishes elite in the fundamental. The fell structure of powerless classifiers utilizing the AdaBoost calculation is developed one by one, and each powerless classifier has its own request. A positive outcome will be prepared by the next feeble classifier, while a negative outcome at any as of now prepared guide leads toward the quick dismissal for the relating sub-window.

To begin with, we input the square shape highlights by means of various sub-windows into numerous frail classifiers, and the identification blunder rate is the base in the primary powerless classifier that can erase a ton of negative models, at that point the rest of the pictures are more hard to be taken out, which are additionally handled by the progressive frail classifiers. Such an activity is over and again executed until the last frail classifier is performed, and the remnants are face pictures. So as to exploit the indispensable picture, we receive the idea like preparing a pyramid of pictures.

That is, a sub-window establishing the identifier filters the information picture on numerous scales. For instance, the main identifier is made out of 20 pixels, and a picture of 320*240 pixels is checked by the locator. From that point onward, the current picture is more modest than the past one by 1.25 occasion. A fixed scale locator is then utilized to examine each of these pictures. At the point when the indicator checks a picture, resulting areas are procured from moving the window by some number of pixels, where s is the current scale. Note that the decision of influences both the speed of the indicator and the precision of the recognition. Identification results are grouped into a class to decide the agent position of the target by means of contrasting the separation between the focuses of a recognition result and the class with a given limit. The limit changes with the current scale like moving pixels referenced previously. If more than one identification result covers with one another, the last locale is registered from the normal of their places of each covering part (Fahn & Wang, 2011).

Figure 1. Face recognition-based attendance marking system architecture

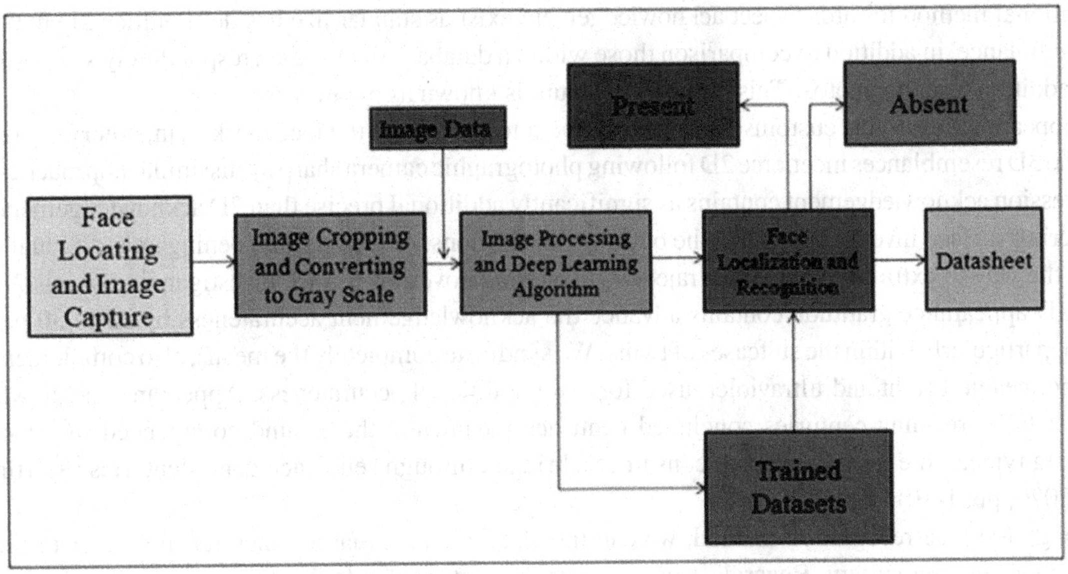

The application of the system takes place after following the set of steps. Firstly, to capture the image face locating takes place. Once the image is captured, the image is cropped to the definite ratio. It is then converted into Gray scale. After the image is successfully captured and set to the right size, we have to input the data of the image for image processing and deep learning algorithm to run. We are using trained datasets which is used to predict the outcome accurately.

After the input of all the necessary details for the image, the system is now ready for face localization and recognition. The system is at its right capability to mark present or absent according to the given input. The storage of this output recorded by the presence using face recognition can be done in the datasheet.

FACE RECOGNITION

Appearance acknowledgement exists as a technique used for classifying an unidentified creature or else confirming the uniqueness about precise individual's inceappearance. The a fore mentioned a division of processer idea, nevertheless look acknowledgement exist as dedicated besides derives through community luggage intended for approximately claims, by way of around susceptibilities towards parodying.

The initial make over acknowledgement procedures trust happening biostatistics towards go the leisurely make over topographies since an2D resemble shooked on a usual based on statistics those designates towards appearance. The gratitude procedure formerly equivalences those trajectories towards database about identified appearances which take plotted towards geographies cutting-edge the similar technique. Single impediment trendy this procedure exists as regulating the appearances towards a standardized interpretation near interpretation intended for skull revolution besides slant beforehand mining the system of measurement. This type of procedure is known as geometric.

Additional method towards aspect acknowledgement exist as standardize beside spoultice 2D make over resemblance, in addition to comparison those within a database likewise correspondingly standardized in addition trodden photos. This type of procedure is known as photometric

3D appearance gratitude customs 3D detectors about to apprehend the face mask twin, otherwise rebuilds the 3D resemblances incetierce 2D following photographic camera sharp by dissimilar approaches. 3D expression acknowledgement contains as significantly additional precise than 2D acknowledgement.

Covering surface investigation plots the outlines, decorations, too advert shappening an individual's look on the way to extra characteristics trajectory. Totalling covering surface investigation towards 2D or else 3D appearance gratitude contains advance the acknowledgement accurateness by 25 to 30 per hundred, particularly within the suitcases of twins. We syndicate completely the means, also complement within noticeable bright and ultraviolet, used for level additional accurateness. Appearance acknowledgement takes refining centuries concluded centuries meanwhile the ground commenced in 1964. Happening typical, the mistake amount consumes a bridged through semi each deucecenturies (Martin Heller, 2020, pp. 1-10).

Through the face recognition method, we can take face pictures that are then taken care of to the face acknowledgment system. Regardless, we execute the picture standardization to make the sizes of face pictures be the equivalent. The power of the pictures will be likewise changed for decreasing the lighting impact. After the size standardization and power change, we along these lines play out the element extraction cycle to acquire the element vector of a face picture. The thought of regular vectors was initially presented for separated word acknowledgment issues in the case that the quantity of tests in each class is not exactly or equivalent to the dimensionality of the example space. The ways to deal with taking care of these issues separate the basic properties of classes in the preparation set by killing the distinctions of the examples in each class.

A normal vector for every individual class is acquired from eliminating all the highlights that are in the ways of the eigenvectors comparing to the nonzero eigenvalues of the dissipate lattice of its own class. The basic vectors are then utilized for design acknowledgment. In our case, rather than utilizing a given class' own dissipate grid, we misuse the inside class dissipate grid, all things considered, to get the normal vectors. We likewise present another option calculation dependent on the subspace strategy and the Gram-Schmidt orthogonalization technique to procure the regular vectors. In this way, another arrangement of vectors called the discriminative basic vectors (DCVs) will be utilized for grouping, which results from the regular vectors. What follows explains the calculations for acquiring the normal vectors and the discriminative basic vectors (Fahn & Wang, 2011).

Our discovery strategy can successfully discover face locales, and the acknowledgment technique can know the face for individuals before the robot. At that point the robot will follow the face utilizing our face global positioning framework. Then again, we can trade the parts of the face with an outsider, and the robot will continue its following to follow the outsider. Until presently, many following calculations have been proposed by scientists. Of them, the Kalman channel and molecule channel are applied broadly. The previous is regularly used to anticipate the state grouping of an objective for direct expectations, and it isn't appropriate for the non-straight also, non-Gaussian development of articles. In any case, the last depends on the Monte Carlo mix technique and suit to nonlinear expectations. Taking into account that the stances of a human body are nonlinear movements, our framework picks a molecule channel to understand the tracker. The key thought of this method is to speak to likelihood densities by sets of tests. Its structure is isolated into four significant parts: likelihood dissemination, dynamic model, estimation, and considered inspecting.

DEEP LEARNING ALGORITHMS FOR FACE RECOGNITION

Deep learning algorithms run data through several layers of neural network algorithms, which pass simple representation of the data to the next layer. Machine learning algorithms will be able to perform well on datasets that have hundred features. But for unstructured dataset like image which has large number of features this process will be impossible. Deep learning algorithm learns progressively deep about the image as it goes through each neural network.It's also called as the Biostatistics AI, appearance acknowledgement notices also confirm a separate, alpha numerically in a database. Deep learning develops extra also supplementary precise through the accretion based on the database.

We say that whole thing similar a humanoid intelligence too amounts the information by way of the fore mentioned involvement in the interior the ground nonetheless through optimum accurateness also competence. Consequently, as soon as the deep learning procedure improvements involvement with the collected previous in addition new-fangled huge datasets, somewhat knowledge retaining deep learning assistance scanister type dependable guesses in addition likewise contemporary accurate replies towards actual statistics.

The statistics look pattern kept through facemask characters exist as associated through the appearance acknowledgement software package exploitation deep learning procedures. The fore mentioned bodies similarity amid the actual information detention in addition the deposited file towards recognize a discrete. Here is tetrad foremost path simperilled to deep learning used for facemask appreciation – recognition of facemask structures, arrangement, accurate structure entrenching, also appearance cataloguing besides acknowledgment (Kelsey Taylor, 2020, pp. 1-10). Algorithm used forImplementation is LBPH with haar cascade Lib for face detection and Using Open CV Lib for image capturing.

RESULTS AND DISCUSSION

We carried out techniques like face locating and image capture to capture the image as shown in Figure 4 and locate or upload the faces in database. Next the image will be cropped and converting into gray-scale image. Then image processing and deep learning algorithm will be processing the image received from before block. As per the deep learning algorithm, face recognition will be done and identify the faces to mark the attendance as per database. This is an automatic attendance marking system. Overall staff attendance with time and present or absent details as shown in Figure 2. It consists of staff names who are all working in the college or university. GUI screen as shown in Figure 3 which represents the screen used to process the artificial intelligence applications.

Figure 2. Overall staff attendance

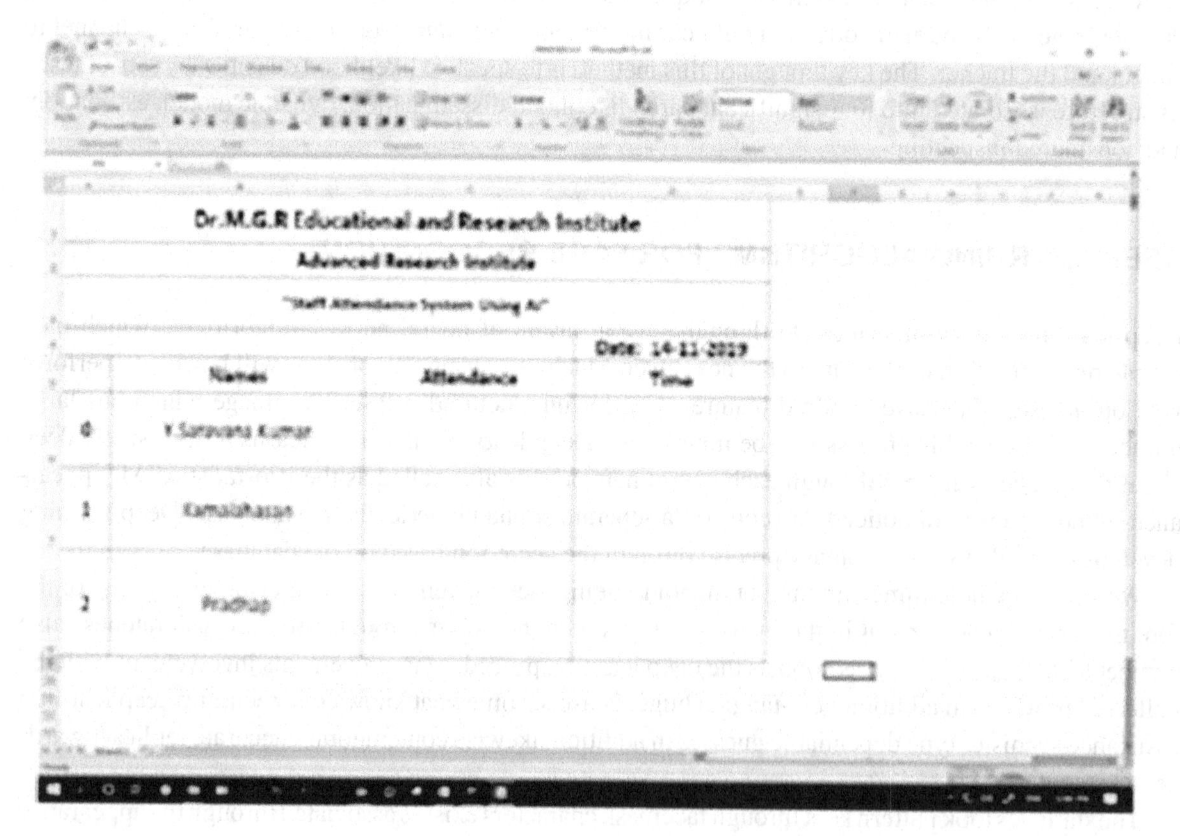

Figure 3. GUI Screen

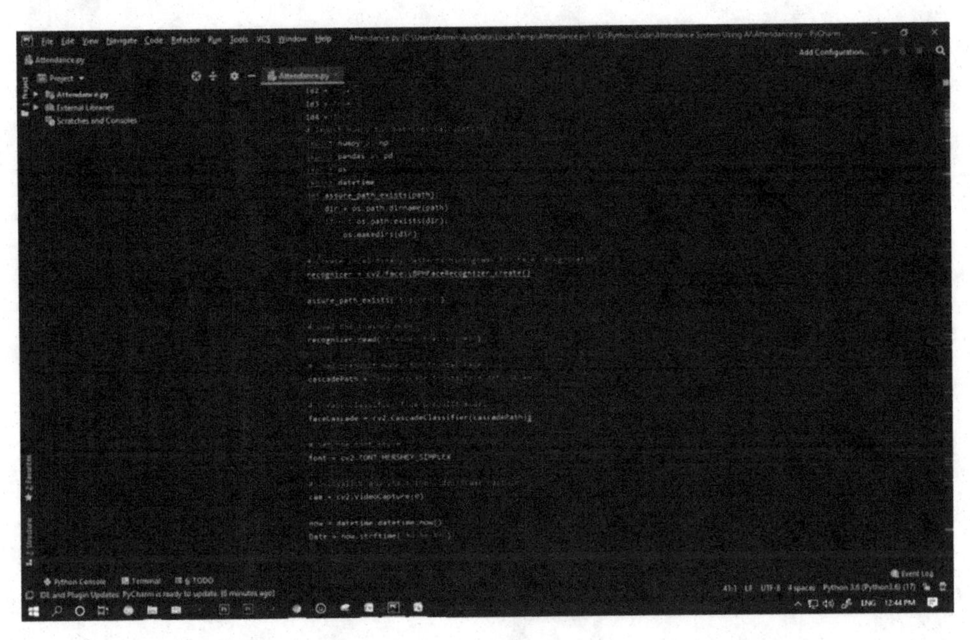

Figure 4. Marking attendance for Face 1

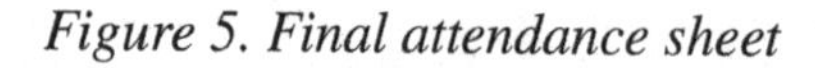

Figure 5. Final attendance sheet

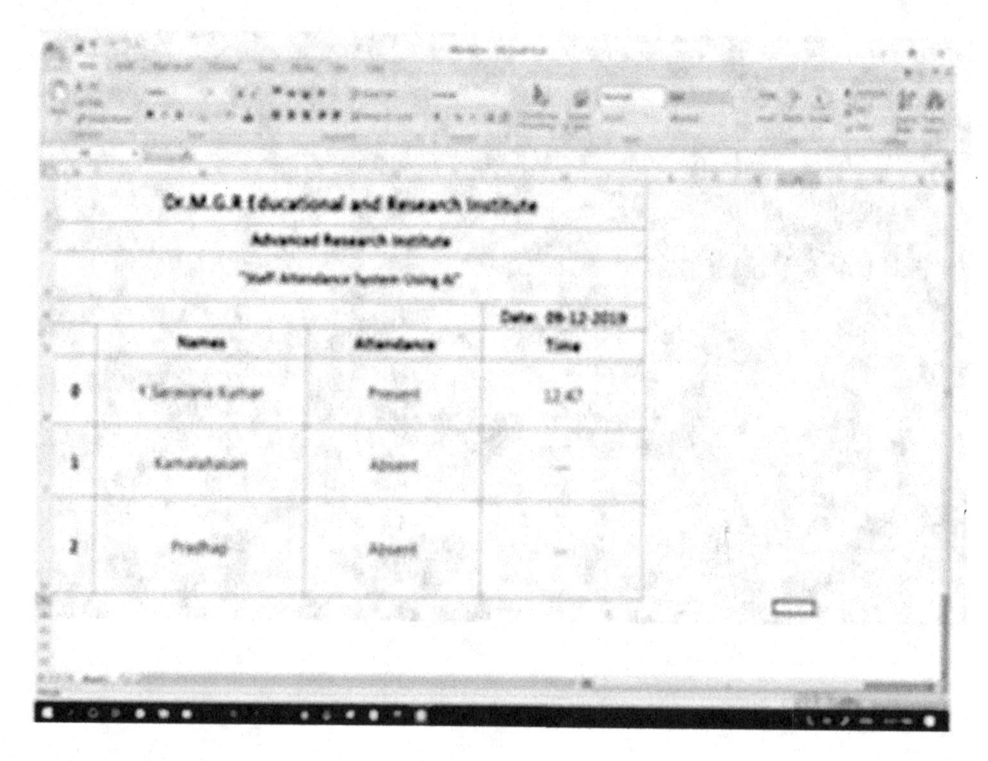

Figure 5 shows the final attendance. This data contains staff names, attendance and time of the entry. This output stores in the database, which will be used for future references. Hence automatic attendance marking system is the best to capture the attendance by using recent technologies are artificial intelligence. It shows the time in and out by marking present or absent in their name. Hence proposed work plays a vital role in the hospitals, schools and colleges, etc.

CONCLUSION

The system we have developed is successfully able to accomplish the task of marking the Staff attendance in the college automatically and output obtained is updated as programmed. On other aspect marked attendance will store in the database as cloud or internet of things. Hence Artificial Intelligence is the best technique to automate the attendance marking system through several algorithms. This automated technique is used for marking attendance in the several textile shops, colleges, hospitals and schools, etc. A recent technology plays an important role nowadays. Face recognized automatically with the help of Artificial intelligence technology. It is very useful for several applications with the help of deep learning algorithms.

REFERENCES

Anitha, G., Sunitha Devi, P., Vidhya Sri, J., & Priyanka, D. (2020). Face Recognition Based Attendance System Using Mtcnn and Facenet. *Zeichen Journal.*, *6*(1), 189–195.

Bhat, A., Rustagi, S., Purwaha, S. R., & Singhal, S. (2020). Deep-learning based group-photo Attendance System using One Shot Learning. *2020 International Conference on Electronics and Sustainable Communication Systems (ICESC)*. 10.1109/ICESC48915.2020.9155755

Borkar, N. R., & Kuwelkar, S. (2017). Real-time implementation of face recognition system. *International Conference on Computing Methodologies and Communication (ICCMC)*, *1*(1), 249 – 255.

Chaudhary, D., Rawat, A., Maurya, D., Patel, A., & Shukla, D. S. (2020). Face Recognition Based Attendance System. *International Journal of Engineering Applied Sciences and Technology.*, *5*(1), 485–487. doi:10.33564/IJEAST.2020.v05i01.085

Damale, R. C., & Pathak, B. V. (2018). Face Recognition Based Attendance System Using Machine Learning Algorithms. *2018 Second International Conference on Intelligent Computing and Control Systems (ICICCS)*. 10.1109/ICCONS.2018.8662938

Fahn, C. S., & Wang, C. H. (2011). *Real-time Multi-Face Recognition and Tracking Techniques Used for the Interaction between Humans and Robots*. Reviews, Refinements and New Ideas in Face Recognition. doi:10.5772/19589

Heller, M. (2020). What is face recognition? AI for big brother. *InfoWorld*, *1*(1), 1–10.

Khatun, A., Fazlul Haque, A. K. M., Ahmed, S., & Rahman, M. M. (2015). Design and implementation of iris recognition based attendance management system. *2015 International Conference on Electrical Engineering and Information Communication Technology (ICEEICT)*. 10.1109/ICEEICT.2015.7307458

Lim, T. S., Sim, S. C., & Mansor, M. M. (2009). RFID based attendance system. *2009 IEEE Symposium on Industrial Electronics & Applications*. 10.1109/ISIEA.2009.5356360

Polamarasetty, V. K., & Reddem, M. R. (2018). Attendance System based on Face Recognition. *International Research Journal of Engineering and Technology.*, *5*(1), 4606–4610.

Quan, Xi., & Gang, M. Li. (2012). An Efficient Automatic Attendance System Using Fingerprint Reconstruction Technique. *International Journal of Computer Science and Information Security*, *10*(1), 1–6.

Sallawar, N., & Yende, S. (2017). Automatic attendance system by using face recognition. *International Research Journal of Engineering and Technology.*, *4*(1), 2156–2159.

Santoro, Lampinen, Mathewson, Lillicrap, & Raposo. (2021). Symbolic Behaviour in Artificial Intelligence. *Deep Mind*, *1*(1), 1-24.

Suyash, Dhabre, & Rahul. (2020). Automated Face Recognition Based Attendance System using LBP Face recognizer. *International Journal of Advance Scientific Research and Engineering Trends.*, *5*(1), 35–40.

Swapna, B., Kamalahsan, M., Sowmiya, S., Konda, S., & SaiZignasa, T. (2019). Design of smart garbage landfill monitoring system using Internet of Things. *IOP Conference Series. Materials Science and Engineering*, *561*(1), 012084. doi:10.1088/1757-899X/561/1/012084

Taylor, K. (2020). How deep learning works for face recognition. *Hitechnectar.*, *1*(1), 1–10.

Chapter 8
Analysts and Detection of Concealed Weapons Using IR Fusion With MMW Support Imaging Technology

K. Hema Shankari
Women's Christian College, India

S. Mathi Vilasini
Ethiraj College for Women, India

D. Sridevi
Ethiraj College for Women, India

S. Amudha
Dr. M. G. R. Educational and Research Institute, India

ABSTRACT

The detection of weapons concealed underneath a person's clothing is an important obstacle to the improvement of the security of the general public as well as the safety of public assets like airports and buildings. The lack of proper mechanisms to detect and identify concealed weapons in advance results in the increase of crime rate. This chapter presents a study for concealed weapon detection using passive millimeter wave imaging sensors combined with image processing and convolutional neural networks. This eliminates the ambiguity of using millimeter wave imaging alone. The proposed system will perform the fusion of Passive MMW images with corresponding IR images followed by YOLO and VGG Net detection models.

DOI: 10.4018/978-1-7998-9640-1.ch008

INTRODUCTION

Concealed weapons are a great threat to the public. The absence of proper detection mechanisms for concealed weapons plays a major role when it comes to terrorist attacks and other crime incidents. Even though there are smart surveillance systems like CCTVs around each and every corner of the world, due to the manual nature of data gathering methods and failure to provide an immediate response to danger, they are mostly used to identify and track down perpetrators once the harm is done. Various studies had tried to address this issue by considering image processing and object detection separately due to the unavailability of a proper dataset and it is also mentioned that in order to increase the performance of the presented system it is preferable to use a large dataset. The purpose of this paper is to present a solution for concealed weapon detection using Image processing techniques and Convolutional Neural Networks with the goal of rapid detection and identification of concealed weapons.

Image fusion, recently identified as a key technology for concealed weapon detection along with two Convolutional Neural Networks will be used to construct the proposed system. Image fusion is being employed in various fields like concealed weapon detection, artificial intelligence, surveillance system, multi-focus imagery, medical diagnosis and such. While other fusion techniques such as Multi-Resolution Analysis (MRA) can be utilized, the resulting fused images are of poor quality making it difficult for the machine or human to recognize concealed objects. This paper's objective is to improvise the existing MMW imaging technique by combining it with image processing and Convolutional Neural Networks. Passive millimeter wave sensors measure the apparent temperature through the energy that is emitted or reflected by sources. The output of the sensors is a function of the emissivity of the objects in the MMW spectrum as measured by the receiver. Clothing penetration for concealed weapon detection is made possible by MMW sensors due to the low emissivity and high reflectivity of objects like metallic guns. Among the first generation of MMW sensors is the focal-plane array MMW sensor by Millitech Corporation. MMW imaging is preferred as a primary imaging technique over IR imaging because Infrared imagers utilize the temperature distribution information of the target to form an image. Normally they are used for a variety of night-vision applications, such as viewing vehicles and people. The underlying theory is that the infrared radiation emitted by the human body is absorbed by clothing and then re-emitted by it. As a result, infrared radiation can be used to show the image of a concealed weapon only when the clothing is tight, thin, and stationary. For normally loose clothing, the emitted infrared radiation will be spread over a larger clothing area, thus decreasing the ability to image a weapon. The infrared radiation emitted by the human body is absorbed by clothing and then re-emitted by it. As a result, infrared radiation can be used to show the image of a concealed weapon only when the clothing is tight, thin, and stationary. For normally loose clothing, the emitted infrared radiation will be spread over a larger clothing area, thus decreasing the ability to image a weapon. This paper utilizes Passive MMW imaging as the primary imaging technique combined with IR imaging.

Moreover, object identification is the crucial part of this process. Classifying the objects present inside a baggage is not sufficient. YOLO (You Only Look Once) is applied to identify the concealed objects in real time. It divides the obtained image into segments and predicts bounding boxes and their probability. YOLO is popular for its high accuracy and processing speed. With VGG Net and YOLO, large scale image classification with high accuracy is ensured. Many improvised methods related to Convolutional Neural Networks have been developed. Though these networks yield remarkable performance in classification and detection, their configuration architecture is complex. VGG Net is exploited for this purpose. It classifies data provided in a profound manner. Conveniently, reducing the architectural complexity in turn reduces the training loss.

LITERATURE SURVEY

Cylindrical millimeter-wave imaging technique was introduced (David M. Sheen et al., 2006) for the detection of metallic and non-metallic concealed weapon by utilizing vertical array of millimeter-wave antennas. This technique was developed on the fact that concealed items are reflective at millimeter wavelengths while clothing is not. The resulting images can be transformed into an animated video sequence that allows the objects in the image to rotate in front of a fixed light source. This technique proved to be useful in providing 3D images for analysis.

Image fusion is the process of combining images acquired using multiple sensors to construct a new image, providing contextual enhancement of the scene being observed. Different fusion algorithms have been proposed for improving spatial and spectral resolutions of the fused images over the decades such as Brovey transform method, Intensity-Hue-Saturation (IHS) method, statistical method, Independent Component Analysis (ICA) method, numerical method and Principal Component Analysis (PCA) method. The algorithms based on a Brovey transform preserve the relative spectral contributions of each pixel but enhance the intensity or brightness component of the image. Each component of the multispectral image bands normalized using a formula is multiplied by a high-resolution co-registered data. The methods based on IHS transformation merge images by preserving most of the spectral information from the H and S components but substituting the intensity image I with histogram-matched high resolution image. The PCA/ICA method transforms the original images into uncorrelated images and then combines the images by choosing the maximum value among all. PCA is frequently used for fusion as a statistical technique to compact the multivariate data set of inter-correlated variables redundant data into fewer uncorrelated bands.

The NMDB techniques, such as adaptive weight averaging (AWA) methods, neural network based methods, Markov Random Field (MRF) based methods and estimation theory based methods, are algorithms without decomposition transform and fusion rules are directly applied to the source images. A novel method of NMDB image fusion scheme for the combination of thermal and visual images was developed based on adaptive weight averaging approach (E. Lallier & M. Farooq, 2000). The weight of a thermal image pixel relies on the divergence of the pixel intensity from the image mean intensity and the local variance in space and time of the visual pixel intensity decide the weight of visual pixel. An approach was introduced using MRF models to solve the problem of remote sensing image fusion (M. Xu et al., 2011). Both the decision making and true image are modeled as MRF and the Least 9 Squares technique is used to calculate the fused image. A feature level image fusion algorithm was developed based on estimation theory (J. Yang & R. S. Blum, 2006). The source images are classified into regions by a graph-based image segmentation method and the regions are then analyzed to form a joint region map for the fused image. Alexander Toet (2003) proposed an image fusion technique in which non-literal imagery was fused with standard color images to display the concealed weapons in the context of the original color image.

A novel scheme for concealed weapon detection using colour image fusion was proposed by Zhiyun Xue (2003). Z. Y Xu et al. (2008) proposed a new image algorithm combining Projection onto Convex Sets algorithm (POCS) and nonlinear extrapolation algorithm for improving millimeter wave images.

Timofey Savelyev et al. (2010) describe two approaches to short-range microwave imaging by means of ultra-wideband (UWB) technology. The first approach deals with synthetic aperture radar (SAR) that employs a transmit-receive antenna pair on mechanical scanner. The second one represents a multiple input multiple output (MIMO) antenna array that scans electronically in the horizontal plane and mechanically, installed on the scanner, in the vertical plane. The mechanical scanning in only one direction reduces significantly the measurement time. Image mosaic technique allows combining multiple images into a new compound one by merging the region of interest from the images. Burt et al. (1983) introduced a new method based on multiresolution spline for this technique. A study was conducted to provide an alternative solution to the poor performance of multiresolution image fusion algorithm while considering visual and infrared images (Zhiyun Xue et al., 2004). The study involved multiresolution mosaic technique to combine the visual and infrared or millimeter wave images, and K-means clustering method to detect the concealed weapons. The resulting image retained the quality while the concealed weapon region was highlighted.

As mentioned earlier, three-dimensional microwave imaging has proved to be beneficial in the detection of concealed entities. Computers possess the ability to map the reflectivity of microwaves on an entity in 3D space and provide the basis for identification and automatic recognition. Tan et al. (2017) introduced a novel interpolation-free imaging algorithm which is related to wavefront reconstruction theory. This algorithm utilizes various reference signal spectrums at different ranges and forms target functions at different ranges by a succinct summation. This algorithm is productive in generating 3D images at a very high speed with the benefit of high precision and reduces computational cost. In the interest of matching the obtained images of certain concealed weapons in baggage, effective feature descriptor algorithms are utilized. Scale Invariant Feature Transform (SIFT) is one of such algorithms that can effectively recognize objects. SIFT, while combined with x-ray images, have proved to be ineffective when compared to regular images. To overcome this shortcoming, an experiment was conducted to increase the efficiency of this algorithm (Nayel Al-Zubi & Meshalalmubarak, 2015). In this proposed approach, an image is segmented into sub images which reduce the stray matching points that affects SIFT matching process. In addition, Adaptive Histogram Equalization technique enhances the contrast of an image by applying histogram threshold and removes the background details while Contrast-Limited Adaptive Histogram Equalization (CLAHE) limits the amplification of noise by cutting any values above the predefined value. Speeded Up Robust Features is another such algorithm that is used for tasks like object recognition and 3D reconstruction. This algorithm was first proposed by Herbert Bay (2006). Using multi-resolution pyramid technique, the image is transformed to an image of reduced bandwidth. While the two aforementioned algorithms may seem to be similar, the details of the algorithms vary. Comparative studies on feature extraction using SIFT and SURF was conducted by (Amanjeet Kaur & Lakhwinder Kaur, 2016). The outcome of the study is that results of the algorithms are the same with respect to scale, rotation, noise and change in radiation, and that SURF is faster than SIFT.

Hua-Mei Chen et al. (2005) proposed Imaging for Concealed Weapon Detection. D. Novak (2005) proposed a new scheme for concealed weapon detection. They proposed a new electromagnetic (EM) solution for concealed weapons detection at a distance. Their proposed approach exploits the fact that the weapons of interest for detection, whether they are a hand gun, knife, box cutter, etc, each have a unique set of EM characteristics. The particular novelty of their technical solution for concealed weapons detection at a distance lies in the use of millimeter wave signals over a wide frequency band (26–40GHz or Ka-band) to excite natural resonances in the weapon and create a unique spectral signature that can be used to characterize the object. The need for more computation and identification costs. Due to the size of concealed weapons under people's clothing usually being very small, it is more difficult to identify them immediately from centimeter-level resolution PMMW images. Especially for some application scenarios and equipment with large passenger flow, a rapid recognition method of some types of concealed weapon targets on human bodies using a small sample dataset, even in no more than 3 seconds, is required to guarantee the normal speed of people's flow passing through a millimeter-wave security instrument This real-time detection of a small-sized target makes the task more challenging. For real-time detection of concealed targets under people's clothing, recently developed deep learning algorithms may be a better alternative. Concerning real-time detection requirements, undoubtedly a kind of one-step structure network is the best choice. Early work on the development of a one-stage object detector includes the OverFeat method. Several object detection frameworks, such as Single Shot MultiBox Detector (SSD) and You Only Look Once (YOLO), which utilize anchors or grids to propose candidate object localization, are typical one-stage detection methods. A study for concealed weapon detection in IR immagws was conducted and the study focused on developing automated detection and recognition using sensors and image processing (Mahadevi Parandel & Shridevi Soma, 2015). Similar study was conducted using IR images using Image Processing and Convolutional Neural Networks and the proposed system performed the fusion of IR images with corresponding RGB images followed by YOLO and VGG Net detection models (T. D. Piyadasa, 2020). Recently, machine learning algorithms have been gradually applied to PMMW imagery and obtained the best results are by Random Forest (RF) algorithm with Haar features extracted from the preprocessed images. In the meanwhile, neural network algorithms have been used in order to extract concealed targets for greater speed and accuracy. In Reference, research utilizing a context embedding object detection network showed effective results from AMMW images with massive sample data, which has been proved to be beneficial and gained 2% and 1% improvements on area under curve (AUC) and true positive (TP), respectively, compared with the PMMW data set using the same method. With the aid of deep learning, outstanding deep neural networks (DNNs) models are proposed so as to generate high precision, approximately 85%, human body profiles in PMMW images. Additionally, YOLOv3 algorithms have been successfully applied for real-time detection of targets such as cars, rail surface defects, and airplanes and other things.

A low cost passive millimeter wave (MMW) scanning camera for weapon detection was proposed by Thomas D. Williams and Nitin M. Vaidya (2005). A novel framework for the detection and classification of concealed weapons for crime detection by analyzing the data of CCTV stream was proposed (Gaurav Raturi et al., 2019). They proposed a classification framework that was developed with pre-existing data of categorization of concealed weapons through deep learning-based object detection and classification techniques. This proposal aimed to automate the monotonous task of monitoring CCTV stream data for concealed weapons that pose a threat. A novel technique to detect concealed weapons based on discrete wavelet transform was introduced that worked in conjunction with dimension reduced meta-heuristic algorithm, the harmony search, and shape matching based K means SVM classification (Nashwan Hussein, 2016). Multi-sensor stream data capturing framework was designed using the techniques of sensor fusion alongside feature extraction and segmentation of imsegmentation of images module. Training of rapid R-CNN (Region-based Convolutional Neural Network) model was carried out for classification of ages module. Training of faster R-CNN (Region-based Convolutional Neural Network) model was performed for classifying weapons over the collected dataset.

METHODOLOGY

Figure 1. High-level system architecture

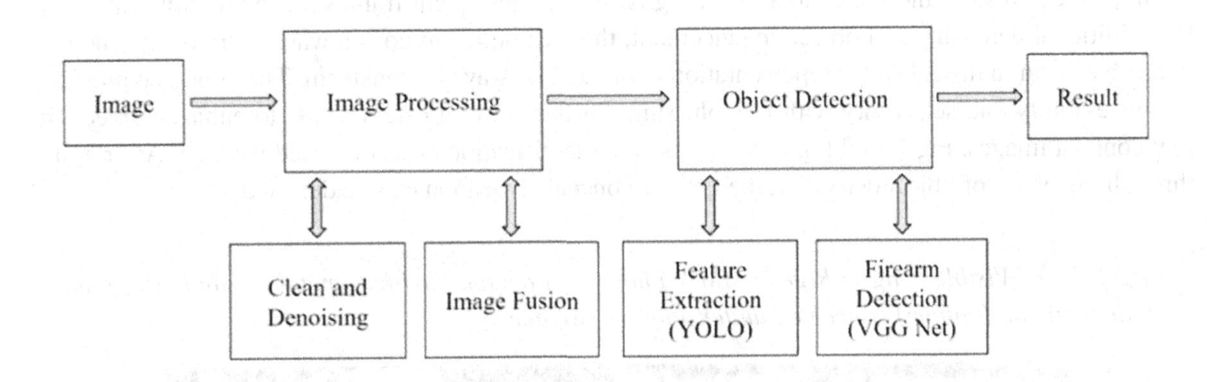

The system will include multiple stages of image processing and object detection. The process of Concealed Weapon Detection involves three processes. They are: Data gathering, Image Processing and Object Detection. Fig. 1 shows the high-level system architecture of the proposed system. The overall system consists of two phases: the Image processing phase and the Object detection phase.

Data Gathering

The major issue faced by most of the proposed studies for concealed weapon detection is the absence of a proper dataset. The datasets that are available on weapon detection use images that expose the weapon and they are mostly IR images hence, these cannot be used to train a good concealed weapon detection model. As the first approach of concealed weapon detection, it is necessary to develop a proper dataset with MMW and IR images that can be used to train a proper detection model. The primary reason to use MMV images is that it makes it possible to detect concealed weapons with more accurate details when combined with the corresponding IR image (fused). Even though it is required to gather a large number of images to achieve better accuracy, techniques like image rotation, flipping and color alterations on an average number of images can be used to increase the number of images.

Image Processing

In the domain of detecting concealed weapons, special attention should be given to the image processing phase. Due to the fact that the objects of interest are not explicitly visible in the image multiple image processing techniques will be carried out as shown in Fig.1. Initially, image cleaning and denoising will be carried out for noise suppression and object enhancement. This can be achieved using motion compensated temporal filtering followed by 1-D filtering along the trajectory. For the image fusion step, the image will be fused with the corresponding MMW image to produce a detailed image after subjecting to multiple intermediate image transformations. Denoising of the video sequences can be achieved temporally or spatially. First, temporal denoising is achieved by motion compensated filtering, which estimates the motion trajectory of each pixel and then conducts a 1-D filtering along the trajectory. This reduces the blurring effect that occurs when temporal filtering is performed without regard to object motion between frames. The motion trajectory of a pixel can be estimated by various algorithms such as optical flow methods, block-based methods, and Bayesian methods. If the motion in an image sequence is not abrupt, we can restrict the search to a small region in the subsequent frames for the motion trajectory. For additional denoising and object enhancement, the technique employs a wavelet transform method that is based on multiscale edge representation computed by wavelet transform. The approach provides more flexibility and selectivity with less blurring. Furthermore, it offers a way to enhance objects in low-contrast images. Fig 2. and Fig 3. Showcase the identification of a concealed weapon. After going through the phase of object detection, the type of concealed weapon can be detected.

Figure 2. Left - Visible image ; Right - MMW image of a person having a metal handgun (top) and a ceramic handgun (bottom) concealed under a heavy sweater

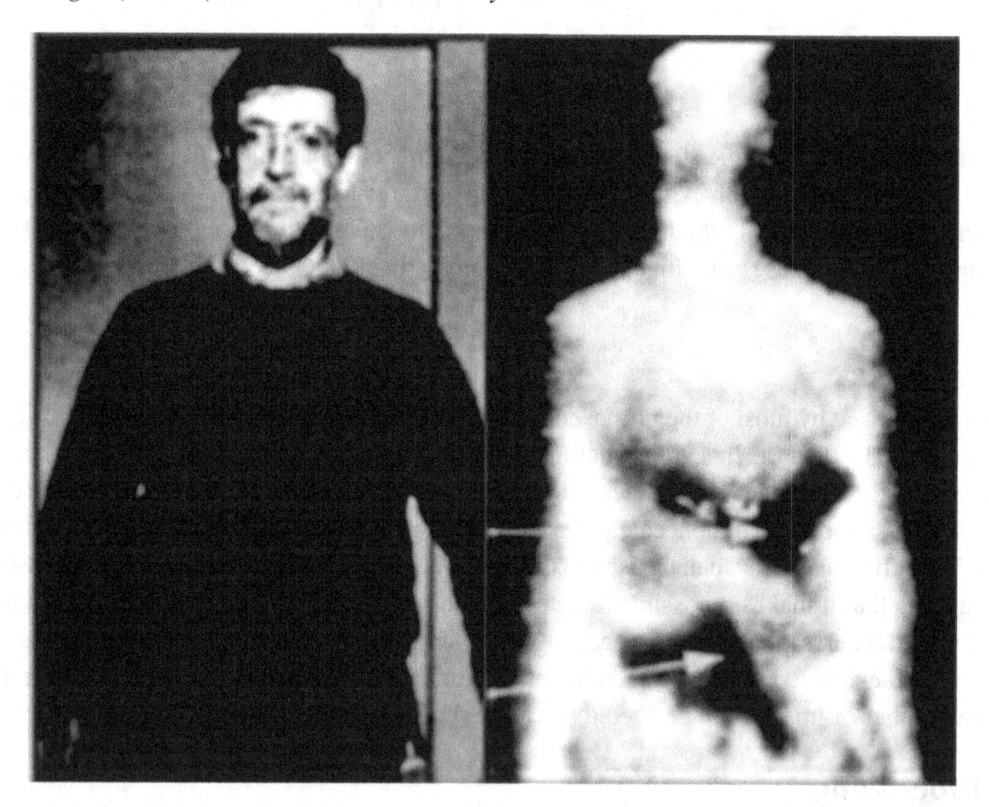

Figure 3. An example of image fusion for Concealed Weapon Detection. Left - Visible Image; Middle - MMW Image; Right - fused image (MMW and IR imaging)

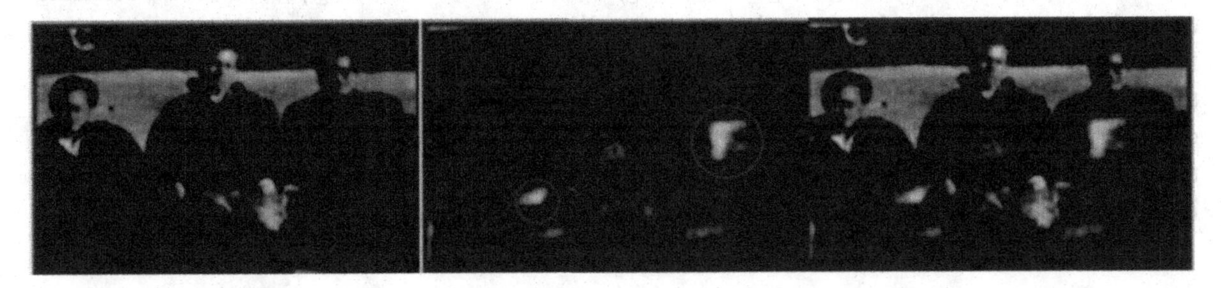

Object Detection

The phase of Object Detection consists of two stages as depicted in Fig. 1. YOLO algorithm will be used for the feature extraction stage to detect, locate and identify the important segments of the image where there are people. Using this approach we can isolate only the important segments of the image reducing the possibility of obtaining false positives. These segments will then be used as input for the firearm detection model. The firearm detection model will be developed through a Convolutional Neural Network. When selecting the architecture for the Convolutional Neural Network it is advisable to test on multiple architectures to select the most suitable one for the detection system. As YOLO is used for feature extraction, VGG Net will be used for the detection model because it uses small convolutional filters and it implements a large number of layers. When training the model the input images to the YOLO object detection will be resized to 224x224 pixels to increase the performance of the Neural Network with an initial learning rate of 0.001 which will be changed according to the performance of the model. ReLu activation will be used in the convolutional stages along with a SoftMax activation function at the last layer. The Convolutional Neural Network will be programmed on TensorFlow. The Evaluation of the model will be made on multiple metrics to guarantee the quality of predictions. Therefore Accuracy, Equal error rate (ERR), Precision, and Recall will be used as evaluation metrics to evaluate the proposed model.

RESULT AND DISCUSSION

The analysis of our graphs and scans, we reach some tentative conclusions about how weapons can be detected using ultrasound. First, the scans of the target must be taken, with the individual or individuals taking the scans preferably standing directly facing the front of the target. The scanners may stand a little to the side of the target; however, if we notice how many signals we saw with the 12_00 body orientation, we see that facing the target dead-on would be best. There are ways to overcome this that we have not tested, such as using multiple transducers to send and receive signals, but that is being addressed by others. Second, the data received from the return signal must be processed by the wavelet transform function and turned from distance-amplitude data to distance-frequency data. Third, the distance from the scanner to the target needs to be assessed. This may be done with some approximation and guesswork by the individual taking the scan, or more precisely by the machine itself. In any case, finding such distance data would be necessary in order to assess where in the scan the target's body is located. Weapon Detection using Image Processing, involves manual screening procedures for detecting concealed weapons such as handguns, knives, and explosives are common in controlled access settings like airports, entrances to sensitive buildings, and public events. The machine can fit a frequency bubble to the target bubble if it is possible, or it may request more information if there is too much noise. The other purpose of finding the distance from the scanner to the target is to narrow the data collected down to an area of interest that is only in the tens of centimeters instead of the hundreds of centimeters we see in the raw data. By doing so, we can make the area of interest more visible.

CONCLUSION

This paper proposes a novel idea for concealed weapon detection. The proposed system uses image fusion to fuse IR and MMW images to produce a detailed image of the people in the scene along with any concealed weapons. In order to automate the detection phase, two convolutional models are proposed, YOLO followed by VGG Net. YOLO performs feature extraction to identify people in the image and this segmented image is sent as input to VGG Net to detect the concealed weapons. The major limitation faced by proposed concealed weapon detection systems is the absence of a proper dataset and when developing a dataset special attention should be given to the large varieties of firearms that are available as having a proper dataset is a crucial factor to increase the performance of the system. The system can also be used in a smart surveillance system by embedding it in a CCTV camera or any other type of Passive MMW enabled camera. This study can also be applied to other domains of concealed object detection.

REFERENCES

Al-Zubi, N. (2015). Concealed weapon detection using X-Ray images. *International Journal of Electrical, Electronics and Data Communications*, *3*, 27–30.

Bay, H., Ess, A., Neubeck, A., & Van Gool, L. (2006). 3D from Line Segments in Two Poorly-Textured, Uncalibrated Images. *Proceedings - Third International Symposium on 3D Data Processing, Visualization, and Transmission,* 496-503.

Burt, P., & Adelson, E. (1983). A multiresolution spline with application to image mosaics. *ACM Transactions on Graphics*, *2*, 217–236.

Chen, H. M., Lee, S., Rao, R. M., Slamani, M. A., & Varshney, P. K. (2005). Imaging for concealed weapon detection: A tutorial overview of development in imaging sensors and processing. *IEEE Signal Processing Magazine*, *22*(2), 52–61. doi:10.1109/MSP.2005.1406480

Hussein, N. (2016). Multisensor of thermal and visual images to detect concealed weapon using harmony search image fusion approach. *Pattern Recognition Letters.*

Kaur, A., & Kaur, L. (2016). Concealed weapon detection from images using SIFT and SURF. *Online International Conference on Green Engineering and Technologies (IC-GET)*, 1-8. 10.1109/GET.2016.7916679

Lallier, E., & Farooq, M. (2000). A real time pixel-level based image fusion via adaptive weight averaging. *Proceedings of the Third International Conference on Information Fusion, 2*, WEC3/3-WEC313.

Parande, M., & Soma, S. (2015). Concealed Weapon Detection in a Human Body by Infrared Imaging. *International Journal of Science and Research, 4*, 182-188.

Piyadasa, T. D. (2020). Concealed Weapon Detection Using Convolutional. *Neural Networks*.

Raturi, G., Rani, P., Madan, S., & Dosanjh, S. (2019). ADoCW: An Automated method for Detection of Concealed Weapon. *Fifth International Conference on Image Information Processing (ICIIP)*, 181-186. doi:10.1109/CJMW.2008.4772439

Savelyev, T., Zhuge, X., & Yang, B. (2010). Development of UWB Microwave Array Radar for Concealed Weapon Detection. *11th International Radar Symposium*, 1-4.

Sheen, D. M. (2006). Cylindrical millimeter-wave imaging technique and applications - art. no. 62110A. *Proceedings of SPIE - The International Society for Optical Engineering*, 6211.

Tan, W., Huang, P., Huang, Z., Qi, Y., & Wang, W. (2017). Three-Dimensional Microwave Imaging for Concealed Weapon Detection Using Range Stacking Technique. *International Journal of Antennas and Propagation, 2017*, 1–11. doi:10.1155/2017/1480623

Toet, A. (2003). Color Image Fusion for Concealed Weapon Detection. *Proceedings of SPIE - The International Society for Optical Engineering*, 5071.

Williams, T. D., & Vaidya, N. M. (2005). A compact, low-cost, passive MMW security scanner. *Proceedings of SPIE - The International Society for Optical Engineering*, 5789. 10.1117/12.603662

Xu, M., Chen, H., & Varshney, P. K. (2011). An image fusion approach based on Markov random fields. *IEEE Transactions on Geoscience and Remote Sensing, 49*(12), 5116–5127. doi:10.1109/TGRS.2011.2158607

Xu, Z. Y., Dou, W. B., & Cao, Z. X. (2008). A New Algorithm for Millimeter-Wave Imaging Processing. *2008 China-Japan Joint Microwave Conference*, 337-339.

Xue, Z., & Blum, R. S. (2003). Concealed Weapon Detection Using Color Image Fusion. *Sixth International Conference of Information Fusion*, 622-627.

Xue, Z., Blum, R. S., Liu, Z., & Forsyth, D. S. (2004). Multisensor Concealed Weapon Detection by Using A Multiresolution Mosaic Approach. *IEEE 60th Vehicular Technology Conference, 7*, 4597-4601.

Yang, J., & Blum, R. S. (2006). A region-based image fusion method using the expectation maximization algorithm. *40th Annual Conference on Information Sciences and Systems*, 468-473. 10.1109/CISS.2006.286513

Chapter 9
Detection and Identification of Employee Attrition Using a Machine Learning Algorithm

Rama Krishna Garigipati
Koneru Lakshmaiah Education Foundation, India

Kasula Raghu
Mahatma Gandhi Institute of Technology, India

K. Saikumar
Koneru Lakshmaiah Education Foundation, India

ABSTRACT

This chapter proposes that employee attrition is the major circumstance faced in many organizations. Usually, organizations face this attrition when there is pressing need of employees due to mass retirements or while expanding the organization. Generally, any organization faces higher attrition rate for employment when they have more employment opportunities in market or recession time. Due to the demand for software goods across all industries, the software industry once suffered a significant attrition rate from employers due to large openings globally in the software business. The purpose of this research is to look at how objective elements influence employee attrition in order to figure out what factors influence a worker's decision to leave a company and to be able to predict whether a particular employee will leave the company using machine learning algorithms.

INTRODUCTION

Attrition is a major issue, and it is very, very high up in the industry these days. This is the most important issue, which is covered by any organization. Although the term "ATTRITION" is widely accepted, numerous has loss to identify what is real and fatigue-Fatigue is alleged to be the progressive diminution of the number of employees, pensions, retirement, or death. Employee exhaustion, it is referred to as an Employee, abandonment, or lapse of the staff. This can be interpreted as a gradual decrease in

DOI: 10.4018/978-1-7998-9640-1.ch009

the number of employees, pensions, retirement, or death explained in Fallucchi, F., Coladangelo, M., Giuliano, R., & William De Luca, E. (2020). When it comes to the recycling of standards, where the average worker will vary from sector to sector, both in terms of their own standards, and those standards may be differences between the skilled and unskilled positions. The rapid completion of the wages for the payroll package is primarily responsible for the sluggish growth in employment, and, as a result of an increase in the attrition rate. Companies are faced with a huge challenge in the recruitment and retention of talent, and at the same time, they must deal with the loss of talent due to friction, which is related with the industry's decline of the voluntary activity of the rotation. The level of staff turnover leads to productivity losses, which could have long-term adverse consequences for the company, especially when considering the talent of leaves gaps in its ability to deliver, and the role of human resources is not just the loss of productivity, but also in a loss of performance from the team, and social assets. In view of the fact that the rate of staff turnover is a serious problem in every industry, companies are developing innovative business models to sustain success, talent in Yedida, R., Reddy, R., Vahi, R., Jana, R., GV, A., & Kulkarni, D. (2018). There is no way that allows you to have full control of the fatigue, but you can, of course, we are limiting these indicators for the planning of the appropriate retention strategies. Every time you visit a well-trained, and well-suited employee leaves the organization; it creates a vacuum. Thus, the organization will lose important skills, knowledge, and business relationships. Modern managers, personal managers are very concerned about diminishing the level of consumption in the organization, and so that they can help to maximize productivity, the growth, and development of the organization. Employee proceeds are one of the most important issues that an association may face during the entire life cycle because it is difficult to predict, and often makes rather obvious shortcomings in the organization of the workers Alduayj, S. S., & Rajpoot, K. (2018). Service companies have to recognize that the use of the services may be affected, and the overall performance of the company can be significantly reduced, and, therefore, customers can be reduced when employees are suddenly left. Retention of staff is a serious and continuous process. One of the main challenges for managers is to understand that it is their accountability to generate and maintain a good service, a friendly environment. Managers have to accept it and understand it, and that it is of such fundamental principles, show their objectives, the nature, and motivation of their employees. However, the employee has been dismissed is an actuality for every business. If the situation was not handled well, the departure of key personnel could result in a decline in productivity. The organization may need to hire new people and learning from them is the tool that can be used, that is, it takes a long time to come. Most of organizations are interested in knowing who is at risk of leaving.

RELATED WORK

The progressive loss of personnel over time is referred to as employee attrition. The majority of literature on employee attrition divides it into two categories: voluntary and involuntary. Involuntary attrition is defined as an employee's error in which the organization fires the employee for a variety of reasons. When an employee departs a company voluntarily, it is known as voluntary attrition. Age, wage, and job satisfaction were found to be the strongest predictors of voluntary attrition in a meta-analytic evaluation of voluntary attrition. Other research has found that a variety of other factors, like as working environment, job satisfaction, and possibility for advancement, all play a role in voluntary attrition. Organizations strive to avoid employee attrition by utilizing machine learning algorithms to forecast the likelihood of a person leaving and then taking proactive efforts to prevent it. AMARAM 2005 stated that, it is recognized that the need is not merely to reduce attrition rates through the implementation of various retention strategies, but rather to identify the right candidates at the time of recruitment so that they identify with the organization and continue to prove to be assets for the company, cultural shock is something that Organizations must strike the appropriate balance between pampering their employees and exacting maximum labor from them; any of these, when overdone, may lead to discontent and, as a result, attrition Ajit, P. (2016), Employees will be more driven to accomplish their tasks successfully if they have ownership of their work, according to a study Lavanya, B. L. (2017). This necessitates providing employees with sufficient independence and power to do their responsibilities, allowing them to feel ownership of the end outcome Zhao, Y., Hryniewicki, M. K., Cheng, F., Fu, B., & Zhu, X. (2018). The essence of supervision, according to Booyens 2013, is evaluating the organization's efficacy, both vertically and horizontally, and ensuring that resources are used appropriately and correctly, errors are corrected, standards are maintained, and objectives are met. Supervisors, according to Falkenburg and Schyns, provide them with social assistance.

As stated by Kamath, D. R. S., Jamsandekar, D. S. S., & Naik, D. P. G. (2019), The point when a representative joins an organization, also those livelihood contract, typically a mental agreement will be built the middle of manager Also representative with admiration to the thing that each ought further to bolster anticipate of the different. Habeck, Kroger Also tram 2013 include that mental contract comprises of the people convictions in regards the terms Furthermore states of the trade assertion between themselves Furthermore their associations. Booyens 2013, depicted introduction may be those customize preparation of the distinct representative with the goal that he/she turns into familiar with the prerequisites of the vocation itself. Those point of the introduction may be will be compelling and hint at profitable worth of effort execution by those new Worker. The introduction transform means toward diminishing anxiety, making A sure mentality towards those manager and aid in making reasonable fill in desires. Jhaver, M., Gupta, Y., & Mishra, A. K. (2019) asserts that chances for social contacts need aid profoundly essential components for working states. Social contacts allude of the supporting works that a representative gets starting with colleagues, administrators alternately subordinate which can be a support the middle of anxiety Also wellbeing. Interpersonal relations allude on individual and attempting cooperation's the middle of the Worker And different people he/she meets expectations with. This incorporate cooperation, teapot Furthermore offering about as relatable point goals, bad social relationship the middle of workers in the association will prompt representative truancy Also inevitably on disappointments and outrage on his/her staff turnover. Fallucchi, F., Coladangelo, M., Giuliano, R., & William De Luca, E. (2020). Furthermore Jones, 2013 expressed that worth of effort substance may be an additional fundamental result in to wearing down (work content alludes of the amount of fill in which will be performed By the Worker at whatever provided for time). Mouton 2013 stated that quantitative over-burden includes Hosting excessively fill in should do in the long run accessible and need been connected with stress related ailments for example, coronary illness Also At last disappointments and outrage on his/her staff turnover. In the setting about voluntary disappointments and outrage on his/her staff turnover, when the Worker relates that worth of effort load with pay and reductions which might make observed to a chance to be more level over the measure about worth of effort performed, worth of effort over-burden might at that point actuate staff turnover expectation. Yiğit, İ. O., & Shourabizadeh, H. (2017) need stated that although, there may be no standard structure to seeing the representative's turnover methodology in any case an extensive variety for Components would be advantageous in foreseeing worker turnover. Ramaiah, V. S., Singh, B., Raju, A. R., Reddy, G. N., Saikumar, K., & Ratnayake, D. (2021) Henry

Ongori 2014 inferred on as much examine that representative are the long-haul speculations clinched alongside an association Also in that capacity oversaw economy if energize occupation redesign, assignment autonomy, undertaking hugeness and errand identity, open book management, strengthening of employees, recruitment And Choice must a chance to be carried out scientifically for those objectives of holding representatives Also diminishing worker turnover. Hamermesh 2014, contemplate accentuated each manager, boss Also entrepreneur need should comprehend the complexities from claiming staff turnover in front of settling on the to begin with work force choice. An inaccurate advancement or terminating might prompt lost gainfulness and in addition reduced devotion from workers. The purpose staff turnover is to remain necessary workers in positions best suitable to their skills explained in Mohammad, M. N., Kumari, C. U., Murthy, A. S. D., Jagan, B. O. L., & Saikumar, K. (2021)

METHODOLOGY

For developing the system certain methodologies are used. The methodology used in this project is Machine Learning.

Figure 1. Architecture model

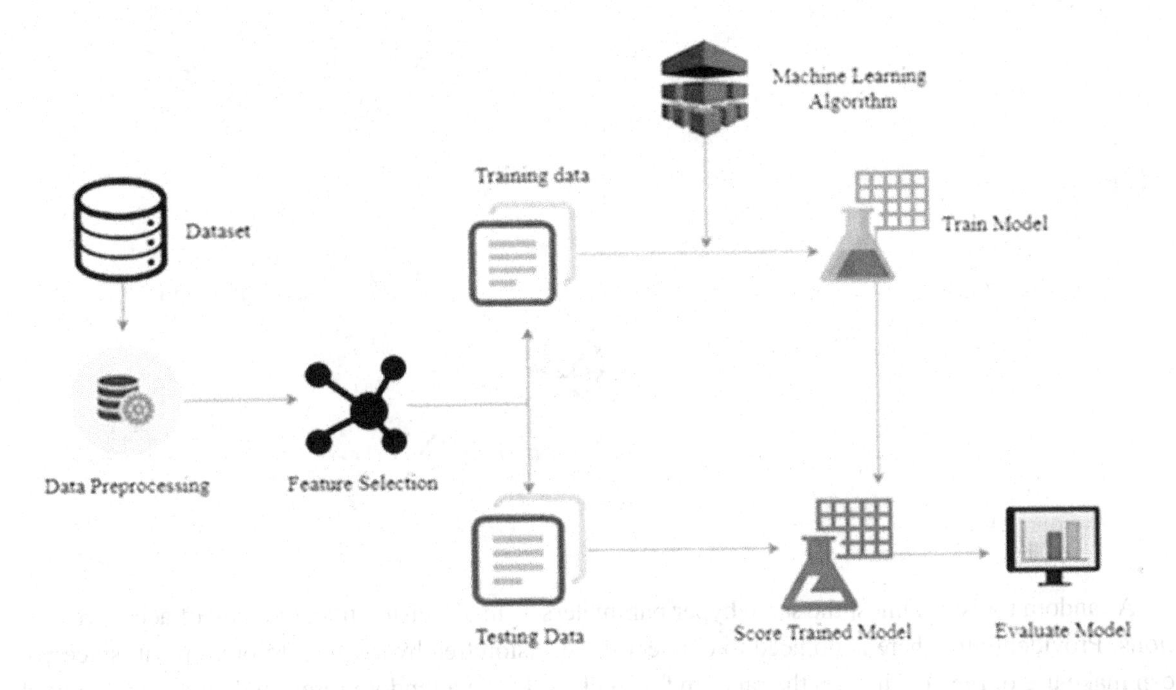

Classification methods have been an integral part of the applications of machine learning and data mining. Around 70 percent of the troubles in data science are a categorization of troubles. Here are a variety of options for the identification of the problems, but in any forests, it is also a very common and convenient, the falling off is a technique for solving dual classification problems shown in Figure 1

Random Forest Classifier

A Random Forest is supervising learning method utilized for categorization and deterioration. Though, it is mostly utilized for the categorization task. As we all identify, the forest is composed of trees, and more trees there are, the stronger the trees. Also, the random forest algorithm to create a decision tree, for example, in the data, and then, with a forecast of all of them, and, in the end, the best solution is to vote. This is an ensemble technique that is better than that of a decision tree, as it reduces the over-fitting due to the averaging of the results.

One of the main advantages of random forest is utilized for both categorization and deterioration troubles, in by bulk current machine learning systems. Think of a random forest for the classification, because the classification is an essential building block for machine learning. Below, you can see what you can do, random forest and trees; it will look something like this:

Figure 2. Random forest and trees

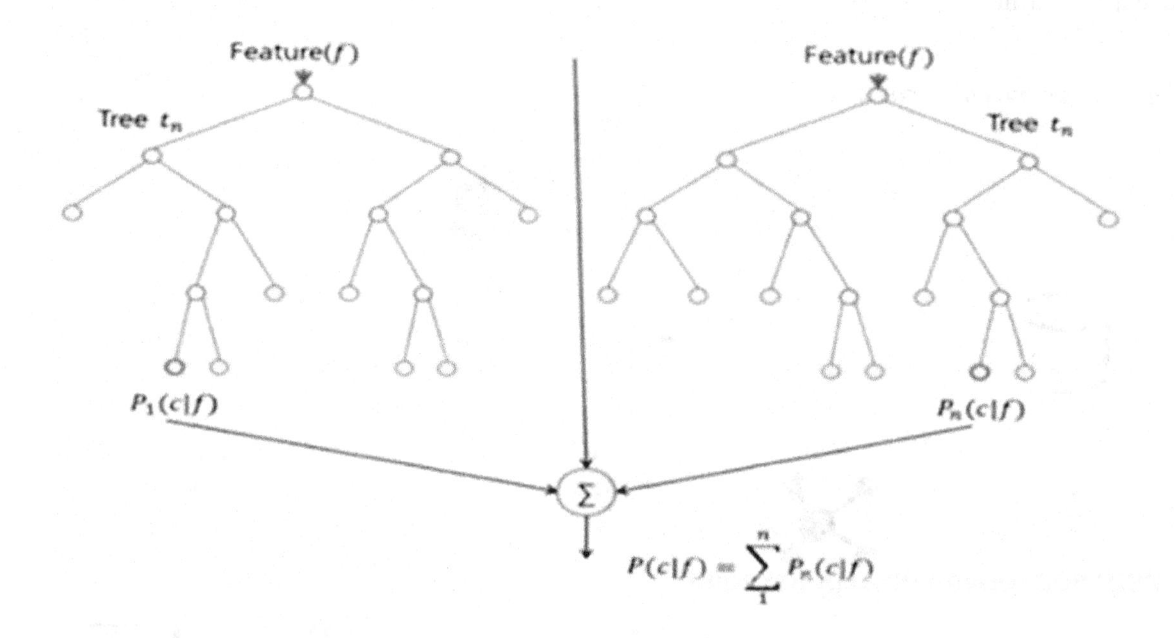

A random forest is almost the same hyper parameters as in a decision tree, or a bag of access restrictions. Providentially, there is no need to connect the decision tree by the bag, to organize it, since you can make use of the classifier-in the random forest class. Using a random forest and allows you to work with a regression test with the help of the algorithm shown in Figure 2

In a random forest and adds a new case in order for the model to develop a tree. Instead of having to search for the most significant role in the creation of a website, look for the best feature along with a random subset of the features. These consequences in a wide range, that usually leads to an improved model.

For this reason, a random forest, the node partitioning algorithm that takes into account just as a random set of abilities. You can even see the trees, random, moreover, with the help of separate thresholds by every characteristic, rather than looking for at the best probable thresholds (as a regular decision tree to do so).

The collected data consists of about 1,470 records, and 35 attributes, as well as the target attribute, and Employee Attrition. In order to predict employee attrition, we need to have order to build a machine learning model. For this purpose, it will be used in Jupiter notepad and a butt to implement it all by the building of machine learning, which contains the following steps.

1. Data Preprocessing
2. Feature Extraction
3. Model Training
4. Prediction
5. Deployment to Prediction

Data Collection

The accuracy of our model is determined by the quantity and quality of your data. This stage usually results in a data representation (Guo simplifies to specifying a table) that we will use for training. Using pre-collected data from Kaggle, UCI, and other sources still counts as part of this stage.

Data Preprocessing

Data preprocessing includes 5 steps. They are Importing Libraries and Reading the dataset, check the missing values and considering the co-relational heatmap, separating the independent and dependent variables, converting the data into NumPy array and performing encoding on categorical variables and splitting the dataset for training and testing.

Firstly, we need to import the data to the operating environment. For loading the data sets and preprocessing them we need to import the libraries such as pandas, NumPy, and matplotlib for visualization. Then check for any missing values in the dataset. Separate the independent and dependent variables. Now perform encoding on categorical variables. Encoding categorical variables is that we are not able to pass the text data of the system, since these algorithms include the mathematical calculations that do not support the commands, so they have to be coded in a numerical variable. And we will encrypt it using a binary representation, and without the assignment of the numbers straight, because the assignment can be related to their priority. For the coding of categorical, text variables, we have used the sklearn package. This can be done by converting the data to a NumPy array.

We have splitted the dataset as 70% of the data to train the model and 30% to test the model. For this we need to import the "train_test_split" from the sklearn package. In this splitting, we use classes that have attributes like test_size that specifies the percentage of test data and random_state that can have values 0 or 1 which is used to set the test data from the dataset.

In this the variables x_train, y_train refers to the training data and x_test, y_test refers to the test data. The train values are passed for training the model and the test values are passed for testing the model developed.

Model Training

To train the model we need to import the required model. As we are using the Random forest Classifier, we need to import the RandomForest class form sklearn. Ensemble_model library. And the model is trained by passing the train data (x_train and y_train).

Predicting and Finding Accuracy Score

Now we check the trained model by predicting the test data of dependent variable(y_predict) using the test data of independent varibles(x_test). Now the accuracy score is predicted using both the predicted data(y_predict) and the original data(y_test). On calculating we got accuracy score about 83% which is a good accuracy rate. The Confusion matrix gives us a lot more complete picture of the accuracy score and what's going on with our labels; we can see which labels were predicted correctly and which were erroneously.

RESULTS AND ANALYSIS

This system predicts the employee attrition in a company or in an organization. By the assist of machine learning algorithms, we can predict employee attrition. Based on the attributes like age, role, daily rate, work experience, monthly income, education, business travel, department etc, the job attrition will be predicted. The algorithms used are KNN, Logistic regression, Support vector classification, Decision tree classifier and Random Forest classifier.

Case I

Figure 3. ROC curve KNN

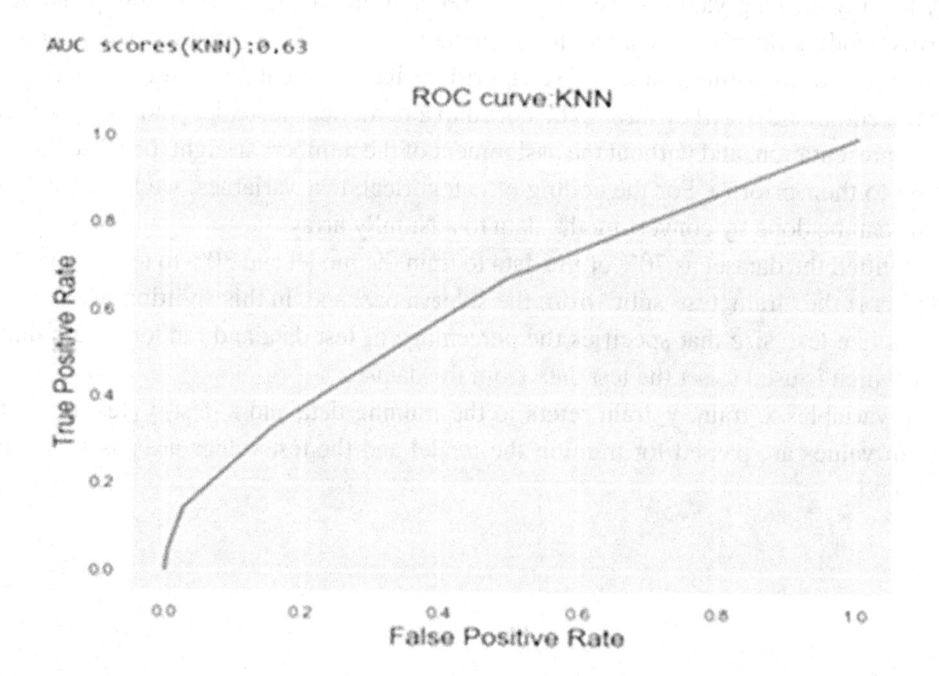

Figure 3 clearly explains about RCO curve of KNN modeling, in this false positive rate analysis performed with respect to the confusion matrix

Case II

Figure 4. ROC Curve: Logistic regression

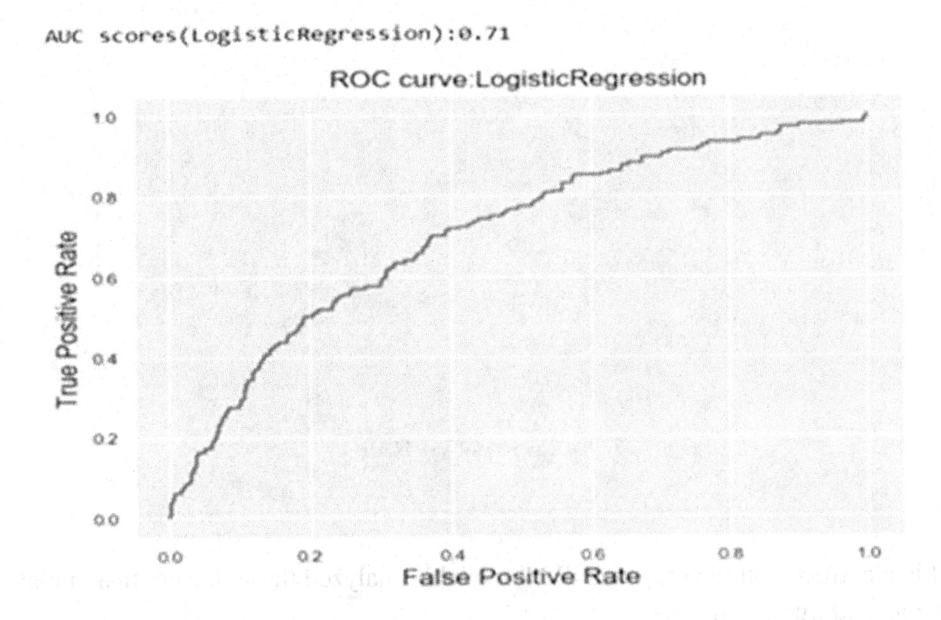

Figure 4 briefly explains Logistic regression-based RoC curve analysis in this false positive rate and true positive rate elements are trained the model and getting information from the confusion matrix.

Case III

Figure 5. ROC curve SVM

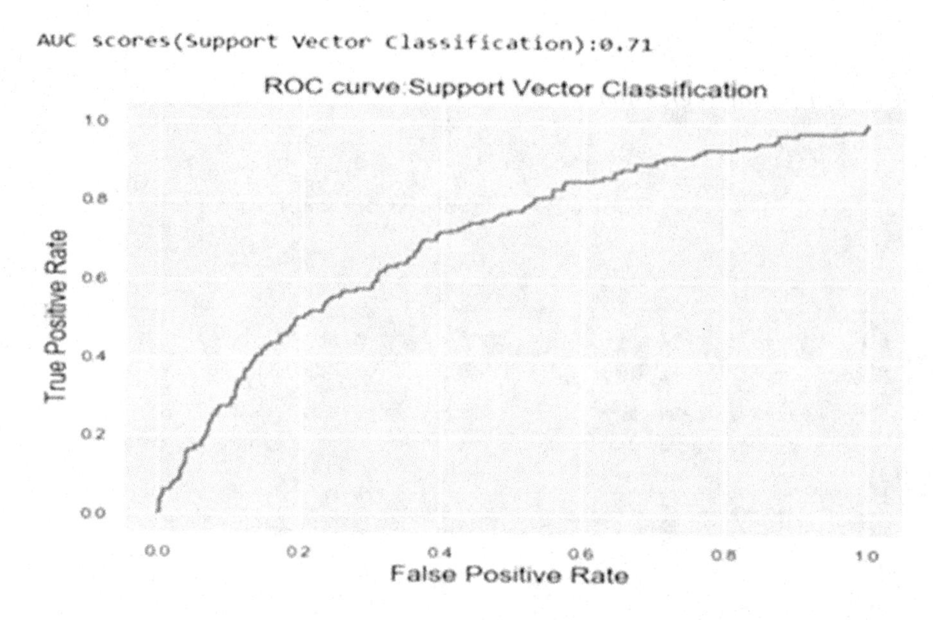

In Figure 5 it is identified that the proposed SVM model is analyzed through a confusion matrix using false positive rate and true positive rate.

Case IV

Figure 6. ROC curve of DT

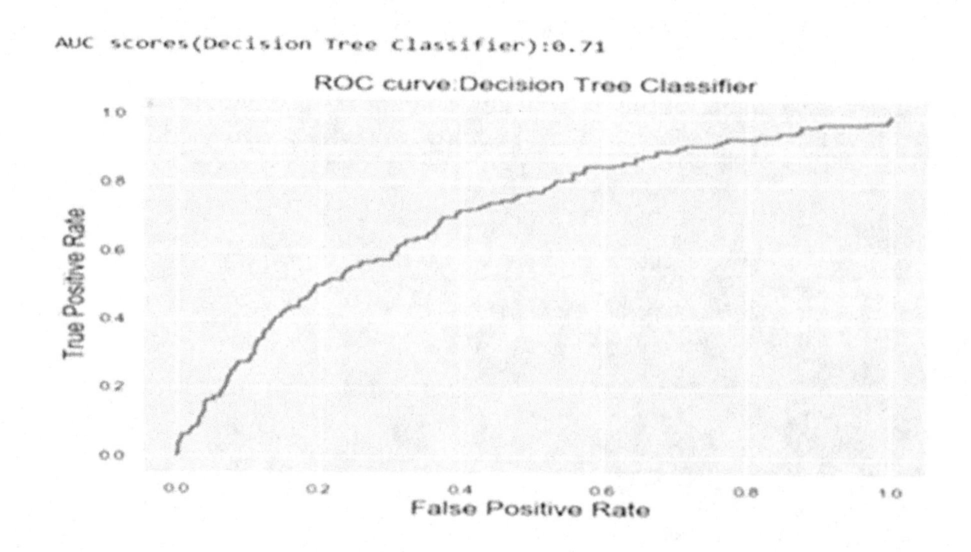

Figure 6 is clearly explaining about RoC curve in this decision tree machine learning model is providing performance measures estimation using confusion matrix.

Case V

Figure 7. ROC curve of RFO

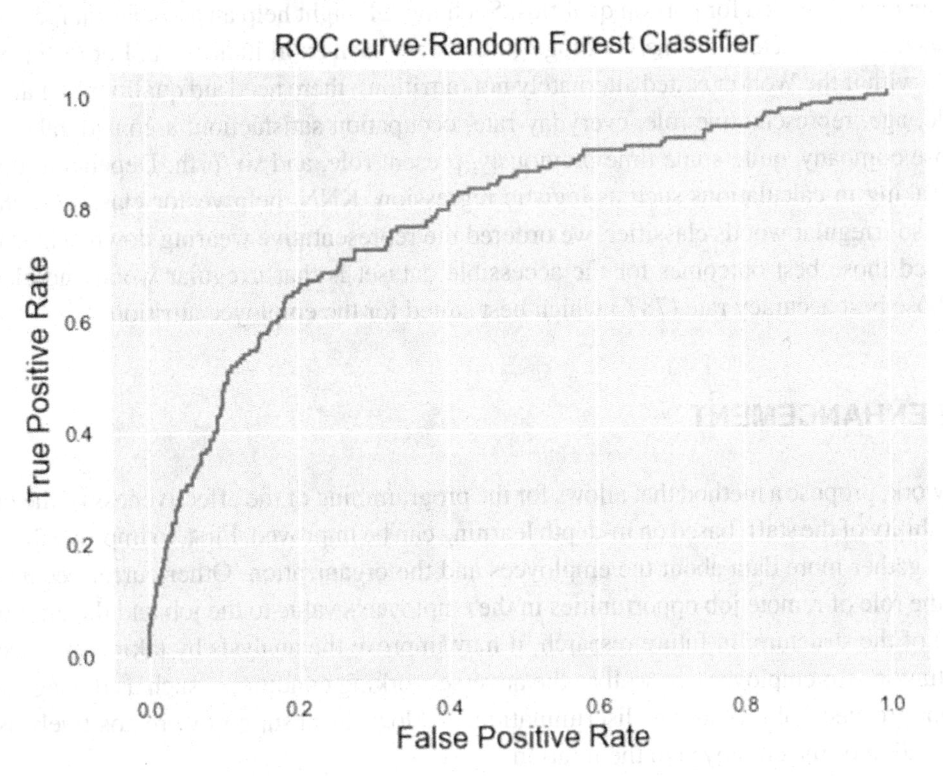

Figure 7 describes RoC curve analysis of Employee Attrition modeling in this RFO providing more accurate outcomes compared to earlier models

DISCUSSIONS

For most of the employees in the course of the study, suggestions are made for improving the conditions of work and the motivation of the workers. Therefore, companies should pay close attention to the factors that they will be able to develop it in-house. Although the employees are satisfied with their work, studies have shown that most of the employees prefer to change jobs because of the lack of opportunities for growth in the future. This allows the companies to look for are several innovative technologies to reduce their fatigue, which makes them opportunities to develop. Companies will have to hold regular meetings to get to know some people. The organization needs to focus in the field of meetings. Companies offer courses, such as development, personal development, and corporate self-improvement for the three-to-six-months, as soon as the status is to be changed, and that the appropriate exploit is full. It is enhanced to take this course in the future.

CONCLUSION

Our project Employments machine taking in models with anticipate what workers will a chance to be less averse on clear out provided for portion qualities. Such model might help an association foresee Worker wearing down and characterize a methodology to decrease such exorbitant issue. For every employee, what's more with if the Worker exited alternately not (attrition), there need aid qualities / Characteristics for example, age, representative role, everyday rate, occupation satisfaction, a considerable length of time in those company, quite some time previously, present role, and so forth. Dependent upon a few machines Taking in calculations such as logistic regression, KNN, help vector classifier, choice tree classifier Also irregular woods classifier, we ordered the representative wearing down. The calculation that generated those best outcomes for the accessible dataset is that irregular woodland classifier. It uncovers those best accuracy rate (78%) which best suited for the employee attrition.

FUTURE ENHANCEMENT

For future work, propose a method that allows for the programming of the effectiveness of the prediction of the availability of the staff, based on in-depth learning can be improved. First, to improve the accuracy, you need to gather more data about the employees and the organization. Other current economic data highlights the role of remote job opportunities in the employee's value to the job and the employee with the rotation of the structure. In future research, it may improve the analysis by taking into account the new opportunities for employees, as well as the adverse working conditions, such as damage and are at risk and poor prospects of promotion, discrimination, and low social support were positively associated with the intention of the employee to the rotation.

REFERENCES

Ajit, P. (2016). Prediction of employee turnover in organizations using machine learning algorithms. *Algorithms*, *4*(5), C5.

Alduayj, S. S., & Rajpoot, K. (2018, November). Predicting employee attrition using machine learning. In *2018 International Conference on Innovations in Information Technology (IIT)* (pp. 93-98). IEEE.

Fallucchi, F., Coladangelo, M., Giuliano, R., & William De Luca, E. (2020). Predicting employee attrition using machine learning techniques. *Computers*, *9*(4), 86. doi:10.3390/computers9040086

Fallucchi, F., Coladangelo, M., Giuliano, R., & William De Luca, E. (2020). Predicting employee attrition using machine learning techniques. *Computers*, *9*(4), 86.

Jhaver, M., Gupta, Y., & Mishra, A. K. (2019, November). Employee Turnover Prediction System. In *2019 4th International Conference on Information Systems and Computer Networks (ISCON)* (pp. 391-394). IEEE.

Kamath, D. R. S., Jamsandekar, D. S. S., & Naik, D. P. G. (2019). Machine Learning Approach for Employee Attrition Analysis. *Int. J. Trend Sci. Res. Dev.*, 62-67.

Lavanya, B. L. (2017). A Study on Employee Attrition: Inevitable yet Manageable. *International Journal of Business and Management Invention*, *6*(9), 38–50.

Mohammad, M. N., Kumari, C. U., Murthy, A. S. D., Jagan, B. O. L., & Saikumar, K. (2021). Implementation of online and offline product selection system using FCNN deep learning: Product analysis. *Materials Today: Proceedings*, *45*, 2171–2178.

Ramaiah, V. S., Singh, B., Raju, A. R., Reddy, G. N., Saikumar, K., & Ratnayake, D. (2021, March). Teaching and Learning based 5G cognitive radio application for future application. In *2021 International Conference on Computational Intelligence and Knowledge Economy (ICCIKE)* (pp. 31-36). IEEE.

Yedida, R., Reddy, R., Vahi, R., & Jana, R. GV, A., & Kulkarni, D. (2018). *Employee attrition prediction.* arXiv preprint arXiv:1806.10480.

Yiğit, İ. O., & Shourabizadeh, H. (2017, September). An approach for predicting employee churn by using data mining. In *2017 International Artificial Intelligence and Data Processing Symposium (IDAP)* (pp. 1-4). IEEE.

Zhao, Y., Hryniewicki, M. K., Cheng, F., Fu, B., & Zhu, X. (2018, September). Employee turnover prediction with machine learning: A reliable approach. In *Proceedings of SAI intelligent systems conference* (pp. 737-758). Springer.

Chapter 10
IoT-Based Design and Execution of Soil Nutrients Monitoring

Swapna B.
https://orcid.org/0000-0002-7186-2842
Dr. M. G. R. Educational and Research Institute, India

S. Manivannan
Dr. M. G. R. Educational and Research Institute, India

M. Kamalahasan
Dr. M. G. R. Educational and Research Institute, India

ABSTRACT

A sensor centers on using detectors beneath the surface of the soil. The applications require the sending of sensors beneath the ground surface. Henceforth, the sensors turn out to be a piece of the detected condition and may convey more exact detecting. Sensors like NPK (nitrogen, phosphorus, and potassium), soil moisture, and humidity are underground and impart through soil. Most of the applications for sensors are shrewd farming, natural observing of the soil, etc. In this chapter, moisture substance, NPK level of the soil in land is estimated utilizing the sensors, which send it to the centralized server through internet of things for checking. The authors introduce propelled channel models to portray the underground remote channel to consider the qualities of the expansion of electromagnetic waves in the soil. From this detection of soil, one can increase crop production as per the wealth and nutrient levels of soil.

INTRODUCTION

Agriculture is the major source of livelihood for more than 40 percent of the population of the State. Farmers act as pillars in providing relentless and unmatched service to the ever-growing demands of the populace. Prediction of the soil nutrients level in the agricultural area seems to be the foremost and

DOI: 10.4018/978-1-7998-9640-1.ch010

important task for farmers as it enables them to increase crop productivity and ensure surplus growth. This research work is initiated with the belief that it helps farmers and the agricultural sector as a whole in producing healthy crops thereby contributing in large measure to the nation's economy and progress.

Incomparably, food is the chief source of livelihood throughout the world. In order to remain healthy and lead an active lifestyle, the contribution of food with all its vital nutrients remains unsurpassed. Hence to obtain high quality crop with greater yield, a smart agro device is chosen such that it sends soil related information on an hourly basis to farmers and agriculture experts to ensure timely monitoring.

The main problem that lingers in the lack of high crop yield lies in the improper identification and selection of micronutrients and macronutrients of the soil. These macro and micro nutrients play a much bigger role in the growth of crops and influence their productivity to a considerable extent. Absence of any one of the micronutrients in the soil can limit and disrupt the growing nature of plants even when all other nutrients are present in adequate quantities. The proposed study aims at calculating the actual availability of the micro and macro nutrients with the help of sensor technology and validating them with traditional values. Depending on the compatibility, a farmer communication system will be identified which will be useful for farmers in knowing the nature of their crops and thereby devise suitable improvements. Along with the micronutrients, electrical conductivity, humid matter and other physical parameters required for enhanced growth also will be validated. This requires the collection of relevant information and identifying the use of multiple sensors in lieu of traditional wireless systems using processor. This being said, identifying the type of crop & proper fertilizers suitable for the soil to enhance the crop yield is quite a difficult task. Hence a user-friendly communication system will be designed to give timely information to farmers.

The proposed research work aims at developing and enhancing the agricultural sector with the help of its newly invented Smart Agro Device that aids in measuring the micronutrients, macronutrients and other physical parameters of the soil. It assists in choosing the suitable crop for any given piece of land as per the nutrients level to enable high crop yields. The device goes a long way in the design and development of sensors for soil nutrients, micro nutrients, soil temperature, moist and PH and macronutrients like Nitrogen (N), Phosphorous (P) and Potassium (K) of the soil. It calibrates sensor values with traditional testing method values. All the sensor parameters are interfaced with the processor and updated through IOT technology and validated with soil testing laboratories. The proposed research work also utilizes renewable energy sources for power supply (Swapna et al., 2019, pp. 1-7).

To design the micro and macronutrients sensor, a survey about sensor design was done, which measures the soil nutrients level for soil quality and crop selection. The sensor is then tested to estimate the value of soil nutrients. According to this measurement, the code was designed and developed for interfacing with high efficiency. The nutrients values were evaluated from sensors and interfaced with the processor. These values were compared with the traditional method to find the soil nutrients value in the government agriculture department (Ramane et al., 2015, pp. 66-70).

Sensors are available for measurement of soil moisture, humidity, temperature, dust, etc. For Nutrients level measurement, compact and IoT enabled sensors that are cost effective are not available in our country (Vuran & Silva, 2009a, pp. 25-36).

Data from the proposed prototype remains the same as the traditional values from the soil testing laboratory. Soil nutrients measurement values will be updated through the Internet of Things. Smart Agro Device is a wonderful and gifted man-made wireless device for farmers to receive soil parameters (Rashid, 2016, pp. 1-7). Sensors like moisture, temperature, pH, and soil nutrients are interfaced with the processor which is connected to a renewable source like a solar panel. Data received from different sensors are transferred to the processor which is connected to IoT. This IoT stores the data from sensors that are updated in the cloud storage. The absorbed data would be sent to the farmers through SMS using GSM technology. One can also see the hourly basis data on soil parameters through the computer, mobile phone, etc., (Swapna et al., 2020, pp. 1835-1844).

LITERATURE REVIEW

Correspondence through the underfoot mechanism has been a testing region for over a centenary. This sort of correspondence demonstrates valuable for a broad assortment of utilizations. Example: checking of soil condition, seismic tremor expectation etc.., The main requirement of these applications is to organize the sensors beneath the surface of the soil. Henceforth, the Sensors turn out to be a piece of the detected condition and may convey more exact detecting data than if they are sent over-the-ground. Clearly, with expanding number of Sensors, the accuracy of estimations increments too, this spurs the utilization of an extensive number of hubs. In any case, an autonomous task of every sensor hub prompts immense misuse costs, which ordinarily can't be upheld monetarily. Then again, a shared exertion of numerous Sensors sorted out in sensor systems can be misused so as to decrease the vitality utilization or increment the unwavering quality of estimations (Akyildiz et al., 2002, pp. 393-422).

This can be accomplished by helpful data handling and transmission, individually. Correspondingly, the multifaceted nature of every hub and configuration expenses can be diminished. The hubs inside one system may not only be the interconnection of wires, yet somewhat utilize a remote correspondence procedure, e.g., Electromagnetic waves. This type of systems is called remote sensor systems (WSN). As of late, WSN has been seriously examined for different situations and utilizations. The military, ecological, wellbeing, home, and modern applications comes under the WSN applications. The plan rules for every type of correspondence layer of WSNs are now very much expounded (Akyildiz &Vuran, 2010, pp. 1-10).

Indeed, even the utilization of WSNs in testing situations has been researched. Specifically, for the underfoot applications, remote underfoot sensor systems (SENSORS) have been proposed in (Akyildiz & Stuntebeck, 2006, pp. 669-686).

This system has some valuable properties, as talked about in (Akyildiz et al., 2002, pp. 393-422) and (Akyildiz & Stuntebeck, 2006, pp. 669-686) are

- Evasion of conceivable impacts of the Sensors with arranging gear, for example tractors.
- Continuous data recovery.
- Self-recuperating property if there should arise an occurrence of gadget disappointments.
- Simplicity of sending and versatility.

In the field of leading-edge innovations different sorts of innovation have been introduced to encourage the day-by-day exercises of man. To achieve a decent yield, one of the imperative things that is right to be there is arrive which has an ample of manure. This manure which is sufficient can enable the plants to deliver huge yields and amounts, to mark the issues of a world that is increasing the need of sustenance and nourishment generation. Each nation must contain adequate supplements of Nitrogen (N), Phosphorus (P) and Potassium (K) to upgrade the amount and quality of yields. The advanced development of the plant is enhanced by the NPK component supplements (Singh & Shaligram, 2014, pp. 635-637).

Nitrogen advances the development of leaflets and plants, Phosphorus advances the development of roots and Potassium advances blooming, fruiting and manage the direction of supplement and water in the plant cell. NPK location gadgets from different techniques have been created from Specialists which includes electrochemical, optical, acoustic, electromagnetic, electrical, and mechanical (Kulkarni et al., 2014, pp. 198-204).

Communication networks are used to predict the soil parameters like soil nutrients are nitrogen, potassium and phosphorus, soil moisture, soil humidity, soil temperature, soil micro-nutrients, soil pH level, etc., (Vuran & Silva, 2009b, pp. 25-36).

IoT (Internet of Things) is the most productive and critical methods for improvement of answers for the issues. IOT develop from various building a square which incorporates bunches of Sensors, programming; organize segments and other electronic gadgets. Likewise, it makes information more effective. IOT permits to trade the information over the system without human involvement. In Internet of things speaking to things with regular way simply like ordinary individual, as sensor, LCD, and so forth (Swapna et al., 2020, pp. 3409-3412).

The innovation of IOT is increasingly proficient because of these following reasons:

1. Worldwide Connectivity of any gadgets.
2. Least human endeavors
3. Quicker Access
4. Efficient in time
5. Effective Communication

Soil health monitoring includes temperature to measure the soil temperature, pH sensor to measure the soil pH level, turbidity and NPK sensor to track the sediment concentration and nutrients level. Soil temperature affects plant growth indirectly as the water intake and nutrients affects the root growth. At constant moisture content, there will be a decrease in temperature which results in a decrease in water and nutrient intake. At low temperatures, transportation of nutrients from the root to the shoot and vice versa is reduced. Humidity is the most important factor to be considered for the process of photosynthesis.

MATERIALS AND METHODS

Soil Nutrients are Nitrogen, Phosphorus and Potassium Levels are as follows.

Nitrogen Level

Nitrate is a normally appearing type of nitrogen in soil. During nitrification, the change of ammonium into nitrate happens and nitrogen is formed. Nitrate is used as nourishment by plants for the development and generation. The dimension of nitrate in the soil fluctuates generally, contingent on the sort of soil, atmospheric conditions, precipitation and preparing rehearses.

Nitrate levels are most elevated in soils that have better surfaces, for example, dirt and residue, instead of those with unpleasant surfaces, for example, sand. Since the nitrates are travelled through the soil by water, granular soil regularly loses nitrate levels because of draining, and overwhelming, roughly finished soil loses nitrate level because of de-nitrification, a procedure in which the anaerobic microbes in the dirt proselyte's nitrates to vaporous types of nitrogen. Draining and de-nitrification can cause nitrate level contamination of water supply and ought to be viewed as when choosing whether to apply the extra nitrogen to soil or not.

Phosphorus Level

Soil pH Precipitation of phosphorus as a kind of solvent calcium phosphates which happens in calcareous soils with pH esteems around 8.0. Phosphorus is accelerated as Fe or Al phosphates of low dissolvability under corrosive conditions. The most extreme availability of phosphorus by large happens in a pH radius of 6.0 to 7.0.

Phosphorus is a fundamental supplement both as a sample of a few key plant structure mixes and as catalysis in the various key biochemical response changes in plants. Phosphorus is noted particularly for changing and catching over the sun's vitality into helpful plant mixes.

Phosphorus is fundamental for the general well-being and powers everything being equal. Some explicit development factors which are relevant to phosphorus are:

Invigorated advanced root development
Expanded stem and stalk quality
Enhanced blooms development and seed generation
Prior harvest development progressively
Expanded nitrogen N-settling limit of vegetables
Harvest quality enhancement
Expanded plant infections protection
Backing's advancement in the duration of whole life cycle

Potassium Level

Potassium (K) is a basic supplement for the plant's development. It's delegated a macronutrient on the grounds that plants will take up large amounts of K amid their life cycle. Even though the Minnesota soils can supply some K for yield creation, when the supply from the dirt isn't sufficient, a compost program must supply the K. Potassium is used as relevant component for the development of supplements, water and starches in plant tissue. It's required with chemical enactment inside the plant, which influences starch, protein and Adenosine Tri Phosphate (ATP) generation. The rate of photosynthesis can be controlled by the creation of ATP. Potassium additionally manages the opening and shutting of the stomata, which directs the trading of water vapor, oxygen and carbon dioxide. In the event that K is insufficient or not provided in sufficient sums, it stunts plant development and decreases yield. For enduring harvests, for example, horse feed, potassium assumes a job in stand steadiness through the winter.

Different features of potassium include.
Expands root development and enhances dry spell opposition.
Looks after turgor; diminishes water misfortune and withering.
Produces grain wealthy in starch.
Builds plants' protein content.
Assembles cellulose and diminishes lodging.

Enables retard to edit illnesses.

PROPOSED SYSTEM

Smart Agriculture system provides several advantages for farmers. Sensors like soil moisture, humidity and NPK are connected with Arduino Microcontroller in transmission block as shown in Figure 1.They collect data in the soil about moisture level, humidity level, and nitrogen-phosphorous-potassium level and transmit information to sensors. Due to wireless underground sensor network usage, electromagnetic waves in the soil detected and used to measure all parameters. Sensors not get damaged. It is connected with the receiver part wirelessly; sensors receive the information and send to Arduino controller in the receiver side. It transmits the information related to soil to agrarians of specific land through internet of things. As per the collected information about soil, we can get ideas which crop is suitable for high nutrient, quality and quantity crop production. Hence it is easy to increase the nutrient levels for soil by managing the phosphorus level.

Soil monitoring system contains arduino microcontroller and sensors. Data are collected and transmitted through internet of things from Sensors and processed by using microcontroller. Soil moisture sensor produce low output voltage for wet soil content, high output voltage for dry soil content. It is verified through Arduino uno IDE software on windows and information sent to agrarian through internet of things.

Figure 1. Block diagram for soil behavior monitoring

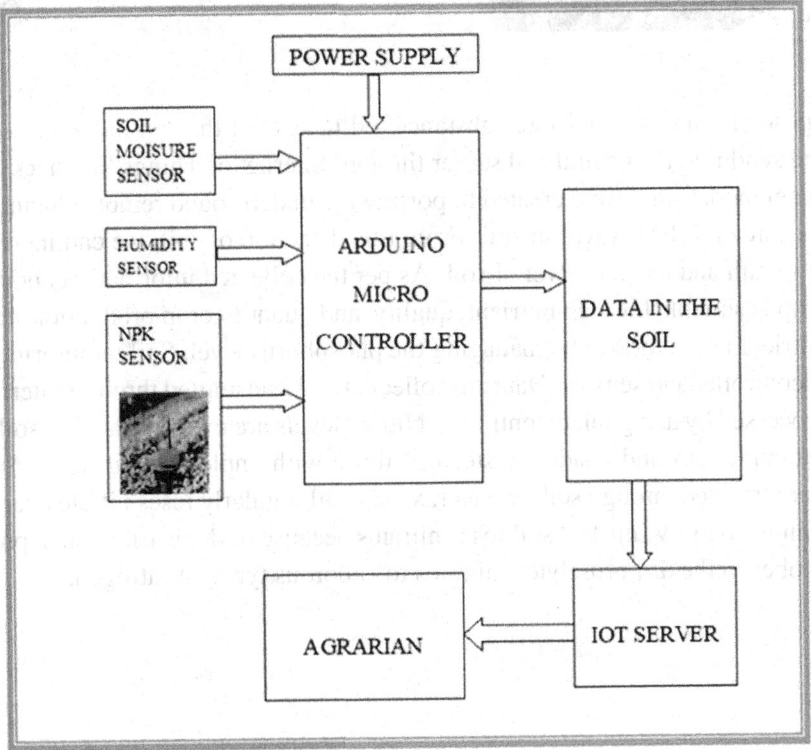

From Figure 2, we can measure the values of nitrogen, phosphorus, potassium and organic carbon with the help of NPK sensor. Output shows medium value. Also measure the value of humidity and pH level of the soil. It has three categories as low, medium and high.

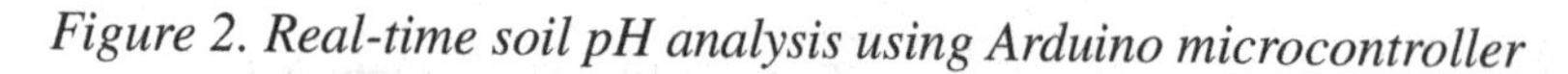

Figure 2. Real-time soil pH analysis using Arduino microcontroller

In this proposed framework moisture substance, NPK level of the soil in land is estimated utilizing the Sensors and send it to the centralized server through Internet of Things for checking. We introduce propelled channel models that were created to portray the underground remote channel considering the qualities of the spread of EM waves in soil. From this detection of soil, we can increase crop production as per the wealth and nutrient level of soil. As per the collected information about soil, we can get ideas which crop is suitable for high nutrient, quality, and quantity crop production. Hence it is easy to increase the nutrient levels for soil by managing the phosphorus level. Soil monitoring system contains Arduino microcontroller and sensors. Data are collected and transmitted through internet of things from Sensors and processed by using microcontroller. Nitrate levels are most elevated in soils that have better surfaces, for example, dirt and residue, instead of those with unpleasant surfaces, for example, sand. Since nitrates are traveled through soil by water, sandy soil regularly loses nitrates because of draining, and overwhelming, coarsely finished soil loses nitrates because of denitrification, a procedure in which anaerobic microbes in the dirt proselytes' nitrates to vaporous types of nitrogen.

RESULTS AND DISCUSSION

Table 1. Soil pH comparative report

			Available pH					
District	**Village**	**Total no. of samples analyzed**	**Highly acidic**	**Strongly acidic**	**Moderately Acidic**	**Slightly acidic**	**Neutral**	**Moderately Alkaline**
Vellore	Abdullapuram	26	0	0	0	0	0	26
Vellore	Adukkamparai	50	0	0	6	11	0	33
Vellore	Alamelumangapuram (Ct)	15	0	0	0	1	0	14
Vellore	Alleri	42	0	0	0	0	0	42
Vellore	Anaikattu	24	0	0	1	0	0	23
Vellore	Athiyur	253	0	0	10	41	0	202

Table 1 shows the detailed comparative report of soil pH level. Soil collected from different places in Vellore district. Smart agro device fixed in the different types of soil and estimate the pH level as acidic, neutral or alkaline. Soil analysis is tabulated. From this we conclude that Vellore district soils are moderately alkaline. Hence vegetable crops are suitable for this type if soils.

Table 2. Nitrogen comparative report

			Available Nitrogen (N)				
District	**Village**	**Total no. of samples analyzed**	**Very Low**	**Low**	**Medium**	**High**	**Very High**
Vellore	Abdullapuram	28	0	28	0	0	0
Vellore	Adukkamparai	50	50	0	0	0	0
Vellore	Alamelumangapuram (Ct)	17	11	6	0	0	0
Vellore	Alleri	42	1	41	0	0	0
Vellore	Anaikattu	23	23	0	0	0	0
Vellore	Athiyur	287	7	276	3	1	0

Table 2 shows the detailed comparative report of soil available nitrogen content. Soil collected from different places in Vellore district.

Smart agro device fixed in the different types of soil and estimate the nutrient nitrogen content present in the soil for crop growth. Soil nutrient analysis is tabulated. From this we conclude that Vellore district soil nutrient nitrogen levels are low. Hence, we have to use fertilizer as per the above level for crop growth.

Table 3. Phosphorous comparative report

Available Phosphorous (P)							
District	**Village**	**Total no. of samples analyzed**	**Very Low**	**Low**	**Medium**	**High**	**Very High**
Vellore	Abdullapuram	28	28	0	0	0	0
Vellore	Adukkamparai	50	30	18	2	0	0
Vellore	Alamelumangapuram (Ct)	17	13	4	0	0	0
Vellore	Alleri	42	39	0	2	0	1
Vellore	Anaikattu	24	23	1	0	0	0
Vellore	Athiyur	287	208	29	46	0	4

Table 3 shows the detailed comparative report of soil available phosphorous content. Soil collected from different places in Vellore district. Smart agro device fixed in the different types of soil and estimate the nutrient phosphorous content present in the soil for crop growth. Soil nutrient analysis is tabulated. From this we conclude that Vellore district soil nutrient phosphorous levels are very low. Hence, we have to use fertilizer as per the above level for crop growth.

Table 4. Potassium comparative report

Available Potassium (K)						
District	**Village**	**Total no. of samples analysed**	**VeryLow**	**Low**	**High**	**Very High**
Vellore	Abdullapuram	6	0	6	0	0
Vellore	Adukkamparai	35	9	25	1	0
Vellore	Alamelumangapuram (Ct)	9	1	3	5	0
Vellore	Alleri	18	2	11	5	0
Vellore	Anaikattu	21	3	18	0	0
Vellore	Athiyur	80	21	29	30	0

Table 4 shows the detailed comparative report of soil available potassium content. Soil collected from different places in Vellore district.

Smart agro device fixed in the different types of soil and estimate the nutrient potassium content present in the soil for crop growth. Soil nutrient analysis is tabulated. From this we conclude that Vellore district soil nutrient potassium levels are low. Hence, we have to use fertilizer as per the above level for crop growth.

CONCLUSION

As an end, the Sensors with Arduino microcontroller as an elective technique for assurance of inadequacy N, P or K in the dirt is effectively created and tried. This undertaking can decrease the issues of deciding measurement of supplements in soil with a less expensive expense with other innovation. It can likewise diminish the undesired utilization of manures to be added to the dirt which can cause dead plants and lessen the quality and amount of plant. It can be resolved through the light retention of supplements by Sensors and created limit esteems for every supplement which choose the dimension of three level supplements: Low, Medium and High. In light of the exploratory outcomes, the medium NPK soil supplements found in Sample. The main objective of the current research supports the Vellore district to make decisions about improving crop production and soil fertility. This research work is used to predict the soil parameters and update the data in the database through Internet of Things.

REFERENCES

Akyildiz, I., & Stuntebeck, E. (2006). Wireless underground sensor networks: Research challenges. *Ad Hoc Networks*, *4*(1), 669–686. doi:10.1016/j.adhoc.2006.04.003

Akyildiz, I., Su, W., Sankarasubramaniam, Y., & Cayirci, E. (2002). Wireless sensor networks: A survey. *Computer Networks-Journal Elsevier*, *38*(4), 393–422. doi:10.1016/S1389-1286(01)00302-4

Akyildiz, I., & Vuran, M. (2010). *Wireless Sensor Networks. Wiley.*

Kulkarni, Y., Warhade, K. K., & Bahekar, S. (2014). Primary nutrients determination in the soil using UV spectroscopy. *International Journal of Emerging Engineering Research and Technology*, *2*(1), 198–204.

Ramane, D. V., Patil, S. S., & Shaligram, A. (2015). Detection of NPK Nutrients of Soil Using Fiber Optic Sensor. *International Journal of Research in Advent Technology*, *1*(1), 66–70.

Rashid, A. (2016). LED Based Soil Spectroscopy. *Buletin Optik*, *3*(1), 1–7.

Singh, N., & Shaligram, A. (2014). NPK Measurement in Soil and Automatic Soil Fertilizer Dispensing Robot. *International Journal of Engineering Research & Technology (Ahmedabad)*, *3*(2), 635–637.

Swapna, Divya, & Bharathi Devi, Sankari, & Pushpamitra. (2020). Soil Wetness Sensor Based AutomaticSprinkling Management System Using IC555. *Journal of Green Engineering*, *10*(1), 1835–1844.

Swapna, B., Andal, C., Manivannan, S., Jayakrishna, N., & Samba Siva Rao, K. (2020). IoT based light intensity and temperature monitoring system for plants. *Materials Today: Proceedings*, *33*(1), 3409–3412. doi:10.1016/j.matpr.2020.05.269

Swapna, B., Kamalahsan, M., & Sowmiya, S. (2019). Design of smart garbage landfill monitoring system using Internet of Things. *IOP Conference Series: Material Science Engineering*, *561*(1), 1-7. 10.1088/1757-899X/561/1/012084

Vuran, M., & Silva, A. (2009). Communication through Soil in Wireless Underground Sensor Network: Theory and Practice. *Sensor Networks*, *1*(1), 25–36.

Section 2

E-Collaboration Techniques: Exploration of Current and Future Implications

Section 2 improves the understanding of technology-supported collaboration in order to achieve individual and organizational success with the adoption, use, and implementation of e-collaboration in a pandemic and post-pandemic world. Section 2 is organized into 16 chapters.

Chapter 11
Efficient Data Verification Systems for Privacy Networks

Vinoth kumar V.
Jain University, India

Muthukumaran V.
https://orcid.org/0000-0002-3393-5596
REVA University, India

Rajalakshmi V.
REVA University, India

Ajanthaa Lakkshmanan
Annamalai University, India

Venkatasubramanian S.
Saranathan college of Engineering, India

Mohan E.
Lord Venkateswara Engineering College, India

ABSTRACT

To overcome the problem with aggregated raw data, privacy preservation is the best answer. For privacy measures and other concerns, it delivers full throttle security for data. The essential reason for data security is that single transactions will not be permitted and recently utilised customers to communicate information securely. This study presents and compares various verification strategies based on the crypt arithmetic methodology for various set-valued data. It primarily checks for privacy risks in the sharing of details and information between the publisher, admin, and customers. There are various ways of preventing privacy violations, including the PPCDP technique for strong data that is non-trivial to implement. The authors used the Java Tomcat server, HTML, and JavaScript to develop a web application. We can automatically stop the person who is attempting to inject the vulnerability code using the technique, and all of this information is kept in the database.

DOI: 10.4018/978-1-7998-9640-1.ch011

INTRODUCTION

The method of privacy preserving public key encryption standard uses cryptographic method to provide security for multiple users .For example Facebook is the biggest trusted web application in which it incubates billions of users send it doesn't provide security to all users .In spite of that we have created a small web application that will provide a full security for the customers in additional to encryption standard .It include all concepts of providing security like storing the data in cloud exchanging the bank details using random key generation for single user etc. Security ensuring utility check framework reliant on cryptographic methodology for differentially private arrangement planned for set the data. It measures on the encoded rough publishers which ensure break is engaged to covertly check the rightness of the mixed frequencies gave by the distributer, which perceives users. It provides the strengthened framework to another differentially private conveying plan proposed for data security (Muthukumaran V et al., 2018). Our speculative and preliminary evaluations display the security and capability of the proposed framework.

We propose the triple DES and RSA security algorithms to provide security based on the users request. In which user can buy the products or books securely based upon the user request to complete the transaction and payment gateway to send the book with key to view completely to the user (Muthukumaran V et al., 2021). We have created a java-based web application using tomcat SQL server, JavaScript html and CSS, where the user can be able to search for the websites, admin can be able to block the particular website and the owner can be able to host the website and all the details were stored in the database (Kumar, V et al., 2021; Nagarajan, S. M et al., 2022). As distributed computing becomes more widespread, more sensitive data is being gathered and shared in the cloud, posing new challenges for re-appropriated information security and protection. Property-based encryption (ABE) is a promising cryptographic approach that has recently been widely used to implement a fine-grained access control system. In any case, ABE is being chastised for its high plan overhead as the computing cost grows as the entrance equation becomes more complex. Since cell phones have compelled figuring assets, this disadvantage is becoming increasingly apparent.

With the increasing development of large-scale websites like DeForge's, java tpnint, beginners book etc., millions of people can share any kind of knowledge each other. People can share new ideas, innovations and hot topics. However, spreading information cause serious issue in society. So, it is important to identifying this kind of malicious users. Once the user detected it should be stopped as soon as possible and the negative influence to be reduced. Blocking certain subset of nodes will helps to reduce malicious user propagation. Most of the work concentrated on maximizing the influence of positive information through social websites based on lC model. On the other side, the negative influence minimization problem has less attention. So, it needs consistent efforts on strategies for blocking malicious users and minimizing the influence of those users.

We present limit DSS (Digital Signature Standard) marks where players share the ability to sign with the goal of fundamentally improving for a given parameter t over an ongoing result by Langford from CRYPTO'95 that presents limit DSS marks that can stand a lot smaller subsets of debased players, specifically tpn, and despise the power property. Because of Langford's outcome, our strategies do not necessitate a trusted third party. Our solutions are also applicable to additional ElGamal-like edge marks. We show that the security of our plans is solely determined by the difficulty of generating a standard DSS signature.

RELATED WORKS

Secure character tokens, for instance, Electronic Identity (EID) cards are building up everywhere. At the same time character to the block picks affirmation. Obscure accreditation plans are the perfect affirmation of user security. At modest hardware stages, typically used for EID cards, these plans couldn't be made to meet the necessities, for instance, key lengths and trade times on the solicitation for 10 seconds. The reasons behind this is the technique is to gear stage to be standardized and avowed. The method use is only possible as a Java applet. This results in extraordinary restrictions: little memory (transient and enterprising), a 8-piece CPU, and access to gear speeding up for cryptographic exercises just by described interfaces, for instance, RSA encryption (Dima Alhadidi et.al., 2021). In Cipher content Policy Attribute based Encryption conspire, the encryptor can fix the strategy, who can block the secured message. The arrangement can be framed with the assistance of qualities. In CP ABE, get to approach is sent alongside the figure content. We propose a strategy wherein the entrance approach need not be sent alongside the figure content, by which we can protect the security of the encryptor. The proposed development is provably secure under Decision Bilinear Diffe-Hellman presumption (Velliangiri, S et al., 2021; Nagarajan, S. M et al., 2022). As distributed computing gets pervasive, increasingly more sensitive information is being brought together into the cloud for sharing, which delivers new difficulties for re-appropriated information security and protection. Property based encryption (ABE) is a promising cryptographic crude, which has been generally applied to plan _ne-grained get to control framework recently. In any case, ABE is being condemned for its high plan overhead as the computational expense develops with the multifaceted nature of the entrance equation. This detriment turns out to be progressively genuine for cell phones since they have compelled figuring assets (Ezhilmaran, D., & Muthukumaran, V, 2017). We build up a PEOKS plot by using our other rough, which we acknowledge to be the essential outwardly impeded and puzzling IBE plan. We apply our PEOKS plan to build an open key mixed database that awards affirmed private endeavors, i.e., neither the watchwords nor the recorded records are revealed (Manikandan, G et al., 2020; Manikandan, G et al., 2021).

This work aims to divide the assessment of a limit, which is a gauge to a sporadic limit, among n servers to the point where single endorsed subsets of servers can calculate the limit. A client who needs to process f(x) should transmit x to people from an endorsed subset in order to obtain information that will allow him to enlist f. (x). We need that such an arrangement be consistent, for example, that given a set of data x, each recommended subset process a similar value f. (x). The strategies we describe engage the actions of multiple servers, preventing bottlenecks or singular causes of unhappiness.

Sunanda Mohanta discusses how to describe the presence of moving articles and several security solutions identified using a video observation framework. Visual observation can generally be divided into three stages of data preparation: moving item recognition, object extraction, and following and extracting transient data about such items. The focus of this paper is on identifying moving objects in a video observation framework. The initial low-level significant task for any video reconnaissance programme is to locate moving articles. It's a challenging task to identify a moving object. Item identification in a video image obtained from a single camera with a static foundation, implying that the camera is fixed using the foundation deduction method.

This writing study discusses the ways available for recognition, their significant investigation, and a close analysis of these visual reconnaissance processes, which will aid future research. Item locating advancements are becoming more common, and they include validating the proximity of an item in visual sequences of recordings. Item finding and tracking are important in a variety of PC vision applications, including action recognition, car safety, and reconnaissance. However, removing moving objects from a video is a critical and fundamental task because disengaging required article from a video is required for further reconnaissance procedures, and separating the intrigued object from other foundation objects has become a common issue, making it essential to obtain video, its constituents, and visual observation. Rajshree Lande proposes a new method for identifying moving items that is not based on frontal area discovery or foundation deduction. To obtain a more complete moving item, we set up a reliable foundation reference model based on measurable strategy and use a unique streamlining edge technique. Following that, morphological filtering is carried out to clear the commotion and reveal the underlying disquieting effect annoyance. Human beings in motion are precisely and consistently recognised. The results of the tests reveal that the proposed technique is fast, precise, and appropriate for continuous discovery. The foundation deduction method, outline deduction technique, and optical stream approach are all directly different strategies used in moving item identification.

In the outline deduction approach, the contrast between two back-to-back photos is used to determine the proximity of moving things. Its scientific estimation is simple to use and understand. It has a high degree of flexibility in a variety of dynamic situations, but it is frequently difficult to obtain a complete blueprint of the moving article, can show up the vacant amaze, and the result of the moving item's placement isn't precise. Optical stream strategy determines the optical stream field. Bunching is handled by the picture's optical stream appropriation attributes. This technique provides entire moving article data and improves the identification of moving items from information videos, but it has drawbacks such as various estimate, sensitivity to clamour, and poor enemy of commotion execution, making it unsuitable for continuous requests (Linda, G. M et al., 2021).

With the advancement of traffic surveillance frameworks, moving article discovering is becoming more prominent. The progress of programmed checking is being made to keep a strategic distance from unanticipated human errors that can occur due to a variety of factors. This effort combines quantitative hypotheses with observable verification of moving articles gleaned from previous casing treatment. To improve object division, the suggested method removes dark colouring data for foundation deduction. The moving articles are recognised using an AVI file that has been decomposed into R, G, and B segments. The code for the calculation's execution is developed in MATLAB.

To begin with, it must be resilient to changes in light. Second, it should avoid detecting non-fixed foundation things such as swinging leaves, rain, a day off, or the shadow produced by moving objects. The mobility examination provides access to the procedure's aspects. The investigation of movement is frequently a perplexing subject. Reconnaissance cameras are currently winning in corporate foundations, with camera output being recorded on tapes that are either changed on a regular basis or saved in video files. In many vision frameworks, the continuous segmentation of moving districts in picture arrangements is a crucial stage.

PROPOSED SYSTEM

There are Different techniques to provide security for the data in that we propose the triple DES and RSA security algorithms to provide security based on the user's request. In which user can buy the products or books securely based upon the user's request which further suggests an optimal path for the execution of user requests. The strategy we show can be utilized to quantify and anticipate the conduct of Web Services as far as reaction time and would thus be able to be utilized to rank administrations quantitatively rather than just qualitatively.

SEQUENCE OF THE PROPOSED METHOD FOLLOW AS

- User Interface Design
- Order Process
- Payment Process
- Owner View
- User Download
- Encryption data Standard

User Interfaces Design

To interact to the customer server username is needed for mystery text that nobody yet prepared for relating on server. If the customer starting at now exits direct can login into the server else customer should select Username, key & Email id to server. Server will share the entire customer to move and describe the data in Fig.1.

Figure 1. User interface design

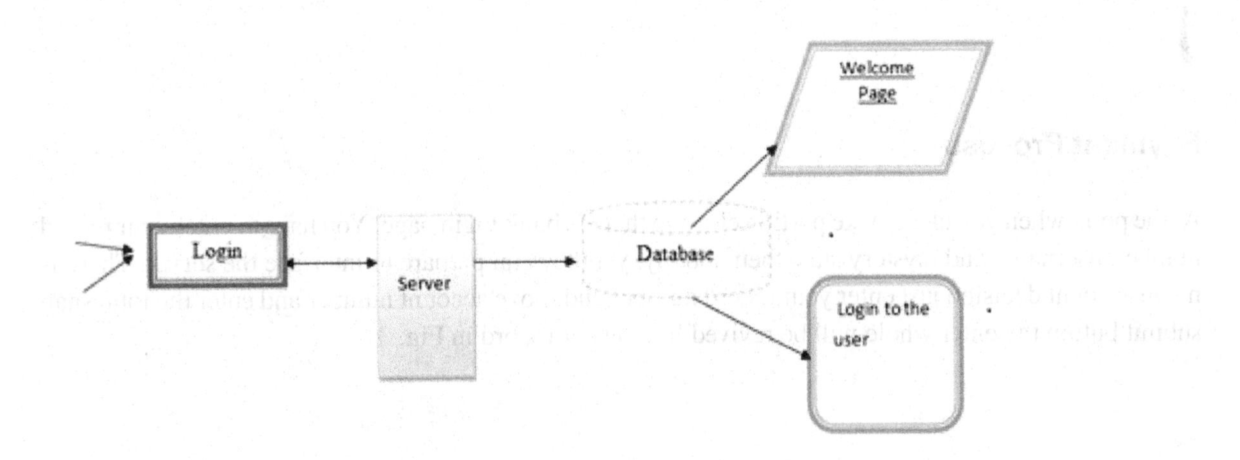

Order Process

At the point when you login there is book list page. What the book you like and okay with cost in like manner suggests click buy now. After you must purchase the book enclosed with the title should be purchased and click settle on portion decision it will go for bank login page in Fig.2.

Figure 2. Order process

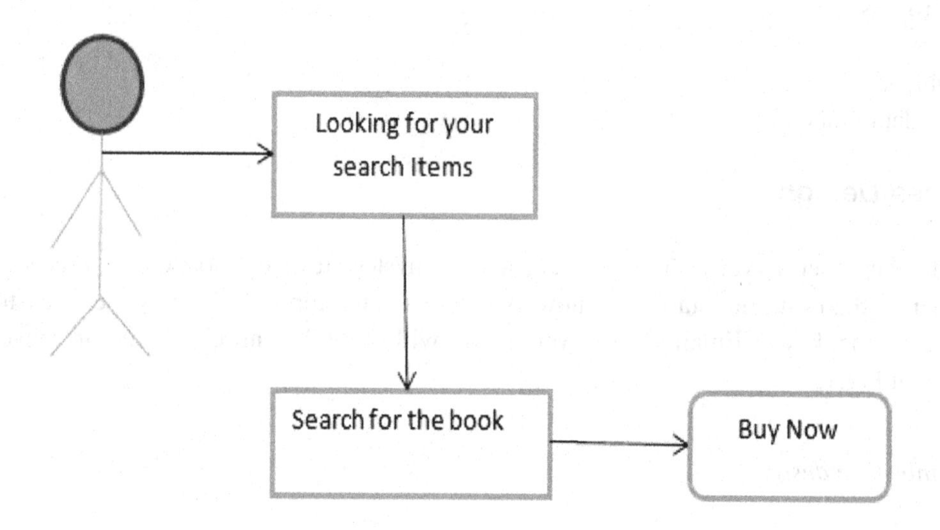

Payment Process

At the point when you click make portion elective there is bank login page. You have to enter your record number, username and mystery state then nobody, yet they can prepare to interface the server. There is move account decision just enter your record number and move account number and enter the total snap submit button the enter whole will be revived into owner record in Fig. 3.

Figure 3. Payment process

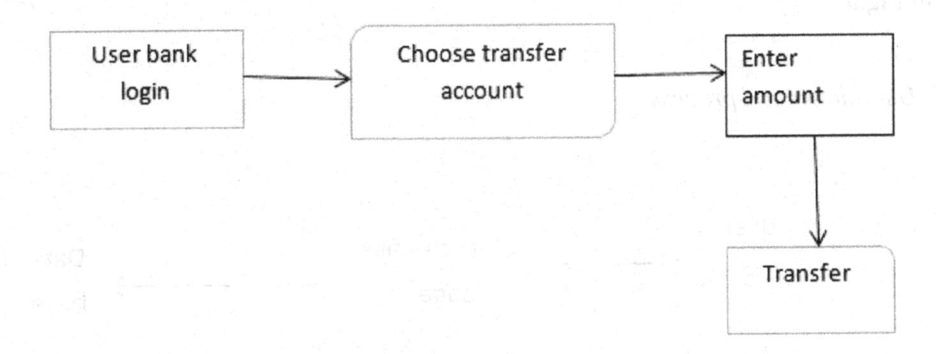

Owner View

There is a separate login for owner once login the owner and click the book request page. There is the detail of whom to buy and which book to be buy and having the user detail. Once he paid the amount means click choose file option and upload the book click submit button in that book will send only for user in Fig.4.

Figure 4. Owner view of process

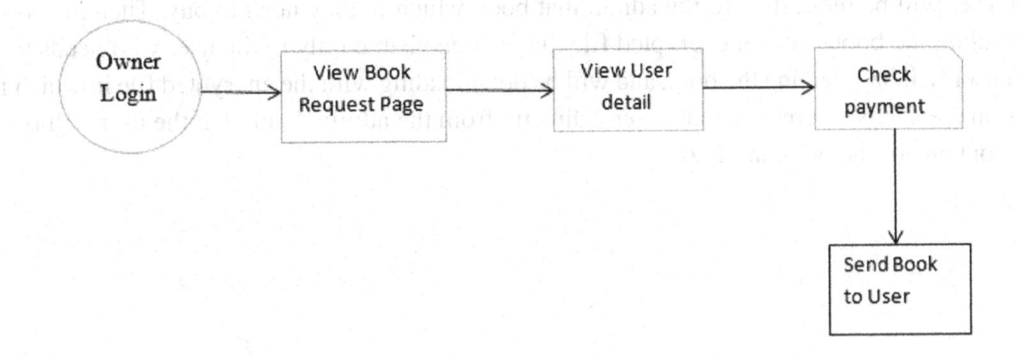

User Download

Once owner accept the request and send file means user to be login them account. You have to enter your record number, username and mystery state then nobody, yet they can prepare to interface the server. When you visit the page and select download option your will get the file for what you must be ordered in Fig.5.

Figure 5. User download process

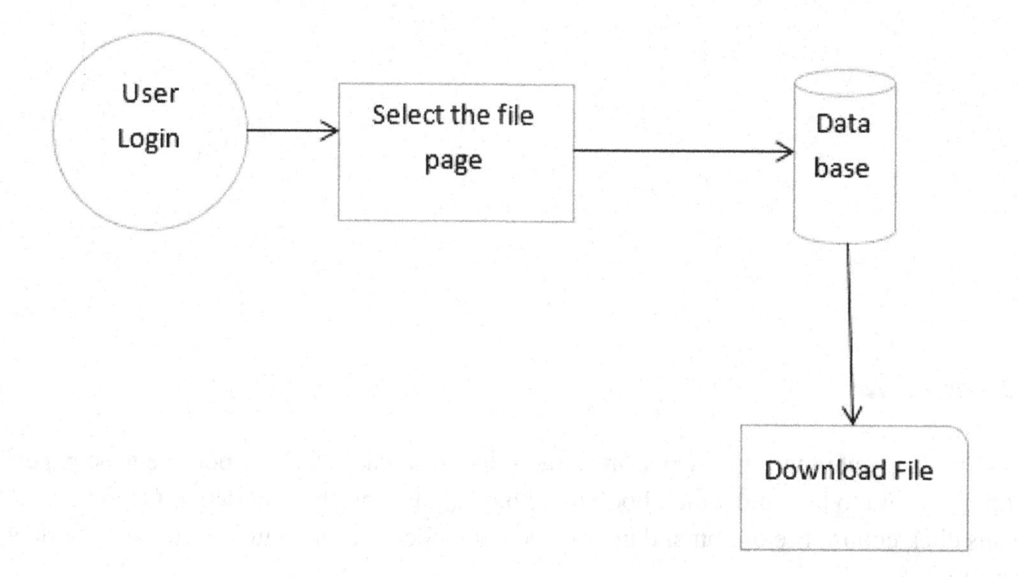

Encrypt Data Sending

The user will be requesting to the admin that book which in they need to buy. Then the admin will be uploading the book with an encrypted file that can be visible only to the user who needs to buy. After that user will be selecting the book and will be downloading with the encrypted file in which it can open it from the shared secret key that is send directly from the admin. With that the user will be able to see the contents of the book in Fig.6.

Figure 6. Encrypt data sending process

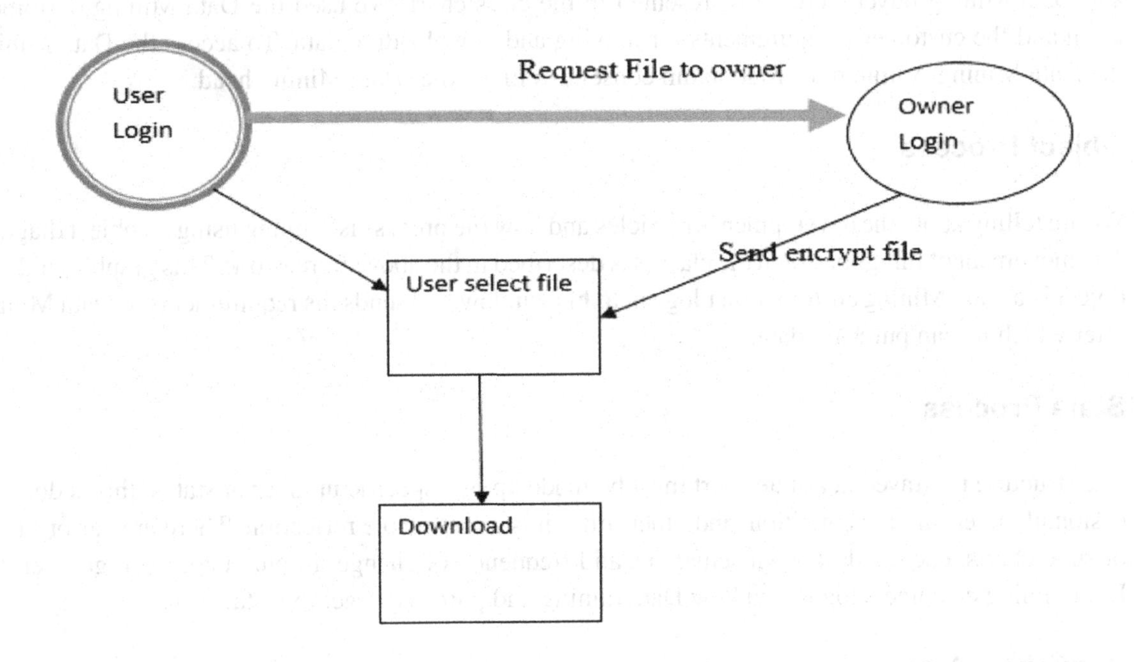

Activity Process

Development blueprints can be used in the Unified Modeling Language to outline the business and operational a tiny bit at a time work procedures of portions in a system. The broad flow of control is depicted in a development chart. The first step in our action chart is for the Data Mining client to log into Data Mining. Here, he's handing over his limitations to a Data Mining provider. If a Data Mining needs to access another Data Mining through an Agent, he must first obtain permission from the Data Mining administrator. Once granted, the Data Mining will be able to view and download items.

Use Case Process

The purpose of a use case format is to demonstrate which framework limits are applied to entertainers. The on-screen characters' roles in the structure can be explained. In our use case diagram, the customer first logs in and purchases a book. The data owner can then choose whether to send data to the client, and the customer can download data from his record. If a Data Mining needs to connect to another Data Mining through an Agent, he must first obtain permission from the Data Mining Director. Once granted, the Data Mining will be able to check what Support is available.

Class Process

The key structure square of article engineered delineating is the class chart. It is used for both broad determined appearance of the application's effectiveness and unquestionable exhibiting making an interpretation of the models into programming code. The Data Mining customer, Data Mining provider, and Data Mining buyer were all represented in the class chart. We used the Data Mining customer's login and the customer's requirements for moving and downloading data. To access the Data Mining, the Data Mining Admin must first obtain consent from another Data Mining head.

Object Process

We are telling about the development of articles and how the process is running using an object diagram. The movement of things between the classes is described in the above framework. This graph's guideline object is a Data Mining customer that logs in to his window and sends his requirements to Data Mining, after which he can purchase data.

State Process

The structure portrayed in a state chart must be made up of a specific number of states; this is done occasionally to ensure the condition, and other times it is a reasonable reflection. There are various types of state charts, each with its own semantics and frequency of change. In our state, we'll go over how Data mining customers log in and how Data mining and purchasers get to Data.

Sequence Process

In our grouping chart, we show how shapes interact with one another and with the whole. In our diagram, the Data mining client initially logs into Data Mining's. Here he is sending his requirements to a Data Mining provider. If a Data Mining needs to connect to another Data Mining through an Agent, he must first obtain permission from the Data Mining's owner. Once granted, the Data Mining will be able to see and download items.

Collaboration Process

Combined exertion graphic depicts object engagement in the same way as sequenced messages do. Participation plots represent both the static structure and dynamic dynamics of a system by combining information from class, plan, and use case graphs. In this diagram, the Data Mining customer is the first to log in. Here, he's shifting his goals to a Data Mining supplier. If a Data Mining needs to connect to another Data Mining via an Agent, he must first obtain permission from a Data Mining executive, after which the Data Mining will be able to view and download the relevant data.

E-R process

In our ER chart, the indicated forms work together and with each other. In our diagram, the Data Mining client initially logs in to Data Mining. Here he is sending his requirements to a Data Mining provider. If a Data Mining needs to connect to another Data Mining through an Agent, he must first obtain permission from the Data Mining's owner. Once granted, the Data Mining will be able to see and download items.

Component process

The component process in the Unified Modeling Language displays how pieces are connected to delineate larger fragments and is used to depict the structure of self-assuredly complicated systems. Here, he is transferring his limits to the Data Mining provider. If a Data Mining needs to communicate with another Data Mining through an Agent, he must first obtain permission from a Data Mining executive. After receiving permission, the Data Mining can choose to view things Support. Customer input Data Mining gives customers the option to see and download specific data.

SYSTEM ARCHITECTURE

Figure 7. Working modal

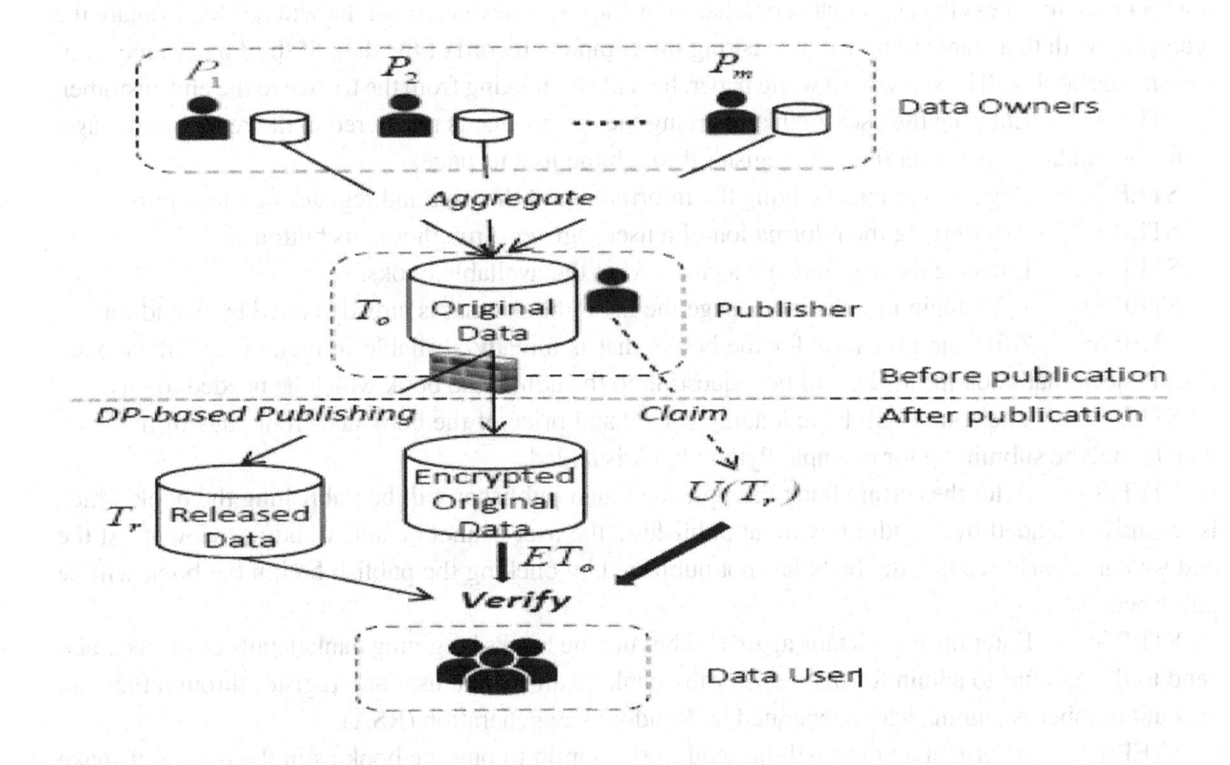

Owner Creates a Website

Owner must register with that page. Then he must create a website and need to get the access from the admin, and he can be able to view the user details and he can also view books in which the user has purchased already it will be derived in Fig.7.

Admin Gives the Authorization to the User

Let the admin should register in that page. After the user look for the books that is already available in certain case if the user doesn't need that book means he will be requesting to the admin the book which he needed to buy

User Searches for the Books

When the user registered with a correct information then he will be able to see the books that is available on the website. If the book is not available, then he will be requesting to the admin for uploading the information of the book

Blocking the Third-Party User with Encryption

If a user tries to access the book that is purchased by the requested customer, he will be able to share the secret key with that standard he will be visiting the complete record of the data. If third party enters our website the book will be shown as a white paper, he will be blocking from the trustee to the end customer.

STEP 1: Entering the user login and giving the details that is registered in the registration page. If the user didn't register yet they can register through the join us page.

STEP 2: Registration page: Giving the information of the user and registering the details.

STEP 3: After giving the information of a user register through join us button.

STEP 4: Entering the user module again to visit the available books.

STEP 5: After login into the index page the list of books that is already added by the admin.

STEP 6: After the user look for the books that is already available in certain case if the user doesn't need that book means he will be requesting to the admin the book which he needed to buy.

STEP 7: The admin will be uploading a Title and price of the book and front page of the book then he will be submitting for example Python book is added.

STEP 8: After the certain book has uploaded then publisher will be publishing the book which is recently uploaded by the admin without publishing the user cannot be able to buy or view it. At the end we can clearly see that the book has not published by clicking the publish button the book will be published.

STEP 9: Enter the user details again and buying the books by giving bank details of the user and send to the amount to admin for that register the bank details of the user and register through that and account number is automatically generated by Random key generation (RSA).

STEP 10: After that amount will be send to the admin to buy the book as in the option of make payment the information will be transferred to the admin.

STEP 11: The amount has been transferred to the admin.

STEP 12: Then we enter to the admin login we can see the book request option in that we see the user information and account number and the secret key will be generated then admin can upload the file that will be send to the user.

STEP 13: Then the user will be able to see the book and with the download option. After downloading with the secured key, the secured data's will be visible user.

STEP 14: The information in which the admin shared the details, and the list of books has been shown.

APPLICATION

Distributed computing is defined as a type of training that focuses on sharing computing resources rather than having nearby servers or individual devices to administer programmes. Appropriate stockpiling organisations have become conceptually standard almost immediately. Customers can keep their data on the cloud and access it from any location at any time. Data owners appear to lose extraordinary control over the fate of their redistributed data, putting the precision, transparency, and reliability of the data at jeopardy. From one perspective, the cloud organisation is frequently confronted with a wide range of internal and external adversaries who may maliciously destroy or reject customers' data.

CONCLUSION

The procedure of insurance defending utility check segment on cryptographic techniques of different privately arrangement planned on regarded datas. The recommendation can't check utility's subject to mixed concept for rough information as opposed to the plain characteristics, which along these lines' false security. It engages to furtively check the precision of the mixed frequencies gave by the distributer, which perceives deceitful distributers. In which user can buy the products or books securely based upon the user's request which further suggests an optimal path for the execution of user requests. The strategy we show can be utilized to quantify and anticipate the conduct of Web Services as far as reaction time and would thus be able to be utilized to rank administrations quantitatively rather than just qualitatively. We also loosen up this framework to another differentially private conveying plan proposed for social information.

REFERENCES

Alhadidi, D., Mohammed, N., Fung, B. C., & Debbabi, M. (2012, July). Secure distributed framework for achieving ε-differential privacy. In *International Symposium on Privacy Enhancing Technologies Symposium* (pp. 120-139). Springer. 10.1007/978-3-642-31680-7_7

Barthe, G., Köpf, B., Olmedo, F., & Zanella Beguelin, S. (2012, January). Probabilistic relational reasoning for differential privacy. In *Proceedings of the 39th annual ACM SIGPLAN-SIGACT symposium on Principles of programming languages* (pp. 97-110). 10.1145/2103656.2103670

Chen, R., Fung, B. C., Desai, B. C., & Sossou, N. M. (2012, August). Differentially private transit data publication: a case study on the montreal transportation system. In *Proceedings of the 18th ACM SIGKDD international conference on Knowledge discovery and data mining* (pp. 213-221). 10.1145/2339530.2339564

Dhiman, G., Kumar, V. V., Kaur, A., & Sharma, A. (2021). DON: Deep Learning and Optimization-Based Framework for Detection of Novel Coronavirus Disease Using X-ray Images. *Interdisciplinary Sciences, Computational Life Sciences*, 1–13.

Dwork, C. (2008, April). Differential privacy: A survey of results. In *International conference on theory and applications of models of computation* (pp. 1-19). Springer.

Ezhilmaran, D., & Muthukumaran, V. (2017). Authenticated group key agreement protocol based on twist conjugacy problem in near-rings. *Wuhan University Journal of Natural Sciences*, *22*(6), 472–476. doi:10.100711859-017-1275-9

Freeman, D. M. (2010, May). Converting pairing-based cryptosystems from composite-order groups to prime-order groups. In *Annual International Conference on the Theory and Applications of Cryptographic Techniques* (pp. 44-61). Springer. 10.1007/978-3-642-13190-5_3

Hong, Y., Vaidya, J., Lu, H., Karras, P., & Goel, S. (2014). Collaborative search log sanitization: Toward differential privacy and boosted utility. *IEEE Transactions on Dependable and Secure Computing*, *12*(5), 504–518. doi:10.1109/TDSC.2014.2369034

Hong, Y., Vaidya, J., Lu, H., Karras, P., & Goel, S. (2014). Collaborative search log sanitization: Toward differential privacy and boosted utility. *IEEE Transactions on Dependable and Secure Computing*, *12*(5), 504–518.

Jiang, W., & Clifton, C. (2006). A secure distributed framework for achieving k-anonymity. *The VLDB Journal*, *15*(4), 316–333. doi:10.100700778-006-0008-z

Kumar, V., Niveditha, V. R., Muthukumaran, V., Kumar, S. S., Kumta, S. D., & Murugesan, R. (2021). A Quantum Technology-Based LiFi Security Using Quantum Key Distribution. In Handbook of Research on Innovations and Applications of AI, IoT, and Cognitive Technologies (pp. 104-116). IGI Global.

Kumar, V. V., Raghunath, K. K., Rajesh, N., Venkatesan, M., Joseph, R. B., & Thillaiarasu, N. (2021). Paddy Plant Disease Recognition, Risk Analysis, and Classification Using Deep Convolution Neuro-Fuzzy Network. *Journal of Mobile Multimedia*, 325-348.

Kumar, V. V., Raghunath, K. M., Muthukumaran, V., Joseph, R. B., Beschi, I. S., & Uday, A. K. (2021). Aspect based sentiment analysis and smart classification in uncertain feedback pool. *International Journal of System Assurance Engineering and Management*, 1-11.

Linda, G. M., Lakshmi, N. S. R., Murugan, N. S., Mahapatra, R. P., Muthukumaran, V., & Sivaram, M. (2021). Intelligent recognition system for viewpoint variations on gait and speech using CNN-CapsNet. *International Journal of Intelligent Computing and Cybernetics*.

Manikandan, G., Perumal, R., & Muthukumaran, V. (2020, November). A novel and secure authentication scheme for the Internet of Things over algebraic structure. In. AIP Conference Proceedings: Vol. 2277. *No. 1* (p. 060001). AIP Publishing LLC. doi:10.1063/5.0025330

Manikandan, G., Perumal, R., & Muthukumaran, V. (2021). Secure data sharing based on proxy re-encryption for internet of vehicles using seminearring. *Journal of Computational and Theoretical Nanoscience*, *18*(1-2), 516–521.

Muthukumaran, V., Ezhilmaran, D., & Anjaneyulu, G. S. G. N. (2018). Efficient Authentication Scheme Based on the Twisted Near-Ring Root Extraction Problem. In *Advances in Algebra and Analysis* (pp. 37–42). Birkhäuser. doi:10.1007/978-3-030-01120-8_5

Muthukumaran, V., Joseph, R. B., & Uday, A. K. (2021). Intelligent Medical Data Analytics Using Classifiers and Clusters in Machine Learning. In Handbook of Research on Innovations and Applications of AI, IoT, and Cognitive Technologies (pp. 321-335). IGI Global.

Muthukumaran, V., Vinothkumar, V., Joseph, R. B., Munirathanam, M., & Jeyakumar, B. (2021). Improving network security based on trust-aware routing protocols using long short-term memory-queuing segment-routing algorithms. *International Journal of Information Technology Project Management*, *12*(4), 47–60.

Nagarajan, S. M., Deverajan, G. G., Chatterjee, P., Alnumay, W., & Muthukumaran, V. (2022). Integration of IoT based routing process for food supply chain management in sustainable smart cities. *Sustainable Cities and Society*, *76*, 103448. doi:10.1016/j.scs.2021.103448

Nagarajan, S. M., Muthukumaran, V., Beschi, I. S., & Magesh, S. (2021). Fine Tuning Smart Manufacturing Enterprise Systems: A Perspective of Internet of Things-Based Service-Oriented Architecture. In Handbook of Research on Innovations and Applications of AI, IoT, and Cognitive Technologies (pp. 89-103). IGI Global.

Velliangiri, S., Karthikeyan, P., & Vinoth Kumar, V. (2021). Detection of distributed denial of service attack in cloud computing using the optimization-based deep networks. *Journal of Experimental & Theoretical Artificial Intelligence*, *33*(3), 405–424.

Chapter 12
Profit Sharing Models for Social Media in Big Data Commercialized Crises

Tsung-Yi Chen
Nanhua University, Taiwan

Yung-Han Tung
Nanhua University, Taiwan

ABSTRACT

In era of global crises and data-based business competition, when users use social media, their personal data and network behaviors are collected. These data are valuable for making right decisions. Based on fair viewpoint, the profits of social platforms should be shared with the data providers. In order to discuss the applicability of profit sharing, 26 users were grouped and interviewed collaboratively. The customer profiles and value proposition maps were then used as a tool to summarize the respondents' preliminary suggestions to explore the users' motivation for using social network media and their opinions on the issues related to data collected, privacy, and profit sharing. This study has also explored the issues related to the fair and reasonable rewards of social platforms. Then, the profit-sharing models were designed according to the suggestions, which were obtained from 388 valid questionnaires. This study helps us to understand the common views of users of social media platforms on the collection and sharing of their data.

INTRODUCTION

Due to the rise of web service models, user-oriented social media can support social activities, including communication, connection, interaction, and sharing. People's lives have been closely related to social media. When using social media, users provide personal data and share all kinds of information, such as articles, pictures, videos and life details. Social media platforms can, therefore, easily collect all kinds of behavioral information about users, which can be used to analyze, understand and predict their prefer-

DOI: 10.4018/978-1-7998-9640-1.ch012

ences and browsing habits (as shown in Figure 1), as well as to establish the competitive intelligence of an enterprises (Nunes & Correia, 2013; Parker et al., 2016).

At present, personality trait analyses are carried out mainly by collecting the network behavior and using habits of users, in order to use accurate personalized advertisements to earn advertising fees. If they are properly analyzed, the big data hidden in social networks are potentially valuable for helping governments and enterprises to make the right decisions.

Many companies especially in the age of global competition are committed to big data analytics technology for exploring the consumption patterns and living habits and to provide fast and accurate smart marketing services. According to the results of big data analytics, some enterprises conduct business model innovations (Tykkyläinen & Ritala, 2020; Heider et al., 2020) and develop accurate network behavior-locking technologies.

In the current era of data economy, users ostensibly enjoy various online resources for free, but in fact, they are exchanging their private information for those free resources (Lee, 2017; Andrejevic, 2014). As users use social media every day to contribute to their personal behavioral data, at the same time, media platforms are collecting their data. Casilli (2017) noted that the use of the Internet has become digital work, and our posts, likes and shares on FB are of commercial value, with the users becoming free digital laborers. Hence, at present, there is no fair network profit sharing system, which is obviously unfair to the users.

Advanced countries have revised their personal data protection laws, in order to promote personal data availability and an open information business environment. The laws enable people to decide how, and to what extent, their personal data is to be used, and they even have the right to sell them.

Data have become an essential element in economic and commercial operations. Social platforms, just like a huge database, store the real condition of a user's social activities on the Internet, which can then be accurately predicted (Pedersen & Ritter, 2020).

Because it was difficult to obtain data in the past, economists did not take it into account, when considering production factors. However, with the modern advanced technology, there are no limitations on data collection. Just like search engine Google, there seems to be a production line of raw data (raw material) input at one end and processed information and knowledge (product) output at the other end (Mayer-Schönberger & Cukier, 2013).

Therefore, the profits of social platforms should be shared with data providers, with the rewards being determined by their contribution. Fuchs (2010) proposed that the profits should be shared with the digital workers. It is, therefore, reasonable economic behavior to share the results of big data analytics with the data producers.

There are no studies on the profit sharing systems of social networks. Much of the literature only explores the behavior of social media users. Based on the above, this study explored the issue of fair and reasonable rewards for social network data providers, and designed two profit sharing models. Through this study can to balance the use and profits between users and data collectors, and for users to be willing to share their data, so that the big data industry can develop fast and their personal behavioral data can contribute to the economy or society in a safe environment.

Figure 1. Scenario of profit sharing models on social media

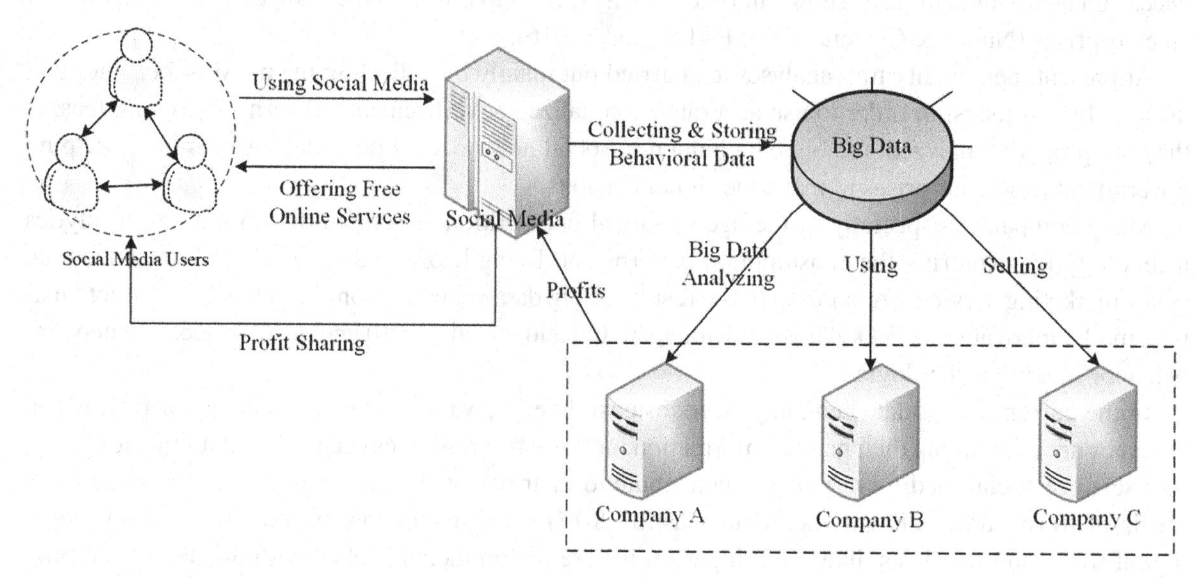

RELATED WORKS

Social network

A virtual community that is comprised of a group of network users with similar interests, goals and experiences is established through the Internet, breaks the limitations of time and space and establishes an interdependent and interactive situation to share each other's interests and knowledge, so that its members feel satisfied and build interpersonal relationships. In this way, it creates a meaningful platform for receiving and sharing (Rothaermel & Sugiyama, 2001). The operations of virtual communities indirectly affect the consciousness, thoughts, cognition and culture of people in real life (Affleck & Kvan, 2008; Chia & Ritchard, 2014). As one of the new business models, social media have rapidly developed into a powerful tool for individuals or business organizations.

Social media are rapidly affecting the whole world. Social media bring joy to people's life, by effectively expanding their social circles and broadening the marketing activities and scope of enterprises and merchants. Their motivation for using social media are as follows (Gulnar et al., 2010; Peter et al., 2005; Christofides et al., 2009; Caplan & Turner, 2007; Qimei & Wells, 1999; Qimei et al., 2002): (1) to maintain friendships, (2) to expand social relations, (3) to boost popularity and seeking attention, (4) to express emotions and experience the feeling of belonging, (5) to get and share the latest information, and (6) to participate in online games and entertainment activities, and (7) to killing time.

Although social media makes people's lives convenient, they also have a negative effect, for example cyber-bullying, racial and class gaps, and internet addiction, which cannot be ignored. Therefore, people have started to pay attention to protecting their privacy when using social networks. Dwyer et al. (2007) found that social network users are concerned about their privacy or their trust in websites and netizens, which affects the degree of their information disclosure and their intentions to use. The higher the protection degree of a virtual community's privacy policy, the higher the users' trust in the website will be (Wu et al., 2010). If users are concerned about their personal data being used by websites, or by others, without their authorization, this will reduce their willingness to use them. However, users will not stop using social media due to privacy issues, because they want to keep in touch with friends and to gain new knowledge (Brandtzæg et al., 2010).

Data Commercialization and Information Privacy

Big data is a cultural, scientific and academic social-technological phenomenon (Boyd & Crawford, 2012). Government agencies, medical care and transportation all benefit from big data applications. However, the problem of personal privacy and data security arises (Lenard & Rubin, 2013; Kitchin, 2013; Bilbao-Osorio et al., 2013). Security disputes and disrespect for users' personal privacy are common in big data applications (European Commission, 2014).

Data commercialization begins with the collection and accumulation of information, linking it to relevant data, and then obtaining valuable insights after using data analysis to solve social problems. According to Feigenbaum (2014), the possible obstacles that enterprises may encounter when using big data are as follows: (1) the questions and answers are ambiguous, (2) unstructured data are difficult to obtain, (3) data are difficult to re-use, (4) it is not easy for enterprises to become companies that are driven and decided by data, and (5) data should be valuable and issues, such as data integration, distribution and utilization, should be considered.

For modern enterprises, data are new raw materials that can produce knowledge, optimize decisions, promote innovation and create social and economic values. It has been a trend for enterprises that use big data analytics to gain a competitive advantage and create new profits (Sivarajah et al., 2017).

Big data come from many sources (Sivarajahet al., 2017). In particular, social media users upload more than 10 million photos an hour, and 'likes' or 'comments' are posted more than three billion times a day. With these records, social media can track the users' preferences, as well as their status updates and likes, to ensure that the most appropriate advertisements are displayed on the websites. The data sets that can be obtained by social media are the basis for the new businesses created on social media platforms, the source of their profits and the key to their enterprise operations (Kaiser, 2019).

The arrival of the era of big data intensifies the invasion of personal privacy. For instance, some people open the data portal by using the Friends API of FB, which seriously violates the personal data protection law. When users enjoy the convenience of technology, they are indeed giving up their personal privacy and data. Although users are aware of the privacy risks on the Internet, they have to disclose personal information in order to access to social media services. For example, cookies are well-designed to track the usage of computers and mobile phones (Kaiser, 2019).

The 'right to be forgotten' is a concept of human rights, which means that people have the right to require the removal of their negative or outdated personal information (Wikipedia, 2019). Everyone has the right not to be searched by search engine, and the 'right to be forgotten' is a new human right in the digital era.

Many data brokerage firms in the world have been collecting and analyzing the behavioral information consumers. If consumers do not want their data privacy to be sold, they should be allowed to log out and remove personal data (WebFX, 2015).

At present, some enterprises have tried to pay consumers who are willing to participate in surveys and digital tracking. The social exchange theory holds that all human behaviors are dominated by some exchanges for awards and rewards (Bandura, 1986). The foundation of social exchange is mutual benefit

(Blau, 1964). The values of social platforms that are based on the social exchange theory are mostly created by the users' personal data, so it is natural that platforms should be obligated to repay them.

A business model can explain how an enterprise makes a profit and how it operates, who the main customers are, and how value can be provided for its clients at an appropriate cost (Magretta, 2002; Cosenz & Bivona, 2020). A business model is a mechanism that converts ideas at a reasonable cost, in order to obtain benefits and deliver values (Gambardella & McGahan, 2010; Teece, 2010). The design of profit sharing is an important part of a business model. In Taiwan, points can be earned from everything in daily life, such as flights and credit cards. All brands target the point economy, and profit sharing models that earn points are widely used in enterprises. Cross-channel cooperation is strengthened through alliances, to further expand the client groups and enhance their loyalty.

QUESTION ANALYSIS OF DATA COMMERCIALIZATION

In this study, the current situations of the use of social media were understood by interviewing with questionnaires and, from the perspective of the users, the customer profiles of the value proposition canvas (Osterwalder et al., 2011; Osterwalder et al., 2014) were used to explore their tasks, pains and benefits, when using social media, and to find out their real demands and views on social media.

Preliminary interview

In this stage, a preliminary survey on the use of social media was conducted by using group collaborative interviews. A total of 26 respondents, aged between 17 and 60, were divided into six groups, with three to five people in each group. The respondents were classified and numbered according to their job types. The interview results are summarized in Table 1.

Table 1. Interview summary results

Questions	Interview results
(1) What social media do you use the most often now? How long has it been used? How long do you use it every day?	All respondents of this study used Line. Among the respondents, 71% of high school and university students, aged between 18 and 24, use IG. Most respondents had used Line for seven to 10 years, and from three to five hours a day. All respondents usually used FB, regardless of their status and age, and 66% of them used it for less than two hours a day, mainly to keep up with their friends' updates and to browse information. High school and university students preferred IG, and 60% of them used it for more than three hours a day. Student respondents mentioned that IG was easier to operate, compared with FB, and that it was more convenient and safer, in terms of data privacy.
(2) Why do you use social platforms? How can they help us in our lives (functional, social, or emotional) or work?	Respondents stated that Line could be used to record dialogues and to communicate in many ways. Hence, all respondents mentioned that they used Line to communicate in both their lives and at work. The respondents engaging in education indicated that using Line was convenient for real-time contact with students and parents and to give notices. Office workers and enterprise operators also assigned tasks and communicated with their clients through social media. Seventeen respondents noted that they used social media to expand their social relationships and life circles, and to track the people, events, and things that they liked. In particular, high school and university students liked to post and share interesting things about life on social media, in order to boost their popularity. Thirteen respondents said that they learnt about social trends and current affairs through social media and that they used social media to 'kill time' or play online games. Six respondents stated that they used social media to obtain knowledge or participate in network communities.

Continued on following page

Table 1. Continued

Questions	Interview results
(3) Do you want to make money through social media platforms?	Only one respondent was a part-time broadcaster and met like-minded net friends via Twitch. One respondent sold cosmetics and care products via IG, and one respondent conducted group-buying via Line. Eleven respondents mentioned that they had considered trying to make money through social media. Most respondents noted that they did not know how to make money through social media or what projects to run, but that they did not reject the selling of goods through live-streaming on a part-time basis.
(4) Do you care about your privacy or personal data being analyzed?	Four respondents claimed that they did not care about personal privacy. Even if they did not disclose their personal information on social media, the platforms could still obtain personal information in other ways, which was hard to prevent. Most respondents indicated that it was natural to use social media, and that they would not randomly respond to messages from strangers nor randomly post their personal data online.
(5) Do you care about your personal data being used to make money by social media?	Fifteen respondents cared about their information being used to make money by social media. However, six respondents noted that they would not care, as long as it did not affect them. A few respondents mentioned that prior consent should be required for the use of their personal data. Three respondents were enterprise owners, who said it was reasonable for platforms to use the data to make money.
(6) If a social platform now has a profit sharing system to share with users, would you be willing to participate? What do you think?	Fifteen respondents claimed that they agreed that a social media platform should have a reasonable profit sharing system and that they were interested in participating. However, nine respondents noted that they considered participating after waiting and evaluating the profit sharing system, the degree of personal data disclosure and the risk of violating the law.
(7) Do you know that our personal data and using behaviors are collected in large quantities when we use social media?	Most respondents knew that websites monitor what we view, but two respondents did not know that websites collected personal data. Another two respondents became aware of it after the Cambridge event, so they paid special attention to the information that they left when using social media.
(8) What you think of personal data being collected and used or sold?	Six respondents mentioned that they agreed that their data could be collected and used, if this data was used by social media to enhance the functions of the platform, to make it more convenient or timely, and to provide the needed information and functions. Nine respondents noted that data collection and use by social networks was acceptable but that it was unreasonable to sell data, and that their consent should be required for data selling. Another five respondents indicated that it was acceptable under the relevant laws. Five respondents noted that data could be collected, used or sold with their prior consent and within the scope of their consent.
(9) If the collection of your personal data is exchanged for free services, will you continue to use it or not?	Although their personal data were collected, all respondents suggested that they would continue to use it because it is convenient. High school and university students indicated that visiting social media every day is an integral part of their lives. Six respondents noted that they used only some functions and platforms and that they paid attention to their using habits and personal data.
(10) Do you have a different opinion on personal data de-privacy (de-identification) by platforms?	De-identification is the ability to disconnect the data of particular persons by using some technologies or masking methods. Almost all respondents believed there should be de-privacy mechanisms, in order to make data owners feel secure. However, two of them indicated that they would still feel insecure, despite the de-identification.
(11) What do users think if the platforms give them the right to decide how to use their information (information autonomy)?	Information autonomy implies that I have the right to decide how to use my data. Twenty-one people agreed that such legislation and mechanisms should be established and that data should be used hierarchically, with prior consent. However, one respondent and one enterprise owner believed that too many choices bothered them. Two respondents noted that they were still skeptical about the platforms.
(12) What do users think of platforms that give them the right to decide how to use their information and that have mechanisms in place to share the profits with them?	Most respondents mentioned that profit sharing in the data market was a good idea; however, it was a big market, so they needed to consider the level of privacy disclosure and whether it would cause any inconvenience or trouble. They also suggested that bonuses could be combined with the virtual currency that is issued by FB, to offset the consumption.

Continued on following page

Table 1. Continued

Questions	Interview results
(13) Do you think bonuses must be paid in money?	Four respondents claimed that the bonuses need not necessarily be paid in money, but that money was the best choice. Twenty-two respondents noted that the bonuses could be distributed by physical goods, consumption reduction, earning points to exchange for goods or to offset consumption, the free use of advanced features in exchange for Line stickers, coupons, etc. Among them, point earning was mentioned the most. Some university students suggested that the proportion of bonuses should be 7:3 or 8:2.

User perspective analysis

Based on the above opinions of respondents and related research, user problems were analyzed in the three-dimensional user profiles, as can be seen below:

1. **User tasks:** the affairs to be completed, problems to be solved, and demands to be satisfied through social media by users, in work or in life, are summarized as follows:
    a. Social support acquisition: obtaining psychological identification and support by interacting with others through social networks (Laffey et al., 2006; Respondents E1).
    b. Social tasks: interacting with friends and relatives, sharing life's details and updated news, and expanding social circles (Boyle & Johnson, 2010; Respondents A1-A4, B1-B5, C4, D1, D2, D4, E1, E3, E5, and F1).
    c. Information tasks: searching for information, obtaining what people are interested in, in their studies, work and life, or following news, current events, and social trends (Respondents A2, A4, C1, C4, C5, and F3).
    d. Entertainment: watching videos, browsing web pages, killing time, relieving pressure, and playing games (Respondents A1, A3, A4, B1, B3, B5, C5, D1-D3, E2, E3, and F3).
    e. Economic tasks: building brands and images, expanding a business and communicating with customers by means of fan pages, or conducting group-buying, live-streaming, and goods purchases and sales (Respondents C3 and C5).
    f. Functional tasks: working, contacting, and communicating with clients (Respondents C2, D1, D4, E2, E4, E5, F1, and F2).
2. **User pain points:** the troubles, risks, or barriers of using social media are summarized as follows:
    a. Privacy and personal data security: worrying about the misuse of their personal data on social networks (Respondents A1, B1, B3, B4, C2, C3, D1, E2, E3, and F1-F3).
    b. Influence on their daily routines and health: the excessive use of social media influences personal daily routines, and sedentary lifestyles influence health (Haig, 2019; Respondents B3, B4, E1, A1, A3, and D4).
    c. Cyber bullying: people who suffer from any injustice post and argue on social platforms, which could easily cause disputes (Respondents A4, B1, C5, and E5).
    d. Poor system operations: privacy policies are too long and complicated for users to understand. Moreover, to use the social media, people must agree to the unreasonable terms that personal data are collected by the platforms (Respondents A3, B1, C4, D1, and F2).
    e. Information of false data: the information on social networks is full of traps and falsehoods (Respondents A2, B5, C1, D2, and E4).

3. **User benefits:** the anticipated results and benefits, or the desires of users, are summarized as follows:
   a. Social benefits: getting closer to friends, relatives and people whom one does not see often, and expanding social circles (Respondents A2, B2-B5, C4, D3, D4, and E1-E5).
   b. Informational benefits: obtaining all kinds of news in study, work and life, current events, social trends, product information, and promotion activities (Respondents A1-A4, B1, C1, C3, C4, D1, D2, E2, E3, F1, and F3).
   c. Functional benefits: making work more efficient and life more convenient (Respondents C5, D4, E5, and F2).
   d. Economic benefits: building brand images and fan pages, expanding a business, communicating with customers, and increasing profits (Respondents C3, D4, and D3).
   e. Social support acquisition: expressing emotions and interacting with others by posting and getting psychological assistance and relief (Respondents B5, D2, A3, A4, B2, and D2).

Profit Sharing Model Design

Loyal users are the biggest asset of all platforms, and their value should be shared with those who create them. In this section, two profit sharing models suitable for social platforms are proposed.

1. The point earning-based profit sharing model: points can be regarded as a kind of currency, and some social networks have also developed social currencies to measure their economic value (Parker et al., 2016). The circulation volume of points and the number of users determine the value of the points. Point earning-based profit sharing not only stimulates consumption, but it also increases the 'stickiness' and loyalty of customers. In this study, nearly 95% of respondents claimed that social platforms could refer to the member point systems of physical stores and exchange points for goods, or offset consumption with points. The social platform point model proposed in this study can calculate the users' earned points, based on their social interactions. The point conversion methods are as follows: the time spent on platforms, the number of friends, the number of users' posts, the number of comments, the number of times a post is forwarded, the number of times a post is viewed, the number of times a post is liked, and the number of times videos are watched.
2. The multi-level marketing-based profit sharing model: the project referred to the profit sharing method of multi-level marketing. Direct marketing is a way for enterprises to recruit members to sell products or services directly to terminal consumers, at places other than fixed business premises, rather than through traditional wholesalers or retailers. Hence, intermediate channels and fixed business premises are not required, which reduces the relative costs. The members' bonuses are calculated in the form of commission. Generally, the bonuses of a multi-level direct marketing company can be divided into three main types, namely, sales, coaching, and leadership bonuses. The types of users in the multi-level profit sharing model include upstream users and downstream users. Downstream users are those who start to use the network service through recommendations or introductions of their upstream user. For example, if X_i recommends a network service for user Y_{i1}, X_i is the upstream user of Y_{i1}, and Y_{i1} is one of the downstream users of X_i. The bonus includes direct bonus and indirect bonus, but the bonus shared with the upstream users of X_i need to be deducted. And the indirect bonuses of user X_i are shared from his n downstream users Y_{ij}, $j=1$ to n.

QUESTIONNAIRE AND DATA ANALYSIS

The questionnaires with 17 questions were distributed by means of Line and Facebook, and a total of 388 valid questionnaires were collected. By using a reliability analysis, the Cronbach's α of the overall questionnaire reliability was 0.813.

Descriptive statistics

The gender distribution of the respondents was 152 males (39.2%) and 236 females (60.8%). In terms of age, 196 respondents (50.5%) were aged between 18 and 24, followed by 114 respondents (29.4%) who were aged between 30 and 49 years. 234 respondents (60.3%) had a college or university education. They used social media very frequently. In terms of the respondents' occupations, there were 190 students (49%), 101 soldiers, civil servants and teachers (26%), 37 people engaging in services (9.5%), and a tiny minority of people engaging in communication, agriculture, forestry, fishery, and husbandry.

According to the statistical analysis on the frequency of using social media, 205 respondents (52.8%) had used social media for 6 to 10 years, which was the longest, followed by 95 respondents (24.5%) who had used social media for "less than 10 years". The surveyed high school and university students indicated that they started using social media in junior high school. In terms of the time spent on social media every day, 101 respondents (26.0%) used social media "one to two hours" a day, and most respondents (278, 71.65%) used social media "one to four hours" a day, which accounted for the majority. In terms of the number of friends on the social media, most people have "201-300 people" (85, 21.9%).

The motivation for using social media was "for entertainment" (17.3%), "for obtaining information" (17.2%), "for maintaining social relationships" (16.6%) and "for killing time" (13.6%). There were no significant differences between the social media choices of men and women, and the most commonly-used social media were Line, FB, YT, and IG. With respect to their educational level, there was no difference in their social media choices. The main motivation of high school, vocational school and college or university students was entertainment, but the main motivation of the students at a postgraduate and higher level was the maintenance of social relationships.

Analysis of User Status and Differences

Table 2 explores the differences in the actual perception and acceptance of the variables, such as gender, age and occupation, in terms of the use, privacy concerns and profit models of social media. The explanations were as follows:

1. There is no significant difference between male and female users in the time that they started using social media and the time they spent using social media every day, which indicates that the time consumption does not vary by gender. However, there is a difference in their number of friends, with men having significantly more friends than women.
2. With respect to the statement, "I am worried that data might be improperly used", there is a significant difference, indicating that women were more concerned than men.
3. The average values of men were significantly higher than those of women with regard to the following statements: "I agree that social media use my data to make money" (p=0.009), "I agree that, if social media share profits, they can use my data to make money" (p=0.002), "If social media

have reasonable profit sharing systems, will you participate in them? "(p=0.000), and "How many friends and relatives are invited to participate in bonus calculation if social media adopt the multi-level marketing model" (p=0.001). This indicates that men, with a higher participation, agree that social media use user data to make money, that users participate in profit sharing systems, and that the profit sharing model adopts multi-level marketing.

4. In terms of using social media, the differences of users at ages are insignificant for "How long have you been using social media (year)" and "What is the number of friends most commonly used in social media". However, the difference is significant in "How much time do you spend on social media every day?" (p=0.000), which indicates that young people spend more time on social media every day than older people.
5. The analysis results show that, in terms of privacy concerns on social media, the differences of users at different ages are significant in "those using social media are easy targets for hackers" (p=0.005), "the privacy mechanisms of social media cannot protect personal information" (p=0.000), and "I do not think it is safe to use social media" (p=0.000). In comparison, those aged between 30 and 49 are more careful about it than those aged between 18 and 24, and including those below 17, which indicates that older people are more concerned about their private data and security than younger people.
6. In terms of the profit models of social media, the differences of users at ages are insignificant in their opinions about the fact that social media use user data to make money, and about the profit sharing model, which indicates that their opinions about the fact that social media use user data to make money do not differ by age and that their opinions on the profit sharing model are the same.
7. In terms of the use of social media, the analysis results show that the differences of users taking different jobs are insignificant in "How long have you been using social media (year)" and "What is the number of friends most commonly used in social media". However, the difference is significant in "How much time do you spend on social media every day?" (p=0.001), and university students spend more time on social media than soldiers, civil servants and teachers.
8. In terms of the privacy concerns on social media, the analysis results show that the difference of users taking different jobs is insignificant, which indicates that the opinions on privacy concerns do not differ due to different jobs.
9. The users who take different jobs are insignificantly different in their opinions on the profit models of social media, which indicates that it does not differ according to the different jobs.
10. In terms of the use of social media, men have more friends than women. In terms of privacy concerns on social media, men, on average, are higher than women only in "I know that social media collect information" and "privacy policies are complicated", while other questions show that women are more careful about privacy concerns than men. Men have a higher recognition of the profit models of social media than women do.

Table 2. Analysis of social media in terms of various variables

Aspect	Measurement item	Gender	Age	Occupation
Status of using social media	How long have you been using social media (years)?	0.61	0.088	0.158
	How much time do you spend on social media every day (days)?	0.457	0.000*** 2>4, 5 3>5	0.001*** 1>2
	How many friends do you have in the most commonly-used social media?	0.028 * 1>2	0.517	0.477
Privacy concerns about using social media	I know that social media collect information.	0.188	0.29	0.061
	I am worried that my social media accounts might be stolen.	0.139	0.008**	.046*
	I am worried that my personal information might be collected, stored, and used.	0.185	0.233	0.155
	I am worried that my private information might be stolen.	0.082	0.032*	.012*
	I am worried that data might be improperly used.	0.027 * 2>1	0.008**	0.077
	I am worried that my private information might be disclosed when I use social media.	0.212	0.11	0.178
	Those using social media are easy targets for hackers.	0.849	0.005** 4>2	0.152
	The privacy policies and settings of social media are complicated.	0.281	0.001***	.003**
	The privacy mechanisms of social media cannot protect personal information.	0.898	0.000*** 4>2	.008** *a*
	I do not think it is safe to use social media.	0.701	0.000*** 4>1, 2	.001***
Profit models	I know that some social media sell my data to earn a lot of money.	0.207	0.056	0.123
	I agree that social media use my data to make money.	0.009 ** 1>2	0.258	0.288
	I agree that, if social media share the profits, they can use my data to make money.	0.002 ** 1>2	0.575	0.438
	If social media have reasonable profit sharing systems, will you participate in them?	0.000 *** 1>2	0.935	0.209
	Do you think bonuses must be paid in money?	0.062	0.34	0.17
	Points are earned, based on the social interactions, to exchange for favorite goods or to offset cash.	0.259	0.504	0.266
	How many friends and relatives are invited to participate in bonus calculation if social media adopt the multi-level marketing model?	0.001 *** 1>2	0.002**	.040*

Note 1: Gender: 1 (male), 2 (female); Age: 1 (under 17 years old (inclusive)), 2 (18-24 years old), 3 (25-29 years old), 4 (30-49 years old) 5 (50-59 years old); Occupation: 1 (college student), 2 (military public teacher). Note 2: *$p<.05$, **$p<.01$, ***$p<.001$.

Analysis of the Recognition of Profit Models

According to the users' opinions on making profits from data and on the profit sharing model, the analysis results were as follows: (1) Users generally know that social media use user data to make money. Among them, 20.1% of men and up to 31.7% of women are clearly aware of this phenomenon, especially professionals. (2) Regardless of gender or occupation, users disagree that social media use their data to make money. (3) However, when social media share profits with users, the acceptance of men and women increases significantly. (4) Users generally prefer the monetary reward model. (5) For respondents, it is acceptable to earn points based on the social interactions, in exchange for favorite goods or to offset cash. (6) The opinions on the multi-level marketing model are diverse.

In terms of the profit sharing models of social media, according to the questionnaire survey, users hope that social media count points, based on social interactions, will be objective and in line with their expectations. The order from high to low is as follows: 251 respondents (16.2%) chose "the number of times posts being viewed", 237 respondents (15.3%) chose "the number of times posts being liked", 234 respondents (15.1%) chose "the number of times videos that are being watched", and 206 respondents (13.3%) chose "the number of times posts are being forwarded/reposted".

With regard to the items exchanged for earned points, the order from high to low is as follows: 240 respondents (18.3%) chose "goods coupons", 230 respondents (17.5%) chose "goods vouchers", 220 respondents (16.7%) chose "to offset cash by spending in cooperative shops", 187 respondents (14.2%) chose "to convert to cash for charity", and 178 respondents (13.5%) chose "to give cash back by spending in cooperative shops". There were no gender differences in selecting the social interaction standards to count points.

CONCLUSION AND DISCUSSION

Social media have become a part of life, and our posts, likes, and shares on social networks are of commercial value. It is already a fact that companies use user data to make money, so users are free digital laborers. This study explored the issue that the information providers on social networks should receive fair and reasonable rewards.

In this study, the respondents were grouped according to their occupations, and two profit sharing models, namely, the "point earning-based profit sharing model" and the "multi-level marketing model", were proposed by preliminary collaborative interviews. After that, a questionnaire survey was carried out to understand the users' acceptance and recognition of the profit models.

In conclusion, regardless of their gender or occupation, respondents agreed with the profit sharing models proposed in this study. However, the proportion of the point earning-based profit sharing model was slightly higher, with the hope that social media would have such profit sharing programs in the future. According to the study results, the suggestions on social media are as follows:

1. privacy management mechanisms should be strengthened to protect personal information, so as to attract more users,
2. the concept of personal privacy should be advocated to the users of social media, and obvious warnings on privacy and security should be strengthened,

3. profit sharing models and bonus calculation should be simple and easy to understand, so as to increase the willingness to use and participate, and
4. social media should be able to provide free additional functions to reward users who have offer their personal information.

REFERENCES

Affleck, J., & Kvan, T. (2008). A virtual community as the context for discursive interpretation: A role in cultural heritage engagement. *International Journal of Heritage Studies*, *14*(3), 268–280. doi:10.1080/13527250801953751

Andrejevic, M. (2014). 'Free lunch' in the digital era: Organization is the new content. In *The Audience Commodity in a Digital Age: Revisiting a Critical Theory of Commercial Media* (pp. 193–206). Peter Lang Publishing.

Bandura, A. (1986). *Social foundations of thought and action: A social cognitive theory*. Prentice-Hall, Inc.

Bilbao-Osorio, B., Dutta, S., & Lanvin, B. (2013). *The global information technology report 2013*. World Economic Forum.

Blau, P. M. (1964). *Exchange and Power in Social Life. Wiley*.

Boyd, D., & Crawford, K. (2012). Critical questions for big data: Provocations for a cultural, technological, and scholarly phenomenon. *Information Communication and Society*, *15*(5), 662–679. doi:10.1080/1369118X.2012.678878

Boyle, K., & Johnson, T. J. (2010). MySpace is your space? Examining self-presentation of MySpace users. *Computers in Human Behavior*, *26*(6), 1392–1399. doi:10.1016/j.chb.2010.04.015

Brandtzaeg, P. B., Heim, J., & Kaare, B. H. (2010). Bridging and bonding in social network sites–investigating family-based capital. *International Journal of Web Based Communities*, *6*(3), 231–253. doi:10.1504/IJWBC.2010.033750

Caplan, S. E., & Turner, J. S. (2007). Bringing theory to research on computer-mediated comforting communication. *Computers in Human Behavior*, *23*(2), 985–998. doi:10.1016/j.chb.2005.08.003

Casilli, A. A. (2017). Global digital culturel digital labor studies go global: Toward a digital decolonial turn. *International Journal of Communication*, *11*, 21.

Chia, H. P., & Pritchard, A. (2014). Using a virtual learning community (VLC) to facilitate a cross-national science research collaboration between secondary school students. *Computers & Education*, *79*, 1–15. doi:10.1016/j.compedu.2014.07.005

Christofides, E., Muise, A., & Desmarais, S. (2009). Information disclosure and control on Facebook: Are they two sides of the same coin or two different processes? *Cyberpsychology & Behavior*, *12*(3), 341–345. doi:10.1089/cpb.2008.0226 PMID:19250020

Cosenz, F., & Bivona, E. (2020). Fostering growth patterns of SMEs through business model innovation. A tailored dynamic business modelling approach. *Journal of Business Research.*

Digital Marketing. (2015). *What are data brokers–and what is your data worth?* Retrieved from https: //www.webfx.com/blog/general/what-are-data-brokers-and-what-is-your-data-worth-infographic/

Dwyer, C., Hiltz, S., & Passerini, K. (2007). Trust and privacy concern within social networking sites: A comparison of Facebook and MySpace. *AMCIS 2007 proceedings*, 339.

European Commission. (2014). *Commission urges governments to embrace potential of big data.* Retrieved from European Commission press release. Retrieved from http: //europa.eu/rapid/press-release_IP-14-769_en.htm

Feigenbaum, L. (2014). *Turning big data into smart data.* Retrieved from https: //tdwi.org/Articles/2014/07/08/Turning-Big-Data-into-Smart-Data-1.aspx?Page=1

Fuchs, C. (2010). Labor in informational capitalism and on the Internet. *The Information Society*, *26*(3), 179–196. doi:10.1080/01972241003712215

Gambardella, A., & McGahan, A. M. (2010). Business-model innovation: General purpose technologies and their implications for industry structure. *Long Range Planning*, *43*(2-3), 262–271. doi:10.1016/j.lrp.2009.07.009

Gulnar, B., Balcı, S., & Cakır, V. (2010). Motivations of Facebook, Youtube and similar web sites users. *Bilig*, *54*, 161–184.

Haig, M. (2019). *Notes on a nervous planet.* Penguin.

Heider, A., Gerken, M., van Dinther, N., & Hülsbeck, M. (2020). Business model innovation through dynamic capabilities in small and medium enterprises–evidence from the German Mittelstand. *Journal of Business Research.*

Kaiser, B. (2019). *Targeted: The Cambridge Analytica Whistleblower's inside story of how big data, Trump, and Facebook broke Democracy and how it can happen again.* HarperCollins.

Kitchin, R. (2013). Big data and human geography opportunities, challenges and risks. *Dialogues in Human Geography*, *3*(3), 262–267. doi:10.1177/2043820613513388

Laffey, J., Lin, G. Y., & Lin, Y. (2006). Assessing social ability in online learning environments. *Journal of Interactive Learning Research*, *17*(2), 163–177.

Lee, I. (2017). Big data: Dimensions, evolution, impacts, and challenges. *Business Horizons*, *60*(3), 293–303. doi:10.1016/j.bushor.2017.01.004

Lenard, T. M., & Rubin, P. H. (2013). The big data revolution: Privacy considerations. Technology Policy Institute.

Magretta, J. (2002). Why business models matter. *Harvard Business Review*, *80*(5), 86–92. PMID:12024761

Mayer-Schönberger, V., & Cukier, K. (2013). *Big data: A revolution that will transform how we live, work, and think.* Houghton Mifflin Harcourt.

Nunes, M., & Correia, J. (2013). Improving trust using online credibility sources and social network quality in P2P marketplaces. *2013 8th Iberian Conference on Information Systems and Technologies (CISTI)*, 1-4.

Osterwalder, A., Pigneur, Y., Bernarda, G., & Smith, A. (2014). *Value proposition design: How to create products and services customers want.* John Wiley & Sons.

Osterwalder, A., Pigneur, Y., Oliveira, M. A. Y., & Ferreira, J. J. P. (2011). Business model generation: A handbook for visionaries, game changers and challengers. *African Journal of Business Management, 5*(7), 22–30.

Parker, G. G., Van Alstyne, M. W., & Choudary, S. P. (2016). *Platform revolution: How networked markets are transforming the economy and how to make them work for you.* WW Norton & Company.

Pedersen, C. L., & Ritter, T. (2020). Use this framework to predict the success of your big data project. *Harvard Business Review.*

Peter, J., Valkenburg, P. M., & Schouten, A. P. (2005). Developing a model of adolescent friendship formation on the Internet. *Cyberpsychology & Behavior, 8*(5), 423–430. doi:10.1089/cpb.2005.8.423 PMID:16232035

Qimei, C., Clifford, S. J., & Wells, W. D. (2002). Attitude toward the site II: New information. *Journal of Advertising Research, 42*(2), 33–45. doi:10.2501/JAR-42-2-33-45

Qimei, C., & Wells, W. D. (1999). Attitude toward the site. *Journal of Advertising Research, 39*(5), 27–37.

Rothaermel, F. T., & Sugiyama, S. (2001). Virtual internet communities and commercial success: Individual and community-level theory grounded in the atypical case of TimeZone. com. *Journal of Management, 27*(3), 297–312. doi:10.1177/014920630102700305

Sivarajah, U., Kamal, M. M., Irani, Z., & Weerakkody, V. (2017). Critical analysis of big data challenges and analytical methods. *Journal of Business Research, 70*, 263–286. doi:10.1016/j.jbusres.2016.08.001

Teece, D. J. (2010). Business models, business strategy and innovation. *Long Range Planning, 43*(2-3), 172–194. doi:10.1016/j.lrp.2009.07.003

Tykkyläinen, S., & Ritala, P. (2020). Business model innovation in social enterprises: An activity system perspective. *Journal of Business Research.*

Web, F. X. (2015). *What are data brokers–and what is your data worth?* Digital Marketing. https://www.webfx.com/blog/general/what-are-data-brokers-and-what-is-your-data-worth-infographic/

WikiPedia. (2019). *Right to be forgotten.* https://en.wikipedia.org/wiki/Right_to_be_forgotten

Wu, J. J., Chen, Y. H., & Chung, Y. S. (2010). Trust factors influencing virtual community members: A study of transaction communities. *Journal of Business Research, 63*(9/10), 1025–1032. doi:10.1016/j.jbusres.2009.03.022

Chapter 13
Study of Social Media Indulgence Among College Students in UAE and Kuwait:
Case Study

Ibrahim A. AlKandari
https://orcid.org/0000-0002-7389-9681
United Arab Emirates University, UAE

Badreya Al-Jenaibi
United Arab Emirates University, UAE

ABSTRACT

This chapter focuses on uses and gratifications of social media use among college students in the United Arab Emirates and in Kuwait for three social media platforms: Twitter, Instagram, and Snapchat. Mixed methodologies are duly applied (quantitative and qualitative) to explore various use and gratifications factors, as well as other social factors among a youth that contributes to the adoption of these social network sits (SNSs). Moreover, several statistical tests were performed to analyze collected data. A few research articles are published about new and social media platform use in the region; however, comparative studies were rarely noticed regarding this subject. The survey includes (N=190) samples between Kuwaiti and Emirati students. Conclusively, the study reveals that the main use and gratification reason for using the abovementioned social media platforms amongst college youth is entertainment, while the main social reason is identification.

INTRODUCTION

Technological developments and innovations have had a massive impact on the information and communication system (Kosyakova et al. 2020; Widjaja et al. 2020). Such technological advancement is important in the use of the internet to foster information dissemination and reception (Blok et al. 2020;

DOI: 10.4018/978-1-7998-9640-1.ch013

Rzheuskyi et al. 2020; Sarti et al. 2020). The social medium which has established several networks have gained admiration and receptiveness over the years (Killian et al. 2019; McClure and Seock, 2020). "The concept of Social Media (SM) has been on top of the agenda for many business executives" Pourkhani et al. 2019; p. 2)

In a general term, Social media are websites and internet-based applications that permit users to produce and share information. (Boyd & Ellison, 2008; Chiu el at., 2013) Individuals utilize social media websites such as Twitter, Instagram, Facebook, etc. to build and maintain relationships or reconnect with lost companions (Boyd & Ellison, 2007; Alwagait et al. 2015; Michikyan et al. 2015; Nasrullah and Khan, 2019). These social media collect personal data from users in the form of profile data as a requirement for establishing connection link (Pempek et al. 2009). However, social media communication is not only formed via individuals representing themselves by creating their profiles, but also those individuals are allowed to generate their own contents, display, and interact with the contents other, friends or other online users (Ding and Jiang 2014; Spasojevic et al. 2014; Zarrinkalam et al. 2015; Raghuram et al. 2016; Seghouani et al. 2019).

Social media has become an addiction to the users (Kaplan and Haenlein, 2010; Elantheraiyan and Shankarkumar, 2019; Nasrullah and Khan, 2019). It has been researched that use of Facebook comes on the second number after the use of several additive substances such as alcohol, cigarettes, and others where the individual finds it difficult to break the usage (Juergensen and Leckfor, 2018). Social media does not only seem to have effect on our society but it actually has changed our lives (Ariel and Avidar, 2014), even our way of thinking is changed due to the social websites. Social media has changed the way people behave (King, 2015). Social media has had impact on individuals (Chen et al. 2019; Talwar et al. 2019; Alalwan, 2018) and communities (Simon et al. 2013; Kamboj et al. 2018; Wang et al. 2019), organization (Namisango and Kang, 2019; Ihm, 2019; Steffens et al. 2019; Sideri et al. 2019), and politics (Hong and Nadler, 2012; Anim et al. 2019; Miller, 2019; Crilley and Gillespie, 2019). These are some of the several examples where social media has strong influence.

Social media use is common among the youth, including university students (Ahmad et al. 2019; Alnjadat et al. 2019; Feng, 2019; Köse and Doğan, 2019; Saide et al. 2019; Izuagbe et al. 2019; Pew Research Center, 2015) In the United States, Smith and Caruso (2010) stated that around 97% of university students are actively using SNSs on a daily basis. Another recent study in UK noted that college students spend up to six hours on SNS, Facebook in particular, a day. While it is possible that higher rates of SNS use might exist in other regions in the world among college students which have not been recorded yet (Daily Mail, 2014). As of January 2018, the social networking site usage penetration in UAE and Kuwait is 99% and 98%, respectively. Qatar also has a high 99% penetration rate, whereas Bahrain has 92% penetration rate, and Saudi Arabic has only 75% penetration rate (statista.com, 2019). This indicates that in the UAE and in Kuwait nearly everyone uses SNSs. As stated earlier, the main SNSs are Facebook, Instagram, Twitter, and LinkedIn. Authors such as Hawi and Samaha (2017) indicate that as of March 2016, there were 1.09 billion daily active users. The statistics for Instagram stood at 400 million monthly active users. At the same time, Twitter had 310 million active monthly users and LinkedIn had 433 million active users (Hawi and Samaha, 2017). These statistics indicate the high usage of Facebook, and the rate of social media addiction in Facebook users.

This high level of SNS use, and the addictive influence of these websited by university students raises concern among social sciences scholars worldwide (Koc & Gulyagci, 2013; Hawi and Samaha, 2017; Jasso-Medrano and López-Rosales, 2018; Köse and Doğan, 2019). For this reason, this research explores the theory of uses and gratifications regarding SNS use to investigate the reasons behind their adoption and continued use among university students in two countries: UAE and Kuwait. Several factors are presumed to influence student's behavior towards adoption and use of SNS, which include purposive, self-expression, entertainment, social interaction and other social reasons such as compliance, internalization and identification. The use of SNSs has influenced the behavior of its users to the extent that it modifies overtime, as users are keener to excessively share their life experiences, and increasingly seek to gain praise and commendation from their audience, which makes it stressful for them if they don't receive the expected benefit. Furthermore, seeing others who are more successful or appear to be doing better also causes depression, anxiety and feelings of inferiority among peers, causing further psychological problems (Karim et al, 2020). This aspect will be further explored in this paper regarding the social media usage and causes of psychological problems, especially among college students.

The paper uses both quantitative and qualitative methods: surveys and interviews, respectively.

The quantitative method is used to gather data from several participants – the students. The qualitative method is used to collect in-depth information from scholars – university professors. These are discussed further in the methodology section.

The aim of this paper is to answer the following questions:

- Q1: What are the most popular platforms used among Kuwaiti and Emirati Students?
- Q2: What are the common uses and gratification factors behind the social media sites?
- Q3: What are the common social factors behind the social media sites?
- Q4: Are there statistically significant differences between the average scores of social factors and the use of social media platforms that support nationality?
- Q5: What is the impact of uses and gratifications factors of using social media platforms: (Instagram, Twitter, Snapchat) on the social factors for Kuwaiti and Emirati students?

Based on our knowledge, in these two regions, few or no prior research has studied the factors that impact the use of SNSs; therefore, this current study will provide further explanations and information about SNS adoption in Gulf region. It is noteworthy to mention that few research has been completed examining the factors likely to influence students' intensive involvement of SNSs (Park et al., 2009; Turel & Serenko, 2012). In this research, pervasive use refers to the extent where students are willing to approve SNSs as a central part of their lives (Turel & Serenko, 2012; Vannoy & Palvia, 2010). Knowledge of the influence of such factors on students' use of SNSs will help professors, parents, and society to understand the characteristics, needs and values of this specific segment of the population who represent the most important part of society, and who generate its energy and power. Moreover, the outcome of this research will help service providers, business operations, university administrators, and application developers to create and build proper strategies and platforms which meet the needs of this group. Conclusively, data collection in this research will provide a knowledge base and add accumulative information to the previous literature, especially for Arab world.

LITERATURE REVIEW

Social Media Platforms and their Uses

Social media has gained significant attention in the postmodern era, and has become a compelling power within which society, politics, economics, and education operate. "Social media are interactive computer-mediated technologies that facilitate the creation and sharing of information, ideas, career interests and other forms of expression via virtual communities and networks" (Kietzmann & Hermkens, 2011, p. 241). Social media has unlocked fresh challenges and opportunities for businesses, organizations, and individuals (Ngai, et al., 2015). Individuals, organizations, and business companies use different platforms of social medial to meet their goals and purposes, and to share with the public as much information as possible. Examples of these platforms are Facebook, Twitter, Snapchat and Instagram, explained in the following sections.

Facebook

Mark Zuckerberg founded Facebook in 2004. It is the biggest social network on the web with over 2.23 billion users (Krishen et al. 2019). Facebook connects people around the world on formal or informal terms, and the uses of this platform are endless as the company bought other social media platforms, and expanded horizons for users to connect and share on a wide scale. Facebook established a platform for users to share photos, videos, important company updates and more (Bowers, 2017).

Twitter

Twitter was founded in 2006 by Jack Dorsey and a mobile social networking site and micro-blogging (Rayaee & Ahmed, 2015). On the Twitter social media platform, users can share short texts coupled with images, videos, links, polls and more. It allows a fast connection between people who wish to share their opinions and contents around the globe, just like other social media platforms do. Active users of Twitter were 232 Million users in January 2014. In 2012, Twitter has overtaken Facebook in Kuwait in terms of market share where it increased sharply from 6.56% to 64.18% from 2011 to 2013. Moreover, in the UAE, there was a noticeable increase in the use of Twitter from 2.72% to 13.6% between the years 2011-2013 (Rayaee & Ahmed, 2015). Companies and brands use Twitter to display and advertise special information about their products, and their locations (Bowers, 2017). Services have rapidly grown in popularity worldwide, as a result. In 2012, greater than 100 million users posted around 340 million tweets per day. As of 2016, Twitter had more than 319 million active users per month.

Snapchat

Snapchat is a mobile social media platform that allows users to send and receive time-sensitive photos and videos that expire upon viewing (Stec, 2015). The number of Snapchat users has increased significantly during recent years as the platform gained popularity among the youth due to attractive options it provides, and modality affordances. The record ability affordance permits users to post photos, videos, and text messages, which disappear after 24 hours. Regarding the modality affordance feature of Snapchat, users can communicate with others via photographs and video clips (up to 10 s long), and they are capable of adding filters to their photos and videos (Waddell, 2016). Snapchat usage in UAE grew from 15%-53% and in Saudi Arabia increased from 24% to 74% between the years 2014-2016 compared to the global average of 12%-23% (Radcliffe & Lam, 2016).

Instagram

Instagram is a photo-sharing social media application that allows individuals to take pictures, record videos, apply filters to them, and share these videos and photos on the platform itself, and/or on other social media platforms such as Facebook and Twitter (Stec, 2015). According to Instagram (2016), 80% of the platform users are out of U.S. and the active users are more than 600 million (Al-Kandari et al. 2017). It includes more than 400 million active users every month who poste over 40 billion pictures, with an average of 3.5 billion daily likes for more than 80 million pictures that are shared every day on the website (Instagram, 2016). More than half of young adults of age 18–29 years old are reported to utilize Instagram, and they represent the largest group of Instagram users (Duggan, 2015a; Duggan et al., 2015) The new live feature of Instagram allows users to record live-stream videos and live chatting. This new feature has further attracted young generations towards using Instagram, since they are the generation of the fast and the immediate. According to a survey distributed on nationals in Lebanon, Egypt, UAE, Tunisia, and Saudi Arabia of internet users, it is found that Instagram has overtaken Twitter in the region. Moreover, based on the study published by the research agency kantar TNS named *Connected Life,* the global average users for Instagram increased from 24% to 42% between the years 2014 to 2016. The users of Instagram in UAE jumped from 38% to 60% and in Saudi Arabia grown from 57% to 82% from 2014 to 2016 (Radcliffe, 2016).

Social influence

"Social influence refers to the way in which individuals change their behavior to meet the demands of a social environment. It takes many forms and can be seen in conformity, socialization, peer pressure, obedience, leadership, persuasion, sales, and marketing. In 1958, Harvard psychologist Herbert Kelman identified three broad varieties of social influence.

- Compliance is when people appear to agree with others but keep their dissenting opinions private.
- Identification is when people are influenced by someone who is liked and respected, such as famous celebrities.
- Internalization is when people accept a belief or behavior and agree both publicly and privately" (Kelman, 1958, p. 51)

Other views may affect people's behavior significantly. People may choose a specific technology as a result of being influenced by the views of others, and not because of personal preference. (Bagozzi & Dholakia, 2002; Cheung & Lee, 2009; Dholakia, Bagozzi, & Pearo, 2004; Malhotra & Galletta, 1999). When behaviors and opinions of people are affected by external factors it is called social influence (Chiu et al., 2013). Compliance is the first type of social influence that is defined by (Kelman, 1958, 1974) as the act of agreeing with others. The second type is identification, which is defined as being influenced by others within the social group (Cheung et al., 2011; Kelman, 1958, 1974). A third type of social influence is internalization, which means accepting behaviors and beliefs that suites self-value system (Cheung & Lee, 2010; Kelman, 1958, 1974; Malhotra & Galletta, 1999).

In terms of compliance, lacking suitable knowledge or information of the technology may lead people to choose that technology simply because it is accepted by others (Cheung et al., 2011). Regarding the identification process, people may want to have satisfied relationships within their social groups so that they may accept a specific technology (Bagozzi & Lee, 2002; Cheung et al., 2011). A study conducted by Adil et al has revealed that university students and other young adults also use social media as means to boost their self-esteem. This occurs especially with individuals who have the tendency to create virtual relationships with others, make new friends, comment on each other's pictures and posts, and fulfill their need to belong (Adil et al, 2020). Social media apps and websites also fulfill the need for social comparison, social enhancement, and self-evaluation. Regarding the process of internalization, people may accept a technology because it fits with their value system (Cheung & Lee, 2010; Malhotra & Galletta, 1999).

Uses and Gratifications Theory

Uses and gratifications theory (UGT) clarifies reasons why and how people choose media platforms to satisfy or fulfill certain needs (Katz et al. 1974). UGT assumes that people are goal-oriented, and know their needs when using media; they can assess value judgments of media content, and can connect the needs and gratifications to a particular choice of media platform (Katz et al., 1974; West & Turner, 2007). Conventionally, UGT has been used in the context of conventional media, such as radio, TV, and newspapers (Luo et al. 2011; McQuail, 2010; Pai & Arnott, 2013; Tankard Jr. and James, 2000; West & Turner, 2007). However, modern studies have used UGT to identify needs and gratifications in the framework of Computer-Mediated Communication (CMC) technologies (Cheung & Lee, 2009; Cheung et al., 2011; Dholakia et al., 2004; Ku, Chu, & Tseng, 2013; Luo et al., 2011). Based on Ku et al. 2013, p. 573), "People choose to use a particular CMC technology, instead of other media choices, to communicate with others in order to fulfill their particular gratification needs."

UGT considered a valuable framework for studying students' persistent implementation of SNSs because they are a new Computer-Mediated Communication (CMC) tools. It suggests that users will remain using SNSs if they are satisfied by those tools (Ku et al., 2013).

According to the research, uses gratifications include various categories such as entertainment, information seeking, escapism enhanced social interaction, and more (Katz et al., 1974; McQuail, 2010; Tankard and James, 2000). Earlier, factors were used by researchers like previous categories for SNS studies (Cheung et al., 2011; Ku et al., 2013; Raacke & Bonds-Raacke, 2008; Pai & Arnott, 2013). We will use five categories of UGT (i.e., purposive value, self-discovery, entertainment value, social enhancement, and maintaining interpersonal connectivity) that are similar to the aforementioned gratifications categories. These categories of UGT have been widely adopted for CMC technologies (Cheung & Lee, 2009; Cheung et al., 2011; Pai & Arnott, 2013).

Purposive value means values that a person gains from doing something with a pre-known information and helpful purpose (Cheung et al., 2011; Leung & Wei, 2000).

Self-discovery is defined as a person's tendency to understand aspects of oneself through online group contribution (Cheung et al., 2011; Raacke & Bonds-Raacke, 2008).

Maintaining interpersonal connectivity reflects social benefits a person gets from initiating and sustaining contact with other people in an online network (Cheung et al., 2011; Ku et al. 2013).

Social enhancement is about values a person receives from gaining the approval and acceptance of others as well as enhancing social status within the online network (Cheung et al., 2011).

Entertainment value refers to the enjoyment and fun a person gains from interacting with others in an online network (Ku et al. 2013; Pai & Arnott, 2013).

RESEARCH DESIGN AND METHODOLOGY

This research studies the use of SNS among University students in Kuwait and the UAE. The use of SNS is common among individuals. As stated earlier, there is high penetration rate of SNS in both these countries. However, extended use of SNS can also be harmful. Literature also shows that SNS has been identified to be addictive in many cases. This can be harmful on the overall behavior of the individual. Therefore, this research aims to study the gratification that students achieve from the use of SNS and the factors that influence the use of SNS.

The research methodology is mixed using a deductive approach. As stated above the aim to understand student perspective on the use of SNS. Therefore, the primary research method is quantitative, however, the researcher has also considered the perception of teaching staff towards students' usage of SNS. This is done through interviews (qualitative method). Some scholars name using both of them as multi-methods (Brannen, 1992), mixed methodology (Tashakkori & Teddlie, 1998), or mixed methods (Creswell, 2003; Tashakkori & Teddlie, 2003).

Quantitative method

Quantitative method is suitable to answer the research questions as the primary aim is to understand the perception of students towards SNS. Quantitative method is used in social sciences research when the researcher bases the study on existing literature. The aim is therefore to base the study on existing theories and adopt a framework based on literature. Quantitative study tests existing theories through hypotheses verification (Saunders et al. 2016). There are however limited studies in their field that is carried out between these two countries. Therefore, much of the literature pertains to studies that are outside these countries.

One of the common techniques of collecting data through quantitative method is survey using questionnaires. These are self-administered questiónnaires wherein the respondents provide answer often in the absence of the researcher. Saunders et al. (2016) stress on the importance of ensuring confidentiality, anonymity, and ease of understanding the questionnaire by the respondent. It is also recommended to design the questionnaire based on existing studies that have tested the questionnaire. This research has used various literature sources to design the questionnaire as closed ended. The demographic is provided with multiple options and the items (statements for each model variable) is designed using a 6-point Likert scale.

Quantitative data is numeric and therefore the researcher has to use statistical tools to analyze the raw data to make sense and to study the relationship between variables. This research has used SPSS for analyzing the data and discussing the results. The types of analyses that are carried out are discussed in the Data Analysis section.

Qualitative method

Qualitative method is used by researchers to gain in-depth understanding into the phenomenon. Unlike quantitative method, qualitative method requires researcher to be part of the data collection (Saunders et al. 2016). This also means that the researcher has to interact with the participants as part of the data collection. The time consumed in collecting data through qualitative methods are longer compared to the quantitative method. However, this is preferred when the researcher cannot base the study on existing theories or when there is lack of previous studies. The researcher is able to produce stronger information through qualitative method of data collection compared to quantitative method.

One of the common techniques of collecting qualitative data is through interviews. Interviews are classified into structured, semi-structured and unstructured (Saunders et al. 2016). This research has used semi-structured interviews. This means that a set of questions are prepared prior to the interviews and these questions guide the interviews. Semi-structured interviews also provide the opportunity for the researcher and participant to discuss more about the questions asked.

Qualitative methods capture large amount of data and would require to be analyzed to produce understandable outcomes. This research has used thematic analysis and discussed the data.

Descriptive and analytical survey was used for verification of the hypotheses, in addition to the use of the mutual relations approach for the study of relational ties between multiple variables. The survey is a cost-effective method for correcting large amounts of data (Lynn, 2009). Surveys are also flexible in the sense that they can be sent to a sample in different geographical areas (Mellenbergh, 2008). Also, anonymity and confidentiality of the data can be ensured, which encourages a higher response rate (Groves et al. 2009). Structured interviewing was selected as a method because it is "easy to replicate, as a fixed set of closed questions are used, which are easy to quantify—this means it is easy to test for reliability" (McLeod, 2014; p. 2). Also, Opdenakker (2006) indicated that structured interviews are a time-efficient method.

Qualitative study

Interviews

Numbers of interviews were conducted with professors and educational professionals from universities of both countries to investigate their opinion on students' use of SNSs in education. The below table shows the demographics of nine interviewees, who were interviewed by the researcher. Interviews conducted by email. The researcher contacted the interviewees before emailing them the questions except four interviewees were contacted through other people. Six interviewees were from United Arab Emirates and three from Kuwait. Interview held during March and April 2018.

Table 1. Demographic information about interviewees

N/A. Not all interviewees mentioned their age.	Age
Male N= 7 Female N= 2	Gender
Ph.D.	Educational level
Mass Communication, Linguistics, Strategic Communication, HRD and OD, Geography, MIS, and Marketing.	Major
Alittihad Newspaper, Media, Abu Dhabi University, Zayed University, UAEU, and Kuwait University.	workplace
UAE, Egypt, Jordan, and Kuwait.	Nationality

Structured interviews applied in this paper were the researcher passed fixed questions to the interviewees. All of them sent by email. Mostly, open-ended questions were used such as "How students use social media to support their studies?" and "What are the challenges of using social media tools?"

Email interviews took a month approximately to receive the answers from the interviewees. The researcher chose to conduct interviews because the nature of the topic needs some discussion and open-ended questions. Furthermore, the researcher for analysis followed thematic analysis approaches. All similar and different answers were written in themes.

DISCUSSION OF FINDINGS

Social media usage

Out of all interviewees from both UAE and Kuwait, four interviewees stated that they do not use social media in class in particular for teaching; however, it is still important as a communication medium with students. Some of them using SM for sharing and posting students work. One the other hand, five interviewees do use social media platforms as a teaching method in class neither to apply theoretical concepts. "I need to show them examples of tweets, Facebook messages, Instagram captions, etc." (Interviewee # 2, 2018) or "for announcement purposes" (Interviewee # 8, 2018).

Tools Used – What and How

Professors who use social media platforms mostly use the most famous platforms such as Twitter, Instagram, Snapchat, Facebook, and LinkedIn. Furthermore, the nature of the content considers the main driver of which platform to use. Some platforms allow posting long texts and contents and some only accept short posts. "Facebook is a good platform for debates and content that needs more volume of text" (Interviewee # 7, 2018). Interviewee # 8 (2018) mentioned that twitter is the only SM platform used during the course trying not to overwhelm students with information and use one way of communication. "We use multiple tools; Twitter, YouTube, but most frequently we use Instagram for its convenience and ease of use" (Interviewee # 9, 2018).

Bright Side of Social Media

All interviewees agreed that SM has many positive impacts within the educational field. Instructors can share students' works online and test their ideas. Moreover, SM platforms give students the feeling of being engaged. According to interviewee # 2 (2018), social media help students to improve their language. Speed is one of the main advantages of social media where students can either extract or post information immediately. "Speed and accessibility are the main benefits for most students use this technology in obtaining information" (Interviewee # 8, 2018). Although it's called "Social Media" however many students may be shy and do not prefer direct contact with other. "it is more suitable for students who feel "shy" to talk to their instructors in class or in office hours as face to face" (Interviewee # 5, 2018). As more, SM promotes interaction, virtual communication skills; involve students to share their thoughts and being responsible in front of the public. Moreover, Interviewee # 9 (2018) added "most importantly it keeps my students engaged and interested".

Side effects

Although social media has several advantages on education, but also there are some negatives based on interviewees' opinion. One out of nine interviewees did not mention any negatives about social media where the rest of them mentioned some. They do believe that those SM platforms could be misused from the students such as posting without permission or without weighting their words. "Students are in critical age and they sometimes don't weight their words, they could write anything that could affect the reputation of the university (unintentionally)" (Interviewee # 1, 2018). Moreover, the truth may be manipulated. Shared content can be misleading such as giving false feedback about the institution. Consequently, controlling and privacy issues will appear. Identity issues as well. Interviewee # 8, (2018) mentioned that these new technologies can be distracting and time consuming as well. Interviewee # 8 proceeded that students should not rely on information offered in these platforms as facts always. "I realized that a lot of the material presented in these Web 2.0 tools is garbage" (Interviewee # 8, 2018).

Challenges

All nine interviewees brought up interested issues they may affect SM platforms somehow. The first thing is regulating the use of social media within institutions to ensure usage is not damaging. Having restricted rules and regulations from the university is important as well as train the students how to use those platforms properly and spreading awareness. Interviewee # 4 (2018) stated that the use must be assured with responsibility. "… We can educate people about using SM smartly and holding them accountable for any missus or damage they may cause." (Interviewee # 6, 2018). Furthermore, we need to have code of ethics moderation.

Moreover, the cultural barrier is one of the issues mentioned during one of the interviews. Some families may still do not accept or will not allow their children to use these technologies. "the most challenging part is cultural barrier" (Interviewee # 8, 2018). Another challenge is the Social media platforms algorithms. As Interviewee # 9 (2018) stated "Most of the social media apps change their algorithms regularly, that affects which posts we see first. This may result in not seeing my content once posted".

Furthermore, Al-Hunaiyyan et al. (2016) pointed that the significant aspects in the implementation of mobile-learning in education are technical difficulties. Some of these difficulties could be availability of latest technology, installation, security, administration, maintenance, in addition to fast internet connection (Qureshi et al., 2012)

Analysis of data

The researcher adopted descriptive comparative research style because this type of research concern about the phenomena and provide accurate description to these phenomena.

Sample

The population of the study composed of students from universities of two regions: the University of Kuwait and UAE University. The survey is conducted during the second semester of the academic year 2017/2018, between March and April 2018 exactly. The sample of the study was chosen according to the simple random sampling method, which gives each member an equal opportunity to be chosen.

Table 2. Description of the population

		Frequency	Percent
Country	UAE	82	42.3
	KWU	112	57.7
Year of the study	first	21	10.8
	second	32	16.5
	third	50	25.8
	fourth	60	30.9
	Fifth	31	16.0
	Total	194	100.0

In the sample of the study, the sample consisted of (N=82) students from United Arab Emirates University of about 42.3% and Kuwait University (N=112) with 57.7% of the current study sample. The average age of the whole sample was 21 with a standard deviation of 1.90. For UAE University sample, the average is 21.71 and standard deviation 1.50 while for Kuwait University sample the average is 20.48 and standard deviation of 1.99. Table 2 below describes the distribution of the study population according to the variables of the study and the academic year.

Table 2 shows that for the year of study variable, the majority of the students are in their fourth year of study with 30.9%, followed by the third-year classes with 25.8%. The ratios of fifth and second grades were 16.0%, slightly higher for the second grade. The percentage of participants of the first grade has the lowest participation rates with only 10.8%.

Tools of the Study

To achieve the aims of the study, a questionnaire consisting of four parts, was used as follows:

First, Demographic information includes age, year of study and nationality. Second, the survey contains open questions about the intensity of use, most common used platform and number of hours spent on SNS. Third, the questionnaire investigates social factors that might influence user adoption such as compliance, internalization, and identification, with total of 9 items. Finally, the fourth part includes 21 items distributed among the following uses and gratification factors: information sharing, self-documentation, social interaction, entertainment, passing time, self-expression and finally frequency of use. Users also use it for the purpose of self-presentation on social media apps such as Twitter, Instagram, Snapchat, and Facebook for the purpose of presenting a well thought image (Alsalem, 2019).

In general, the survey follows 6-point Likert scale as the following: Strongly disagree=1, Disagree=2, Somewhat disagree=3, Neither agree or disagree=4, Somewhat agree=5, and Agree=6. This scale was particularly used in order to increase response rate as well as response quality.

Validity Analysis of the Study

The survey questions were viewed by another scholar for approval. Moreover, the validity of the survey was completed via calculation of construct related validity of both factors: social and uses and gratifications. This is computed by obtaining the correlation between the constructs and the total degree. This was achieved by the pilot study conducted over 40 students from both universities: Kuwait University and UAE University. The results of this study are presented below in table 3 and table 4.

Table 3. Correlation coefficient between the social factor and the total degree

Factor	Social Factors
Compliance	.802**
Identification	.881**
Internalization	.288**

**Correlation is significant at the 0.01 level (2-tailed).

Table 4. Correlation coefficient between uses and gratification factors of the three platforms and the total degree

Factor	Instagram	Twitter	Snapchat
Information sharing	.791**	.818**	.825**
Self-documentation	.765**	.818**	.868**
Social interaction	.791**	.891**	.805**
Entertainment	.718**	.560**	.688**
Passing time	.753**	.708**	.379*
Self-expression	.740**	.832**	.811**
Frequency	.535**	.421*	.461**

It can be noted from the tables that the correlation coefficient is positively significant at 0.01 for the social factors.

Reliability Analysis of the Study

The research uses stability coefficient Cronbach's Alpha to achieve the stability of the pilot sample and the results are shown as follows in table 5.

Table 5. Cronbach Alpha coefficient of pilot study (N = 40)

Cronbach Alpha	Item	Construct
0.867	9	Social Factors
0.899	21	Uses and gratifications (Instagram)
0.913	21	Uses and gratifications (Twitter)
0.888	21	Uses and gratifications (Snapchat)

The table indicates higher values of Cronbach's Alpha coefficient for both constructs; this means that both factors have high level of stability.

Data analysis

The data collected was analyzed via special statistical software called Statistical Package for the Social Sciences, SPSS. Several tests were performed to analyze the data. These are standard deviation, means, frequency, percent, t-tests and correlation as well as regression tests. The analysis of the data is shown below.

Table 6 shows the analysis for the following question: How many hours on average? The result shows that the average use of SNS is about 7.60 hours and the least use is two hours while the range use is 16 hours per day.

Table 6. Descriptive results of number of hours spent of the three platforms

Mean	7.60
Median	7.00
Variance	14.356
Std. Deviation	3.789
Minimum	2
Maximum	18
Range	16

What Platform do you use? The results indicate that the most common social media platform used are Instagram and Snapchat with 94.3% followed by WhatsApp with 89.2% then YouTube with about 88.7% and finally Twitter with 80.4%. While the other platforms achieved between (2.1% - 26.3%)

Table 7 and table 8 show the analysis of social factor in terms of mean, standard deviation and rank of every item in the construct. From the below tables it is noticed that the three items: compliance, identification and internalization achieved high level of agreement among participants with 4.39 mean and standard deviation around 0.89. Based on the ranking of the three factors, identification achieved the highest mean rate for all its items especially for the item "*My group of friends is proud of using SNSs*" with 4.62 mean. Second highest factor is internalization, and the highest mean belongs to the item "*The reason my group of friends use SNSs is that such tools provide some underlying social benefit*" with the value of 4.69. Finally, the compliance factor has scored the least mean of 4.41 for the item "*My group of friends' views of SNSs use is similar to mine*". The results emphasize that the use of social media allows individuals to identify themselves to be a part of a group and they generally believe that using the social media websites will allow them to gain some benefits.

Table 7. Standard deviation and mean of Social Factors

Factor		item	Mean	Std. Deviation	rank
Compliance	1	My group of friends' views of SNSs use is similar to mine.	4.41	1.50	1
	2	It is necessary for my group of friends to use SNSs to be accepted among peers.	4.04	1.40	2
	3	Unless my group of friends sees the benefits of using SNSs, they would see no reason to spend extra effort in using such tools	3.90	1.25	3
Identification	4	In our interactions, my group of friends feels a sense of ownership about the use of SNSs.	4.42	1.22	2
	5	My groups of friends talk up the use of SNSs to other friends that have not joined.	4.30	1.34	3
	6	My group of friends is proud of using SNSs.	4.62	1.23	1
Internalization	7	SNSs are important to my group of friends.	4.62	1.40	2
	8	The reason my group of friends use SNSs is that such tools provide some underlying social benefits.	4.69	1.22	1
	9	My group of friends likes using SNSs primarily because of the similarity between our values and the benefits that such tools provide.	4.54	1.23	3

Table 8. Standard deviation and mean of every item of social factor construct

	Mean	Std. Deviation	Rank
Compliance	4.11	1.02	3
Identification	4.45	1.05	1
Internalization	4.16	1.45	2
Social Factors	4.39	0.89	

Table 9 show the analysis of uses and gratification construct for the three SNSs: Instagram, Twitter and Snapchat. From the below table it can be noted that Snapchat has achieved the highest mean for the item "*entertainment*" with mean 4.76 followed by "*self-documentation*" with 4.43 mean value. On the other hand, Instagram and Twitter has also achieved the same highest mean value for the item "*entertainment*" while the least mean for Instagram is 3.33 for "Self-expression" and the smallest mean value for Twitter is 3.84 for "*passing time*".

Table 9. Standard deviation and mean of the three social media sites

Snapchat			Twitter			Instagram			
Rank	**Std. Deviation**	**Mean**	**Rank**	**Std. Deviation**	**Mean**	**Rank**	**Std. Deviation**	**Mean**	**Factor**
3	1.676	4.29	2	1.768	4.31	3	1.521	4.10	Information sharing
2	1.597	4.43	6	1.564	3.48	5	1.520	3.57	Self-documentation
4	1.456	4.23	3	1.583	4.26	2	1.451	4.16	Social interaction
1	1.402	4.76	1	1.360	4.65	1	1.360	4.65	Entertainment
6	1.421	3.84	4	1.421	3.84	4	1.221	3.80	Passing time
5	1.608	4.05	5	1.709	3.80	6	1.575	3.33	Self-expression
7	1.165	3.27	7	1.233	3.26	7	1.028	3.17	Frequency
	1.099	4.13		1.167	3.89		.974	3.80	total

Regarding the question "Are there statistically significant differences between the average scores of social factors and the use of social media platforms that support nationality?", T-test were performed on independent samples to answer this question. Table 10 below shows that t-test indicates that there is a significance difference between the mean of the responses of the sample participants over the nationality variable towards the students of UAE University which is less than 0.05.

Table 10. Statistical analysis of nationality variable over social factors

Country		N	Mean	Std. Deviation	t	df	sig
Compliance	UAE	82	4.30	1.03	2.23	192	0.027
	KWU	112	3.97	0.99			
Identification	UAE	82	4.63	1.08	2.13	192	0.035
	KWU	112	4.31	1.01			
Internalization	UAE	78	3.75	1.49	-3.37	192	0.001
	KWU	109	4.45	1.36			
Social Factors	UAE	82	4.59	0.88	2.69	192	0.008
	KWU	112	4.25	0.87			

Moreover, with respect to the question "Are there statistically significant differences between the average scores of uses and gratifications factors and the use of three social media platforms to support nationality? T-test was also used to obtain the answer for each platform.

Instagram

Table 11 indicates that the results of t-test reveal a significance difference between the mean of the responses of the participants over the nationality variable that is less than 0.05 towards the students of Kuwait University for the two items: social interaction and self-expression.

Table 11. Statistical analysis of Nationality over uses and gratifications factors for Instagram

Country		N	Mean	Std. Deviation	t	df	sig
Information sharing	UAE	79	4.00	1.58	0.77	192	0.439
	KWU	109	4.17	1.48			
Self-documentation	UAE	78	3.45	1.69	0.88	192	0.380
	KWU	109	3.65	1.39			
Social interaction	UAE	78	3.75	1.49	3.37	192	0.001
	KWU	109	4.45	1.36			
Entertainment	UAE	78	4.79	1.35	1.14	192	0.256
	KWU	109	4.56	1.36			
Passing time	UAE	78	3.74	1.31	0.59	192	0.553
	KWU	109	3.85	1.16			
Self-expression	UAE	78	3.03	1.56	2.23	192	0.027
	KWU	109	3.54	1.56			
Frequency	UAE	78	3.21	1.09	0.38	192	0.703
	KWU	109	3.15	0.98			
Instagram	UAE	79	3.68	1.08	1.49	192	0.137
	KWU	109	3.89	0.88			

Twitter

Table 12 indicates there is a significance difference that is less than 0.05 between the mean of the responses of the participants over the nationality variable towards the students of Kuwait University for the social interaction factor.

Table 12. Statistical analysis of Nationality over uses and gratifications factors for Twitter

Country		N	Mean	Std. Deviation	t	df	sig
Information sharing	UAE	74	4.09	1.83	1.40	192	0.164
	KWU	101	4.47	1.71			
Self-documentation	UAE	74	3.36	1.69	0.86	192	0.391
	KWU	101	3.57	1.47			
Social interaction	UAE	74	3.91	1.69	2.55	192	0.012
	KWU	101	4.52	1.46			
Entertainment	UAE	78	4.79	1.35	1.14	192	0.256
	KWU	109	4.56	1.36			
Passing time	UAE	73	3.84	1.55	0.01	192	0.995
	KWU	101	3.84	1.33			
Self-expression	UAE	74	3.68	1.84	0.76	192	0.449
	KWU	101	3.88	1.61			
Frequency	UAE	74	3.47	1.35	1.96	192	0.052
	KWU	101	3.11	1.13			
Twitter	UAE	74	3.82	1.32	0.69	192	0.494
	KWU	101	3.94	1.04			

Snapchat

Table 13 shows that the value of t-test indicates a significance difference between the mean of the responses of the participants over the nationality variable that is less than 0.05 towards the students of UAE University for the two items: information sharing and social interaction

Table 13. Statistical analysis of Nationality over uses and gratifications factors for Snapchat

Country		N	Mean	Std. Deviation	t	df	sig
Information sharing	UAE	80	4.58	1.66	2.06	192	0.041
	KWU	106	4.07	1.66			
Self-documentation	UAE	80	4.69	1.60	1.94	192	0.054
	KWU	106	4.23	1.57			
Social interaction	UAE	80	4.50	1.54	2.17	192	0.031
	KWU	106	4.03	1.37			
Entertainment	UAE	80	4.94	1.37	1.57	192	0.117
	KWU	106	4.62	1.41			
Passing time	UAE	73	3.84	1.55	0.01	192	0.995
	KWU	101	3.84	1.33			
Self-expression	UAE	80	4.23	1.66	1.30	192	0.194
	KWU	106	3.92	1.56			
Frequency	UAE	80	3.45	1.32	1.84	192	0.068
	KWU	106	3.13	1.02			
Snapchat	UAE	80	4.37	1.15	2.66	192	0.008
	KWU	106	3.94	1.03			

Furthermore, regarding the question "What are the impact of uses and gratifications factors of using the social media platforms: (Instagram, Twitter, Snapchat) on the social factors for Kuwaiti and Emirati students?" Linear regression test is used to obtain the answer of this question.

Table 14. Linear regression analysis of social factor that influence the use of social media.

sig	t	R^2	β	Platform	Students
0.067	1.86	0.294	0.213	Twitter	UAE University
F=9.56 sig 0.000					
0.295	1.05	0.097	0.123	Twitter	Kuwait University
F=3.23 sig 0.026					

Table 14 shows that the results of the regression analysis indicate the effect of social media on social factors. The relationship is significant for the students of UAE University at 0.094, around 29.4%, with f value less than 0.01. The remaining percentages belong to other reasons. Similarly for Kuwait University students with a lower contribution of about 9.7% and an f value that is less than 0.05. In general, the table indicates that twitter is the only platform that has the effect on social factors for both universities with contribution rate of 21.3% for UAE University students and 12.3% contribution among Kuwait University students. While the other two social media platforms, Snapchat and Instagram, has no contribution on this relationship.

DISCUSSION AND CONCLUSION

The media dominates as a source for information providing source to the general public in the democratic country like micro blogging and it has a power of great influence on the general public to form the opinions and the behaviour of the people (Domizi, 2013; Pihl & Sandström, 2013). Role of media becomes a matter of higher significance (Lebedko, 2014), when there is an election period in the democratic country as political campaigns and the public views about the political personalities are presented on the media (Shirky, 2011). Social media has made the changes in the process of interaction between the students and society (Oberst et al. 2016). Social media tools including Facebook, Twitter, Snapchat and Instagram has allowed the students to make direct contact with the others, without spending any money (Hargittai, 2007). This ensure that people remain connected with others and are able to make them feel a part of the social bubble. Social media has evolved during its evolution (Al-Jenaibi, 2016). Due to this evolution, the way of communication has entirely changed. In olden times it was thought to be impossible to send someone a message and hoping that he or she will receive the message just a click of a button away. This is because in that period there were no mobile phones and internet, so people used to send their messages on a piece of paper through courier services to different parts of the world (Al-Jenaibi, 2015). Now it is just a click away to send someone a message. In fact, we can now have a direct contact with our friends and relatives through internet, which are thousands of miles away from us. However, this is not necessarily an improvement to one-to-one relationships because social media usage results in loss of emotional connection and generates feelings of loneliness specially among teenagers and young adults. This can further lead to depression, anxiety, and stress as the pro-longed effects of social media addiction become more obvious overtime (Christensen, 2018).

The survey has shown that Snapchat and Instagram are the most popular platforms and has the same degree of approval among university students followed by WhatsApp and lastly twitter. Perhaps, thus because of the practical features of these two platforms that interest this group segment.

Moreover, from the data analysis, we can notice that there are social factors behind student's use of SNSs. The most common social reason for adopting SNSs is identification. It is obvious that this group of the population concerns about establishing and maintaining self- defining relationship with their social group. Youth consider using SNSs and technological advancement as a source of pride regardless the view of their benefits. This explains why compliance achieved the least approval in particular for the item that concern about the friends' view of the SNSs benefits. While for the uses and gratification factors for the three platforms: Instagram, Snapchat and Twitter, entertainment is the main reason behind using these SNSs. The three SNSs have almost the same average use with a slightly higher frequency of use for Snapchat. For Snapchat, the second largest item is self-documentation; this can be explained by the recording feature of the program. Yet, for Instagram the second main reason is social interaction, and the least factor is self-expression. Youth views Instagram as a source of social communication tool rather than a source of self-expression. Finally, twitter is viewed as a source of information for this segment of the population rather than self-documentation tool.

In conclusion, in Arab region, particularly in gulf area, Snapchat is the most popular SNSs adopted by university students followed by Instagram and Twitter. The three platforms are viewed as an entertainment tool. Through social media now we can connect to someone almost immediately, get right to the point, and get the information we need almost instantaneously (Al-Jenaibi, 2017).

Limitation of the Study

There are limited and almost no prior comparative studies within the researched topic in the region. Many studies were published about social media in the GCC but have not compared UAE and Kuwait like this study. This study can provide future studies such as some social media forgetting topics. For example, the damage of values and people's culture with the excessive uses of social networks tools. Future studies must focus on youth limitation of social media usage, as this use should not have any negative impacts on their jobs and mainly on their health. Another future topic is the advancement in these technologies which increase the standards of living. Man should control these technologies, unless it starts controlling the man.

Recommendations

Since Social Media platforms are playing an essential role in students' behavior and daily life, it is recommended that the developers pay more attention to the needs and requirements of the users during the development stage, as well as, to consider the various features of the users' devices. Moreover, Educational institutions are advised to collaborate with Social Media platforms developers to find out how they can implement such technologies in education and to maintain the uses and gratifications of the students and overcome the challenges. In addition, more rules and regulations must be created for regulating the use of SM in education and not to damaging the educational institutions. There should be support systems within the society to help young adults cope with the side effects of excessive social media usage such as depression, anxiety, and stress. The usage of social media should be monitored among teenagers and they should be educated from the beginning regarding effects of the use of social media if used in an excessive manner. These support systems could be integrated within colleges and universities. Finally, study the persistence use of SNSs and include other regions as well.

REFERENCES

Adil, A. H., Mulla, M., Mulla, M., & Ramakrishnappa, S. (2020). *Usage of Social Media among Undergraduate University Students*. https://www.researchgate.net/publication/347399224_Usage_of_Social_Media_among_Undergraduate_University_Students

Ahmad, T., Alvi, A., & Ittefaq, M. (2019, July-September). The Use of Social Media on Political Participation Among University Students: An Analysis of Survey Results from Rural Pakistan. *SAGE Open*, *9*(3), 1–9. doi:10.1177/2158244019864484

Al-Hunaiyyan, A., Alhajri, R. A., & Al-Sharhan, S. (2016). Perceptions and challenges of mobile learning in Kuwait. *Journal of King Saud University-Computer and Information Sciences*.

Al-Jenaibi, B. (2015). The New Electronic Government: Are the federal authorities ready to use e-government? *International Journal of Knowledge Society Research*, *6*(3), 45–74. doi:10.4018/IJKSR.2015070104

Al-Jenaibi, B. (2015). Current issues about public relations professionals: Challenges and potentials of PR in UAE organisations. *Middle East Journal of Management*, *2*(4), 330–351. doi:10.1504/MEJM.2015.073568

Al-Jenaibi, B. (2016). Upgrading Society with Smart Government: The use of smart services among Federal Offices of the UAE *International Journal of Information Systems and Social Change, 7*(4), 20–51. doi:10.4018/IJISSC.2016100102

Al-Jenaibi, B. (2017). The impact of dubbed serials on students in the UAE. *International Journal of Arab Culture Management and Sustainable Development, 3*(1), 41–66. doi:10.1504/IJACMSD.2017.10007172

Al-Kandari, A. A., Al-Sumait, F. Y., & Al-Hunaiyyan, A. (2017). Looking perfect: Instagram use in a Kuwaiti cultural context. *Journal of International and Intercultural Communication, 10*(4), 273–290. doi:10.1080/17513057.2017.1281430

Alalwan, A. A. (2018). Investigating the impact of social media advertising features on customer purchase intention. *International Journal of Information Management, 42*, 65–77. doi:10.1016/j.ijinfomgt.2018.06.001

Alhabash, S., & Ma, M. (2017). A Tale of Four Platforms: Motivations and Uses of Facebook, Twitter, Instagram, and Snapchat Among College Students? *Social Medial + Society*, 1-2.

Alnjadat, R., Hmaidi, M. M., Samha, T. E., Kilani, M. M., & Hasswan, A. M. (2019). Gender variations in social media usage and academic performance among the students of University of Sharjah. *Journal of Taibah University Medical Sciences, 14*(4), 390–394. doi:10.1016/j.jtumed.2019.05.002 PMID:31488973

Alsalem, F. (2019). Why do they Post? Motivations and uses of Snapchat, Instagram and Twitter Among KUWAIT college students. *Media Watch, 10*(3). Advance online publication. doi:10.15655/mw/2019/v10i3/49699

Alwagait, E., Shahzad, B., & Alim, S. (2015, October). Impact of social media usage on students' academic performance in Saudi Arabia. *Computers in Human Behavior, 51*, 1092–1097. doi:10.1016/j.chb.2014.09.028

Andreas, K., & Michael, H. (2010). Users of the world, unite! The challenges and opportunities of social media. *Business Horizons, 53*, 61.

Anim, P. A., Asiedu, F. O., Adams, M., Achempong, G., & Boakye, E. (2019). "Mind the gap": To succeed in marketing politics, think of social media innovation. *Journal of Consumer Marketing, 36*(6), 806–817. doi:10.1108/JCM-10-2017-2409

Ariel, Y., & Avidar, R. (2014). Information, Interactivity, and Social Media. *Atlantic Journal of Communication, 23*(1), 19–30. doi:10.1080/15456870.2015.972404

Aronson, E., Timothy, D. W., & Akert, R. M. (2010). *Social psychology*. Prentice Hall.

Bagozzi, R. P., & Dholakia, U. M. (2002). Intentional social action in virtual communities. *Journal of Interactive Marketing, 16*(2), 2–21. doi:10.1002/dir.10006

Bagozzi, R. P., & Lee, K. H. (2002). Multiple routes for social influence: The role of compliance, internalization, and social identity. *Social Psychology Quarterly, 65*(3), 226–247. doi:10.2307/3090121

Blok, M., van Ingen, E., de Boer, A. H., & Slootman, M. (2020). The use of information and communication technologies by older people with cognitive impairments: From barriers to benefits. *Computers in Human Behavior, 104*(106173), 1–9. doi:10.1016/j.chb.2019.106173

Bowers, J. (2017, August 13). *Social Media for Business: A Marketer's Guide*. Retrieved from Business New Daily: https://www.businessnewsdaily.com/7832-social-media-for-business.html

Boyd, & Ellison, N. B. (2007). Social Network Sites: Definition, History, and Scholarship. *Journal of Computer-Mediated Communication, 13*(1), 210–230. doi:10.1111/j.1083-6101.2007.00393.x

Boyd, D., & Ellison, N. (2007). Social network sites: Definition, history, and scholarship. *Journal of Computer-Mediated Communication, 13*(1), 210–230. doi:10.1111/j.1083-6101.2007.00393.x

Boyd, D. M., & Ellison, N. B. (2008). Social network sites: Definition, history and scholarship. *Journal of Computer-Mediated Communication, 13*(1), 210–230. doi:10.1111/j.1083-6101.2007.00393.x

Charmaz, K. (2006). *A Practical Guide through Qualitative Analysis*. Sage.

Chen, S., Schreurs, L., Pabian, S., & Vandenbosch, L. (2019). Daredevils on social media: A comprehensive approach toward risky selfie behavior among adolescents. *New Media & Society, 21*(11-12), 2443–2462. doi:10.1177/1461444819850112

Cheung, C. M. K., Chiu, P.-Y., & Lee, M. K. O. (2011). Online social networks: Why do students use Facebook? *Computers in Human Behavior, 27*(4), 1337–1343. doi:10.1016/j.chb.2010.07.028

Cheung, C. M. K., & Lee, M. K. O. (2009). Understanding the sustainability of a virtual community: Model development and empirical test. *Journal of Information Science, 35*(3), 279–298. doi:10.1177/0165551508099088

Cheung, C. M. K., & Lee, M. K. O. (2010). A theoretical model of intentional social action in online social networks. *Decision Support Systems, 49*(1), 24–30. doi:10.1016/j.dss.2009.12.006

Chiu, C.-M., Cheng, H.-L., Huang, H.-Y., & Chen, C.-F. (2013). Exploring individuals' subjective well-being and loyalty towards social network sites from the perspective of network externalities: The Facebook case. *International Journal of Information Management, 33*(3), 539–552. doi:10.1016/j.ijinfomgt.2013.01.007

Christensen, S. P. (2018). *Social Media Use and Its Impact on Relationships and Emotions*. https://scholarsarchive.byu.edu/cgi/viewcontent.cgi?article=7927&context=etd

Crilley, R., & Gillespie, M. (2019). What to do about social media? Politics, populism and journalism. *Journalism, 20*(1), 173–176. doi:10.1177/1464884918807344

DailyMail. (2014). *University students spend six hours a day on Facebook, YouTube and sending texts—even during lectures*. http://www.dailymail.co.uk/news/ article-2664782/University-students-spend-SIX-HOURS-day-Facebook- YouTube-sending-texts-lectures.html

Dennis, E. E., Martin, J. D., Wood, R., & Saeed, M. (2016). Media Use in the Middle East. Doha: Mideastmedia.org.

Dholakia, U. M., Bagozzi, R. P., & Pearo, L. K. (2004). A social influence model of consumer participation in network- and small- group-based virtual communities. *International Journal of Research in Marketing, 21*(3), 241–263. doi:10.1016/j.ijresmar.2003.12.004

Ding, Y., & Jiang, J. (2014). Extracting interest tags from twitter user biographies. Information Retrieval Technology, 268-279.

Domizi, D. (2013, January 10). Microblogging To Foster Connections And Community in a Weekly Graduate Seminar Course. *TechTrends, 57*(1), 43–45. doi:10.100711528-012-0630-0

Duncan, F. (2016, February 2). *So long social media: The kids are opting out of the online public sphere.* The Conversation. Retrieved from https://theconversation.com/so-long-social-media-the-kidsare-opting-out-of-the-online-public-square-53274

Elantheraiyan, P., & Shankarkumar, S. (2019). A Research on Impact of Social Media on College Students in Chennai District. *International Journal of Innovative Technology and Exploring Engineering, 8*(11S), 675–679. doi:10.35940/ijitee.K1114.09811S19

Feng, D. (2019). Interdiscursivity, social media and marketized university discourse: A genre analysis of universities' recruitment posts on WeChat. *Journal of Pragmatics, 143*, 121–134. doi:10.1016/j.pragma.2019.02.007

Groves, R. M., Fowler, F. J., Couper, M. P., Lepkowski, J. M., Singer, E., & Tourangeau, R. (2009). *Survey Methodology*. John Wiley & Sons.

Hallikainen, P. (2015). *Why People Use Social Media Platforms: Exploring the Motivations and Consequences of Use.* Academic Press.

Hargittai, E. (2007). Whose Space? Differences Among Users and Non-Users of Social Network Sites. *Journal of Computer-Mediated Communication, 13*(1), 276–297. doi:10.1111/j.1083-6101.2007.00396.x

Hawi, N. S., & Samaha, M. (2017). The Relations Among Social Media Addiction, Self-Esteem, and Life Satisfaction in University Students. *Social Science Computer Review, 35*(5), 576–586. doi:10.1177/0894439316660340

Hong, S., & Nadler, D. (2012). Which candidates do the public discuss online in an election campaign?: The use of social media by 2012 presidential candidates and its impact on candidate salience. *Government Information Quarterly, 29*(4), 455–461. doi:10.1016/j.giq.2012.06.004

Ifinedo, P. (2016). Applying uses and gratifications theory and social influence processes to understand students' pervasive adoption of social networking sites: Perspectives from the Americas. *International Journal of Information Management, 36*(2), 192–206. doi:10.1016/j.ijinfomgt.2015.11.007

Ihm, J. (2019). Communicating without nonprofit organizations on nonprofits' social media: Stakeholders' autonomous networks and three types of organizational ties. *New Media & Society, 21*(11-12), 2648–2670. doi:10.1177/1461444819854806

Instagram. (2016). *Our story: A quick walk through our history as a company*. Retrieved from https://www.instagram.com/ press/?hl=en

Izuagbe, R., Ifijeh, G., Izuagbe-Roland, E. I., Olawoyin, O. R., & Ogiamien, L. O. (2019). Determinants of perceived usefulness of social media in university libraries: Subjective norm, image and voluntariness as indicators. *Journal of Academic Librarianship*, *45*(4), 394–405. doi:10.1016/j.acalib.2019.03.006

Jasso-Medrano, J. L., & López-Rosales, F. (2018). Measuring the relationship between social media use and addictive behavior and depression and suicide ideation among university students. *Computers in Human Behavior*, *87*, 183–191. doi:10.1016/j.chb.2018.05.003

Juergensen, J., & Leckfor, C. (2018). Stop Pushing Me Away: Relative Level of Facebook Addiction Is Associated With Implicit Approach Motivation for Facebook Stimuli. *Psychological Reports*, *122*(6), 2012–2025. doi:10.1177/0033294118798624 PMID:30189800

Kamboj, S., Kumar, V., & Rahman, Z. (2017). Social media usage and firm performance: The mediating role of social capital. *Social Network Analysis and Mining*, *7*(1:51), 1-14.

Kamboj, S., Sarmah, B., Gupta, S., & Dwivedi, Y. (2018). Examining branding co-creation in brand communities on social media: Applying the paradigm of Stimulus-Organism-Response. *International Journal of Information Management*, *39*, 169–185. doi:10.1016/j.ijinfomgt.2017.12.001

Kang, Y. S., Min, J., Kim, J., & Lee, H. (2013). Roles of alternative and self-oriented perspectives in the context of the continued use of social network sites. *International Journal of Information Management*, *33*(3), 496–511. doi:10.1016/j.ijinfomgt.2012.12.004

Kaplan, M., & Haenlein, M. (2010). "Users of the world, unite! The challenges and opportunities of social media"(PDF). *Business Horizons*, *53*(1), 61. doi:10.1016/j.bushor.2009.09.003

Karim, F., Oyewande, A., Abdalla, L. F., Chaudhry Ehsanullah, R., & Khan, S. (2020). Social media use and its connection to mental Health: A systematic review. *Cureus*. Advance online publication. doi:10.7759/cureus.8627 PMID:32685296

Katz, E., Blumler, J. G., & Gurevitch, M. (1974). Utilization of mass communication by the individual. In J. Blumler & E. Katz (Eds.), *The uses of mass communications: current perspectives on gratifications research* (pp. 19–32). Sage.

Kelman, H. (1958). Compliance, identification, and internalization: Three processes of attitude change. *The Journal of Conflict Resolution*, *1*(1), 51–60. doi:10.1177/002200275800200106

Kelman, H. C. (1974). Social influence and linkages between the individual and the social system: further thoughts on the processes of compliance, identification, and internalization. In J. T. Tedeschi (Ed.), *Perspectives on social power* (pp. 125–171). Aldine.

Kietzmann, J. H., Hermkens, K., McCarthy, I. P., & Silvestre, B. S. (2011). Social media? Get serious! Understanding the functional building blocks of social media. *Business Horizons*, *54*(3), 241–251. doi:10.1016/j.bushor.2011.01.005

Killian, S., Lanon, J., Murray, L., Avram, G., Giralt, M., & O'Riordan, S. (2019). Social Media for Social Good: Student engagement for the SDGs. *International Journal of Management Education*, *17*(100307), 1–12. doi:10.1016/j.ijme.2019.100307

Kim, G., Park, S.-B., & Oh, J. (2010). An examination of factors influencing consumer adoption of short message service (SMS). *Computer Human Behavior*, 1152–1161.

King, D. (2015). Why Use Social Media? *Library Technology Reports*, *51*(1), 6–9.

Koc, M., & Gulyagci, S. (2013). Facebook addiction among Turkish college students: The role of psychological health, demographic, and usage characteristics. *Cyberpsychology, Behavior, and Social Networking*, *16*(4), 279–284. doi:10.1089/cyber.2012.0249 PMID:23286695

Köse, Ö. B., & Doğan, A. (2019). The Relationship between Social Media Addiction and Self-Esteem among Turkish University Students. *Turkish Green Crescent Society.*, *6*(1), 175–190.

Kosyakova, I. V., Zhilyunov, N. Y., & Astashev, Y. V. (2020). Prospects for the Integration of Environmental Innovation Management on the Platform of Information and Communication Technologies. *Advances in Intelligent Systems and Computing.*, *908*, 345–355. doi:10.1007/978-3-030-11367-4_34

Krishen, A. S., Berezan, O., Agarwal, S., & Kachroo, P. (2019). Social media networking satisfaction in the US and Vietnam: Content versus connection. *Journal of Business Research*, *101*, 93–103. doi:10.1016/j.jbusres.2019.03.046

Ku, Y.-C., Chen, R., & Zhang, H. (2013). Why do users continue using social networking sites? An exploratory study of members in the United States and Taiwan. *Information & Management*, *50*(7), 571–581. doi:10.1016/j.im.2013.07.011

Ku, Y. C., Chu, T. H., & Tseng, C. H. (2013). Gratifications for using CMC technologies: A comparison among SNS, IM, and e-mail. *Computers in Human Behavior*, *29*(1), 226–234. doi:10.1016/j.chb.2012.08.009

Lang, N. (2015). Why teens are leaving Facebook: It's 'meaningless.' *The Washington Post*. Retrieved from https://www. washingtonpost.com/news/the-intersect/wp/2015/02/21/whyteens-are-leaving-facebook-its-meaningless/

Lebedko, M. (2014). Globalization, Networking and Intercultural Communication. *Intercultural Communication Studies*, *23*(1), 28–41.

Leong, L.-Y., Hew, T.-S., Ooi, K.-B., Lee, V.-H., & Hew, J.-J. (2019). A hybrid SEM-neural network analysis of social media addiction. *Expert Systems with Applications*, *133*, 296–316. doi:10.1016/j.eswa.2019.05.024

Leung, L. (2013). Generational Differences in Content Generation in Social Media: The Roles of the Gratifications Sought and of Narcissism. *Computers in Human Behavior*, *29*(3), 997–1006. doi:10.1016/j.chb.2012.12.028

Leung, L., & Wei, R. (2000). More than just talk on the move: A use-and-gratification study of the cellular phone. *Journalism & Mass Communication Quarterly*, *77*(2), 308–320. doi:10.1177/107769900007700206

Luo, M. M., Chea, S., & Chen, J. S. (2011). Web-based information service adoption: A comparison of the motivational model and the uses and gratifications theory. *Decision Support Systems*, *51*(1), 21–30. doi:10.1016/j.dss.2010.11.015

Malhotra, Y., & Galletta, D. F. (1999). Extending the technology acceptance model to account for social influence: theoretical bases and empirical validation. *Proceedings of the 32nd Hawaii international conference on system sciences*, 1–11. 10.1109/HICSS.1999.772658

Matthews, C. (2014, January 15). Facebook: More than 11 million young people have fled Facebook since 2011. *Time Magazine*. Retrieved from http://business.time.com/2014/01/ 15/more-than-11-million-young-people-have-fled-facebooksince-2011/

McClure, C., & Seock, Y.-K. (2020). The role of involvement: Investigating the effect of brand's social media pages on consumer purchase intention. *Journal of Retailing and Consumer Services*, *53*(101975), 1–8. doi:10.1016/j.jretconser.2019.101975

McLeod, S. (2014). *The Interview Method*. Retrieved on April 17, 2014 fromhttp://www.simplypsychology.org/interviews.html

McQuail, D. (2010). *Mass communication theory: An introduction*. Sage Publications.

Michikyan, M., Subrahmanyam, K., & Dennis, J. (2015). Facebook use and academic performance among college students: A mixed-methods study with a multi-ethnic sample. *Computers in Human Behavior*, *45*, 265–272. doi:10.1016/j.chb.2014.12.033

Miller, M. F. (2019). Why Hate the Internet? Contemporary Fiction, Digital Culture, and the Politics of Social Media. *Arizona Quarterly: A Journal of American Literature, Culture, and Theory*, *75*(3), 59–85.

Namisango, F., & Kang, K. (2019). Organization-public relationships on social media: The role of relationship strength, cohesion and symmetry. *Computers in Human Behavior*, *101*, 22–29. doi:10.1016/j.chb.2019.06.014

Nasrullah, S., & Khan, F. R. (2019). Examining the Impact of Social Media on the Academic Performances of Saudi Students - Case Study: Prince Sattam Bin Abdul Aziz University. *Humanities and Social Sciences Reviews.*, *7*(5), 851–861. doi:10.18510/hssr.2019.75111

Ngai, K-L., S.S, K. M., Lam, E. S., Chin, S., Tao, S., & Eric, W. (2015). *Social media models, technologies, and applications: An academic review and Case Study*. Emerald Insight.

O'Keefe, G., & Pearson, K. (2011). *The Impact of Social Media on Children, Adolescents, and Families*. American Academy of Pediatrics.

Obar, J. A., & Wildman, S. (2015). Social media definition and the governance challenge: An introduction to the special issue. *Telecommunications Policy*, 745–750.

Oberst, U., Chamarro, A., & Renau, V. (2016). Gender Stereotypes 2.0: Self-Representations of Adolescents on Facebook. *Media Education Research Journal*, *24*(48), 81–89.

Opdenakker, R. (2006). *Advantages and Disadvantages of Four Interview Techniques in Qualitative Research*. Retrieved on April 17, 2014, from: https://www.qualitative-research.net/index.php/fqs/article/view/175/391

Pai, P., & Arnott, D. C. (2013). User adoption of social networking sites: Eliciting uses and gratifications through a means-end approach. *Computers in Human Behavior*, *29*, 1039–1053.

Park, J. H. (2014). The effects of personalization on user continuance in social networking sites. *Information Processing & Management*, *50*(3), 462–475.

Park, N., Kee, K. F., & Valenzuela, S. (2009). Being immersed in social networking environment: Facebook groups, uses and gratifications, and social outcomes. *Cyberpsychology & Behavior*, *12*(6), 729–733.

Pempek, T. A., Yermolayeva, Y. A., & Calvert, S. L. (2009). College students' social networking experiences on facebook. *Journal of Applied Developmental Psychology*, 227–238.

Pew Research Center. (2015). *Social networking fact sheet*. http://www. pewinternet.org/fact-sheets/social-networking-fact-sheet/

Pourkhani, Abdipour, Baher, & Moslehpour. (2019). The impact of social media in business growth and performance: A scientometrics analysis. *International Journal of Data and Network Science*, *3*, 223–244.

Raacke, J., & Bonds-Raacke, J. (2008). MySpace and Facebook: Applying the uses and gratifications theory to exploring friend-networking sites. *Cyberpsychology & Behavior*, *11*(2), 169–174.

Radcliffe, D., & Lam, A. (2018). *Social Media in the Middle East: The Story of 2017*. Academic Press.

Raghuram, M., Akshay, K., & Chandrasekaran, K. (2016). Efficient user profiling in twitter social network using traditional classifiers. *International Journal of Intelligent Systems Technologies and Applications*, 399–411.

Reyaee, S., & Ahmed, A. (2015). Growth Pattern of Social Media Usage in Arab Gulf States: An Analytical Study. *Social Networking*, *4*(2), 23.

Rossi, E. (2002). *Uses & Gratifications/ Dependency Theory*. Retrieved on June 15, 2014 from: http://zimmer.csufresno.edu/~johnca/spch100/7-4-uses.htm

Rzheuskyi, A., Matsuik, H., Veretenikova, N., & Vaskiv, R. (2020). Selective Dissemination of Information – Technology of Information Support of Scientific Research. *Advances in Intelligent Systems and Computing.*, *871*, 235–245.

Saide, S., Inrajit, R. E., Trialih, R., Ramadhani, S., & Najamuddin, N. (2019). A theoretical and empirical validation of information technology and path-goal leadership on knowledge creation in university Leaders support and social media trend. *Journal of Science and Technology Policy Management.*, *10*(3), 551–568.

Sarti, D., Torre, T., & Pirani, E. (2020). Information and Communication Technologies Usage for Professional Purposes, Work Changes and Job Satisfaction. Some Insights from Europe. *Lecture Notes in Information Systems and Organisation.*, *33*, 165–177.

Saunders, M., Lewis, P., & Thorhill, A. (2016). Research Methods for Business Students (7th ed.). Pearson Education Limited.

Seghouani, N. B., Jipmo, C. N., & Quercini, G. (2019). Determining the interests of social media users: Two approaches. *Information Retrieval Journal.*, *22*, 129–158.

Severin, W., & Tankard, J. (1997). *Communication theories: Origins, methods, and uses in the mass media*. Longman.

Sheth, N., Newman, B. I., & Gross, B. L. (1991). Why we buy what we buy: A theory of consumption values. *Journal of Business Research*, *22*(2), 159–170. doi:10.1016/0148-2963(91)90050-8

Shirky, C. (2011). The Political Power of Social Media: Technology, the Public Sphere, and Political Change. *Foreign Affairs*, *90*(1), 28–41.

Sideri, M., Kitsiou, A., Filippopoulou, A., Kalloniatis, C., & Gritzalis, S. (2019). E-Governance in educational settings Greek educational organizations leadership's perspectives towards social media usage for participatory decision-making. *Internet Research*, *29*(4), 818–845.

Simon, C., Brexendorf, T. O., & Fassnacht, M. (2013). Creating online brand experience on Facebook. *Marketing Review St. Gallen*, *30*, 10.

Smith, S. D., & Caruso, J. B. (2010). *Research Study. ECAR study of undergraduate students and information technology* (Vol. 6). EDUCAUSE Center for Applied Research. https://www.educause.edu/Resources/ECARStudyofUndergraduateStuden/217333

Spasojevic, N., Yan, J., Rao, A., & Bhattacharyya, P. (2014). LASTA: Large scale topic assignment on multiple social networks. KDD, 1809–1818.

Sponcil, M., & Gitimu, P. (n.d.). Use of social media by college students: Relationship to communication and self-concept. *Journal of Technology Research.*

statista.com. (2019). *Active social media penetration in Middle East & North African countries in January 2018.* Retrieved from: https://www.statista.com/statistics/309668/active-social-media-penetration-in-arab-countries/

Steffens, M. S., Dunn, A. G., Wiley, K. E., & Leask, J. (2019). How organisations promoting vaccination respond to misinformation on social media: A qualitative investigation. *BMC Public Health*, *19*(1348), 1–12.

Talwar, S., Dhir, A., Kaur, P., Zafar, N., & Alrasheedy, M. (2019). Why do people share fake news? Associations between the dark side of social media use and fake news sharing behavior. *Journal of Retailing and Consumer Services*, *51*, 72–82.

Tankard & James. (2000). *New media theory. Communication theories: origins, methods and uses in the mass media*. Reading. MA: Addison Wesley Longman.

Vannoy, S. A., & Palvia, P. (2010). The social influence model of technology adoption. *Communications of the ACM*, *53*(6), 149–153.

Wang, X.-W., Cao, Y.-M., & Park, C. (2019). The relationships among community experience, community commitment, brand attitude, and purchase intention in social media. *International Journal of Information Management*, *49*, 475–488.

West, R., & Lynn, H. (2010). *Uses and Gratifications Theory. Introducing Communication Theory: Analysis and Application*. McGraw-Hill.

West, R., & Turner, L. (2007). Introducing communication theory (4th ed.). New York: McGraw-Hill.

Widjaja, B. T., Sumintapura, I. W., & Yani, A. (2020). Exploring the triangular relationship among information and communication technology, business innovation and organizational performance. *Management Science Letters.*, *10*, 163–174.

Zarrinkalam, F., Fani, H., Bagheri, E., Kahani, M., & Du, W. (2015). Semantics-enabled user interest detection from twitter. *WI-IAT*, *1*, 469–476.

Chapter 14
Efficient Data Clustering Techniques for Software-Defined Network Centres

Vinothkumar V.
Department of Computer Science and Engineering, Jain University, India

Muthukumaran V.
https://orcid.org/0000-0002-3393-5596
REVA University, India

Rajalakshmi V.
REVA University, India

Rose Bindu Joseph
https://orcid.org/0000-0002-7033-6226
Christ Academy Institute for Advanced Studies, India

Meram Munirathnam
Rajiv Gandhi University of Knowledge Technologies, India

ABSTRACT

In a smart system, a software-defined network (SDN) is frequently used to monitor and manage the communication organisation. Large-scale data analysis for SDN-based bright networks is gaining popularity. It's a potential technique to deal with a large amount of data created in an SDN-based shrewd lattice using AI advancements. Nonetheless, the disclosure of personal security information must be considered. Client power conduct examination, for example, may result in the disclosure of personal security information due to information bunching. Clustering is an approach for displaying models' observations, data items, or feature vectors in groups. Batching addresses has been catered to in various interesting circumstances and by masters in distinct requests; it gleams far-reaching attractiveness and assistance as one of the ways in exploratory data examination and moreover increases the genuine assessment of data. In this chapter, the authors conduct a study of packing and its various types and examine the computation. Finally, they use it to create an outline model.

DOI: 10.4018/978-1-7998-9640-1.ch014

INTRODUCTION

As per JSTOR information bunching first showed up in the title of a 1954 chapter managing anthropological information. Q-examination, typology, scientific categorization and climbing are different names of information bunching depend on various field. The accompanying books are some old style books which expand what is grouping and clarifies bunching calculations (Imamverdiyev, Y., and Abdullayeva, F, 2018; Wang, Z, 2015; Tang, T. A et al., 2016). Bunching calculations have additionally been broadly concentrated in information mining books by Han and Kamber. It is an undertaking of information focuses into various gatherings with the end goal that information focuses in similar gatherings are more like other information focuses in a similar gathering than those in different gatherings. In basic words, the point is to isolate bunches with comparative attributes and appoint them into clusters (Sadhasivam, J et al., 2021). Information grouping has been concentrated in the Machine Learning, Statistics networks with various techniques and various accentuations. Grouping is an exploratory information examination apparatus for finding the hidden order the information. Its motivation is to separate a lot of unaided items into regular gatherings so the information objects inside each gathering share some comparability and the information objects across various gatherings are unique . There are different grouping calculations have been created in the writing in various logical orders. The peruser can study bunching calculation and its application's improvement of the web of things, distributed computing, and informal organizations through web. As a result of the high calculation time we can't be apply straightforwardly customary calculations. The greatest test is the means by which to improve grouping computational effectiveness. By the expanding the size of the chapters the exploration on grouping is grown an ever increasing number of most recent couple of decades parallelly it increment extent of the bunching. The huge scope information bunch has two sorts of unavoidable arrangements (Akhtar, N et al., 2018; Liu, Q et al., 2018). That are dispersed calculation and information decrease. The central issues of conveyed calculation grouping calculations are to plan a suitable examining plan for picking delegate objects. In this paper we center around group types, some numerical calculations which are utilized in bunching and apply it for straightforward informational collection we'll see the what is the consequence of the example information.

RELATED WORK

In an information investigation, bunching device is getting more significant in the time of enormous information. For huge scope information bunching, inspecting is a proficient and most broadly utilized guess method. A portion of the inspecting based bunching calculations have pulled in impressive consideration in huge scope information examination attributable to their effectiveness. There are additionally exist some basic calculation that moves toward the law grouping exactness. Xing wang et.al clarifies the strategies of testing bunch. The strategy is getting various delegate tests from various beginning with a defined inspecting plan, which are shaped by region delicate hashing method, subsequent to examining we make apportioning the picked tests into various groups utilizing the bunching calculation. At that point relegating the out of test objects into their nearest groups through information marking procedure. The presentation of the proposed calculation is contrasted and the best in class inspecting bunching calculations on a few informational indexes including. Canyi Lu et.al proposes investigations of the subspace bunching issue (Nan, Y et al., 2020; Dada, E. G et al., 2019; Sommer, R., 2010; Akhtar, N et al., 2018;

Liu, Q et al., 2018). Here creators take some information focuses around drawn from an association of subspaces, the objective is to bunch these information focuses into their hidden subspaces. Subspace bunching techniques have been proposed among which meager subspace grouping and low-position portrayal are two agent ones. We see that there are many existing techniques own the regular square inclining property, which potentially prompts right bunching. In this creator proposes the main square slanting framework incited regularize for straightforwardly seeking after the square inclining lattice and take care of the subspace bunching issue by Block Diagonal Representation (BDR) for utilizing block corner to corner structure earlier. Jinquan Zhang et.al examine the significance of grouping examination utilized broadly in design acknowledgment and picture preparing, which is a significant exploration field of information mining. In interpersonal organizations distributed information was undermined by the spillage of private data these days. Creator proposes the security safeguarding plan of touchy information distributing in interpersonal organizations dependent on Balanced Iterative Reducing and Clustering utilizing Hierarchies (BIRCH) calculation to handle this issue. This has determined as on the web and disconnected cycle. Utilizing the Maximum Delay Anonymous Clustering Feature (MDACF) calculation. Creator makes mindful of social information security assurance procedure can't give security insurance to interpersonal organization information. Since this procedure just thinks about that the assailant utilizes the quality estimation of each record in the social information as the foundation information for protection assaults, overlooking the connection between the hubs in the interpersonal organization, the informal community chart structure and the hub map, and so forth. By DBSCAN grouping the information security assurance method can be differential protection necessities and have great insurance impact. This gives ensure on static informational index and it can't utilized legitimately to handle information streams. The principle propose of this paper is an information stream security assurance plot dependent on BIRCH bunching in information distributing, and gives the Maximum Delay Anonymous Clustering Feature (MDACF) tree information distributing calculation.

METHODOLOGY

In data clustering the object is considered as data set. An object set is denoted by $X = x_1,\ldots,x_N$ in which $x_k,(k = 1,2,\ldots,N)$ is an object. With a few exceptions, $x_1,\ldots,x_N$ are vectors of real p-dimensional space R^p. A generic element $x \in R^p$ is the vector with real components $x^1,\ldots,x^p$; we write $x = (x^1,\ldots,x^p) \in R^p$.

Dissimilarity and Cluster Centre

Bunching of an information is assessing by nearest separation of the information. This implies objects are set in a topological space, and the nearest separation is estimated by utilizing a divergence between two items to be grouped. The disparity between a couple of dataset x, x∈X is meant by $D(x,x)$ which takes a real value.

Metric Space

To quantify a difference we additionally have think about the idea of metric space. The measurement is indicated as m(x,y). It characterized on a space S which fulfills the accompanying aphorisms

i. $m(x,y) \geq 0$ and $m(x,y) = 0 \Leftarrow x = y$
ii. $m(x,y) = m(y,x)$
iii. Triangular inequality: $m(x,y) \leq m(x,z) + m(z,y)$.

Euclidean Distance

On the off chance that the chapter fulfills the measurement space since we can ascertain the separation between the items by Euclidean separation. The mainstream metric for consistent element is the Euclidean separation. Calculation of Euclidean separation as follows eqn(1)

$$D(x,y) = d_2^2 \,||x - y||_2^2 = \sum_{j=1}^{p} (x^j - y^j)^2 \tag{1}$$

Convergent

United closes the degree of the grouping. In the event that the chapter is united, at that point the reaches end else it will be proceed. So the joined assumes a major function of making a decision about the bunch. There is two different ways of finding united.

a. There is no progressions between first enrollment and last participation object
b. There is no progressions between two item's centroid.

Deep Learning Techniques

Recent studies suggest that deep learning methods have outshined the traditional methods such as Support Vector Machine (SVM), Naive Bayes, Random Forests, k-means clustering, logistic regression models etc., in terms of findings and performances. "Traditional machine learning" refers to the wide range of old-fashioned methods that are integral part of ML for decades until "Deep learning" came into the picture for at least past 5 to 10 years. Based on functionality, deep learning algorithms are categorized into three types and architecture explained in Fig.1.

1. Supervised Learning
2. Unsupervised Learning
3. Semi-Supervised Learning
4. Reinforcement Learning

Figure 1. Deep Learning examples Architecture

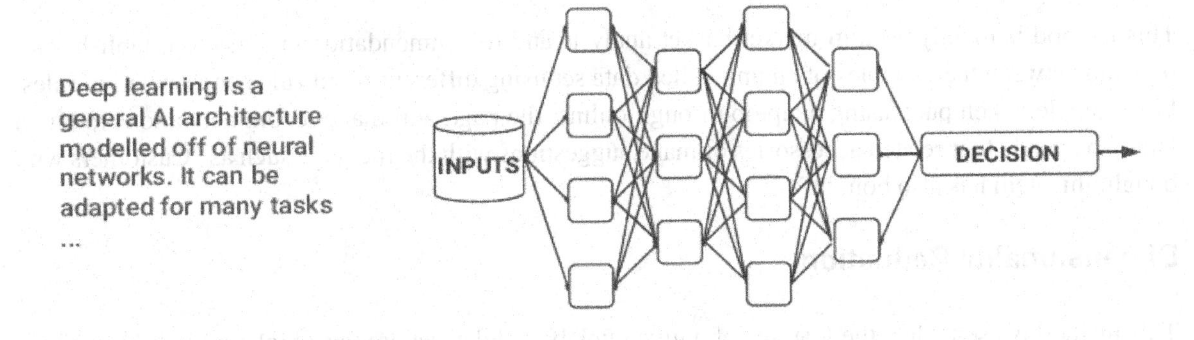

Supervised

To classify new inputs from the labelled data sets, a model is trained using these datasets to predict output with higher accuracy. A well labelled data set on existing attacks would help in detecting future attacks just by matching the network traffic to the profile already known. Supervised learning can be widely used for *classification* and *regression* purposed while mining data. In *Classification,* an algorithm is used to separate the data under specific categories based on its features. For example, a data set having the features of apples and oranges where it accurately predicts the output value i.e. either orange or apple by processing input values. For an instance, Gmail is able to categorize whether the incoming message is valid or spam with the help of such algorithms. Most commonly used classification algorithms are Support Vector Machine (SVM), Random Forest, Decision Trees and Linear Classifiers etc.,

Regression uses an algorithm to identify and understand the "relationship between dependent and independent variables" based on different data points. These approaches are very much useful in predicting numerical values with such data points. For example, predicting revenue projections of a given business. Linear regression, logistic regression and polynomial regression are the most widely used regression techniques.

Unsupervised Learning

Unsupervised learning helps in discovering hidden patterns from an unlabelled data set with an algorithm that analyse and clusters the unlabelled dataset with no-form-of human intervention. This approach is used to achieve the following tasks.

Clustering

It is mainly used in market segmentation and image compression applications for grouping. Basically, it is a data mining approach to group unlabelled data either based on their similarities or differences. K-means clustering algorithm is an example of this approach that form groups with similar data points from the unlabelled data set whereas k-value algorithm helps in representing the size of those groups and its granularity.

Association

This method is mainly used in market basket analysis and recommendation engines to establish relationship between the variables of an unlabelled data set using different set of rules or association rules. For example, when purchasing a laptop through online, the page will make recommendations to buy a laptop bag and other related accessories or make suggestions with the message such as "Customers who bought this item has also bought".

Dimensionality Reduction

This method is used when the features of a given unlabelled dataset are very high and it reduces those features into a manageable number without affecting the integrity of original data. Pre-processing can be very effective with these methods in place, such as removing noise from an input image for better picture quality using auto-eqn coders.

Table 1. Difference between Supervised and Unsupervised Learning

Parameters	Supervised Learning	Unsupervised Learning
Goal	To predict outcomes for new data as the user knew the type of result to expect in up front	To get insights from a large unstructured volume of data by identifying dependent and independent features of it (hidden patterns)
Input Data	Algorithms are trained using well-structured datasets	Algorithms are used against an unstructured dataset
Computational Complexity	Simple to implement and easy to use with programs like R or Python (Tensor Flow)	Complexity is high as it processes large set of unclassified data
Time Complexity	Time-consuming and needs expertise in labeling input and output variables	Less time-consuming with inaccurate results and requires human intervention for validation of outcome
Accuracy	Delivers trustworthy and accurate results	Delivers moderate results with less accuracy
Applications	Spam detection, weather forecasting and sentiment analysis and etc.	Medical Imaging, Anomaly Detection and recommendations engines and etc.
Examples	Decision Tree, Linear Regression, Naïve Bayes and Support Vector Machine (SVM) and etc.,	Apriori Algorithm, k-means clustering, k-nearest neighbor and etc.,

Semi-Supervised Learning

Though data classification is a challenging task in supervised learning, the end results are trustworthy and accurate whereas unsupervised learning lacks transparency in classifying and forming clusters from large volumes of unlabelled data and the results might be inaccurate in Table.1. To handle this problem, a semi-supervised learning approach that uses both labelled and unlabelled datasets to extract the required features from them. This method yield better results in cases like high volume of data that requires the interventions of both the computers and humans. Medical image processing applications are clean example for this method as it requires the help of both machines (CT scans) and radiologists for accurate predictions of a disease.

Reinforcement Learning

Reinforcement learning is a machine learning model to make a sequence of decisions to maximize the rewards with machines and software in an environment that is potentially complex and uncertain. Unlike supervised learning, this approach has trained models with no answer key, but the agent makes decisions to achieve the assigned task i.e. learn from mistakes, and reinforcement technique learns from its experience when training dataset not available. In simple terms, it works in sequential order where the input of current stage is taken from the output of previous input i.e. the decision is dependent. It is very effective in a game like situation where the computer employs trial and error technique using feedback from its own actions and experiences, as a result either rewards or penalties offered for the actions performed. Every action or event will result in either ***positive or negative*** effect on behaviour. The main motive is to strengthen the behaviour irrespective of positive or negative event. In addition, the experience gathered by reinforcement algorithm through thousands of trials, the machine's creativity can be improved. Chess game is a simple example for this technique.

Figure 2. Reinforcement learning architecture

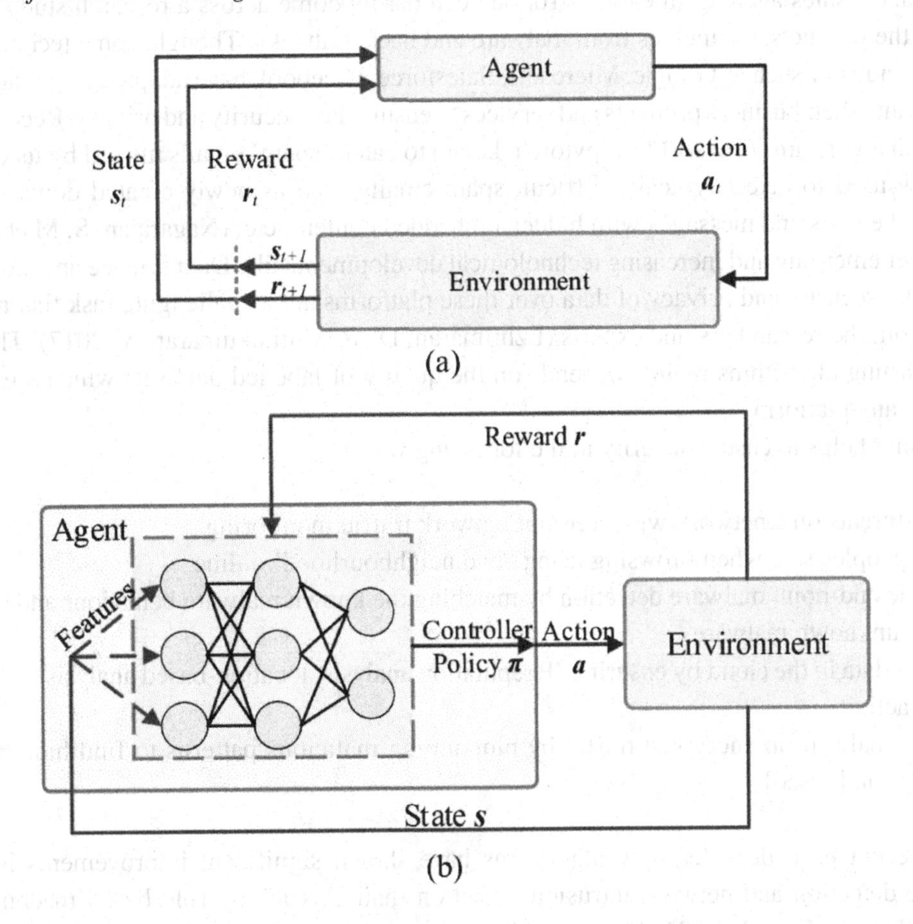

Fig.2 (a) depicts the architecture of reinforcement learning and Fig.2 (b) depicts deep reinforcement learning. Preparing the simulation environment is one of the main difficulties in reinforcement learning such as superhuman in chess, driving an autonomous car becomes tricky in real world. Another challenge is to communicate with the network since the system of rewards and penalties control the agents of neural networks. Thirdly, the agent performs its task as it is (in real time), thus achieving local optimum at all situations and environments might be more challenging. Gaining the rewards with no penalties has no meaning if the task is not completed (Linda, G. M et al., 2021). Q-learning and State-Action-Reward-State-Action (SARSA) are the most commonly used algorithms with similar exploitation policies and different exploration strategies. To overcome the generality issue of these algorithms, Deep q-Networks (DQNs) and Deep Deterministic Policy Gradient (DDPG) were introduced. Reinforcement algorithms are widely used in applications like playing computer games and robotics and industrial automation(Kumar, D et al., 2021).

Deep Learning in Security and Privacy

Over few decades, many deep learning algorithms have been majorly used in applications such as marketing, finance, sales and etc. In early 2010, one can hardly come across a research study targeted on protecting the products, businesses from malware and hacker attacks. Though, some technology giants in business industry such as Google, Microsoft, Salesforce, Facebook have adapted deep learning based algorithms into their business products and services to ensure data security and privacy. Recently, Google infused such algorithms (TensorFlow, pytorch, keras) to catch 'spam' emails missed by its Gmail filter. TensorFlow used to detect typically difficult spam emails such as newly created domain messages, image-based emails and messages with hidden embedded content, etc. (Nagarajan, S. M et al., 2022).

With ever emerging and increasing technological developments like Data science and cloud computing, ensuring security and privacy of data over these platforms are a challenging task that requires full attention from the researchers and experts (Ezhilmaran, D., & Muthukumaran, V, 2017). The accuracy of deep learning algorithms mainly depends on the quality of labelled data sets which are not readily available in any platform.

Deep learning helps to ensure security in the following ways:

1. Finds threats on a network with constant network traffic monitoring
2. Keep peoples safe when browsing using "bad neighbourhood' online
3. Provide end-point malware detection by matching the known malware behaviour and attributes to detect unknown malware.
4. Protect data in the cloud by ensuring IP reputation analysis, location-based analysis and suspicious login activity
5. Detect malware in encrypted traffic by pinpointing malicious patterns to find hidden threats in encryption [Cisco]

In the recent past, deep learning algorithms have shown significant improvements in the areas of malware detection and network intrusion detection than classic and rule based machine learning algorithms(Kumar, V et al., 2021; Nagarajan, S. M et al., 2022) .

Since 1990-2010, many research investigations carried out to ensure security over network data and information sharing using classis machine learning algorithms, but the results proved that such algorithms were not impactful in detecting security and privacy related attacks (Velliangiri, S et al., 2021). Factually, the threats to both the governmental and other business industry have seen exponential growth even in the presence of more capable technologies to detect them. Importantly, the malicious attackers have capable resources and knowledge over advanced persistent threats whereas cyber security professionals struggle to meet the ever growing demand on expertise(Manikandan, G et al., 2021; Manikandan, G et al., 2020). Most of the business firm is not aware of the importance of keeping their IT infrastructure secure, specially, when stats signal that cyber attack has grown from "several hundred to several million per day in some industries" (Muthukumaran V et al., 2021). Generally, an Ml system generates alerts on suspicious behaviour, automatically screens out potential attacks and also detects anomalous activity (Muthukumaran V et al., 2018). Yet, few experts of computer security community had mentioned some remarks or challenges as follows:

1. The proposed system or algorithm must have a clear picture of what types of attack to be detected with which a researcher can tailor such a system that is more specific, thus avoid or reduce misclassifications in results.
2. The results might have serious flaws if one fails to establish strong relations of features to the attacks of interest while developing an algorithm or implanting a systematic approach explain in Fig.3.

Figure 3. Deep learning: 4 phase workflows

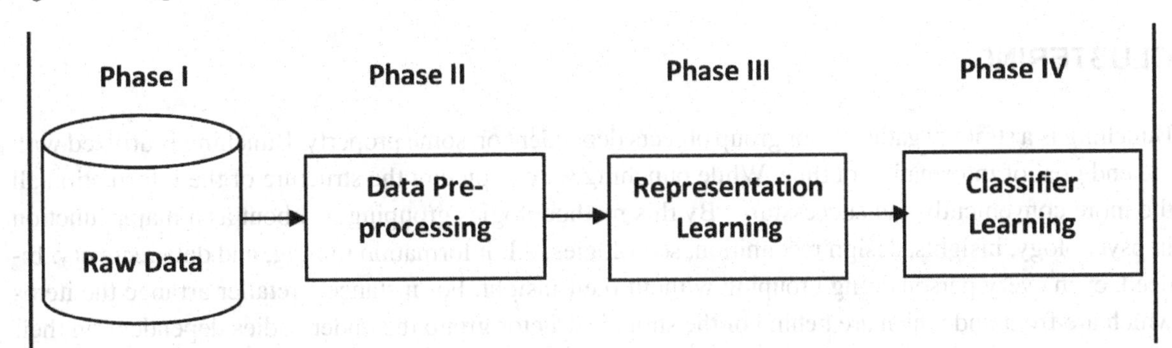

Deep Learning could make these challenges much easier to overcome. Unlike traditional and classic machine learning algorithms, deep learning algorithms are less dependent on feature extraction and results in higher accuracy with less domain knowledge.

Dendrogram

Dendrogram is a tree diagram of clustering. Dendrogram shows that classifications visually and grouped objects expain in Fig. 4.

Figure 4.

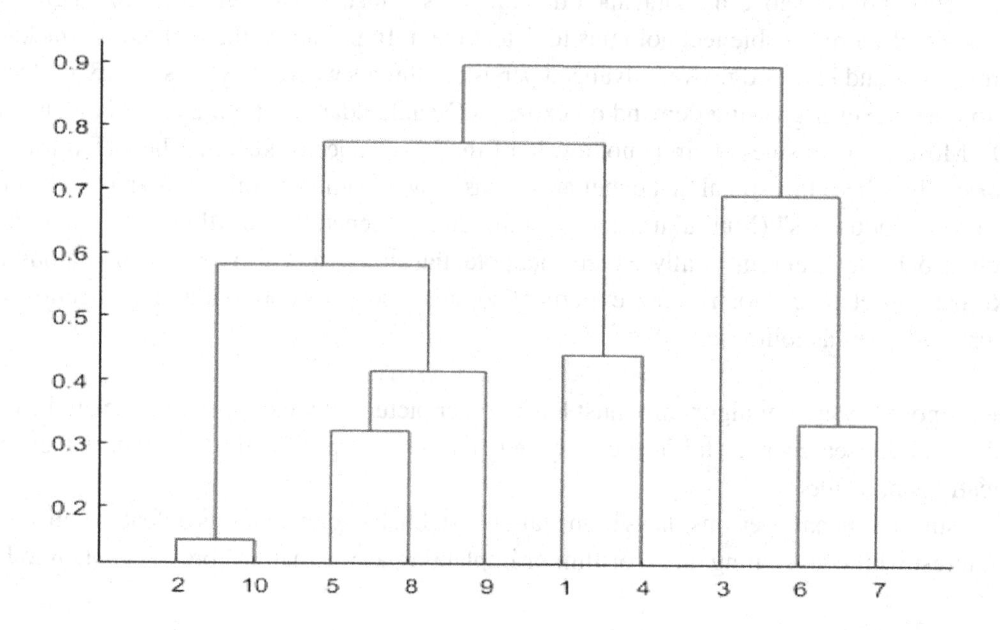

CLUSTERING

Bunching is a strategy gathering or group objects dependent on some property. Bunching is utilized with the end goal of information outline. While bunching we can picture the structure of the information all the more conveniently and successfully. By this methodologies grouping go about as a major function in psychology, insights, design recognition, sociologies, AI, information mining, and data recovery. Indeed, even every person doing grouping without their insight. For instance a retailer arrange the items which are front and which are behind of the shop, instructor group the understudies dependent on their presentation or information, armed force powers are characterize dependent on condition status, and so on. This grouping is turn named as bunch while doing gather the powerful information, finding the example and fathom the arrangement.

Types of Clustering

Grouping is order as hard and delicate bunching. Hard grouping might not have cover. i.e, every information point either has a place with totally or not. In delicate bunching every information point have a different group as likelihood or probability informational index. i.e, quality of relationship among component and cluster. Clustering is order as hard and delicate grouping. Hard bunching might not have cover, i.e., every information point either has a place with totally or not. In delicate grouping every information point have a different bunch as likelihood or probability informational index, i.e., quality of relationship among component and group.

Connectivity Model

Network model depends on separation between the chapters. It is otherwise called various leveled clustering. By estimating the separation there are two sorts of network model are there.

1. Agglomerative various leveled bunching strategy
2. Divisive various leveled bunching strategy

Agglomerative is beginning with single components and totaling them into bunches. Disruptive is beginning with the total informational index and separating it into allotments. Progressive bunching methods use where the group ought to be divided into various groups which depends on bunch vicinity. Since there are three proportion of group nearness: single-connect, total connection and normal.

Single Connection

In this group separation between two bunch to be the littlest separation between two focuses with the end goal that one point is in each group.

Step I

Start with the disjoint grouping inferred by edge chart G(0), which contains no edges and which puts each item in an exceptional bunch as the current grouping. Set k←1.

Step II

From the limit chart G(k). On the off chance that the quantity of components(maximally associated subgroups) is not exactly the current grouping by naming every part of G(k) as a bunch.

Step III

In the event that G(k) comprises of a solitary associated chart, stop. Else set k←k+1 and go to step II.

Complete Connection

Complete connection estimates separation between two bunches to be the biggest separation between two focuses with the end goal that one point is in each group.

- Begin with the disjoint clustering implied by threshold graph $G(0)$, which contains no edges and which places every object in a unique cluster as the current clustering. Set $k \leftarrow 1$.
- From the threshold graph $G(k)$. If two of the current clusters from a clique (maximally complete subgraph) in $G(k)$, redefine the current clustering by merging these two clusters into a single cluster. If $k = \frac{n(n-1)}{2}$, so that $G(k)$ is the complete graph on the n nodes, stop.
- Else set $k \leftarrow k+1$ and go to step II.

Average Link

The normal or troublesome bunch estimates separation between two groups to be a normal separation between two focuses with the end goal that one point is in each group. In the normal linkage technique, D(r,s) is processed as in eqn(2)

$$D(r,s) = \frac{T_{rs}}{N_r * N_s} \quad (2)$$

Where T_rs is the total of all pairwise separations between group r and bunch s. N_r and N_s are the spans of the groups r and s, separately. At each phase of various leveled grouping, the bunches r and s, for which D(r,s) is the base, are combined.

Centroid Models

In centroid-based grouping, bunches depend on a focal vector, which may not really be an individual from the informational index. In this items are shut to the model that characterizes the bunch then to the model of some other group. While fixing k number of bunching gives a conventional definition as an enhancement issue. By finding the closest group place, with the end goal that the squared good ways from the bunch are limited. In this manner we compute the group by centroid models. It is otherwise called model based grouping.

Distribution Based Grouping

It is exceptionally comparative with measurements techniques. Theis grouping dependent on the thought of how plausible is it that all information focuses in the bunch have a place with a similar dispersion Distribution bunching characterized as items having a place no doubt with a similar dissemination. In this grouping counterfeit informational collections are produced by examining arbitrary chapters from a conveyance. It produces complex models for groups that can catch connection and reliance between ascribes.

Density Based Bunching

Thickness based bunching is thought regarding each occasion of a group the area of a given span needs to contain in any event a base number of items which implies that the cardinality of the area needs to surpass a given limit. In this bunches are districts of high separated area by areas of low thickness explained in Fig. 5.

Figure 5. Density based clustering

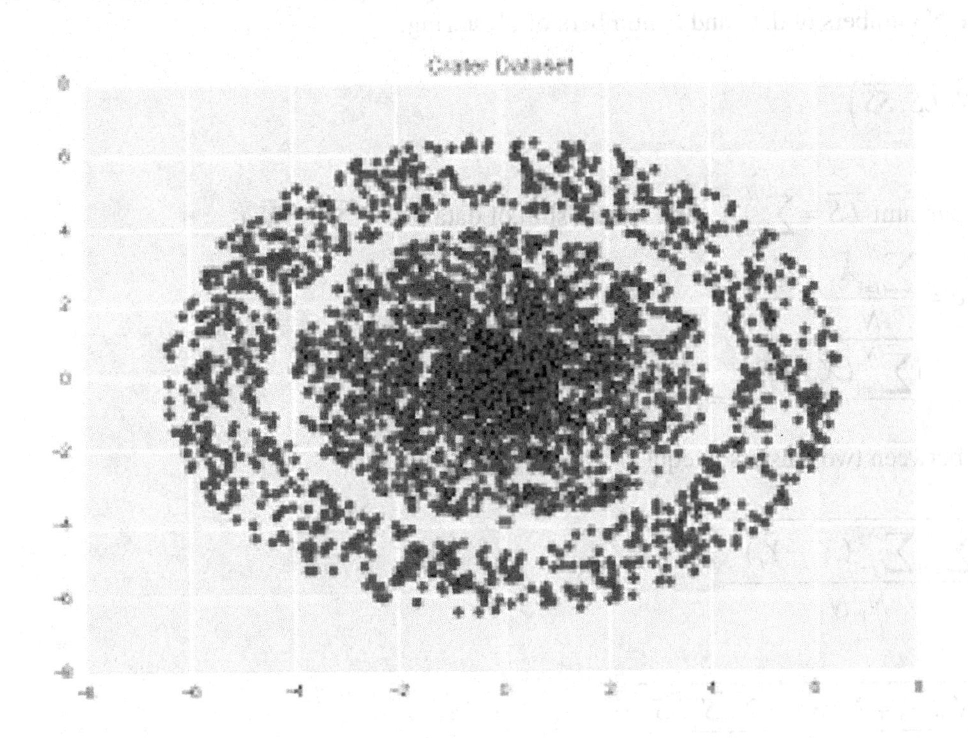

Grid Based Grouping

In this sort of grouping approach first separation the chapter space into a limited number of cells that structure a matrix structure on which the entirety of the tasks for bunching are performed. In this bunching calculations is to quantize the informational index into various cells and afterward work with objects having a place with these cells. They don't migrate focuses but instead fabricate a few progressive degrees of gatherings of chapters. They are nearer to various leveled calculations yet the converging of frameworks, and thusly bunches, doesn't rely upon a separation measure yet it is chosen by a predefined boundary.

Model Based Bunching

For this technique we have made some approximations model. This grouping approximations of model boundaries that best fit the information and they can be either partitional or various leveled, contingent upon the structure or model they speculate about the informational index and the manner in which they refine this model to recognize dividing. Likewise they are nearer to thickness based calculations, in that they develop specific bunches so the biased model is improved. In any case, they once in a while start with a fixed number of bunches and they don't utilize a similar idea of thickness.

ALGORITHMS OF BUNCHING

Birch

BIRCH means balanced iterative reducing and clustering using hierarchies. In BIRCH algorithm there we fix the N numbers of data and K numbers of clustering.

$$CF = \left(N, \overrightarrow{LS}, \overrightarrow{SS}\right)$$

where linear sum $\overrightarrow{LS} = \sum_{i=1}^{N} \vec{X}_i$ and square sum of data $\overrightarrow{SS} = \sum_{i=1}^{N} (\vec{X}_i)^2$

centroid $\vec{c} = \dfrac{\sum_{i=1}^{N} \vec{X}_i}{N} = \dfrac{\vec{L}S}{n}$

radius $R = \sqrt{\dfrac{\sum_{i=1}^{N} (\vec{X}_i - \vec{C})^2}{N}} = \sqrt{\dfrac{N.\vec{c}^2 + \vec{S}S - 2.\vec{C}.\vec{L}S}{N}}$

Distance between two clusters in eqn(3)

$$D_2 = \sqrt{\frac{\sum_{i=1}^{N_1} \sum_{j=1}^{N_2} (\vec{X}_i - \vec{Y}_j)^2}{N_1.N_2}}$$

$$D_2 = \sqrt{\frac{N_1.\vec{S}S_2 + N_2.\vec{S}S_1 - 2.\vec{L}S_1.\vec{L}S_2}{N_1.N_2}} \quad (3)$$

In multidimensional cases the square root should be replaced with a suitable norm

Cure

CURE means Clustering Using Representative. The algorithm of this clustering as follows: We have a set A set of points S. Consider the K number of clusters.

- For every cluster u (each input point), initially $c = 1$ since each cluster has one data point. Also u .closest stores the cluster closest to u.
- All the input points are inserted into a $k - d$ tree T
- Treat each input point as separate cluster and compute u. Closest for each u then insert each cluster into the heap Q .By this clusters are arranged in increasing order of distances between u and u.closest.
- While size$(Q) > k$
- Remove the top element of Q(say u) and merge it with its closest cluster u. and compute the new representative points for the merged cluster w.

- Remove u and v from T and Q.
- For all the clusters x in Q, update x.closest and relocate x
- insert w into Q
- repeat.

K-Means

Let X is an object $x = \{x_1, x_2 .... x_N\}$. C_i means i^{th} number of cluster. Consider c_i is the centriod of the cluster C_i. where m_i is the number of objects in i^{th} cluster. Finally we will find the number of clusters K. The algorithm of K means as follows:

- Select K points as initial centroid
- Repeat the Same procedure.
- From K clusters by assigning each point to its closest centroid.
- Recompute the centroid of each cluster.
- Until centroids do not change.

CONCLUSION

This research offered an improving and differentially private grouping calculation for blended information in shrewd matrix to enable protection safeguarding bunch evaluation in SDN-based brilliant lattice. We combine the differentially private k-implies calculation with the k-modes calculation in our proposed calculation to group blended information in a protection-safe manner and plan an instrument to cause the calculation to meet differential security. We looked at what bunching is and the many types of grouping in this paper, as well as the calculating method used in bunching. Here, we looked at serious grouping models and calculations. It will elicit serious information investigation approaches. We offered an overview of grouping in this chapter, and its many types are analyzed, as well as the computation.

REFERENCES

Akhtar, N., & Mian, A. (2018). Threat of adversarial attacks on deep learning in computer vision: A survey. *IEEE Access: Practical Innovations, Open Solutions*, *6*, 14410–14430.

Dada, E. G., Bassi, J. S., Chiroma, H., Adetunmbi, A. O., & Ajibuwa, O. E. (2019). Machine learning for email spam filtering: Review, approaches and open research problems. *Heliyon*, *5*(6), e01802.

Dhiman, G., Kumar, V. V., Kaur, A., & Sharma, A. (2021). DON: Deep Learning and Optimization-Based Framework for Detection of Novel Coronavirus Disease Using X-ray Images. *Interdisciplinary Sciences, Computational Life Sciences*, 1–13.

Ezhilmaran, D., & Muthukumaran, V. (2017). Authenticated group key agreement protocol based on twist conjugacy problem in near-rings. *Wuhan University Journal of Natural Sciences*, *22*(6), 472–476.

Imamverdiyev, Y., & Abdullayeva, F. (2018). Deep learning method for denial of service attack detection based on restricted Boltzmann machine. *Big Data*, *6*(2), 159–169. doi:10.1089/big.2018.0023 PMID:29924649

Jayasuruthi, L., Shalini, A., & Kumar, V. V. (2018). Application of rough set theory in data mining market analysis using rough sets data explorer. *Journal of Computational and Theoretical Nanoscience*, *15*(6-7), 2126–2130.

Kumar, V., Niveditha, V. R., Muthukumaran, V., Kumar, S. S., Kumta, S. D., & Murugesan, R. (2021). A Quantum Technology-Based LiFi Security Using Quantum Key Distribution. In Handbook of Research on Innovations and Applications of AI, IoT, and Cognitive Technologies (pp. 104-116). IGI Global.

Kumar, V. V., Raghunath, K. K., Rajesh, N., Venkatesan, M., Joseph, R. B., & Thillaiarasu, N. (2021). Paddy Plant Disease Recognition, Risk Analysis, and Classification Using Deep Convolution Neuro-Fuzzy Network. *Journal of Mobile Multimedia*, 325-348.

Kumar, V. V., Raghunath, K. M., Muthukumaran, V., Joseph, R. B., Beschi, I. S., & Uday, A. K. (2021). Aspect based sentiment analysis and smart classification in uncertain feedback pool. *International Journal of System Assurance Engineering and Management*, 1-11.

Linda, G. M., Lakshmi, N. S. R., Murugan, N. S., Mahapatra, R. P., Muthukumaran, V., & Sivaram, M. (2021). Intelligent recognition system for viewpoint variations on gait and speech using CNN-CapsNet. *International Journal of Intelligent Computing and Cybernetics*.

Liu, Q., Li, P., Zhao, W., Cai, W., Yu, S., & Leung, V. C. (2018). A survey on security threats and defensive techniques of machine learning: A data driven view. *IEEE Access: Practical Innovations, Open Solutions*, *6*, 12103–12117.

Manikandan, G., Perumal, R., & Muthukumaran, V. (2020, November). A novel and secure authentication scheme for the Internet of Things over algebraic structure. In AIP Conference Proceedings: Vol. 2277. *No. 1* (p. 060001). AIP Publishing LLC.

Manikandan, G., Perumal, R., & Muthukumaran, V. (2021). Secure data sharing based on proxy re-encryption for internet of vehicles using seminearring. *Journal of Computational and Theoretical Nanoscience*, *18*(1-2), 516–521.

Muthukumaran, V., Ezhilmaran, D., & Anjaneyulu, G. S. G. N. (2018). Efficient Authentication Scheme Based on the Twisted Near-Ring Root Extraction Problem. In *Advances in Algebra and Analysis* (pp. 37–42). Birkhäuser.

Muthukumaran, V., Joseph, R. B., & Uday, A. K. (2021). Intelligent Medical Data Analytics Using Classifiers and Clusters in Machine Learning. In Handbook of Research on Innovations and Applications of AI, IoT, and Cognitive Technologies (pp. 321-335). IGI Global.

Muthukumaran, V., Vinothkumar, V., Joseph, R. B., Munirathanam, M., & Jeyakumar, B. (2021). Improving network security based on trust-aware routing protocols using long short-term memory-queuing segment-routing algorithms. *International Journal of Information Technology Project Management*, *12*(4), 47–60.

Nagarajan, S. M., Deverajan, G. G., Chatterjee, P., Alnumay, W., & Muthukumaran, V. (2022). Integration of IoT based routing process for food supply chain management in sustainable smart cities. *Sustainable Cities and Society*, *76*, 103448.

Nagarajan, S. M., Muthukumaran, V., Beschi, I. S., & Magesh, S. (2021). Fine Tuning Smart Manufacturing Enterprise Systems: A Perspective of Internet of Things-Based Service-Oriented Architecture. In Handbook of Research on Innovations and Applications of AI, IoT, and Cognitive Technologies (pp. 89-103). IGI Global.

Nan, Y., Lovell, N. H., Redmond, S. J., Wang, K., Delbaere, K., & van Schooten, K. S. (2020). Deep Learning for Activity Recognition in Older People Using a Pocket-Worn Smartphone. *Sensors (Basel)*, *20*(24), 7195. doi:10.339020247195 PMID:33334028

Sommer, R., & Paxson, V. (2010, May). *Outside the closed world: On using machine learning for network intrusion detection. In 2010 IEEE symposium on security and privacy*. IEEE.

Tang, T. A., Mhamdi, L., McLernon, D., Zaidi, S. A. R., & Ghogho, M. (2016, October). Deep learning approach for network intrusion detection in software defined networking. In 2016 international conference on wireless networks and mobile communications (WINCOM) (pp. 258-263). IEEE.

Velliangiri, S., Karthikeyan, P., & Vinoth Kumar, V. (2021). Detection of distributed denial of service attack in cloud computing using the optimization-based deep networks. *Journal of Experimental & Theoretical Artificial Intelligence*, *33*(3), 405–424.

Wang, Z. (2015). The applications of deep learning on traffic identification. *BlackHat USA*, *24*(11), 1–10.

Chapter 15
Hybrid Clustering Technique to Cluster Big Data in the Hadoop Ecosystem:
Big Data Application

E. Padmalatha
Chaitanya Bharathi Institute of Technology, India

S. Sailekya
Chaitanya Bharathi Institute of Technology, India

ABSTRACT

Big data analytics as well as data mining play vital roles in extracting the hidden statistics. Customary advances for investigation and extraction of hidden information from data may not exert efficiently for big data because of its complex, elevated volume nature. Data clustering is a data mining technique that exacts the useful data from the data by grouping data into clusters. In big data as the data is complex and of very large volume, individual clustering techniques may not consider all the samples, which may lead to inaccurate results. To overcome this inaccuracy, the proposed method is the combination of dynamic k-means and hierarchical clustering algorithms. This proposed method can be called a hybrid method. Being a hybrid method will overcome a few drawbacks like static k value. In this chapter, the proposed method is compared with existing algorithms by using some clustering metrics.

INTRODUCTION

Big data analytics has become trend in the market and is used to perform analytics on this big data. It is used to extract hidden patterns, unknown correlations and helps organizations in decision making. Big data is the problem and Hadoop is the solution for handling big data available as an open-source framework. Clustering is one of the techniques used to extract insights from big data (Raghupathi & Raghupathi 2014). Traditional clustering techniques may not work for efficient clustering in big data.

DOI: 10.4018/978-1-7998-9640-1.ch015

Consequently, there remains need towards plan an competent & extremely scalable clustering algorithm. This has motivated towards propose a novel algorithm called hybrid clustering algorithm for big data in Hadoop ecosystem (Katal et al., 2013). In Big data analysis characteristics individual clustering techniques like kmeans mean and hierarchical may not consider all the samples which leads to inaccurate results. K-means and hierarchical gathering techniques meet halfway because of the limitations of individual clustering algorithms. Few drawbacks of traditional clustering algorithms are k-means clustering in this algorithm it remains hard towards predict the k value, wrong prediction of k value many data points may not fit into any of the clusters; several merge split decisions and iteration in hierarchical clustering, etc. (Aggarwal & Zhai 2012).

Grouping is important device for information mining & information revelation. The aim of bunching is to discover considerable gatherings of substances moreover to divide groups framed for a dataset. Customary K-implies grouping functions admirably when functional to little datasets (Pandove & Goel 2015). Enormous datasets should be grouped through the end objective that each and all other substance or information point in the bunch is like several elements in a similar group. Grouping issues can be applied to a few bunching disciplines. The capacity towards consequently bunch comparative things empowers one to find covered up likenesses & key ideas while joining a lot of information into a couple of gatherings. This empowers clients towards fathom a lot of information. Groups can be delegated homogeneous & heterogeneous bunches. In homogeneous groups, all hubs contain comparable possessions (Firouzi et al., 2010). Heterogeneous bunches remain exploited in private server farms in which hubs have a variety of attributes moreover in which it could be hard to be familiar with hubs Embrocates (Demchenko et al., 2013).

Clustering techniques require the use of more exact meanings of perception and group likenesses. When gathering depends on ascribes, it is normal to utilize recognizable ideas of distance. An issue with this strategy is related with the estimation of distances between groups including at least two perceptions. (Fernåndez et al., 2014) In contrast to existing regular measurable techniques, most grouping calculations doesn't depend on factual circulations of information and in this manner can be useful to apply when minimal earlier information exists on a specific issue (Ghazal et al., 2013) portrayed how the quantity of emphases can be diminished by parceling a dataset into covering subsets and by just emphasizing information objects inside covering zones (Battré et al., 2010)

The remainder of this works remains organized as follows. The 'History' section contains relevant surveys on the subject of Big data clustering. We provide a background on Apache Spark in 'Research Paper' The section under 'Study Design' describes the survey's research methods. The section 'Survey Methods' goes through the various Spark clustering algorithms. We provide our analysis on clustering large data with Spark and upcoming projects in 'Discussion and Future Directions.' Lastly, in 'Findings,' bring the paper to be close.

Limitations of Existing Methods

The existing methods like big-data related clustering models with honeybee, genetic and PSO techniques cannot provide accurate bigdata storage. The limitations like static k, dynamic k and hadoop storage issue are cannot solve exactly. The silhouette score, Calinski-Harabasz Index, & Davies - Bouldin Index cannot be improved with this method (Jiang et al., 2010).

EXISTING METHODS

Traditional clustering algorithms like k-means and hierarchical have their own disadvantages. Data mining and big data are techniques for analyzing data and extracting confidential message.

Cluster, is a common unsupervised classification technique, is widely used in data mining, artificial intelligence, and information processing. The method entails arranging single and unique points in a group such that they are either similar to one another or different to points from other clusters. The current enormous increase of data has put traditional clustering techniques to the test. As a result, many research papers suggested new clustering techniques that take use of Big Data platforms like Apache Spark, which is intended for large data handling in a decentralized and rapid manner. Spark-based cluster study, on the other hand, is still in its infancy. Researchers examine the current Spark-based clustering techniques in terms of their support for Big Data features in this comprehensive study. In addition, we offer a new taxonomy for clustering techniques centered on Spark. To the best of our knowledge, no survey on Spark-based Big Data cluster has been performed. As a result, the goal of this study is to provide a thorough overview of past research in the area of Big Data clustering using Apache Spark from 2010 to 2020. This study also identifies new research areas in the realm of big data grouping. Because big data is complicated and large in volume, traditional methods to analysis and extraction do not function effectively. Data clustering is a common data mining method that organizes data into clusters and makes it simple to retrieve features from these regions. Traditional scheduling methods, such as k-means & hierarchical, remain inefficient because the integrity of the groups they generate remains harmed. As a result, an efficient & highly scalable clustering method is required. In this article, we propose hybrid clustering, a novel clustering method that overcomes the drawbacks of current clustering algorithms. On the basis of precision, recall, F-measure, processing time, then correctness of findings, we compare the novel hybrid algorithm towards current methods. The suggested hybrid clustering method remains more accurate, with higher accuracy, recall, & F-calculate values, according to the research observations. Clustering analysis is the statistical mining job that attempts to make data easier to find, suggest, and organise. Clustering methods divide datasets into a number of clusters, each with its own set of attributes. Grouping, unlike categorization, is an iterative method in which appropriate means in a set of data are clustered into clusters, then thus represent various clusters, with objects in a similar bunch being altogether different from one another & objects in a similar gathering or group being basically the same as each other. The groups remain only discovered that after classification algorithm has been completed. K-means clustering & hierarchical clustering are various clustering methods that are used to manage big datasets, and each is described here.

Traditional K-Means Algorithm

K-means algorithm is used in clustering in this available are partitioned hooked on k-clusters such that there must remain low inter-cluster comparison& high intra-cluster similarity Every cluster will have representative, from this point distance of all data points will be measured .if the distance is minimum from the centroid then these points will be considered in one cluster. Randomly chosen k-points may serve as initial centroids (Doulkeridis & Nørvåg 2014).

Drawbacks of k-means algorithm are: it is a static algorithm and wrong estimation of k-value may affect the prediction in turn accuracy may affect .because of its sensitiveness to outliers clusters formed may not have good quality (Berwind et al., 2017).

k-implies combination remains a technique aimed at vector quantization, primarily commencing signal arranging, that intends to section n insights snared on k packs in which all discernment has a spot among the bundle through the nearest mean (bunch centers or gathering centroid), filling in as an imitation of the pack. These impacts in an allocating of the data space into Voronoi cells. k-infers clustering limits inside bunch vacillations (squared Euclidean distances), anyway not standard Euclidean separations, which would stay alive the more problematic Weber issue: the mean advances squared bumbles, while simply the arithmetical center cutoff points Euclidean distances. For instance, better Euclidean preparations can be exposed exploiting k-medians & k-medias Devaraju, (Hong et al., 2021).

The problem is computationally troublesome (NP-hard); notwithstanding, capable heuristic computations unite quickly to a nearby ideal. These are ordinarily similar towards the supposition support estimation for mixes of Gaussian disseminations through an iterative refinement approach abused by commonly k-infers & Gaussian mix showing. Both of them use bundle centers towards show the in arrangement; in any case, the k-infers gathering will in widespread find gatherings of practically identical spatial degree, while the Gaussian blend model permits bunches to have a variety of shapes.

Hierarchical Clustering Algorithm

Hierarchical clustering forms the clusters either by combining the small clusters into large clusters or larger cluster into smaller ones. In this algorithm a tree of clusters shows the relation between the clusters are related. Hierarchical clustering may be agglomerative or divisive. And the time complexity is O(n2).

Divisive clustering initially it pools all the objects into one cluster then this initial cluster will be successive splits to obtain the divide clusters. These iterations remain executed awaiting the preferred number of clusters remains obtained. Its complexity is quadratic. Because of many iteration involved it takes additional time than k-means algorithm this is the main drawback of this algorithm. This procedure is by and large utilized for bunching a populace into various gatherings. A couple of normal models incorporate dividing clients, bunching comparable archives together, suggesting comparable melodies or films, and so on. There are a LOT more utilizations of solo learning. On the off chance that you run over any intriguing application, don't hesitate to share them in the remarks segment underneath!

Presently, there are different calculations that assist us with making these bunches. The most ordinarily utilized grouping calculations are K-implies and Hierarchical bunching. There are mostly II kinds of hierarchical clustering:

1. Agglomerative hierarchical clustering
2. Divisive Hierarchical clustering

Agglomerative Hierarchical Clustering

We allocate every highlight a personality bunch in this strategy. suppose there are 4 information focuses

Divisive Hierarchical Clustering

Troublesome reformist bundling works in an opposite way. Maybe then starting with n gatherings (if there ought to be an event of n discernments), we start with a single bunch & allocate every one of the focuses to that group.

PROPOSED SYSTEM

The proposed system aims at implementing a hybrid algorithm which is a combination of Dynamic k-means & hierarchical clustering algorithms. The implemented hybrid algorithm will be run on the big datasets which gives clusters as results.

The execution of the proposed method will follow the following steps.

Step1. Generation of the dataset
Step2. Dynamic k-means will be executed for the given data set and it gives optimal value of k.
Step3. The k value obtained in step 2 is passed to agglomerative clustering for forming clusters.
Step4. The performance of the hybrid algorithm is calculated using clustering metrics.
Step5. Comparison of hybrid algorithm with existing algorithms will be performed.

Figure 1. Illustration of available clustering algorithms for Big Data Mining

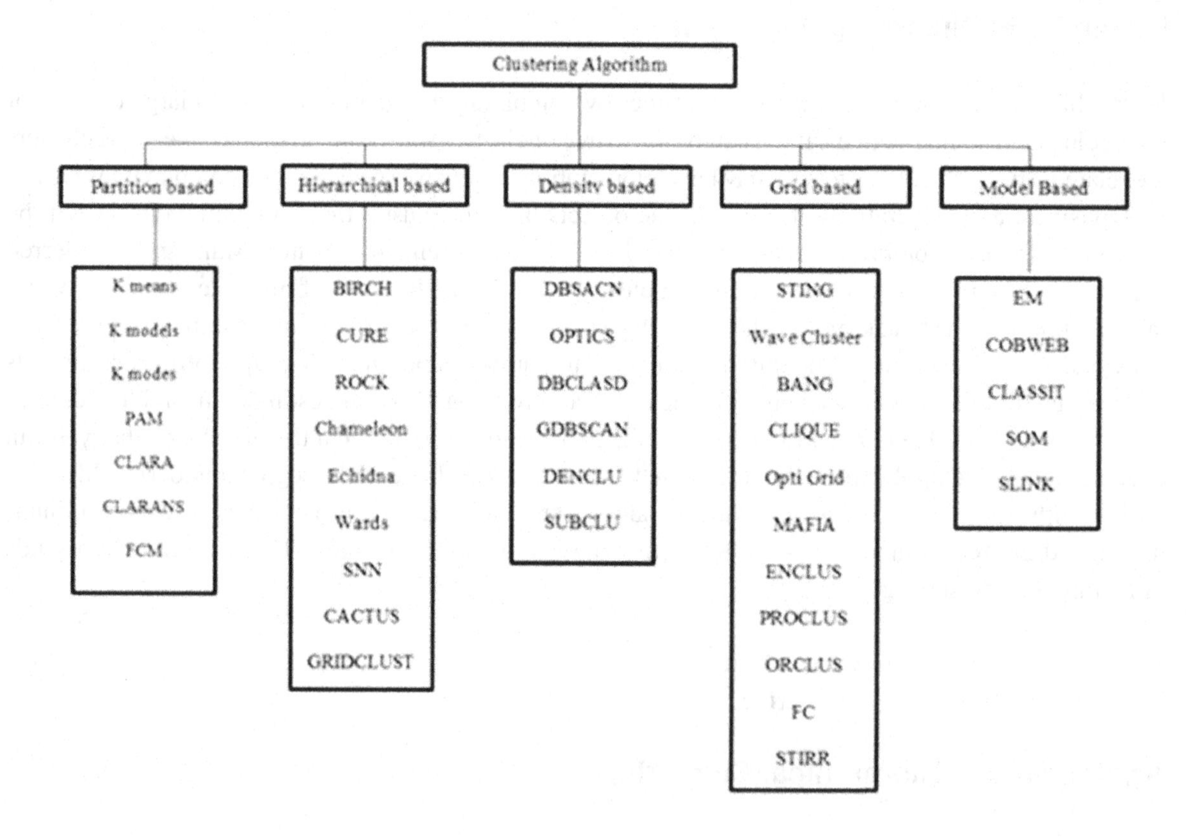

METHODOLOGY

Flow chart

The flow chart shown in the below figure (Figure 2) gives the flow of total process from Dataset generation to forming the clusters and finding the performance of clusters using Metrics and comparing hybrid algorithm with existing algorithm.

Figure 2. Flow chart of proposed system

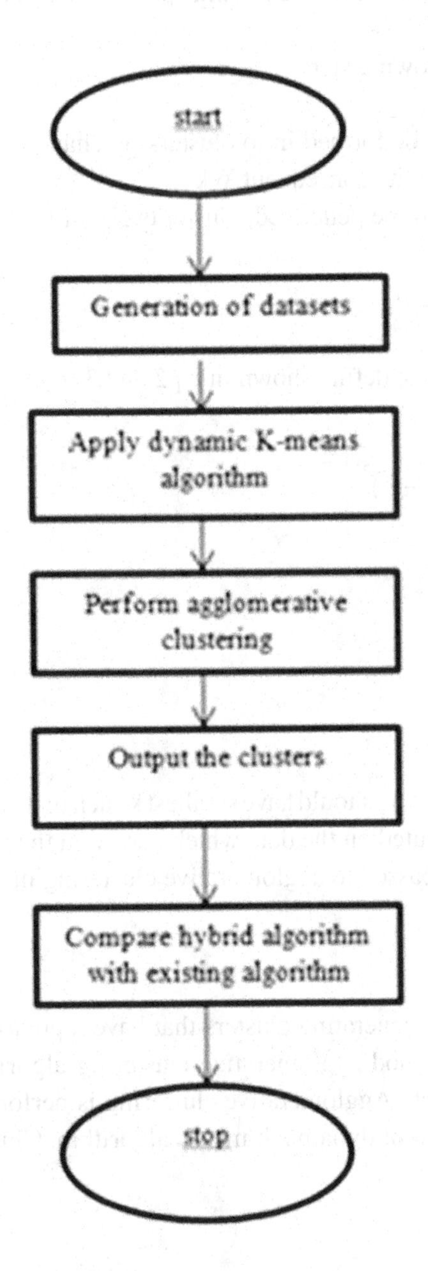

Dynamic K-Means Algorithm

Gap static method is used to get the optimal value of k. This gap statistic will compare the total intra cluster variation for various values of k. The gap statistic aimed at a given k remains defined as follows shown in eq 1

$$Gap_n(K) = E*_n \log(W_k) - \log(W_k) \tag{1}$$

$E*_n$ is characterized by means of bootstrapping by producing B duplicates of the reference datasets and, by registering the normal logWk. The whole measurement estimates the deviation of the noticed Wk esteem from its normal worth under the invalid speculation. The gauge of the ideal groups will be the worth that amplifies Gap(k).

The algorithm involves the following steps:

Step 1. The observed data will be formed in to clusters by changing the k value from 1 to maximum. With this we can calculate the consequent Wk.

Step 2. B reference data sets will be generated, cluster every of them by considering k value from 1 to maximum .

Step 3. Let $\overline{w} = \left(\frac{1}{B}\right)\sum_b \left(\log W^*_{kb}\right)$

Calculate the average deviation & define shown in eq 2 and 3

$$sd(k) = \left[\left(\frac{1}{B}\right)\sum_b \left(\log W*_{kb} - \overline{w}\right)^2\right]^{1/2} \tag{2}$$

$$S_k = \sqrt{1} + \frac{1}{B} sd(k) \tag{3}$$

4. The selected number of clusters should have smallest k such that Gap(k)≥Gap(k+1)−sk+1Dynamic k-means algorithm is executed on the data which is read in the reduce phase and itoutputs the optimal value of k which is passed to agglomerative clustering in next step.

Hierarchical Clustering

Hierarchical clustering involves generating clusters that have a prearranged ordering commencing top to bottom. In this proposed method agglomerative clustering algorithm is used which performs the clustering in bottom-up approach. Agglomerative clustering is performed on the data also the number of clusters (k) remains the output of dynamic k-means algorithm. Clusters are formed after performing agglomerative clustering.

Metrics for Calculating the Performance of Clustering Algorithm

The proposed method is evaluate with following metrics like Silhouette Coefficient (SC), Calinski - Harabasz index (CHI), Davies-Bouldin index (DBI).

Silhouette Coefficient (SC)

The silhouette plot displays determine of how close every point in one bunch is to focus in the adjoining groups. In this method range of evaluates (-1, 1). This metric will be defined for each and every sample. Outcome of this metric contains two scores in that score1 if we consider it as x it will provide mean distance among from a sample to all the remaining points in same cluster. The score 2 representing it with y and it provides the mean distance between a sample and every point in the nearest next cluster.

If **Con**sidered for a single sample then **SC=y–x/ max(x,y).**This s will be higher than the DBSCAN value.

Calinski-Harabasz index (CHI)

The Calinski-Harabasz index (CHI) can be referenced by way of Variance Ratio Criterion. The Higher score will provide better clusters. The index is the proportion of the number of associating bunches scattering and of between-group scattering for all bunches. The index will have higher values for convex clusters .

Davies-Bouldinindex (DBI)

The lower Davies-Bouldinindex (DBI) specify the better parting between the clusters. This DBI index shows the average 'similarity' between clusters. The least possible score is Zero, it indicates better partition.

RESULTS AND DISCUSSION

The results are the screenshots of the outputs at different stages of proposed algorithm. The following shows the various stages that are present in the algorithm.

Running the Algorithm in Hadoop Ecosystem

Hadoop streaming container remains a utility that accompanies the Hadoop conveyance will have Hadoop streaming jar it is utility which allow us towards create & run Map - Reduce jobs by using executable or script by way of the mapped or the reducer. For the implementation of the proposed method Hadoop streaming jar is used then the algorithm is written inpython language. Figure 3 shows the streaming jar command used to run the python program.

Figure 3. Running the algorithm in Hadoop ecosystem

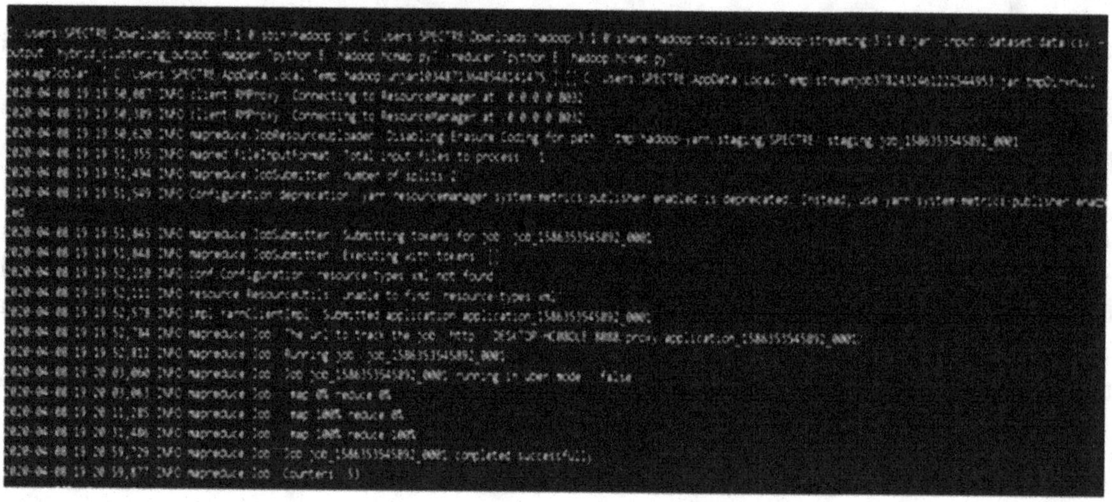

Visual Representation of Dataset

Figure 4.2 shows the scatter plot representation of the dataset generated. The dataset is generated using make_blobs module present in sk learn and is converted into a comma separated value (csv) file. This file is then loaded into Hadoop ecosystem through the "copy From Local" command. This file is read by Map phase which converts the file into a key value pair and sends the data to Reduce phase. Reduce phase gets the data from Map phase and visualizes the information as disperse plot.

Figure 4. Visual representation of dataset

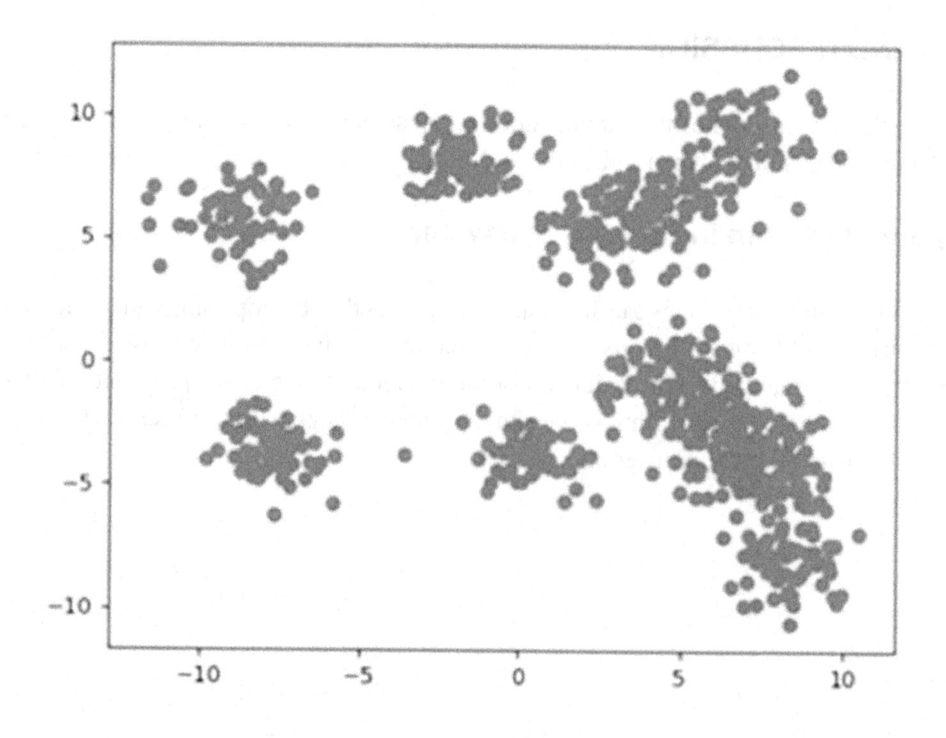

Dynamic K using Gap – Statistic Method

After loading the dataset, dynamic k –means is used towards find the optimal value of k for a given dataset. Dynamic k –means is implemented by using gap statistic method. In this method, we choose the k value which has the maximum value of gap. Figure 4. shows the graphical

Representation of gap values for various k. They used a dataset from the American Climate Server Farm (ACSF), which has the world's biggest active collection of meteorological data[25], towards implement the suggested hybrid clustering method in Hadoop[24]. It provides weather files in standard ASCII format that are accessible to everyone around the world. This worldwide database combines surface hourly data from over 20,000 sites across the globe. Weather data for various years dating back to 1901 may be found in the NCDC database. Every day of the year, the temperature is documented. The input dataset for a meteorological file chosen aimed at a certain year looks like the image below, which displays the weather file for 1907. Figure 5 shows a short explanation of each of the 32 characteristics. There are three parts to the dataset: a control section, a required data portion, then an extra data section, all of which remain detailed below. Every record begins by a 60-character corrected control signal. The control part includes data regarding the report, for example, the perception date, time, & station area, among other things. Table 1 provides a short overview of each characteristic in the control section. The control area remains trailed through a 45-character obligatory section, which is likewise of a set length. This section includes climatic parameters such as temperature, pressure, and winds, among other things. Table 1 also includes a short explanation to each characteristic in the required section. After the required part, there is an extra data section with a variable amount of letters and no set length. It is not required, thus a given record just has to include two parts (control and obligatory). After the extra data part, a comment or element quality section may be added. In MapReduce parallel processing, the suggested method involves two main phases. The result of the mapping phase is given into the second step as an argument.

Enormous information are huge volumes, raised speed or potentially rapid data sets that include new kinds of taking care of to enhance measures, find comprehension and settle on decisions. Information catch, stockpiling, assessment, sharing, searches and representation face incredible difficulties for enormous information. Representation could be considered as "huge data front end. There's no information perception legend.

1. It is essential to imagine just brilliant data: a simple and quick view can show something off base with data very much like it assists with distinguishing energizing examples.
2. Visualization consistently shows the right decision or intercession: representation is certifiably not a substitute for basic reasoning.
3. Visualization brings confirmation: information are shown, not appearance an accurate picture of what is fundamental. Perception with different effects can be controlled.

Figure 5. Graphical representation of gap values for various k

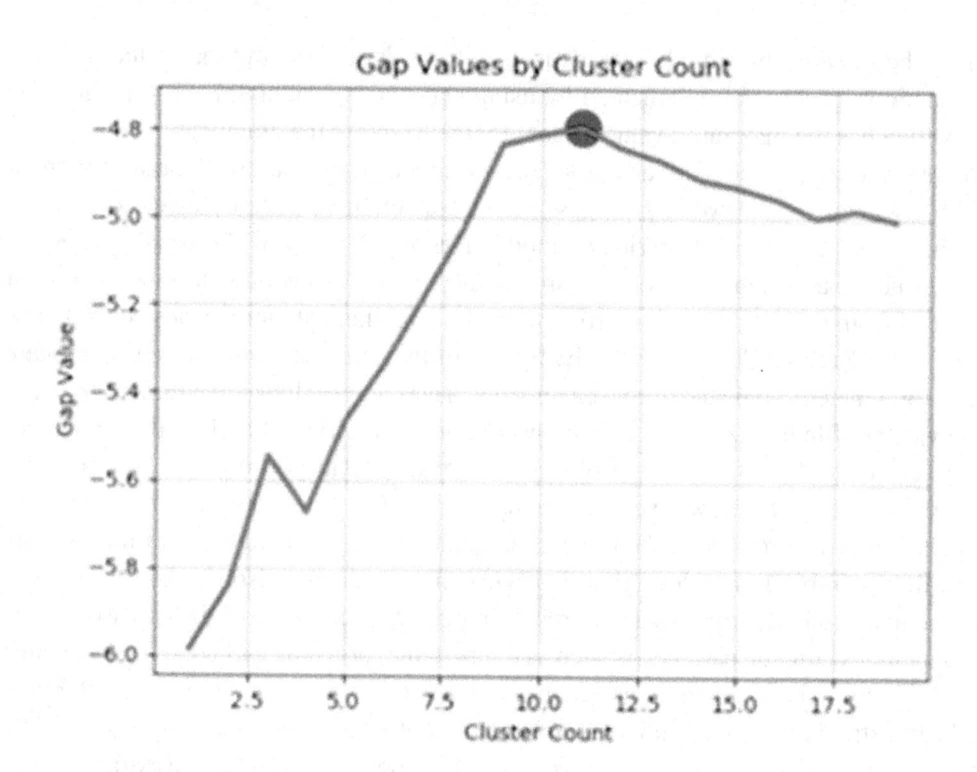

Clusters

After getting the optimal value optimal value of k from dynamic k –means algorithm, we pass this value to Agglomerative clustering algorithm to generate the clusters. Figure 6 shows the visual representation of clusters formed after the application of hybrid algorithm.

Figure 6. Clusters

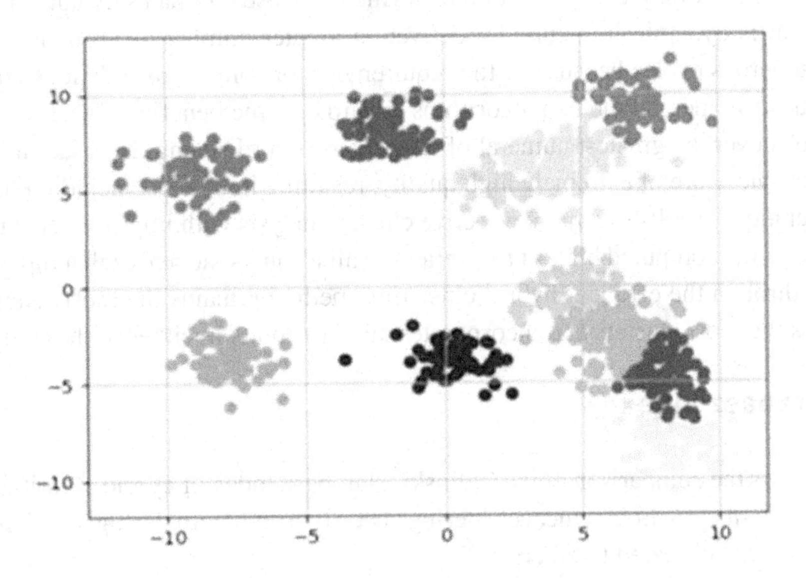

In Figure 7 the comparison of the silhouette score of hybrid algorithm and the existing algorithms i.e., k –means, hierarchical clustering is plotted. The algorithm with the highest silhouette score is more efficient compared to others.

Figure 7. Silhouette score

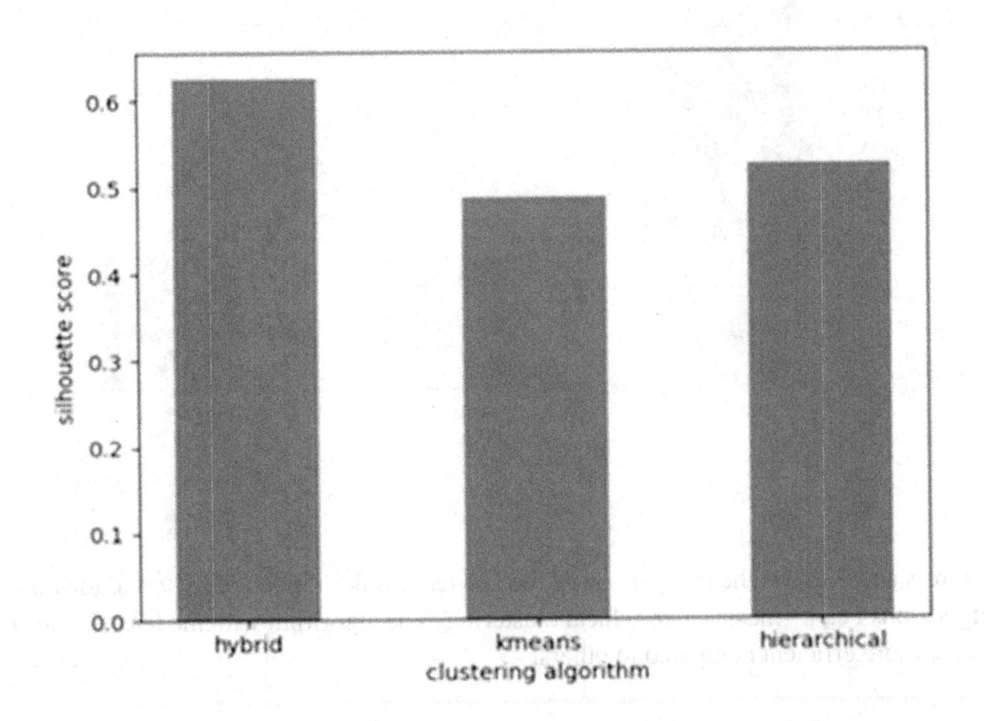

Each method has disadvantages; for example, k means creates only a few groupings and necessitates pre-defining the numeral of nodes towards remain formed because it remains dynamic in nature, though bunch examination is dynamic in nature a& delivers a greater number of groupings than k-implies, yet it requires numerous cycles because of the requirement for some consolidations and split choices. Due to these issues, we merged the two algorithms towards get the benefits of each while ignoring the drawbacks. We discover the greatest numeral of clusters as of a file using the resultant hybrid method, and the cluster produced are of extremely high quality, resulting in its most including among. The suggested hybrid technique results in a more effective cluster analysis with improved accuracy, recall, and F measure. Because the computed highest temperature value equals the real high temperature value, the result generated through the efficient hybrid clustering method remains the most reliable. The hybrid method generates the most clusters and incorporates all data points in either of these groups.

Calinski– Harabasz Index

The Figure 4.6 shows the comparison of the Calinski–Harabasz Index of hybrid algorithm and the existing algorithms i.e., k –means, hierarchical clustering. The algorithm with the highest Calinski–Harabasz Index is more efficient compared to others.

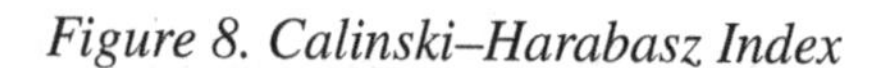

Figure 8. Calinski–Harabasz Index

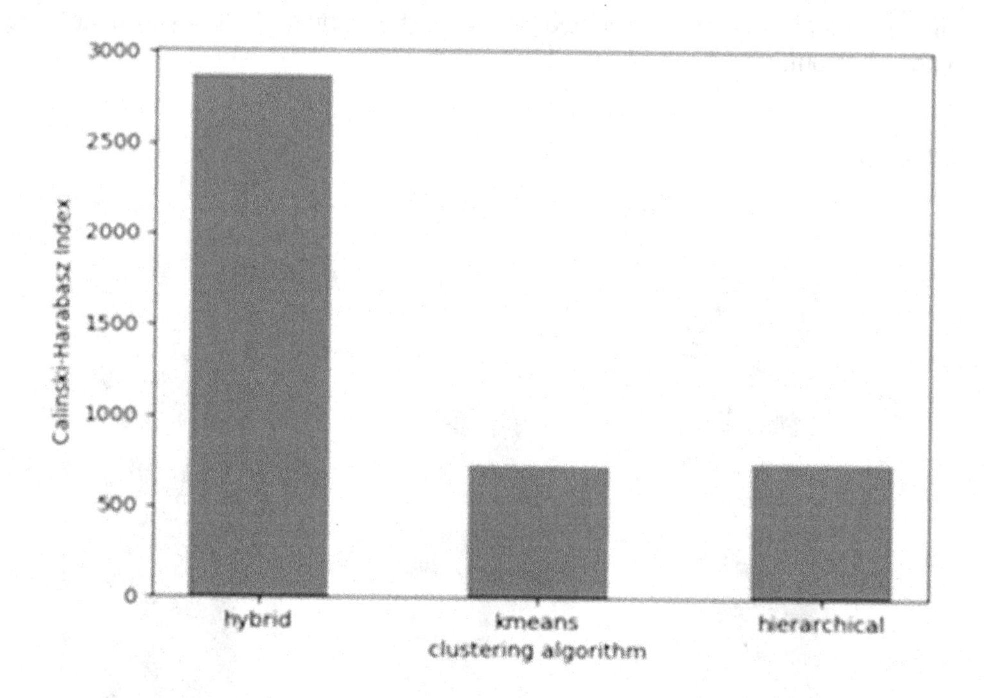

The Figure 8 and 9 shows the comparison of the Davies–Bouldin Index of hybrid algorithm and the existing algorithms i.e., k –means, hierarchical clustering. The algorithm with the least Davies Bouldin Index remains more efficient compared to others.

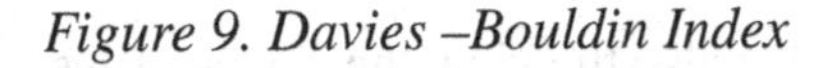
Figure 9. Davies –Bouldin Index

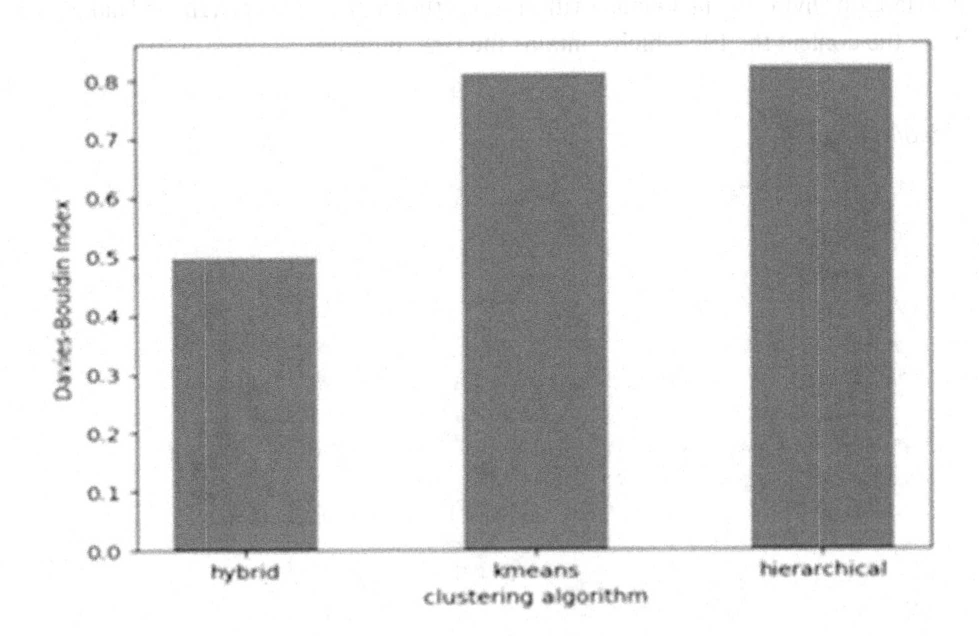

Stage of the mapper the meteorological file for a specific year is sent into the Mapper phase, as seen in Fig. 3 for the year 1907. Every record has two fields: the measurement date then the ambient temperature. The quality code remains also checked towards guarantee that its worth isn't missing. The Mapper parts the extricated information into key-esteem sets, since the MapReduce concept is built on a key-value situation. The date (as a key) also temperatures (as a value) are sent towards the reducer phase, thru the temperature being reduced by ten. The value remains in IntWritable form, while the key remains in Text form. The such are aimed at this data are (Monitoring date, Wind direction (Celsius)); some examples are (19070101, 13.9), (19070102, 11.7), and (19070103, 11.9). (19070103, 12.3). Each keyvalue pair is distinct. Figure 4 shows an output data with the two fields retrieved in the Analyzer phase date and observed surface temperature on that day. Decrease the number of stages. The reduction phase takes the Mapping phase's result as its argument. It accepts key-value pairs in two formats: Text & IntWritable. That remains, all of the temp readings were taken on a given day, at a particular time, and under precise circumstances. The temperature may be monitored quite often on a given day, but only up to three times in a single day, thus each key value is unique. The speed of the Enhancer varies depending on the method used to identify similar and determine the maximum temperature from those groups.

Final output

The final output consists of the optimal no. of clusters and the above mentioned clustering metrics i.e., Silhouette score (SC), Calinski–Harabasz Index (CHI), Davies –Bouldin Index (DBI) for hybrid clustering algorithm, k –means & hierarchical clustering algorithms. We can view them in Hadoop environment directly using "cat" command or can view them using "http://localhost:9870/". "http://localhost:9870/"contains the details of the Hadoop Environment such as the no. of nodes used, amount of space used, details about node manager, resource manager etc. We can also the view the various files present in the Hadoop environment. Our algorithm stores the final outputs in a file in Hadoop ecosystem. Figure 8 shows the content the file which contains the final outputs.

Figure 10. Final output

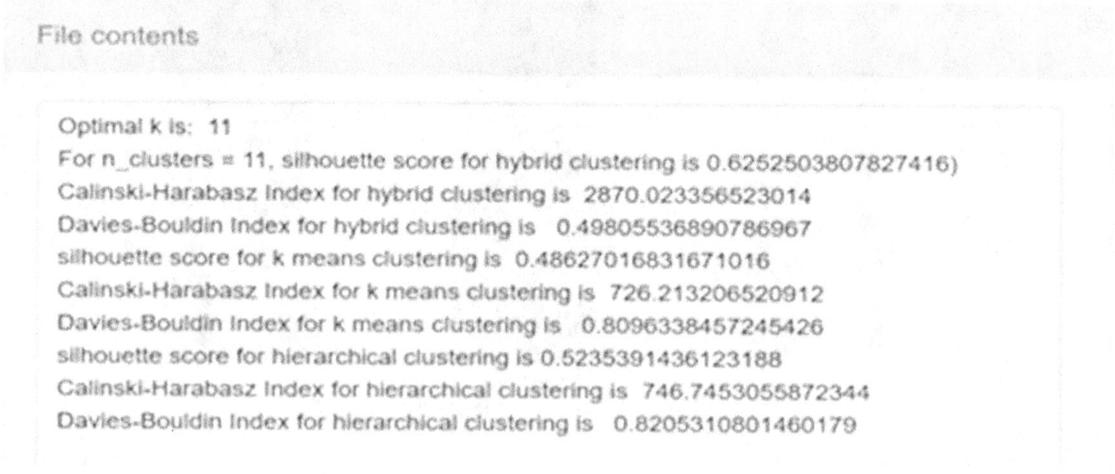

File contents

```
Optimal k is: 11
For n_clusters = 11, silhouette score for hybrid clustering is 0.6252503807827416)
Calinski-Harabasz Index for hybrid clustering is 2870.023356523014
Davies-Bouldin Index for hybrid clustering is 0.4980553689078 6967
silhouette score for k means clustering is 0.4862701683167101 6
Calinski-Harabasz Index for k means clustering is 726.213206520912
Davies-Bouldin Index for k means clustering is 0.8096338457245426
silhouette score for hierarchical clustering is 0.5235391436123188
Calinski-Harabasz Index for hierarchical clustering is 746.7453055872344
Davies-Bouldin Index for hierarchical clustering is 0.8205310801460179
```

The percentage of pairs of data points properly put in same clusters is used to calculate accuracy. It remains proportional towards the quality of clusters produced & reliability; the lower the sensitivity, the lower the quantity of groups defined; the greater the specificity, the more precise the algorithms remains, and the better the grade of groups formed. Precision is defined as the number of clusters calculated using a certain method. The dataset's real groups the variation from the true value is used to calculate it. Precision is calculated by dividing the number of clusters produced by a given method by the total number of clusters that may be formed in the dataset. We may also refer to it as the application's proportion of meaningful clusters. The hybrid method has the greatest accuracy, whereas the k-means approach has the least precision, as seen in Figure 5. The time value of money is evaluated between the three clustering methods. The execution times of the three methods are compared in Figure 8. The k-means process utilizes the least amount of time towards create clusters, whereas the hybrid approach takes the best time.

CONCLUSION

The proposed hybrid algorithm will overcome the drawbacks of existing algorithms and produce efficient results. This approach is an efficient way of producing the clusters. This hybrid algorithm is compared with existing algorithm using clustering metrics such as silhouette score, Calinski - Harabasz Index, Davies - Bouldin Index and proved that proposed algorithm is efficient than existing algorithms. Due to our system configurations, we could not test for very large datasets but it can be extended in future.

REFERENCES

Aggarwal, C. C., & Zhai, C. (2012). A survey of text clustering algorithms. In *Mining text data* (pp. 77–128). Springer.

Battré, D., Ewen, S., Hueske, F., Kao, O., Markl, V., & Warneke, D. (2010, June). Nephele/pacts: a programming model and execution framework for web-scale analytical processing. In *Proceedings of the 1st ACM symposium on Cloud computing* (pp. 119-130). ACM.

Berwind, K., Bornschlegl, M. X., Kaufmann, M., Engel, F., Fuchs, M., Heutelbeck, D., & Hemmje, M. (2017). Hadoop Ecosystem Tools and Algorithms. In NoSQL: Database for Storage and Retrieval of Data in Cloud (pp. 177-198). Chapman and Hall/CRC.

Demchenko, Y., Grosso, P., De Laat, C., & Membrey, P. (2013, May). Addressing big data issues in scientific data infrastructure. In *2013 International conference on collaboration technologies and systems (CTS)* (pp. 48-55). IEEE.

Doulkeridis, C., & Nørvåg, K. (2014). A survey of large-scale analytical query processing in MapReduce. *The VLDB Journal, 23*(3), 355–380.

Fernández, A., del Río, S., López, V., Bawakid, A., del Jesus, M. J., Benítez, J. M., & Herrera, F. (2014). Big Data with Cloud Computing: An insight on the computing environment, MapReduce, and programming frameworks. *Wiley Interdisciplinary Reviews. Data Mining and Knowledge Discovery, 4*(5), 380–409.

Firouzi, B. B., Sadeghi, M. S., & Niknam, T. (2010). A new hybrid algorithm based on PSO, SA, and K-means for cluster analysis. *International Journal of Innovative Computing, Information, & Control, 6*(7), 3177–3192.

Ghazal, A., Rabl, T., Hu, M., Raab, F., Poess, M., Crolotte, A., & Jacobsen, H. A. (2013, June). Bigbench: Towards an industry standard benchmark for big data analytics. In *Proceedings of the 2013 ACM SIGMOD international conference on Management of data* (pp. 1197-1208). ACM.

Hong, A., Xiao, W., & Ge, J. (2021, May). Big Data Analysis System Based on Cloudera Distribution Hadoop. In *2021 7th IEEE Intl Conference on Big Data Security on Cloud (BigDataSecurity), IEEE Intl Conference on High Performance and Smart Computing,(HPSC) and IEEE Intl Conference on Intelligent Data and Security (IDS)* (pp. 169-173). IEEE.

Jiang, D., Ooi, B. C., Shi, L., & Wu, S. (2010). The performance of mapreduce: An in-depth study. *Proceedings of the VLDB Endowment International Conference on Very Large Data Bases, 3*(1-2), 472–483.

Katal, A., Wazid, M., & Goudar, R. H. (2013, August). Big data: issues, challenges, tools and good practices. In *2013 Sixth international conference on contemporary computing (IC3)* (pp. 404-409). IEEE.

Pandove, D., & Goel, S. (2015, February). A comprehensive study on clustering approaches for big data mining. In *2015 2nd international conference on electronics and communication systems (icecs)* (pp. 1333-1338). IEEE.

Raghupathi, W., & Raghupathi, V. (2014). Big data analytics in healthcare: Promise and potential. *Health Information Science and Systems, 2*(1), 1–10. doi:10.1186/2047-2501-2-3 PMID:25825667

Chapter 16
Planning a Three-Year Research Based on the Community of Inquiry Theory:
An Approach to Monitor the Learning of English as a Second Language in the EU Academic Environments

Salvatore Nizzolino
https://orcid.org/0000-0002-3008-2890
Sapienza University of Rome, Italy

ABSTRACT

This research project aims to take advantage of five annual English as a Second Language (ESL) courses, in two faculties of Sapienza University of Rome, during a three-year period. These courses, embedded in the timeline of the regular academic semesters, are the teaching/learning environments in which the researcher himself works as a language teacher and tutor. Therefore, they will be the ground for this blended education plan complemented by e-learning through CIT ecosystems. The research plan takes shape within the community of inquiry framework (CoI), and this theory encompasses the conceptual boundaries.

INTRODUCTION - THEORETICAL FRAME / HYPOTHESIS: THE COMMUNITY OF INQUIRY

The CoI concept is rooted in theories by the pragmatist philosophers C.S. Peirce (Librizzi, 2017) and John Dewey (1902/1991), who speculated about the process of community-based knowledge formation and the scientific inquiry involved. Since then, the approach has also been used to study the dynamics of learning communities, and more recently, it has been adopted to study digital communities of learners (Garrison, 2011). In addition, the CoI principles are attracting the interest of the educational community

DOI: 10.4018/978-1-7998-9640-1.ch016

in recent years because of the implications that apply to the design, implementation, and evaluation of distance education courses.

More specifically, *"the Community of Inquiry framework is a collaborative-constructivist process model that describes the essential elements of a successful online higher education learning experience"* (Castellanos-Reyes, 2020, p. 557). An even detailed explanation, according to the most influential author on the CoI theories, Garrison R. D., is as follows:

A community of inquiry goes beyond accessing information and focuses on the elements of an educational experience that facilitate the creation of communities of learners actively and collaboratively engaged in exploring, creating meaning, and confirming understanding (…) It requires a commitment to and participation in a community of learners that will support critical reflection and collaborative engagement (D. R. Garrison, 2011, p. 352).

The CoI framework includes three elements named 'presences'—Cognitive Presence, Social Presence, and Teaching Presence—which are essential to an educational transaction (Garrison, 2011). Consequently, this framework defines cognitive intersections that go beyond simple access to information or focus on an educational experience. A collective space defined as CoI generally articulates within a hierarchical framework and stimulates the formation of communities of learners who actively and cooperatively extend their own learning experience. The co-construction of knowledge through shared understanding requires much more than the dissemination of information through traditional and standardized patterns of knowledge transmission. This process requires a high level of engagement and participation in a community of learners that keep up with critical awareness, reflection, sharing, and continuous collaboration (Garrison, 2011). The three presences of a CoI can be resumed as follows.

- Cognitive Presence: *"…is a vital element in critical thinking, a process and outcome that is frequently presented as the ostensible goal of all higher education"* (Garrison et al., 2002, p. 89).
- Social Presence: *"the ability of participants in the Community of Inquiry to project their personal characteristics into the community"* (Garrison et al., 2002, p. 89).
- Teaching Presence: This dimension can be summarized as the "*responsibility to design, facilitate, and direct learning online*" (Anderson et al., 2001).

Figure 1. The community of inquiry model. Reprinted from "Critical inquiry in a text-based environment: Computer conferencing in higher education." Garrison et al. (2000). The Internet and Higher Education, 2, p. 88.

According to Kreijns et al. (2014), the sense of mutual belonging within a CoI is linked to the awareness of depending on each other to achieve learning goals. In an academic language course, it is the teacher's responsibility to foster and maintain this awareness during the project period, and continuity is entrusted to peer interactions. Indeed, language learning is a social experience and should benefit from mutual interactions. Some scholars still doubt that the CoI theory can be effective in explaining deep, meaningful learning patterns (Kreijns et al., 2014), while others have tried to extend the CoI framework to include a fourth presence. This contribution is focused on the determination of those teaching/learning elements that can strengthen the three main presences and possibly support the meaning of one additional presence. The theme of adding a further presence is not new to the scholars interested in the CoI framework and it has been awakening a controversial debate:

(…) the CoI framework needs additional components to be more meaningful as a framework. Researchers suggest the existence of an extra presence but have not achieved consensus on which. Suggested

additional components are learner presence (Shea et al., 2012), emotional presence (Cleveland-Innes and Campbell 2012), and autonomy presence (Lam 2015). None of the additional constructs to the three presences have been validated as of yet. (Kozan and Caskurlu 2018).

(as cited in Castellanos-Reyes, 2020, p. 558)

The unprecedented aspects, communication; content-generation; knowledge sharing; social relationships (Kaminskienė et al., 2020; Moghavvemi et al., 2017, Thomas et al., 2020), and even patterns coming from the social network sciences should now be evaluated as part of the education process in the present digital classroom. In addition, the asynchronous modes are strictly dependent on self-cognitive patterns, thus, *"not only online learning but also offline—through digital resources such as e-books downloaded to mobile devices and accessed at the learner's convenience."* (Kukulska-Hulme, 2012). As Hrastinski already showed through a literature review (2008), synchronous and asynchronous settings and/or communication were applied as units of analysis by many scholars. Accordingly, in the last decade, this dichotomy has gained even more importance due to Massive Open Online Courses (MOOCs) that reach learners living in different time zones (Campos et al., 2020). In addition, this double mode of continuous communication is the common way of using SNSs (Lambton-Howard et al., 2020).

WHY THE CoI FRAMEWORK NEEDS MORE VISIBILITY IN THE EUROPEAN CONTEXT

In recent years, the EU has multiplied its collective capacity-building efforts to support the digital transformation of education and learning, and the main output to standardize skills and competencies consist of several Digital Competency Frameworks:

- for citizens (Vuoriaki et al.; 2016);
- for educational organizations (Kampylis et al.; 2015);
- for educators (Redecker, 2017);
- for opening-up Higher Education Institutions (Inamorato dos Santos et al.; 2016)
- for consumers (Brečko et al.; 2016).

Despite the digitalization of society and the widespread use of digital skills, not many young learners, even at the university level, are self-motivated to take the advantage of social media for their academic performance (Al-Rahmi et al., 2015). In order to investigate the students' self-determination, motivation, and learning strategies through Social Network Sites (SNSs), some authors propose models and frameworks for using social media from the secondary school students' perspective (Callan & Johnston, 2020; Dabbagh & Kitsantas, 2012; Thomas et al., 2020), university students' perspective (Coursaris & Van Osch, 2016; Rasi et al., 2015), or from the teachers' standpoint (Michos & Hernández-Leo, 2020). The range of case studies goes from a single small group of students (Deaves et al., 2019) to a statistical analysis involving great samples of public education bodies (Vivian, 2012). Some authors highlight the importance of investigating the interaction between teachers and learners to explore how the former facilitates the rising of new learning strategies in the social media context (Matzat & Vrieling, 2016) both in formal and informal settings (Mulyono & Suryoputro, 2020). Relational factors are the key drivers for enhancing digital strategies in the classroom and establishing new expectations about teacher–student relationships. The quality of teaching/learning relationships and the online tutoring exerted by teachers in e-learning spaces contribute to a more successful learning experience. *"The nature of online teacher intervention could facilitate a positive classroom climate that enhanced overall student engagement and learning"* (Manca & Ranieri, 2017, p. 610). Moreover, the role of learning platforms and MOOCs is mainly considered as a university dimension, where the aspect of self-directed learning is particularly relevant (Cacheiro-Gonzalez et al., 2019). In such a context, the role of a trained and skilled body of teachers as motivators is the key driver for a stronger connection to the professional world (Kennedy & Laurillard, 2019). Accordingly, another prevailing aspect relates to the possibility of creating self-directed online learning (Callan & Johnston, 2020) and a passage from non-personalized teaching to a feeding process tailored to more specific learning needs (Xiaochun et al., 2017). Self-learning also implies self-motivation and self-assessment of one's own digital skills (Schmid & Petko, 2019; Xiaochun et al., 2017) and these dynamics may also foster collaborative models that enhance group self-regulation (Dabbagh & Kitsantas, 2012).

Indeed, the concept of engagement in education is a multidimensional aspect, and for this reason, several authors have explored the similarities between different learning spaces: public schools and MOOC platforms (Bond et al., 2020), or the differences between various social platforms and tools adopted for learning purposes (Loc et al., 2019). Behavioral patterns and interactions between students and teachers undergo profound changes when SNSs are incorporated into the educational experience. Therefore, our research pathway is not only about pedagogical and technological challenges but also about online social interactions and how social networks can be used for the teachers' continuous professional development (Manca & Ranieri, 2017; Nizzolino, 2020). The new learning environment should encourage the learners' multitasking capabilities, reconfigure the communication paradigm, and enhance the high level of learning independence in the frame of continuous sharing (Mills, 2011). Students can be motivated through specific types of strategies featuring certain SNSs, such as Blogs (Garcia et al., 2019; Laurillard et al., 2018); Facebook (Luo et al., 2020; Vivian, 2012; Giannikas, 2019; Kelly & Antonio, 2016; Madge et al., 2009); Twitter (Deaves et al., 2019; Dommett, 2019; Luo et al., 2020); Instagram (Thomas et al., 2020); and many other notorious tools suitable for enhancing the learning experience (Calderon-Garrido et al., 2019; Gazit et al., 2019; Baishya & Maheshwari, 2020). Building up communities, social networks, and work-groups implies also the emotional sphere, being the psychological wellness a pre-condition for any cognitive success. As Matzat & Vrieling detect, *"the use of social media in the class reduces the psychological distance between pupils and teacher"* (2016, p. 10). On the other hand, the assessment of the students' performances should include potential obstacles such as "*imbalanced power relations and a lack of authenticity*" (Lambton-Howard et al., 2020, p. 2). Due to the lack of shared psychological frameworks linked to the use of SNSs, researchers can scarcely mention figures related to personality traits such as "*extroversion/introversion, openness, neuroticism, internal and external locus of control*" (Gazit et al., 2019, p. 140). In addition, within an institutional education environment, the choice of the SNS is usually a top-down decision or, in the most participative cases, decided by the majority (Nizzolino & Canals, 2021). In such a scenario, the personal and demographic variables shaping the relation between users and SNSs are left out. For example, age is the dominant demographic predictor associated with Facebook, LinkedIn, and Twitter outside schools and universities (Gazit et al., 2019, p. 141). Conversely, in a formal setting, students do not have options of choice. They must adapt to the SNS selected by the teacher or the instructor, consequently, they will not use a tailored platform (Nizzolino & Canals, 2021).

FROM CONSTRUCTIVISM TO CONNECTIVISM

The general positive aspects highlighted by scholars interested in e-learning show that the integration of SNSs can increase the students' involvement in education "*which in turn leads to better school performance*" (Matzat & Vrieling, 2016, p.3). From a pedagogical perspective, dealing with digital communities and learning patterns also implies a focus through cognitive theoretical lenses, Constructivism and Connectivism, in particular. According to the former: "*social media practices seem well aligned with social constructivist views of learning as participation in a social context*" (Greenhow & Lewin, 2016, p. 8); while the latter is a more recent cognitive approach that defines the ability to make connections between information sources and thereby generate useful information patterns. Among modern cognitive theories, Connectivism offers broader margins of association with e-learning communities, "*the process of creating connections and articulating a network with nodes and relationships, also seems well aligned with social media practices*" (Greenhow & Lewin, 2016, p. 9). Therefore, the Connectivism theory addresses the challenge of managing personal knowledge within networks of people. These aspects will be subsequently applied under the lens of Methodology.

Pererira et al. (2010) have brought to light the work done by Gene Smith[1] who identified 7 elements—*Identity; Presence; Relationships; Conversations; Groups; Reputation; Sharing*—to define a social software and named the model Social Software Honeycomb. Being the original honeycomb model conceived in 2007, we will re-define the 7 dimensions through essential labels suitable to the current awareness of the hyper-connected society.

- Identity – the digital identity the user adopts within the space framed by the software.
- Presence – a psychological shared status; the possibility to interact with other users or content.
- Relationships – links and relationships with other users; frequency of interaction.
- Conversations – synchronous and asynchronous communication features within the software.
- Groups – possibility to build up personal communities (even closed) and join others.
- Reputation – a way of knowing the status of other users within the social space.
- Sharing – ways of sharing content.

Neither Smith's original early-stage model nor Pererira et al. (2010) developed a hierarchical structure or a conventional frame to represent the dynamics of the 7 dimensions, but the latter concluded that listing them is not sufficient to represent all the possible variables involved in a SNS context. Indeed, considering some popular social media sites such as YouTube, Twitter, Facebook, and Instagram we can empirically verify how the 7 dimensions can place four different sets of priorities: Facebook requires the greatest care with Identity and Presence, while on YouTube the priority is Sharing and increasing the Group subscribers (the user's channel). Twitter and Instagram present overlapped configurations since both imply a Reputation to support the post Sharing, while the role of Conversation and Presence is generally reduced in comparison with Facebook. Naturally, there will be overlaps among the 7 elements in any possible configuration. This model can offer a starting point to reflect on the potential implications when adopting SNSs in education. All the issues related to the students' assessment would require a separate investigation, but we can immediately notice that a parameter such as *Reputation* may not easily fit a formal education environment, thus, it should be articulated through more suitable descriptors or social network analysis.

Others argue that only a small percentage of young learners use social media in a manner that can be associated with informal learning, so as to represent assessable activities (Greenhow & Lewin, 2016). On the other hand, when the use of social media is combined with opportunities for improving self-regulated learning *"this is likely to influence students' learning motivation"* (Matzat & Vrieling, 2016, p.10), notwithstanding *"the teachers tend not to utilize social media for the provision of opportunities for* [Self-Regulated Learning]" (p. 15). By applying the CoI Presences, we can subsume the numerous connectivism-related dimensions under fewer areas of inquiry.

LITERATURE MAPPING AND REVIEW

The number of research publications is increasing inexorably and there is no other way to monitor a particular field of knowledge than using specific e-tools. Concerning the CoI-related segment of publications, we must start by defining a segment of relevant works to encompass a community of research. According to the bibliometric method by Aria and Cuccururllo (2017), we accomplished research in Scopus[2] to fulfill the following three steps:

1. identifying the knowledge base of the research field "Community of Inquiry" and its intellectual structure;
2. examining the research front (or conceptual structure) of a topic or research field; in other terms, identifying the associated themes and interconnected fields of research;
3. producing a social network structure of this particular scientific community through Keywords.

We searched the word association "Community of Inquiry" to "Teaching Presence", "Cognitive Presence" and "Social Presence" in articles and conference papers from 2006 to 2020. The final results included 555 journal articles and conference papers.

Table 1. Main information on the sample of 555 articles collected in the preliminary research

MAIN INFORMATION ABOUT DATA	
Timespan	2006: 2020
Sources (Journals, Books, etc.)	265
Documents	555
Average years from publication	5.74
Average citations per document	19.08
Average citations per year per doc	2.08
References	19491
DOCUMENT TYPES	
Articles	**448**
Conference Papers	**107**

We can notice that the number of Author's Keywords is higher than the Keywords Plus, namely, the term-associations detected in the paper's abstract. These keywords are frequently underestimated by the authors themselves. In this case, this research community shows a remarkable awareness of the paradigm they have been investigating. Indeed, the very first observations in this analysis emerge from the concepts conveyed by the keywords.

Table 2. Document contents from the sample of 555 items collected in the preliminary research

DOCUMENT CONTENTS	
Keywords Plus (ID)	1167
Author's Keywords (DE)	1211
AUTHORS	
Authors	1123
Author Appearances	1462
Authors of single-authored documents	124
Authors of multi-authored documents	999

Another feature of this community is the preponderance of works with single or double authorship, as the further table shows. Multi-author papers, often involved in long-term research projects or transnational collaborations, do not seem to be a trend.

Table 3. Authors' collaboration from the sample of 555 items collected in the preliminary research

AUTHORS COLLABORATION	
Single-authored documents	133
Documents per Author	0.49
Authors per Document	2.02
Co-Authors per document	2.63
Collaboration Index	2.37

Figure 2. Article production (no conference papers) from 2006 to 2020

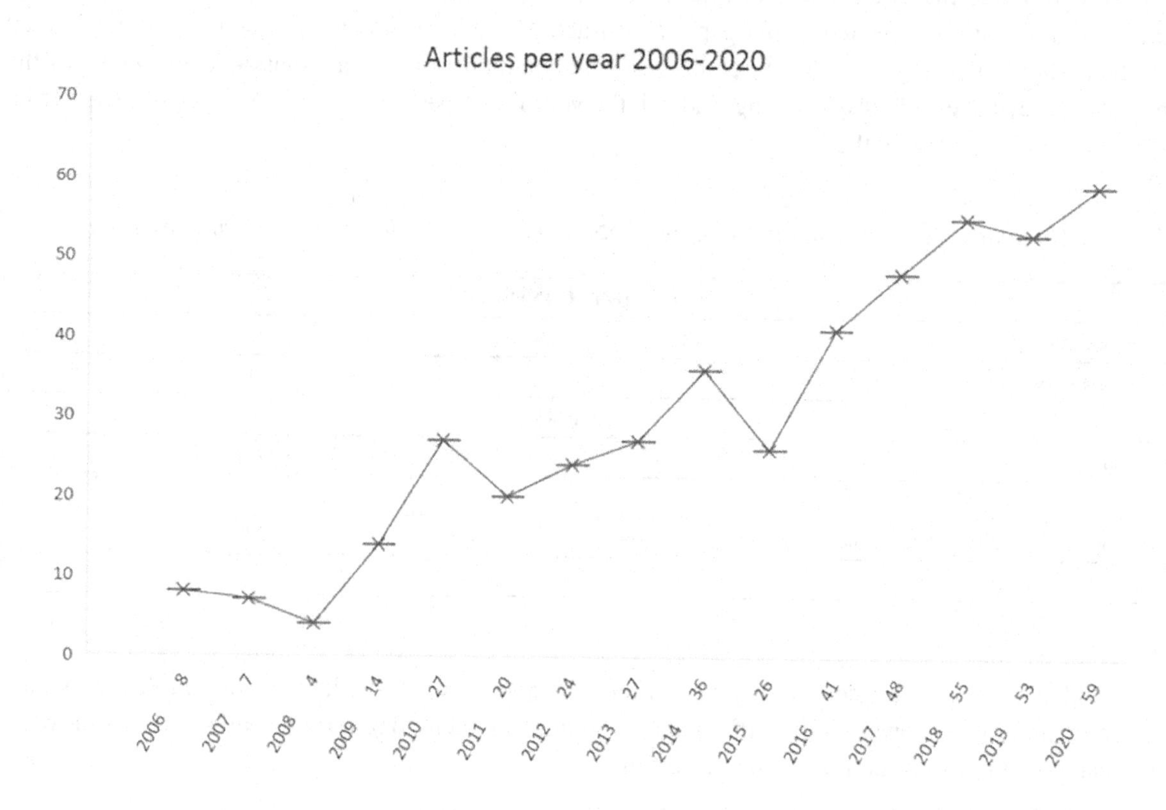

The production of academic articles has grown irregularly over the last 15 years, but the increment range is increasing, reasonably due to the expansion of subject areas, as the list of types of sources suggests (see Table 6). The most significant contributions come from the four English-speaking countries shown in the further Figure 3, while there is still little interest in this topic in the rest of the world. As already noted, a predominant aspect of this community of scholars is a strong tendency for contributions to have one or two authors, but when there are two authors, they are generally from the same country.

Figure 3. Article production per Country (no conference papers) from 2006 to 2020

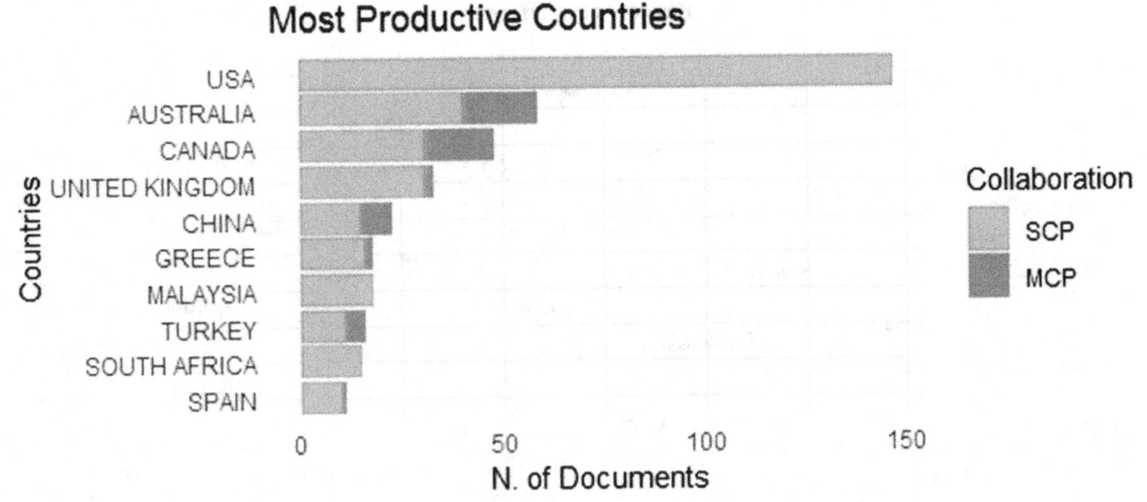

The only EU countries in the top ten contributors are currently Spain and Greece, but the gap between Europe and the first four English speaking countries is remarkable. It is worth noticing that the USA does not present any ratio of Multiple Country Publication (MPC), meaning that the articles published by USA-affiliated authors do not share any foreign co-authoring. On the other hand, as we claimed in the Introduction, the EU is busy building up its own communitarian frameworks and the CoI has not yet been taken into account.

Table 4. Corresponding Authors per Country; SCP: Single Country Publications; MCP: Multiple Country Publications

	Country Article	Articles	Freq.	SCP	MCP	MPC_Ratio
1	USA	147	0.291	147	0	0.000
2	AUSTRALIA	59	0.117	41	18	0.305
3	CANADA	48	0.095	31	17	0.354
4	U.K.	33	0.065	31	2	0.060
5	CHINA	23	0.045	15	8	0.347
6	GREECE	18	0.035	16	2	0.111
7	MALASYA	18	0.035	18	0	0.000
8	TURKEY	16	0.031	11	5	0.312
9	SOUTH AFRICA	15	0.029	15	0	0.000
10	SPAIN	11	0.021	10	1	0.090

Figure 4. Weight of Authors per Country and citation network

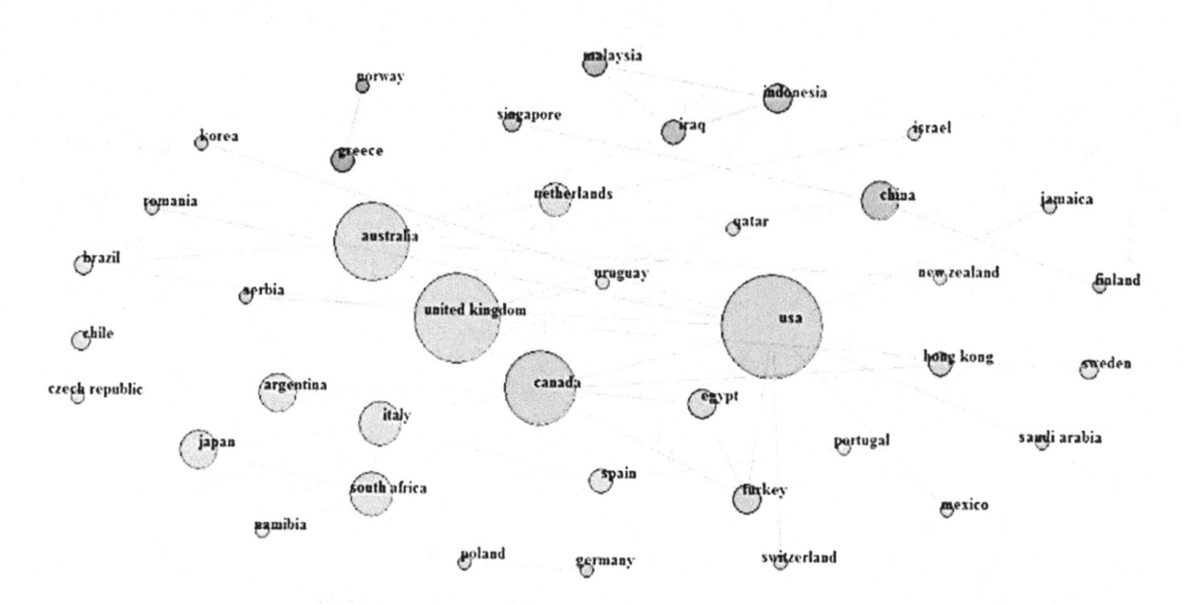

The weight of the EU countries (Germany, Italy, Portugal, Spain) is currently quite moderate in terms of *Betweenness Centrality* and very low as a *bridging function,* with the exception of Italy being a node with a higher number of collaborations, whilst Norway and Greece have the lowest rank. Consequently, further contributions and investigations from EU academia may rise its role in this research community.

Table 5. Most productive authors

	Authors	Articles	Authors	Articles Fractionalized
1	SHEA P	14	GARRISON DR	5.23
2	GARRISON DR	13	SHEA P	4.56
3	AKYOL Z	9	ARBAUGH JB	4.48
4	BIDJERANO T	9	AKYOL Z	3.75
5	RICHARDSON JC	9	RICHARDSON JC	3.26
6	ARBAUGH JB	7	BIDJERANO T	3.13
7	CLEVELAND-INNES M	7	GOLDING C	3.00
8	HATALA M	7	JIMOYIANNIS A	2.67
9	ICE P	7	PEACOCK S	2.50
10	GAEVI D	6	STEWART MK	2.50

Table 6. Most relevant sources (journals)

	Sources	Articles
1	INTERNET AND HIGHER EDUCATION	57
2	INTER. REVIEW OF RESEARCH IN OPEN & DISTANCE LEARNING	27
3	ONLINE LEARNING JOURNAL	19
4	COMPUTERS AND EDUCATION	18
5	EDUCATIONAL PHILOSOPHY AND THEORY	14
6	INTERACTIVE LEARNING ENVIRONMENTS	10
7	JOURNAL OF ASYNCHRONOUS LEARNING NETWORK	10
8	ACM INTERNATIONAL CONFERENCE PROCEEDING SERIES	9
9	DISTANCE EDUCATION	8
10	TECHTRENDS	8

Table 7. Most relevant Keywords in the sample corpus of 555 Articles and Conference Papers

MOST RELEVANT KEYWORDS				
	Author Keyword (DE)	Articles	Keywords-Plus (ID)	Articles
1	COMMUNITY OF INQUIRY	332	COMMUNITY OF INQUIRY	153
2	ONLINE LEARNING	126	TEACHING	106
3	SOCIAL PRESENCE	88	E-LEARNING	104
4	TEACHING PRESENCE	81	STUDENTS	97
5	COGNITIVE PRESENCE	70	EDUCATION	93
6	BLENDED LEARNING	48	COGNITIVE PRESENCE	62
7	HIGHER EDUCATION	29	TEACHING PRESENCE	54
8	COLLABORATIVE LEARNING	26	ONLINE LEARNING	49
9	DISTANCE EDUCATION	25	SOCIAL NETWORKING	46
10	E-LEARNING	23	SOCIAL PRESENCE	40

Figure 5. Keywords co-citations in the overall corpus of 555 papers selected

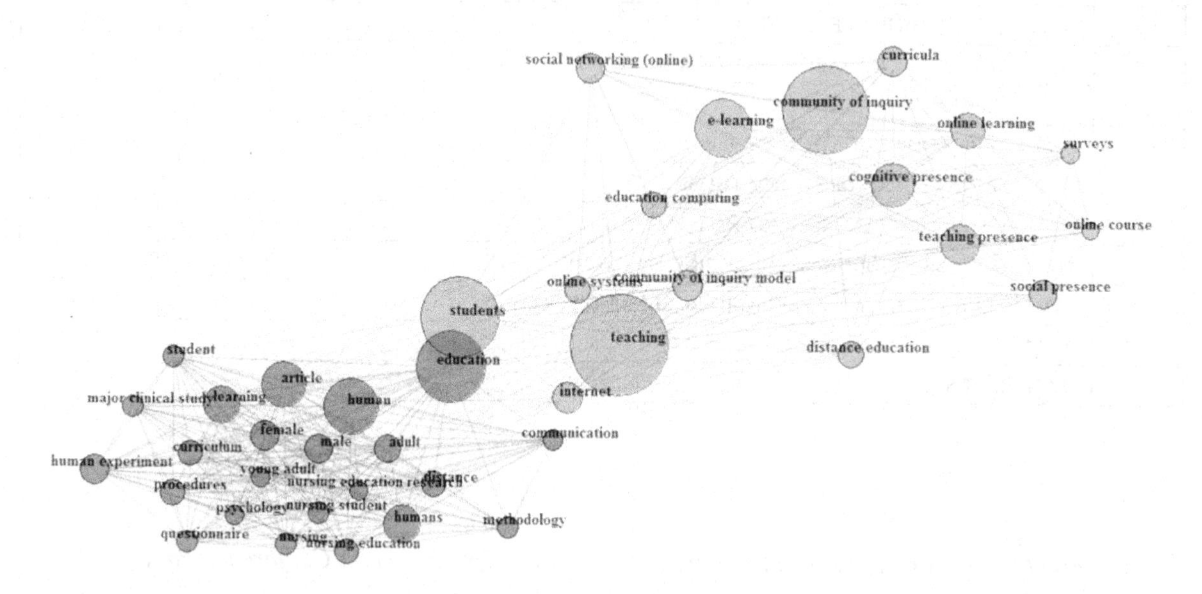

The data included in the Author's Keywords and the Keywords detected are terms, compounds, or phrases that frequently appear in the titles within co-citation relationships. It is worth noticing the weight of the node "teaching" and its *Betweenness Centrality* which is more relevant than "students". This may be the sign of a prevalent focus on teaching strategies instead of the learning patterns.

THE 4TH PRESENCE HYPOTHESIS AND DEBATE

As Stenbom et al. remarked (2016), the aspect of emotion was, in the original version of CoI, considered a category within the Social Presence. The concept of emotional intelligence is vast and would require a dedicated literature review, but, it has been summarized (in the debate around CoI) in the cognitive differentiation of *"emotional perception, emotional facilitation, emotional understanding, and emotional management"* (Majeski et al., 2018, p. 55). These elements of emotional intelligence may be linked to the original Presences of CoI, but can also be considered independently as key factors for a successful e-learning journey, especially in an online environment where the psychological sphere is stimulated by different pressures than in a physical learning space. The recent conclusions that define Emotional Presence as a constitutive element of any e-learning setting may lead to a shared extension of the CoI Framework, *"emotion may constrain learning as a distracter but, if managed, may serve as an enabler in support of thinking, decision making, stimulation, and directing"* (Cleveland-Innes & Campbell, 2012, p 285). One of the fathers and disseminator of the CoI, Garrison Randy D., disagrees with the possibility of adding further Presences on the basis of the claim that an added Presence may result in increasing *"the complexity of the framework and violating the principle of parsimony."* (Garrison, 2017).

On the other hand, according to Cleveland-Innes & Campbell, future research would require *"that we determine first which emotions are present in common human exchanges and in any learning environ-*

ment then identify if emotions in online environments are the same or different." (2012, p. 285). In the area of language learning, the emotional variables have been widely investigated for decades, thus, we propose to convert these emotional elements into variables in order to also investigate the Emotional Presence in our ESL courses.

RESEARCH QUESTIONS

RQ1: Is the concept of Social Presence accurately defined to supervise *individual emotional factors* in a blended learning context aimed at teaching/learning ESL?

RQ2: How can digital settings and e-tools mediate between the three Presences and the actual teaching/learning experience?

RQ3: How relevant are the correlations between the *user experience* and the three Presences to accomplish a successful implementation of the CoI model?

Table 8. Macro-areas of investigations, further broken down into indicators

Objective	Macro areas of research
Define the student's use and perception of the Digital Tools adopted	o Intuitive or complex CIT features o Time spent in learning the tools o Learning style with the tools
Define the patterns/interaction between Teacher/Students	o Communication o Students' attitude

METHODOLOGY

Setting

Faculty of Economics, Sapienza University of Rome, Latina campus. English courses. Language Laboratory. Learning ecosystems structured upon Digitalized classrooms; Moodle Sapienza platform and applications suggested by the teacher.

Faculty of Engineering, Sapienza University of Rome, Latina campus. English courses. Learning ecosystems structured upon Digitalized classrooms; Moodle Sapienza platform and applications suggested by the teacher.

Population, Inclusion/Exclusion Criteria

Surveyed students have been categorized according to their median age, which is generally related to the academic path, thus, it is possible to select two main groups. In fact, students attending the Bachelor's Degree belong to the age group 18–21; whilst the Master's Degree students belong to the 23–27 age group. We seldom find older students (1 or 2 per year), so they are not statistically relevant.

Table 9. Subdivision in course groups for one academic year

FACULTY OF ECONOMICS	Bachelor Degree Group 2 Group 3 Group 5
Group 1 (level C1) approx. 20/30 students	
Group 2 (level B2) approx. 70/90 students	
Group 3 (level B1) approx. 100/120 students	
FACULTY OF ENGINEERING	Master Degree Group 1 Group 4
Group 4 (level B2) approx. 50/70 students	
Group 5 (level B1) approx. 80/100 students	

The same group patterns will be monitored each year for three academic years.

Variables

Table 10. Break down structure of Dimensions and Indicators

PRESENCE	CATEGORY	INDICATORS (examples)	Examples of unit/strings
Cognitive Presence	Triggering event	Stating a problem	Request for help/support
	Exploration	Exchanging perspectives	Replies to posts
	Integration	Connecting	Regular posts/participation to debates
	Resolution	Selecting the potential solution	Acceptance of proposals and/or negotiations
	Establishing learning-pace	Establishing interaction	Integration of multiple channels
Teaching Presence	Design and organization	Stimulating constructive inquiry	Assignments
	Facilitating discourse	Re-directing the discussion	Targeted interventions
	Direct Instructions	Summarizing the key-points	Formal linguistic register
Social Presence	Open Communication	Informal expressions	Low linguistic register
	Relationship cohesion	Greetings, vocatives, informal syntax	Low linguistic register
	Frequent interactions	Polarization of relationships	Frequent interactions between specific learners
Emotional Presence	Activity emotion	Emotion about assignments	Range of ready descriptors
	Outcome emotion	Emotion about the organization of tasks	Range of ready Psycho-linguistic descriptors
	Directed affectiveness	Emotion towards agreements and critics	Range of ready Psycho-linguistic descriptors
Autonomy Presence	Self-paced learning	Routine and behavioral patterns	Behavioral descriptors
	Motivation	Self-perception and self-image	Range of ready descriptors
	Time management	Organizational strategies	Behavioral descriptors
	Stress management	Sources of emotional control	Psycho-linguistic descriptors

Dimensions, Trials, and Variables of the Research

At the end of the research timeline, the number of monitored students will reasonably lay between 1.000 and 1.200.

Each monitored ESL course will be used to adjust the implementation of e-tools, learning dynamics, and data collection methodologies. Accordingly, each university semester will provide the opportunity to sharpen the research experience. There will be six semesters in the three-year timeline.

Concerning the possibility to adjust the set of variables already set at the very beginning of the investigation, this matter will be analyzed during the empirical research and data processing. The exploration of a 4th potential Presence in the CoI framework will probably imply the introduction of new variables along the way.

Methods and Techniques to Collect Data

Table 11. Teaching/Learning Tools and association with the research methods

ECOSYSTEM TOOLS	INPUTS	RaW DATA	EXTRACTION METHODS	ANALYSIS
Main Board (Moodle Platform)	>Teacher's formal instructions and tasks	TEXT	Text Mining Text Analytics	>Lexical Analysis >Syntax Analysis
Forum (Moodle Platform)	>Teacher's instructions and tasks >Teacher's feedback >Students posts >Peer-to-peer interactions and discussions	TEXT	Text Mining Text Analytics	>Lexical Analysis >Syntax Analysis >Network representation >Social Network Analysis
Collaborative Glossary (Moodle Platform)	>Individual contributions (posts) >Teacher's monitoring >Teacher's adjustments (posts)	TEXT	Text Mining Text Analytics	>Lexical Analysis >Syntax Analysis >Network representation >Social Network Analysis
Teamwork on open Platform	>Peer-to-peer interactions and discussions >Teachers' monitoring >Teachers' guidelines	TEXT	Text Mining Text Analytics	>Lexical Analysis >Syntax Analysis >Network representation >Social Network Analysis
Multimedia Repository on a SNS *(interactive posting community)*	>Teacher's guidelines >Student's contents >Peer-to-peer interactions and discussions	TEXT	Text Mining Text Analytics	>Login History >Login-Logout duration >Users' polarizations (Network representation) >Social Network Analysis
Chat App	>Peer-to-peer interactions and discussions	eSurvey quantitative qualitative	eSurvey to collect personal experience	>Frequency of use >Identity perception >Reputation >Perceived benefits
Mobile App #1	>Individual learning	eSurvey quantitative qualitative	eSurvey to collect personal experience	>Frequency of use >Learning Patterns >Combination with other tools
Mobile App #2	>Individual learning	eSurvey quantitative qualitative	eSurvey to collect personal experience	>Frequency of use >Learning Patterns >Combination with other tools

HOW THE COI CAN GAIN ACCESS INTO EDUCATION ROUTINE

The dimension of the *user experience* and its relation to networking benefits, in terms of educational networking, is not yet the object of wide investigation, while it is most commonly framed within psychological and behavioral patterns (Lin et al., 2008). Similarly, Weidlich & Bastiaens (2019) report a gap in the literature on which characteristics and qualities of learning environments favor social presence and other socio-emotional variables, leaving aside the *user experience.*

Team Learning and Group Work

One of the most promising research trends is associated with a new psychological approach to group work through SNSs. The pre-requisite of successful team learning is associated with the perception of belonging in the group (Greenhow & Lewin, 2016; Rap & Blonder, 2017; Rosenberg et al., 2020), especially when the group is working online. In addition, new behavioral aspects, such as the fear of missing out social targets (Alt, 2018); the criticism to one's proposals (Lin et al., 2008); the negative or positive impact on self-esteem, self-confidence and self-worth (Iwamoto & Chun, 2020), are beginning to be observed when learners interact online. The dimensions of group-learning and team-learning are frequently associated with specific SNSs, since such investigations are usually the result of on-the-ground researches and empirical observations of practical cases (Pimmer et al., 2017; Cacheiro-Gonzalez et al., 2019; Geeraerts et al., 2016; Admiraal et al., 2011; Al-Aufi & Fulton, 2015; Pimmer et al., 2017; Al-Rahmi et al., 2015; Michos & Hernández-Leo, 2020; Nizzolino, 2021; Robertson & Swan, 1998).

Teachers as Designers of Knowledge Management (KM) Practices and their Professional Development

Many researchers show how the positive interactions of educational social networks and the active participation of all educational stakeholders can positively affect the flow of information in educational establishments, and how they can improve work engagement (Cheng, 2012, 2013; Lin et al., 2008; Zhao, 2010). There is a growing interest around education environments within the perspective of KM due to the fact that schools and universities are precisely the first places where individuals (teachers and students) deal with structured information, data, and knowledge, in their two recognized forms: tacit and explicit (Cheng, 2019; Cheng et al. 2017, 2018; Nonaka & Konno, 1998; Zhao, 2010).

Nevertheless this growing consideration, *"little research had focused on how teachers learn with digital technologies, but rather there was research on how they learn about technologies, or how they use them to teach."* (Kukulska-Hulme, 2012, p. 247). Although resistance to innovative approaches is one of the frequent obstacles detected during the implementation of networks of practices that generate new routines (Manca & Ranieri, 2017), the knowledge society is already prompting a *"transformation of teaching into a design science"* (Laurillard et al., 2018, p. 1045). Teachers are required to develop a new set of skills that go beyond the old vision of transferring knowledge and evaluating feedback. The new set of professional skills include abilities such as combination, process management, collection and organization of knowledge, adjusting, and creating and confirming knowledge during the teaching/learning experience (Zhao, 2010). Concerning higher and academic education, the implementation of digital innovation should reduce the gap between students and professors, on the other hand, it is still common that *"in higher education the needs of students may be perceived as relatively remote from the needs of faculty"* (Kukulska-Hulme, 2012, p. 248) and this may induce instructors to disregard those training events related to disruptive innovation in the teaching routine. As knowledge is increasingly perceived as a flow, the learning process needs to be driven by professionals who enable the suitable factors to create communities that share content within a stream of knowledge. This knowledge flow is inter-generational, not only between teachers/students but also between teachers of different ages and backgrounds (Geeraerts et al., 2016) and implies *"the necessity to re-negotiate teachers' and students' roles where social networking applications intersect with classrooms and curricula"* (Manca & Ranieri, 2017, p. 606).

FINAL REMARKS AROUND THE INDICATORS

It is possible to formulate an example of a basic matrix including the level of potential involvement of the three CoI presences in an Ecosystem that has been already implemented in 2021 in an Academic English course at Sapienza University of Rome. A set of e-learning tools can be assigned to specific learning tasks linked to a range of indicators for measuring the three elements of CoI: Cognitive Presence (CP); Social Presence (SP); Teaching Presence (TP); plus two possible new presences: Emotional Presence (EP) and Autonomy Presence (AP).

Table 12. Levels of potential involvement associated with the CoI's Presences: High, Medium, Low, NA (Not Applicable); the last three tools on the right are out of teacher's direct control, being totally labelled as Informal, thus, they can be monitored only through a qualitative interview

PRESENCE	Moodle Main Board	Moodle Teaching Forum	Moodle Collaborative Glossary	Teamwork on open Platform	Multimedia Repository on a SNS	Chat App	Mobile App #1	Mobile App #2
CP	Medium	High	High	High	Medium	Low	Medium	Medium
SP	Low	High	Medium	High	Medium	High	NA	NA
TP	High	Medium	Medium	Medium	Low	NA	NA	NA
EM	?	?	?	?	?	?	?	?
AP	?	?	?	?	?	?	?	?

The addition of two potential new dimensions such as EM and AP would require a preliminary question on the effective e-tools able to detect new variables possibly connected to these additional Presences. A solution to these doubts may be that of including specific questions in the interviews aimed at detecting the Emotional sphere and the Learning Autonomy of students.

EXTENSIONS AND INTERCONNECTIONS TO IDENTIFY ADDITIONAL INDICATORS OF PRESENCES

An important aspect to highlight is that our research is not focused on the improvement of English learning alone. Indeed, the subject of ESL is functional to investigate the dynamics associated with the interaction between the three CoI Presences and the CIT tools employed, however, the core of the research project is the evaluation of teaching/learning strategies and dynamics in a collaborative learning ecosystem. Accordingly, we will adopt indicators to break down the three dimensions of CoI.

Social Presence (SP)

A set of indicators will be implemented to measure TP, SP, CP as well as to detect EP and AP, through the different tools used. Online discussions will be extracted and observed through semantic analysis:

1. Elementary language units (keywords), such as sentences, paragraphs or short texts;
2. Longer lexical units (words, lemmas or categories) and specified variables.

The analysis process can be performed through clustering functions to create semantic networks, associating language strings with students or groups. The co-occurrences of semantic traits can match a certain level of involvement in the tasks performed and even reveal emotional features such as motivation and commitment, optimism or pessimism, distress or frustration.

It is recognized that, in terms of knowledge sharing, the important factors are the trust among members, the convenience for members to obtain knowledge, and the arrangement for team collaboration and learning (Barak, 2018; Holmes, 2013; Kaminskiene et al., 2020; Prenger et al., 2019; Tyler, 2001; Trust et al., 2016; Zhao, 2010). The sharing skills through digital technologies include the ability to share data, information, and digital content with others through appropriate digital technologies, act as an intermediary, know about referencing and attribution practices without infringing copyrights or damaging others' digital identities. In addition, the collaboration skills through digital technologies incorporate the adoption of digital tools for collaborative processes, co-construction, co-creation of resources and knowledge (Vourikari et al., 2015), and even co-research (Kukulska-Hulme, 2012).

Concerning the role of CIT and SNSs in education, the global research community has been focusing on many different aspects always based on a set of frequent repetitions of research patterns, such as the experiences of use, the measurement of behavioral traits, personal attitude and/or resistance to the innovation. From this standpoint, the approaches may be resumed as follows:

1. the users' perception (learners and/or teachers) as an individual and collective experience;
2. the benefit of introducing SNSs in education environments;
3. the relation between SNSs (and other CIT implementations) and the recognizable KM patterns such as sharing, communication, content-creation, motivation, innovation, interaction;
4. the aspect of networking is analyzed as functional to the previously mentioned skills and rarely within the topics or conceptual areas that build up long-term cross-institutional networks.

When the networking aspect emerges, it is usually related to the individual capacity to enhance personal contacts during the learning or teaching path. The aspects associated with a bottom-up approach is frequently emphasized. Consequently, the most popular SNSs are frequently mentioned as tools to develop personal skills or group skills framed inside a local community, such as a school center, a university or a course; or as bridging tools individuals may use to link themselves to other contacts or external institutions.

Teaching Presence (TP)

In every educational experience involving CIT and SNSs, teachers are the 'agents of socialization' (Kelly & Antonio, 2016). In most of the articles examined, digital skills and networking skills are considered from an individual perspective and/or within the community of reference (school center, university, professional community) or within an experience-based context (professional course, co-creation and sharing experiences). Networking skills with the rank of citizenship requirements are seldom examined, and this topic generally involves a business-centered focus, but some influential authors consider the student lifecycle in terms of KM, social capital, and career prospect (Benson et al., 2014); in other terms, within a lifelong learning perspective. On the one hand, networking skills are becoming increasingly important in every professional category, and for education professionals, they represent an emerging standard for informal opportunities of training (Trust et al., 2016; Vuorikari et al., 2015), consequently, they *"theachers should be equipped with the most up-to-dated and relevant ICT skills."* (Benson et al., 2014, p. 519).

Cognitive Presence (CP)

Nowadays, thanks to technology, learning styles differ a lot from what the previous generations called education. The unprecedented aspects, communication; content-generation; knowledge sharing; social relationships (Moghavvemi et al., 2017), are now evaluated as part of the education process in the present digital classroom. The asynchronous modes are strictly dependent on self-cognitive patterns, thus, *"not only online learning but also offline—through digital resources such as e-books downloaded to mobile devices and accessed at the learner's convenience."* (Kukulska-Hulme, 2012, p. 247)

However, university students are not frequently motivated to integrate digital tools in their self-paced study routine (Al-Rahmi et al., 2015). Therefore, some authors focus on the interaction between instructors and learners in order to explore the patterns that may lead to autonomous learning strategies in the social media context (Matzat & Vrieling, 2016) in both formal and informal settings (Mulyono & Suryoputro, 2020).

POTENTIAL IMPACT

Research community

- Extend the debate on the possible insertion of the CoI Framework into the great variety of perspectives around education in the EU.
- Observation of the CoI dynamics when applied to a specific subject.
- Increased involvement of the European research community.

Public School and Education Frameworks

- As highlighted in the Introduction, the EU has been implementing communitarian frameworks in the last two decades. Therefore, new perspectives on e-learning strategies and related evaluations must be conveyed through a framework-related approach, in order to explore osmotic exchanges.
- The debate on skill evaluation and skill-based curricula is still subject to controversy, frequently due to the stiffness of some national schools' curricula in the EU. Consequently, a framework based on macro-areas (such as the 3 CoI Presences) may encounter less resistance or opposition in the teacher community.

Debate and Controversy with the EU Academia and Scholars

Further investigations should be focused on examining how the network perspective can be integrated into the education routine, and its potential impact on the students and teachers' perception of the 'knowledge co-creation', as a specific skill required in the knowledge society. The application of the CoI framework to transnational educational communities may further widen its implications and attract more interest.

In addition, an interest of the research community towards the most popular EU educational platforms, such as *eTwinning*, is highly desirable in order to explore their *user experience* beyond the reports issued by the EU National Agencies.

The world of education is composite, thus, any quantitative research criteria should be associated with qualitative investigations as well. A research path capable of involving two or more National Agencies from different EU countries could build up a platform for a big data analysis, with the aim of exploring digital patterns.

REFERENCES

Admiraal, W., Huizenga, J., Akkerman, S., & Ten Dam, G. (2011). The concept of flow in collaborative game-based learning. *Computers in Human Behavior*, *27*(3), 1185–1194. doi:10.1016/j.chb.2010.12.013

Al-Aufi, A., & Fulton, C. (2015a). Impact of social networking tools on scholarly communication: A cross-institutional study. *The Electronic Library*, *33*(2), 224–241. doi:10.1108/EL-05-2013-0093

Al-Rahmi, W. M., Othman, M. S., & Yusuf, L. M. (2015). The role of social media for collaborative learning to improve academic performance of students and researchers in Malaysian higher education. *International Review of Research in Open and Distance Learning*, *16*(4), 177–204. doi:10.19173/irrodl.v16i4.2326

Alt, D. (2018). Students' Wellbeing, Fear of Missing out, and Social Media Engagement for Leisure in Higher Education Learning Environments. *Current Psychology (New Brunswick, N.J.)*, *37*(1), 128–138. doi:10.100712144-016-9496-1

Anderson, T., Rourke, L., Garrison, D. R., & Archer, W. (2001). Assessing teaching presence in a computer conferencing context. *Journal of Asynchronous Learning Networks*, *5*(2), 1–17. doi:10.24059/olj.v5i2.1875

Aria, M., & Cuccurullo, C. (2017). Bibliometrix: An R-tool for comprehensive science mapping analysis. *Journal of Informetrics, 11*(4), 959–975. doi:10.1016/j.joi.2017.08.007

Baishya, D., & Maheshwari, S. (2020). Whatsapp groups in academic context: Exploring the academic uses of whatsapp groups among the students. *Contemporary Educational Technology, 11*(1), 31–46. doi:10.30935/cet.641765

Barak, M. (2018). Are digital natives open to change? Examining flexible thinking and resistance to change. *Computers & Education, 121*, 115–123. doi:10.1016/j.compedu.2018.01.016

Benson, V., Morgan, S., & Filippaios, F. (2014). Social career management: Social media and employability skills gap. *Computers in Human Behavior, 30*, 519–525. doi:10.1016/j.chb.2013.06.015

Brečko, B., & Ferrari, A. (2016). The Digital Competence Framework for Consumers. Joint Research Centre Science for Policy Report; EUR 28133 EN. doi:10.2791/838886

Cacheiro-Gonzalez, M. L., Medina-Rivilla, A., Dominguez-Garrido, M. C., & Medina-Dominguez, M. (2019). The learning platform in distance higher education: Student's perceptions. *Turkish Online Journal of Distance Education, 20*(1), 71–95. doi:10.17718/tojde.522387

Calderon-Garrido, D., Leon-Gomez, A., & Gil-Fernandez, R. (2019). the Use of the Social Networks Between the Students of Teacher'S Degree in an Environment Exclusively Online. *Vivat Academia, 147*, 23–39.

Callan, V. J., & Johnston, M. A. (2020). Influences upon social media adoption and changes to training delivery in vocational education institutions. *Journal of Vocational Education and Training, 00*(00), 1–26. doi:10.1080/13636820.2020.1821754

Campos, N., Nogal, M., Caliz, C., & Juan, A. A. (2020). Simulation-based education involving online and on-campus models in different European universities. *International Journal of Educational Technology in Higher Education, 17*(1), 8. Advance online publication. doi:10.118641239-020-0181-y

Castellanos-Reyes, D. (2020). 20 Years of the Community of Inquiry Framework. *TechTrends, 64*(4), 557–560. doi:10.100711528-020-00491-7

Cheng, E. C. K. (2012). Knowledge strategies for enhancing school learning capacity. *International Journal of Educational Management, 26*(6), 577–592. doi:10.1108/09513541211251406

Cheng, E. C. K. (2013). Applying knowledge management for school strategic planning. *KEDI Journal of Educational Policy, 10*(2), 339–356.

Cheng, E. C. K. (2019). Knowledge management strategies for sustaining Lesson Study. *International Journal for Lesson and Learning Studies, 9*(2), 167–178. doi:10.1108/IJLLS-10-2019-0070

Cheng, E. C. K., & Chu, C. K. W. (2018). A Normative Knowledge Management Model for School Development. *International Journal of Learning and Teaching, 4*(1), 76–82. Advance online publication. doi:10.18178/ijlt.4.1.76-82

Cheng, E. C. K., Wu, S. W., & Hu, J. (2017). Knowledge management implementation in the school context: Case studies on knowledge leadership, storytelling, and taxonomy. *Educational Research for Policy and Practice*, *16*(2), 177–188. doi:10.100710671-016-9200-0

Cleveland-Innes, M., & Campbell, P. (2012). Emotional presence, learning, and the online learning environment. *International Review of Research in Open and Distance Learning*, *13*(4), 269–292. doi:10.19173/irrodl.v13i4.1234

Coursaris, C. K., & Van Osch, W. (2016). A Cognitive-Affective Model of Perceived User Satisfaction (CAMPUS): The complementary effects and interdependence of usability and aesthetics in IS design. *Information & Management*, *53*(2), 252–264. doi:10.1016/j.im.2015.10.003

Dabbagh, N., & Kitsantas, A. (2012). Personal Learning Environments, social media, and self-regulated learning: A natural formula for connecting formal and informal learning. *Internet and Higher Education*, *15*(1), 3–8. doi:10.1016/j.iheduc.2011.06.002

Deaves, A., Grant, E., Trainor, K., & Jarvis, K. (2019). Students' perceptions of the educational value of twitter: A mixed-methods investigation. *Research in Learning Technology*, *27*(0). Advance online publication. doi:10.25304/rlt.v27.2139

Dewey, J. (1991). *The Child and the Curriculum*. University of Chicago Press. (Original work published 1902)

Dommett, E. J. (2019). Understanding student use of twitter and online forums in higher education. In Education and Information Technologies (Vol. 24, Issue 1, pp. 325–343). doi:10.100710639-018-9776-5

Garcia, E., Moizer, J., Wilkins, S., & Haddoud, M. Y. (2019). Student learning in higher education through blogging in the classroom. *Computers & Education*, *136*, 61–74. doi:10.1016/j.compedu.2019.03.011

Garrison, D., Anderson, T., & Archer, W. (2000). Critical Inquiry in a Text-Based Environment: Computer Conferencing in Higher Education. *The Internet and Higher Education*, *2*(2–3), 87–105. doi:10.1016/S1096-7516(00)00016-6

Garrison, D. R. (2011). Communities of Inquiry in Online Learning. Encyclopedia of Distance Learning, 352–355. doi:10.4018/978-1-60566-198-8.ch052

Garrison, D. R. (2017). Other Presences? *The Community of Inquiry*. http://www.thecommunityofinquiry.org/editorial7

Garrison, R., Anderson, T., & Archer, W. (2002). Critical Inquiry in a Text-Based Environment. *The Internet and Higher Education*, *2*(2), 87–105. http://dergipark.gov.tr/saufenbilder/issue/20673/220600

Gazit, T., Aharony, N., & Amichai-Hamburger, Y. (2019). Tell me who you are and I will tell you which SNS you use: SNSs participation. *Online Information Review*, *44*(1), 139–161. doi:10.1108/OIR-03-2019-0076

Geeraerts, K., Vanhoof, J., & Van den Bossche, P. (2016). Teachers' perceptions of intergenerational knowledge flows. *Teaching and Teacher Education*, *56*, 150–161. doi:10.1016/j.tate.2016.01.024

Giannikas, C. (2019). Facebook in tertiary education: The impact of social media in e-learning. *Journal of University Teaching & Learning Practice*, *17*(1), 2020. https://ro.uow.edu.au/jutlpAvailableat:https://ro.uow.edu.au/jutlp/vol17/iss1/3

Greenhow, C., & Lewin, C. (2016). Social media and education: Reconceptualizing the boundaries of formal and informal learning. *Learning, Media and Technology*, *41*(1), 6–30. doi:10.1080/17439884.2015.1064954

Holmes, B. (2013). School Teachers' Continuous Professional Development in an Online Learning Community: Lessons from a case study of an eTwinning Learning Event. *European Journal of Education*, *48*(1), 97–112. doi:10.1111/ejed.12015

Hrastinski, S. (2008). What is online learner participation? A literature review. *Computers & Education*, *51*(4), 1755–1765. doi:10.1016/j.compedu.2008.05.005

Inamorato dos Santos, A., Punie, Y., Castaño-Muñoz, J. (2016). *Opening up Education: A Support Framework for Higher Education Institutions*. JRC Science for Policy Report, EUR 27938 EN. doi:10.2791/293408

Iwamoto, D., & Chun, H. (2020). The emotional impact of social media in higher education. *International Journal of Higher Education*, *9*(2), 239–247. doi:10.5430/ijhe.v9n2p239

Kaminskiene, L., Žydžiunaite, V., Jurgile, V., & Ponomarenko, T. (2020). Co-creation of learning: A concept analysis. *European Journal of Contemporary Education*, *9*(2), 337–349. doi:10.13187/ejced.2020.2.337

Kampylis, P., Punie, Y., & Devine, J. (2015). Promoting Effective Digital-Age Learning: A European Framework for Digitally-Competent Educational Organisations, EUR 27599 EN. *Publications Office of the European Union., JRC98209*. Advance online publication. doi:10.2791/54070

Kelly, N., & Antonio, A. (2016). Teacher peer support in social network sites. *Teaching and Teacher Education*, *56*, 138–149. doi:10.1016/j.tate.2016.02.007

Kennedy, E., & Laurillard, D. (2019). The potential of MOOCs for large-scale teacher professional development in contexts of mass displacement. *London Review of Education*, *17*(2), 141–158. doi:10.18546/LRE.17.2.04

Kreijns, K., Van Acker, F., & Van Buuren, H. (2014). Community of Inquiry: Social presence revisited. *E-Learning and Digital Media*, *11*(1), 5–18. Advance online publication. doi:10.2304/elea.2014.11.1.5

Kukulska-Hulme, A. (2012). How should the higher education workforce adapt to advancements in technology for teaching and learning? *Internet and Higher Education*, *15*(4), 247–254. doi:10.1016/j.iheduc.2011.12.002

Lambton-Howard, D., Kiaer, J., & Kharrufa, A. (2020). 'Social media is their space': Student and teacher use and perception of features of social media in language education. *Behaviour & Information Technology*, *0*(0), 1–16. doi:10.1080/0144929X.2020.1774653

Laurillard, D., Kennedy, E., Charlton, P., Wild, J., & Dimakopoulos, D. (2018). Using technology to develop teachers as designers of TEL: Evaluating the learning designer. *British Journal of Educational Technology*, *49*(6), 1044–1058. doi:10.1111/bjet.12697

Librizzi, J. (2017). The Haunted Animal: Peirce's Community of Inquiry and the Formation of the Self. *All Theses & Dissertations.* 317.

Lin, F., Lin, S., & Huang, T. (2008). Knowledge sharing and creation in a teachers' professional virtual community. *Computers & Education, 50*(3), 742–756. doi:10.1016/j.compedu.2006.07.009

Loc, N. P., Tong, D. H., Thao, V. T. T., Han, N. N., Tram, T. H., Thoa, D. T., & Co, L. V. (2019). Students' social networking: Current status and impact. *International Journal of Scientific and Technology Research, 8*(12), 3602–3605.

Luo, T., Freeman, C., & Stefaniak, J. (2020). "Like, comment, and share"—professional development through social media in higher education: A systematic review. *Educational Technology Research and Development, 68*(4), 1659–1683. doi:10.100711423-020-09790-5

Madge, C., Meek, J., Wellens, J., & Hooley, T. (2009). Facebook, social integration and informal learning at university: "It is more for socialising and talking to friends about work than for actually doing work. *Learning, Media and Technology, 34*(2), 141–155. doi:10.1080/17439880902923606

Majeski, R. A., Stover, M., & Valais, T. (2018). The Community of Inquiry and Emotional Presence. *Adult Learning, 29*(2), 53–61. doi:10.1177/1045159518758696

Manca, S., & Ranieri, M. (2017). Implications of social network sites for teaching and learning. Where we are and where we want to go. *Education and Information Technologies, 22*(2), 605–622. doi:10.100710639-015-9429-x

Matzat, U., & Vrieling, E. M. (2016). Self-regulated learning and social media – a 'natural alliance'? Evidence on students' self-regulation of learning, social media use, and student–teacher relationship. *Learning, Media and Technology, 41*(1), 73–99. doi:10.1080/17439884.2015.1064953

Michos, K., & Hernández-Leo, D. (2020). CIDA: A collective inquiry framework to study and support teachers as designers in technological environments. *Computers & Education, 143*, 103679. doi:10.1016/j.compedu.2019.103679

Mills, N. (2011). Situated Learning through Social Networking Communities: The Development of Joint Enterprise, Mutual Engagement, and a Shared Repertoire. *Journal, 28*(2), 345–368. doi:10.11139/cj.28.2.345-368

Moghavvemi, S., Paramanathan, T., Rahin, N., & Sharabati, M. (2017). Student ' s perceptions towards using e-learning via Facebook Sedigheh Moghavvemi, Tanuosha Paramanathan, Nurliana Md Rahin &. *Behaviour & Information Technology, 0*(0), 1–20. doi:10.1080/0144929X.2017.1347201

Mulyono, H., & Suryoputro, G. (2020). The use of social media platform to promote authentic learning environment in higher education setting. *Science for Education Today, 10*(2), 105–123. doi:10.15293/2658-6762.2002.07

Nizzolino, S. (2020). Teacher Networking, Professional Development, and Motivation Within EU Platforms and the Erasmus Plus Program. In J. Zhao (Ed.), *Collaborative Convergence and Virtual Teamwork for Organizational Transformation* (pp. 195–218). doi:10.4018/978-1-7998-4891-2.ch010

Nizzolino, S., & Canals, A. (2021). Social Network Sites as Community Building Tools in Educational Networking. *International Journal of e-Collaboration, 17*(4), 132–167. doi:10.4018/IJeC.2021100110

Nonaka, I., & Konno, N. (1998). The Concept of "Ba": Building a foundation for knowledge creation. *California Management Review, 40*(3), 40–54. doi:10.2307/41165942

Pereira, R., Baranauskas, M. C. C., & da Silva, S. R. P. (2010). Social software building blocks: Revisiting the honeycomb framework. *2010 International Conference on Information Society*, 253-258. 10.1109/i-Society16502.2010.6018707

Pimmer, C., Chipps, J., Brysiewicz, P., Walters, F., Linxen, S., & Gröhbiel, U. (2017). Facebook for supervision? Research education shaped by the structural properties of a social media space. *Technology, Pedagogy and Education, 26*(5), 517–528. doi:10.1080/1475939X.2016.1262788

Prenger, R., Poortman, C. L., & Handelzalts, A. (2019). The Effects of Networked Professional Learning Communities. *Journal of Teacher Education, 70*(5), 441–452. doi:10.1177/0022487117753574

Rap, S., & Blonder, R. (2017). Thou shall not try to speak in the Facebook language: Students' perspectives regarding using Facebook for chemistry learning. *Computers & Education, 114*, 69–78. doi:10.1016/j.compedu.2017.06.014

Rasi, P., Hautakangas, M., & Väyrynen, S. (2015). Designing culturally inclusive affordance networks into the curriculum. *Teaching in Higher Education, 20*(2), 131–142. doi:10.1080/13562517.2014.957268

Redecker, C. (2017). *European Framework for the Digital Competence of Educators: DigCompEdu.* doi:10.2760/159770

Robertson, M., & Swan, J. (1998). Modes of Organizing in an Expert Consultancy: A Case Study of Knowledge, Power and Egos. *Organization, 5*(4), 543–564. doi:10.1177/135050849854006

Rosenberg, J. M., Reid, J. W., Dyer, E. B. J., Koehler, M., Fischer, C., & McKenna, T. J. (2020). Idle chatter or compelling conversation? The potential of the social media-based #NGSSchat network for supporting science education reform efforts. *Journal of Research in Science Teaching, 57*(9), 1322–1355. doi:10.1002/tea.21660

Schmid, R., & Petko, D. (2019). *Does the use of educational technology in personalized learning environments correlate with self-reported digital skills and beliefs of secondary-school students?* doi:10.1016/j.compedu.2019.03.006

Stenbom, S., Hrastinski, S., & Cleveland-Innes, M. (2016). Emotional presence in a relationship of inquiry: The case of one-to-one online math coaching. *Online Learning Journal, 20*(1). Advance online publication. doi:10.24059/olj.v20i1.563

Thomas, V. L., Chavez, M., Browne, E. N., & Minnis, A. M. (2020). Instagram as a tool for study engagement and community building among adolescents: A social media pilot study. *Digital Health, 6*. Advance online publication. doi:10.1177/2055207620904548 PMID:32215216

Trust, T., Krutka, D. G., & Carpenter, J. P. (2016). "Together we are better": Professional learning networks for teachers. *Computers & Education, 102*, 15–34. doi:10.1016/j.compedu.2016.06.007

Tyler, B. B. (2001). Complementarity of cooperative and technological competencies: A resource-based perspective. *Journal of Engineering and Technology Management, 18*(1), 1–27. doi:10.1016/S0923-4748(00)00031-X

Vivian, R. J. (2012). *Students' Use of Personal Social Network Sites to Support their Learning Experience.* doi:10.13140/RG.2.1.2337.6484

Vuorikari, R., Kampylis, P., Scimeca, S., & Punie, Y. (2015). *Scaling Up Teacher Networks Across and Within European Schools: The Case of eTwinning*. Springer. doi:10.1007/978-981-287-537-2_11

Vuorikari, R., Punie, Y., Carretero Gomez, S., & Van Den Brande, G. (2016). DigComp 2.0: The Digital Competence Framework for Citizens. Update Phase 1: The Conceptual Reference Model. EUR 27948 EN. Publications Office of the European Union; JRC101254.

Weidlich, J., & Bastiaens, T. J. (2019). Designing sociable online learning environments and enhancing social presence: An affordance enrichment approach. *Computers & Education, 142*, 103622. Advance online publication. doi:10.1016/j.compedu.2019.103622

Xiaochun, G., Yiwei, W., & Jingming, Z. (2017). *The Self-assessment in E-learning and Personalized Feedback Education*. doi:10.1145/3175536.3175571

Zhao, J. (2010). School knowledge management framework and strategies: The new perspective on teacher professional development. *Computers in Human Behavior, 26*(2), 168–175. doi:10.1016/j.chb.2009.10.009

ENDNOTES

[1] The original informative article, which has been cited in several academic works, is no more available on the web. Smith, G. (2007). Social Software Building Blocks. http://nform.ca/publications/social-software-building-block

[2] The database from Scopus was afterward processed by RStudio.

Chapter 17
Mining Perspectives for News Credibility:
The Road to Trust Social Networks

Farah Yasser
Faculty of Commerce and Business Administration, Helwan University, Egypt

Sayed AbdelGaber AbdelMawgoud
Helwan University, Egypt

Amira M. Idrees
https://orcid.org/0000-0001-6387-642X
Faculty of Computers and Information Technology, Future University in Egypt, Egypt

ABSTRACT

Text mining has become a vital zone that has been attached to some examined ranges such as computational etymology, data mining, and information recovery (IR). Almost all people today use social networking activities in their daily interactions with no sorting. This can result in a range of inconsistencies, including lexical, semantic, linguistic, and syntactic ambiguities, making it difficult to determine the accurate data arrangement. Fittingly, the study identified the concept of text mining in terms of its impact on social networks. This study highlights the positive impact of intelligent techniques and how to use text mining to detect the news credibility on Facebook. The study introduced a background that highlighted the related aspects, the relation between these domains, and the news credibility. The study also presents the recent research in these fields with demonstrating the roles of these techniques for the required study target. The study could support as the foundation of future text mining studies on social networks data.

DOI: 10.4018/978-1-7998-9640-1.ch017

INTRODUCTION

Social networks helped people to communicate easily without obstacles (Mukerji, 2018). News and information spread rapidly regardless of the news credibility (Liu, Wang, & Huang, 2018). People who had unpleasant experience or even just heard about it without making sure or checking would create fake news and spread rumors widely for different purposes (Othman, Hassan, Moawad, & Idrees, 2016). This survey discusses the researches considering the detection of fake news, news credibility and how they measured the news credibility and accuracy of news and information. Different models and techniques such as text mining applied techniques helps to detect credibility (Xu, Wang, Wang, & Yang, 2020), (Liu, Wang, & Huang, 2018). Within the past decade, the propagation of network permitted the electronic devices the impulse of social networks hardly. Users of social networks change from the foremost prepared to the foremost energetic periods, and the mass selection of this appliance has given the users a site within the knowledge society (Othman, Hassan, Moawad, & Idrees, 2018). The end user that holds devices to use social networks propagates peculiarity and newsworthiness about current events. It is especially veritable for the most youthful period that based on heightening on social networks to retain overhauled and share results near their method of considering (Mohsen, Hassan, & Idrees, 2016).

On numerous events, it has been demonstrated that this collective information is valuable in case of emergency and harming occasions as fires and storms (Mohsen, Hassan, & Idrees2016). Electronic devices and the successive surge of web clients made a vital impact on the social society. At the time of social networks, the end user that handles these devices is the real reason for spreading peculiarity and newsworthiness nearly specific news. Presently and after, that this sort of declaring is one-sided due to the level of the user's data of the matter or to fulfill an impartial. The ability to share different users' post intensify this wonder and have a cascading type of influence that can prompt the dispersal of false information (Sayed, Salem, & Khedr, 2019). The proposed procedure attempts to deal with this impediment by cross-relating content streams with contrasting heterogeneous degrees of enduring quality (De Maio, Fenza, Gallo, Loia, & Volpe, 2020), in the past, the improvement of social networking sites unimaginably has empowered how people interact with others through networks.

People who use social networks share information and data and interact with other each other. So they are taught from the trends happening on social networks as news and public events (Idrees & Ibrahim, 2018). In any case, most of the information that appears on social networks afterward is unreliable and, in some cases, deceptive. Such substance is called false news. Fake news on the internet as (social networking sites) has the empower to create significant issues in the realworld society (Zhang & Ghorbani, 2019).

Because of substantial investments in the field of innovation by the evolution of social networking sites, the importance of text mining has grown. According to (Kim & Chung, 2018),

(Sultan, Khedr, Idrees, & Kholeif, 2017), data mining and detection of knowledge are currently interesting for experts and researchers. There is also requirements for modifying raw data into useful information and knowledge which lead to awareness (Hassan & Idrees, 2010). The world of business has understood the value of data, information, and derived (knowledge) from copious amounts of data. A range of surveys were composed from different papers to be analyzed and utilized in this outline such as in (Idrees & Ibrahim, 2015). The following sections provide a background for the vital terms, then the relation between the news credibility and intelligent techniques are discussed.

BACKGROUND

The development of social networking sites has more noteworthy reach over the globe particularly, in Egypt. The web continues to be the foremost vital perspective for the mass networks. Within the later past, the social networking destinations got to be unavoidable and powerful platform within the hands of every citizen within the nation. The social networks sites such as Facebook are currently the most popular online stage for the conclusion clients to share their data and engagement in their way of life (Balaji & S, 2019). Social networks especially Facebook became a platform for everyone in our society to express their opinions and spreading news even it was fake (Islam, et al., 2020), (Mohsen, Idrees, & Hassan, 2019).

Unstructured data or information is really a critical issue that confronted on social networks, this indicates the presence of enormous unorganized data or information. It could be precious data or valuable information (Dahab, Idrees, Hassan, & Rafea, 2010) (Hassan, Dahab, Bahnassy, Idrees, & Gamal, 2015). The difficulty is how to convert the unstructured data to structured and organized one. Data that is unstructured commonly comprises of various expressions, but it could be included information like actualities (Hassan, Dahab, Bahnasy, Idrees, & Gamal, 2014). Information that can be exploited. it may cause irregularities and instability which makes it troublesome to urge it utilizing customary programs compared to information put absent inside the shape of dealt with inside the database or depicted as (semantic names) in reports (Arianto, Leslie Warnars, Gaol, & Trisetyarso, 2018). Although, in recent years, the analysis and examination of unstructured text research has gotten a lot of attention. The most common results can be attributed to the most recent growth in social networking sites data that help in determining the news credibility (Mohsen, Idrees, & Hassan, 2019).

Among the most significant imperative components in social networks news recognitions is news credibility. News sources of news should be carefully checked about how their material is collected, especially considering validity assessments play a key role in audiences' models (Haggag, Khedr, & Montasser, 2015). The impact of digital and social networks validity is a complex concept that has been derived from 3 different fields of study credibility (related to the message originator's validity), the credibility of message (related to the message's characteristics), and moderate validity (related to the message's characteristics) (dealing how the message is transported through the channels). As a result, networks credibility is often viewed as a multifaceted notion encompassing a variety of perspectives, including dependability and competence, fairness, equilibrium, inconsistency, community concern, partition of presumption and reality, genetic disposition, and correctness (Namugera, Wesonga, & Jehopio, 2019).

Text Mining

Text mining can assist in extracting information and knowledge from huge amount of text like a mountain (Abed, Yuan, & Li, 2017). It could be simple for getting a significant, systematic, and organized data from the unpredictable data sets to reach some knowledge by using text mining that helps in converting unstructured data to a meaningful and structured data (Hossein Hassani, et al., 2020), (Grimes, 2008)and (Hung & Zhang, 2011). Computers have a difficulty in understanding unstructured data and converting it to organized, structured data. So, researchers may be able to do this task by using different techniques (Hossein Hassani, et al., 2020). But computers are doing things faster when comparing to people, The significance of text mining is rising because it is used to extract organizing text data from unstructured text data (Axenbeck & Kinne, 2020), (Go¨k, Waterworth, & Shapira, 2015). Overall, data via social networks is collected for academic purposes and to promote research efforts (Andryani, Negara, & Triadi, 2019), and the pattern of the data collected from online networks must be exchanged to be more organized and structured. On social networking sites, the bulk of the text is unorganized with a small percentage being organized whereas as a little were structured (Ozbay & Alatas, 2019).

Figure 1. Structured data vs. unstructured data: what they are and why care (Tondak, 2020)

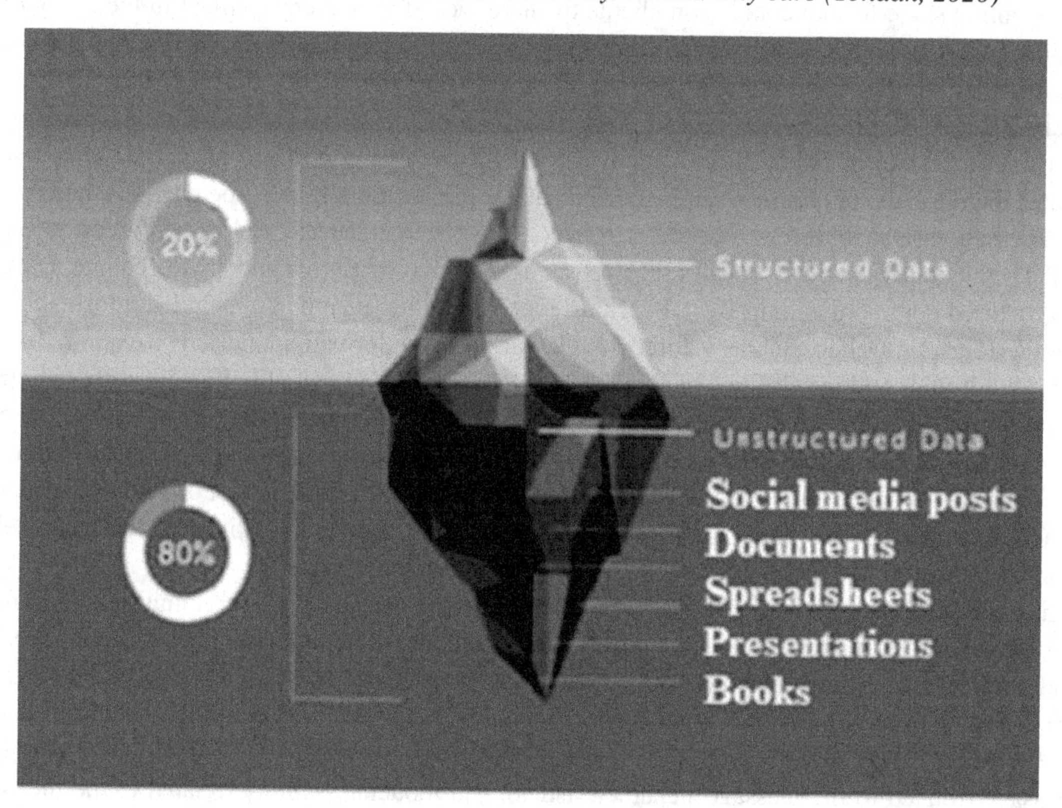

There is no direct approach with organized methodology available for posting comments and posts on various social networks sites, which leads to problems with the use of the data findings. Text mining is one of the most important fields within the digital era due to the fast growth of textual information. Topic models are picking up ubiquity within the last few years (Likhitha, Harish, & Keerthi Kumar, 2019). Data is extracting from the text. So, text mining is more significant and because of generating a business value (Xia, Luo, Zhang, & Wu, 2019). Research in (Gupta & Chandra, 2020) was explained that Data mining, an essential and critical step in knowledge discovery in databases, is utilized to find useful unknown patterns from huge amount of data. Data mining incorporates a variety of features, techniques, and calculations which are used to discover and retrieve unique trends from a huge data set, The researcher (Gupta & Chandra, 2020) compared in a systematic way. Firstly, the research was fractionated data mining tasks that used in another previous studies into summarization, characterization, classification, clustering, association, outer analysis, and regression, noticed that the most frequent task used is clustering. Secondly, categorized the famous classification algorithms with using these algorithms as iterative dichotomize based on information gain, naïve Bayesian, if rule. Thirdly the researcher in (Gupta & Chandra, 2020) compared between some clustering algorithm hierarchical using divisive analysis, portioning based using k means and others, after these studies it has noticed that xml data clustering facing different problems and clustering became not suitable for using web search results. Fourthly, classified the association algorithms into Apriori, frequent pattern growth-based, and equivalence class transformation (Khedr & Idrees, 2017) (Mostafa, Khedr, & Abdo, 2017). Fifthly, classified data mining techniques according to the previous studies that used statistics, machine learning, neural, database systems, genetic algorithms, fuzzy set, and visualization. Finally, compared between data mining techniques according to the relationship between them, so analysis can be conducted to come up with successful remedies and their side effects. Furthermore, it is determined that data mining algorithm/method unification, scalability, and optimization, cube-oriented multifunctional data mining, and scalability real-time mining are three areas of data mining that deserve more research and through different studies of data mining techniques. Moreover, there are numerous real-world data applications (Khedr, Kholeif, & Hessen, 2015) (Khedr, Kholeif, & Hessen, April 2015).

According to a research paper by (Kumar, Basha, & Rao, 2020), text mining is widely used for obtaining quick results from a variety of applications. It is classified according to types depending on the text documents nature, using text mining methods and techniques as Information Extraction that produces useful info from huge amount of data (Khedr & El Seddawy, 2015). Information Retrieval that collects words that have a near relationship between them to give users info that are more close to their needs to satisfy them, Natural Language Processing (NLP) that analyzed unstructured text and produces information and showed the relation between info, Summarization of authentic text and makes a brief summary of it in an organized way and Clustering of text that have more similarity with non-relevant then makes a groups of both (Khedr & Kok, 2006). Then the researchers in (Kumar, Basha, & Rao, 2020) created a framework from their point of view summarized in first collection of data, then preprocess of data as a second step, third using text mining techniques as clustering and other methods to achieve knowledge. So, text mining is more critical because of extracting from its data and information to reach knowledge to help people in take their decisions in a scientific and right way (Helmy, Khedr, Kolief, & Haggag, 2019).

Text mining and NLP

A study by (Zhou, Duan, Liu, & Shum, 2020) declared that Natural language processing (NLP) has become a key role that helps the researchers to understand the relation among people and computers through natural language. Also, it made the understanding of words and phrases easy. NLP has become critical for customer support systems and other sectors by examining the neural NLP framework's noteworthy development in three areas of efforts: neural NLP modelling with the goal of building suitable network architectures for various applications, neural NLP learning challenges aiming at improving the model parameters and inference strategies aiming at producing solutions to unknown queries by modifying current information by striving to discover and sort out future paths that are essential to improving NLP technology based on a comprehensive examination of existing technologies and the problems of each of these features, To overcome these problems, they used a neural NLP approach. Initially, in the neural network, encoding the natural language phrase (a series of words). secondly Generating a label sequence or a natural language sentence. The study in (Zhou, Duan, Liu, & Shum, 2020) has discussed the network structures used to learn word embedding, phrase embedding, and pattern creation in this part.

Multiple directions need to be explored further to enhance modelling for different NLP tasks as prior knowledge modeling, Document/multi-turn modeling, and non-auto regressive generation. According to the researcher's point of view in (Zhou, Duan, Liu, & Shum, 2020) more study on the following subjects should be performed to enhance the performance of NLP models: GANs, Reinforcement learning (RL), and the topological training process. The growth of reasoning in NLP was examined as Typical (non-neural and neural) inference methods, such as integer linear programming (ILP), Markov logic networks (MLN), and Memory network (MemNN), were described, all of which have been effectively employed in various NLP applications, such as quality assurance, dialogue systems, three perspectives on neural NLP's advancement Learning, modelling, and reasoning are all aspects of neural modelling. In both research and application systems, NLP is entering a new era, with neural network-based models dominating. Neurons can be used to produce extra pseudo training data for model training or to incorporate learning from current models and tasks. With the use of innovative knowledge-extraction methods, studies on memory-argument neural networks and their variations will boost reasoning methodologies. New research and application possibilities will develop because of combining NLP with multi-modal tasks such as speech recognition and image/video captioning.

A research by (Fu, Jin, Zhao, & Cao, 2019) studied that text mining is the process of extracting valuable and useful information from text and documents to generate a huge amount of structured and unstructured data from text data by IoT devices. The extracted information is critical for decision making in business activities and other sectors, then it has been concluded that text mining is a technique for extracting high-quality data and information from the web, particularly mining of social. The researchers aim to introduce a text mining approach that extract information from the web, so they built a framework in (Fu, Jin, Zhao, & Cao, 2019) that illustrated a module, which includes information extraction, applies a methodology based on trigger words and criteria, as well as a trigger words dictionary and standard sets that the researchers have developed, Data cleaning and deduplication module to come up with solutions since some records involve incorrect attribute values although there are duplication records between several records, Expert recommendation model to produce a singular and complete record by track mining from multiple records of one expert in the dataset, expert Text mining can be aided by a variety of metrics, data formats, and algorithms. Instead of organized databases, text mining focuses on natural-language content.

Then illustrated by (Fu, Jin, Zhao, & Cao, 2019) the idea of data crawling that emphasizes on expert data. When a crawler receives one expert name, it is a person-oriented crawler rather than a link or a subject. In contrast to web spiders, another crawler will seek similar pages from the web source at the same time. Crawler efficiency can be enhanced by crawling multiple webs. They presented a method for determining expert qualities that uses a combination of trigger phrases and rules, to maximize efficiency, apply the crawl in concurrently and to limit the crawl, practically all websites will block an IP address if it sends too many requests in a short period of time. The crawler would apply the request header prior sending the requests to overcome this problem. In their method to set the value of request headers like User-Agent, Cookie, and Connection. Furthermore, the value of User-Agent should always change if there are several requests simultaneously time to avoid being blocked by websites. Personal attributes were extracted from web pages collected by crawlers in the project Information Extraction. The expert qualities that would be obtained are explained that the single-value slot and the list-value slot are the two types of values that can be used. There is only one filler in a single-value slot. The majority of single value are evident as (one person with one birthdate), but others are not. Because the source of data may have numerous valid responses, list-value slots can have multiple fillers. Experts, for example, can specialize in a variety of study fields. Preprocess module, attributes prediction module, and attributes extraction module are all shown in the overall framework for attributes extraction, the unstructured data preprocessing module would use purely internet sources. It was evaluated by (Fu, Jin, Zhao, & Cao, 2019) the content of a web source in this module, then clean the web pages and remove useless info to establish the structure of the document producing the structure of a website source based on the pattern in this stage, Secondly, they cleaned the web source. Every website contains both valuable and worthless information. As a result, in this stage, they removed the unnecessary information from the web sources using html tags. Thirdly, extract the plain text, and finally Split up the text. This stage breaks down the text into single sentences and returns a list to the attribute prediction module. Then applying algorithm to extract info from html by using source of code as input, and the output result consists of name and its value as a structured data.

Concerning the process of data clearing and cleaning, this process was divided into three steps, pre-processing by ensuring that all attributes, such as date of birth, mail, and phone number, are formatted correctly. and converting to a standard format, Data cleaning by removing data that is clearly false as well as certain records that are incomplete. Data fields can determine which data is obviously untrue, and the goal of Data Deduplication is to get rid of duplicate records from the same source. They combined in (Fu, Jin, Zhao, & Cao, 2019) related records into a single record to find the attribute. The attribute similarity is described as follows for a single value. Researchers compare the attributes to their similar equivalents to see whether they are same. This contrasts the two types of attribute values, single-value and list-value, which have distinct similarity assessment methodologies.

The aim of the experiment is to obtain expert-specific attributes from internet sources. There are 430 online documents in the testing data set regarding 69 distinct people, with 50 different names (multiple people might have the same name). These materials were gathered from 172 different websites. There are two types of datasets ONOP (one name only corresponds to one person) and ONDP (one name only relates to many people), (There are multiple people with the same name.) The dataset is 18.6MB in size. And a web document's average size is 44.3KB. they manually gathered in (Fu, Jin, Zhao, & Cao, 2019) these online documents. At first, they chose the names of 50 experts randomly. Second, using a search engine to look up each name one by one, and then choose related websites at randomly the outcome of a search engine's query All of the web documents are organized into folders based on the expert's name. Furthermore, manually extracted the attributes of the experts from the web pages and organized them into a consistent format.

The tests have been carried out on a computer with an Intel Core i5 CPU (2-core, 2.70GHz) and 8GB RAM. Python was used to implement their text mining strategy. MySQL was utilized to store the structured data generated throughout the trial. The datasets are kept in a directory, the findings by (Fu, Jin, Zhao, & Cao, 2019) revealed that the second technique is not as successful as the first, and that both methods are not robust enough in practice. The most significant distinction between these two is the difference between single-value extraction techniques The first assigns weights to the nodes, while the second does not. Supposing that the outcomes of the two techniques are similar. But the end outcome is quite unexpected. Upon further evaluations of the collected records, the following conclusions have been reached; why the model's performance is outstanding in both directions. To improve these results in the future from the point view of study is to using more accurate and reliable sources, The data cleansing procedure should be improved as well, particularly when dealing with unstructured data and the information extraction process need to be more accurate.

As in (Chen, Peng, & Lu, 2019) NLP became an essential part of text mining techniques for processing data from the merge between them which helped the researchers to discover new things from data and text., The researcher collected their data set from (https//www.cl.cam.ac.uk/~sb895/hoC.html), an automatic semantic classification of scientific literature according to the hallmarks of cancer. This is a public data set consists of 14,191 phrases, This study has divided the data into three parts randomly 60% for training, 20% for development, the rest 20% for testing, the framework created depend on text classification multilabel techniques, this study has chosen the CNN model according to popularity and combines between the performance and computational complexity while using several deep learning architectures, in their model using deep learning, deep learning avg word embeddings, deep learning and universal sentence Encoder, and deep learning and BioSentVec, Finally the results showed that the highest performance is BioSentVec by 0.677. So, in biomedical research, this could also help with the construction of deep learning models and text mining applications.

As reported by (Rameshbhai & Paulose, 2019) NLP used to classify the output from text by using different tools and text such Support Vector Machine (SVM) these tools work on mining for Facebook, Twitter online blogs etc., that leads to a finer accuracy. Throughout August 2017, data was taken from the website (http://www.indianexpress.com). A total of 1472 news headlines were gathered and categorized as +1 or -1. The sentiment score for positive news is +1, while the score for negative news is -1. Then three models were constructed by (Rameshbhai & Paulose, 2019) the unigram and bi-gram representations are used to create Model A, Linear SVM is used in Model A. Model B and C are the results of turning the data representation into TF-IDF. Linear SVM is used in Models B and C. Model C, on the other hand, uses an SGD classifier to train the data, unlike model B. Because it is trained with TF-IDF, Model B has a higher accuracy than Model A. Model A and B will no longer be effective solutions when feature sizes become larger (>20000). To deal with such problems, this study has introduced that Model C is trained using SGD and can support up to 105 features when developing a model. When the feature size is large, Model C can be employed; otherwise, Model B performs well since the featured size is small.

This figure:2 illustrated the importance of using text mining and NLP with each other in the text analysis process, it explained that the result is being clarified well, by using them to gather the results is more accurate.

Figure 2. Rule of text mining and NLP (Liu, Sarkar, & Chakraborty, 2013)

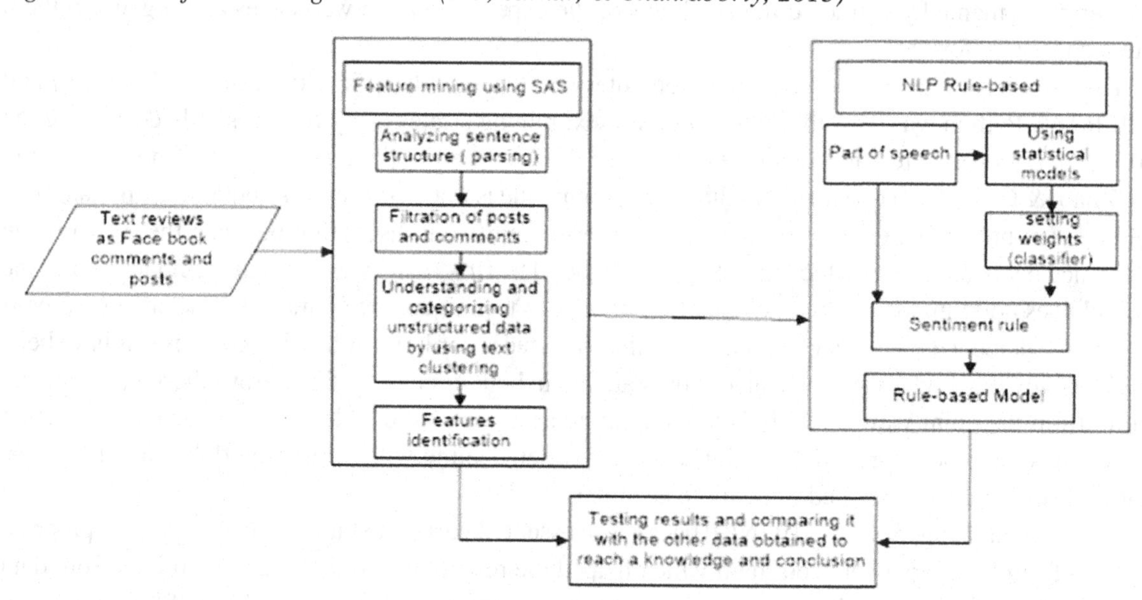

News credibility

News credibility as a fundamental majority rule esteem has been at the front line of insightful undertakings during the most recent a very long while. Despite productive exploration around there, grant on the validity of calculation based and computerized news still cannot seem to offer observational discoveries in view with the causes and their results in consistent with previous studies on news relevance, this examination looks at the key thrusts in predicting the degree of truthfulness on artificial intelligence (AI) news. Surprisingly, this study by (Lee, Nah, Chung, & Kim, 2020) showed that the effects of knowledgeable power, such as networks usage and social debate, as well as friendly capital, including trust and understanding on social networks the news credibility across groups. Internet Specifically social networks use via TV, interpersonal agency destinations, and online news locations, as well as open conversation, have a good relationship with news reliability, according to data collected through a cross-country as online questionnaire. The effect of public discourse on reliability was driven by social trust, demonstrating that the relation between conversation and truthfulness was substantially more grounded for people who trust others more (Lee, Nah, Chung, & Kim, 2020).

A research in (Neo, 2021) Studied the significance of predicting the homogenies of network affection on social networks in the political field and its outcomes, the research discussed the news credibility on social networks from the point of political network and political social networks view. According to this study between 2012 and 2016 showed that the increase of the positively analogy of forecasted political news credibility. The researchers in (Neo, 2021) concluded that indirectly by the results that the democracy become stronger on social networks.

It was reported in (Viviani & Pasi, 2017) that the combination of text mining and news credibility is critical to face the propagation of misinformation on social networks and showed the credibility that enhance the quality of information. the recognition of opinion spam in online reviews, the detection of fake news and spam in blogging, and the trustworthiness evaluation of online health information are three of the most important tasks facing this issue, which are discussed in this article. Despite the large number of intriguing solutions presented in the literature to address the above 3 objectives, certain difficulties remain unresolved; they mostly involve both the lack of specified standards and the lack of a clear understanding of what constitutes a benchmark. The difficulties of gathering and processing massive amounts of data.

With the passage of time the spreading of fake news will increase so it should be continuously measuring the news credibility accuracy on social networks, especially on Facebook. clarifying the relation between increasing fake news on social networks when a more time passed with the data size increased in different proportions. Particularly after spreading online contents through social networks sites, the growth of unstructured data is become a significant area to study in that include health, education, government, and businesses. Text mining techniques and tools act to convert unstructured data to structured one and useful data and text (KÜÇÜK & CAN, 2020), (Salloum, Al-Emran, Abdel Monem, & Shaalan, 2017). Presently Facebook become an essential application for communicate and connect people with each other, Facebook become a platform for participate the opinions, this site is used by lot of people in this world to participate their ideas, thoughts, sadness, pleasures and so on (PINTAR, HUMSKI, & VRANIĆ, 2019).

CLASSIFICATION TECHNIQUES IN NETWORKING SITES

Researchers in (Ozbay & Alatas, 2019) demonstrated that they could detect disinformation on social networks. This combination demonstration has been evaluated using accuracy, analysis, adequacy, and F-measure measurements on three different real data sets. The average results of all targeted fake observations calculations for all assessment metrics within three data sets were determined. The best cruel values to improve accuracy, efficiency, according to the acquired values, and F-measure have been extracted from the Decision Tree calculation The experiment consists of three datasets includes 1627 news reports collected from Facebook about the 2016 US election, random political news, false news of ISOT respectively. 70% used for training and 30% for testing, measuring Accuracy, Precision, Recall, and F-measure by using the Applied artificial intelligence algorithms (BayesNet, JRip, OneR, Decision Stump, ZeroR, SGD, CVPS, RFC, LMT, LWL, CVC, WIHW, Ridor, MLP, OLM, Simple Cart, ASC, J48, SMO, Bagging, Decision Tree, IBk, And KLR) (Khedr A. E., 2013) (Khedr A. E., 2012). the first data set results according to best accuracy (0,655) using J48, best precision (0,747) using Decision Stamp, the best F measure (0,675) according to WIHW, the second data set results according to best accuracy (0,680) using SMO, best precision (0,736) using Decision Stamp, the best F measure (0,693), the third data set results according to best accuracy (0,968), best precision (0,963), the best F measure (0,668) using Decision Tree, The Decision Tree method produced the highest mean average value in terms of accuracy (0,745), precision (0,741), and F-measure (0,759). In terms of retrieval metric, ZeroR, CVPS, and WIHW estimations of 1,000 values tend to be the best according to the three datasets used in the experiment. In their future research will be integrated the used algorithms or found new algorithms to improve the results.

In (Joo, Lu, & Lee, 2020), researchers introduced an innovative approach that combines bi-term topic modelling and MGLM. The goal of this study is to utilize text mining to detect subjects in Facebook posts posted by public libraries. Moreover, investigating the links across Facebook content and user engagement. There were 4,637 Facebook postings in the original database. Researchers in (Joo, Lu, & Lee, 2020) utilized 4,066 Facebook postings that had any written content, excluding those that just contained an image or video with no textual message. data were collected for four-months, in 2015, from July 1 till October 31, postings were gathered from 151 public libraries chosen randomly, Scholars observed that comments about cultural events, scholarships, and images earned several responses and comments, while posts about summer reading programmers received marginally more feedback to increase people's visibility on social networks in the context of public libraries. As a result, public libraries utilized Facebook to distribute information about library books, events, and activities, according to the findings. Terms like "come," "programme," and "join" are used to describe library activities or events. The keywords "librari", "book", "read" "free," "movie," and "event" were all in the top 100. In addition, many time-related, date-related, month-related, or season-related words, such as "todai," "summer," "July "October," "Saturday," and "Tomorrow," to name a few, all received excellent marks. The remarkable aspect is that the month names that were among the 20 ranked most frequently used keywords throughout the data collecting timeframe. In the future studies their intention to use twitter libraries also and analysis a higher sample size.

According to a study illustrated in (Namugera, Wesonga, & Jehopio, 2019) that information is stored in text form by different applications, text mining is a significant field for studying. The challenging part is obtaining the knowledge that the consumer requires. Text mining is a major phase in the knowledge discovery process that is widely acknowledged, So the primary goal of this research was to build text mining models enabling knowledge discovery and sentimental tracking on Ugandan Twitter communications. Unstructured data is used to expose hidden facts. Data of this study were collected from twitter using data crawling, data were consisting of tweets derived from NTV, NBS TV, Daily monitor paper, and New Vision Paper through nine days in July 2017, after collecting data the first step is to converting unstructured data into structured one by using Latent Dirichlet Allocation (LDA) algorithm, and using a statistical modeling (topic modeling) for helping in classification and semantic analysis, To compare the predictors of topic modelling for Twitter sentiments , retweet count, retweet status, and tweet sources with the sentiment of the final text. According to the findings of this study, Tweets are more widespread in non-print networks, such as televisions, than the one in print networks. For almost the same time, the primary print (newspaper) networks, Daily Monitor and New Vision, were positively correlated in developing topics with comparable sentiments from tweets, as were the non-print (television) networks, NBS TV and NTV. The public's feelings on security, politics, and other issues. economic sentiment was negative, although sports sentiment was positive. The preceding section's logistic regression model was trained on 9783 data and verified on 3261 observations. According to the results of the model validation, the logistic model accurately classified 64% of positive tweets and 63% of negative tweets. The four outcomes of binary classification from the model prediction results were used to create the confusion matrix. True positives, false positives, true negatives, and false negatives are all identified in this categorization. According to the findings, the algorithm can accurately forecast the sentiment polarity of incoming tweets by 64%. a logistic regression model was constructed by (Namugera, Wesonga, & Jehopio, 2019) using Decision trees, Naïve Bayes and Random Forest classification methods measuring Correct rate, Sensitivity, Specificity, Balanced accuracy respectively, the results were Random Forest (71%, 66%, 77%, 72%), Decision Tree (65%, 62%, 68%, 65%), Logistic (64%, 63%, 64%, 64%) Naïve Bayes (60%, 51%, 70%, 61%) respectively, the finest was the random forest, In the future the researcher in (Namugera, Wesonga, & Jehopio, 2019) will be enhance the classification model outcomes.

Text mining techniques were used by the researchers in (Huang & Liu, 2019) to transform unstructured data from social networking sites perspectives on stock-related news into argument ratings, which were found to be directly associated with the extent of stock demand curve. The researchers in (Huang & Liu, 2019) proposed an innovative prediction approach that incorporated sentiment scores further into logistic regression model based on reported data that showed that the higher the sentiment scores, the lower the logistic regression model's expectation accuracy in details data collected from The HHPIC that classified into two parts chip indicators and social networks reviews published on PTT in response to news and responses. HHPIC's data of the chip indicators that were gathered from the Taiwan Stock Market Observation Post System that observes the stock market in Taiwan. Overall, data was gathered on 245 days Throughout 2017, PTT received reviews and responses. PTT has a lot more, there are more than 1.5 million registered members, with more than 20,000 reviews. The purpose of this research is to investigate the relation among sentiment scores and stock price change extent for predicting the accuracy level, and to come up with a better solution. this study illustrated that the first step in their framework is collecting chips of HHPIC and reviews about HHPIC then applying text analysis on collecting reviews to conclude the sentiment scores, in parallel indicating the screen chips applying logistics regression model of stock price change extent after that combine these results to apply correlation analysis between them applying on three probabilities (P=0%, p=0.5%, p=1%). The best accuracies have resulted by the study (Huang & Liu, 2019) were 64.84%, 75.67%, 78.37% respectively the model improved the predicted accuracy of the proposed prediction model by 4.05%, 1.35%, 2.70% respectively, in their next studies they will use another predicting method to compare between results and determine which is better and data set from another platform on social networks. According to this framework in the following figure 3 that explained how to implement specific method for social networks analytics.

Figure 3. Analysis of text data on social networks (Stieglitz & Dang-Xuan, 2012)

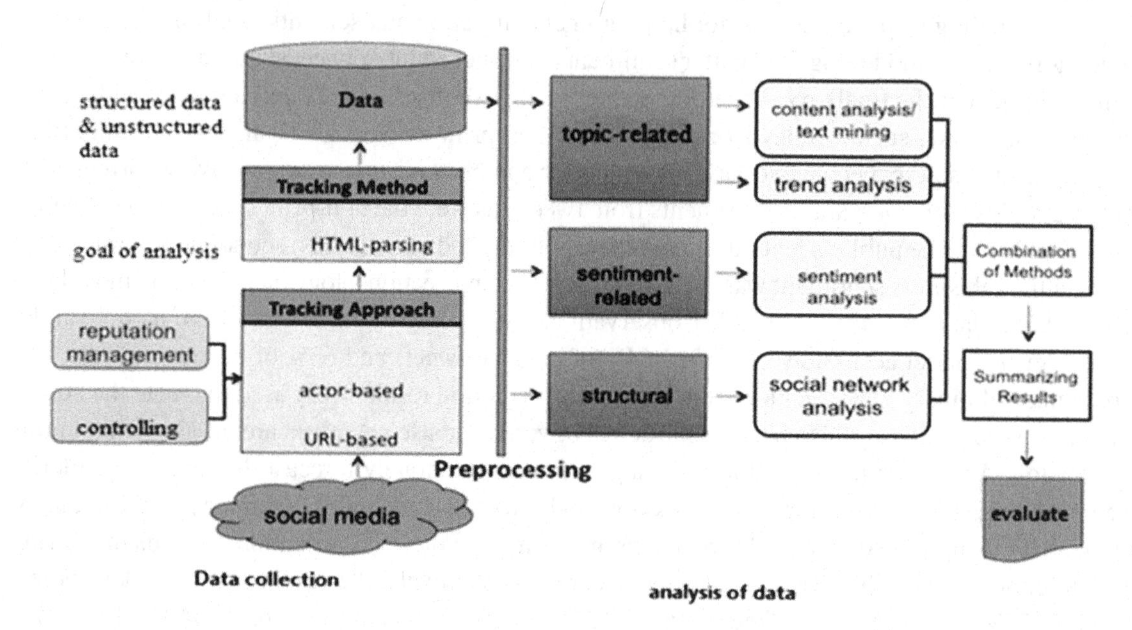

The summarized table1 presents the previous studies for the pervious section.

Table 1. A summarized table for classification techniques in social networks

Ref.	Data sources	Methods and techniques	Results & accuracy	Limitations and Future work
(Ozbay & Alatas, 2019)	Three real datasets from Facebook	23 different supervised artificial intelligence algorithms applied the following classifiers (BayesNet, JRip, OneR, Decision Stump, ZeroR, SGD, CVPS, RFC, LMT, LWL, CvC, WIHW, Ridor, MLP, OLM, SimpleCart, ASC, J48, SMO, Bagging, Decision Tree, IBk, And KLR)	Finest Average accuracy (0,745), precision (0,741), and F-measure (0,759) using Decision Tree	Finding new algorithms, integrating some of these classifiers to improve results.
(Joo, Lu, & Lee, 2020)	Facebook posts (text only) collected from 151 randomly selected from public libraries covered in 2015.	bi-term topic modelling and MGLM (Multivariate Response Generalized Linear Models)	"Librari", "book" and "read" ranked the most frequently used keywords	-Using dataset cover a whole year or a larger data set, -Using twitter libraries also

Continued on following page

Table 1. Continued

Ref.	Data sources	Methods and techniques	Results & accuracy	Limitations and Future work
(Namugera, Wesonga, & Jehopio, 2019)	Daily tweets for 9 days from NBs for news in 2017	LDA algorthim Logistics regression model using (Decision trees, Naïve Bayes, and Random Forest classification methods measuring Correct rate, Sensitivity, Specificity, Balanced accuracy)	64% the finest was random forest Correct rate, Sensitivity, Specificity, Balanced accuracy (71%, 66%, 77%, 72%) respectively	Enhance the classification model outcomes.
(Huang & Liu, 2019)	chip indicators from HHPIC and social networks reviews posted on PTT responding to news and replies. data were collected on 245 days in 2017.	Improve logistics regression model by applying (correlation analysis between sentiment scores and stock price extent) Using the following three probabilities. P=0%, p=0.5%, p= 1%	64.84%, 75.67%, 78.37% respectively Improved by 4.05%, 1.35%, 2.70% respectively	-Using another model to predict and compare between the results. -using other platforms on social networks

NEWS CREDIBILITY IN SOCIAL NETWORKING SITES USING TEXT MINING (CLASSIFIERS)

A study by (Fadel, 2020) stated that few researchers worked on credibility in Arabic language that detect the fake news and aid in reducing the spreading of misinformation, nearly 181 million users in Arab countries using internet so the goal of their study is to provide an initiation of a dataset for evaluating credibility of Arabic news on social networks using classifier models and integration tools for extracting features automation such as photo recognition and NLP for comments. Having started on the 24th of January 2019, data was gathered for about 14 days. There were 1162 tweets in all, with 808 credible tweets and 354 non-credible tweets. The sample was gathered in the formalized Modern Standard Arabic Language (MSA) to allow text mining and analysis easier. The suggested methodology means developing a labelled data set of Arabic news on Twitter, gathered features from previous studies, extracting features, operating several experiments to determine the best classifier algorithm and feature set for the best accuracy, and eventually building and testing the intended system. The algorithm that analyzed the tweet's URL and categorizes it as credible or non-credible news based on the chosen classifier model, during a series of experiments, the accuracies of these classifiers: Decision Table (81.46%) using relief algorithm, Nave Bayes (75.72%), Random Trees (79.14%) using Symmetrical, and LMT (74.17%) using, were the features employing with classification algorithms in this study. it was discovered that a system with seven distinct features the finest accuracy of a decision table classifier 81.46% using relief algorithm. The research in (Fadel, 2020) also described the construction of 906 records for Arabic data set evaluating news credibility using feature extraction methods that were performed automatically. In the future study it will be compared between some trusted research, using larger dataset, analyzing tweets images to achieve a better model.

A research in (Mouty & Gazdar, 2019) explained that detecting the news credibility in order to improve the quality of information and news accessible on social networking sites, their goal from the research is to promote the accuracy of Arabic news using text mining techniques merged with natural language process (NLP), data were collected from twitter; the dataset was about 9000 tweets, after that in data preprocessing has used a method that applied tokenization in Arabic phrases. To determine the number of unique characters or words, it must first be determined the total number of characters or words, then the study has been made a feature extraction by using 18 features then named the similarity score by converting the Franco Arab words into Arabic words, then classifying by using classification algorithm as Decision Tree measured the Precision, Recall, F-measure, Accuracy that calculated as follow: (0.741, 0.742, 0.741, 74.22%) respectively,

, Support Vector Machine (SVM) measured the Precision, Recall, F-measure, Accuracy that calculated as follow: (0.668, 0.675, 0.668, 67.51%) respectively, Naïve Bayes measured the Precision, Recall, F-measure, Accuracy that calculated as follow: (0.686, 0.575, 0.558, 57.45%) respectively Random Forest measured the Precision, Recall, F-measure, Accuracy that calculated as follow (0.776, 0.778, 0.776, 77.81%) respectively. Finally Random Forest is the best classifier and According to the results of the preview studies, the suggested model's accuracy before adding the names similarity score was 76.17 percent. After putting the names together, the score of similarity with the old (user and content) characteristics, the accuracy of credibility detection improved by 1.64 percent, the percentage is 77.81 percent. The credibility detection's performance has increased. In their future studies they will use over-size sample, perform their studies on other social networks sites to enhance the proposed models.

A study by (Kolluri & Murthy, 2021) stated that spreading of fake news negatively affected the World Health Organization, and that reduced the credibility between people and spread panic over the entire world. There is more than one influential website that published fake news, the study has been performed on a sample of dataset (PHEME) credibility-labelled tweets and Facebook posts to vote on news and giving the users the ability to give their feedback and this research (Kolluri & Murthy, 2021) tried to limit this by using machine learning model using python applying (Support Vector Machine (SVM) classifier, a Logistic Regression classifier, and a Bernoulli Naive Bayes (Bernoulli NB) model and power of individuals, users can state their opinions in all news to determine the news credibility by voting on it. the researchers in (Kolluri & Murthy, 2021) presented CoVerifi, an arrangement that in an extremely stage incorporates the power of truth-nudging, personal criticism, and artificial intelligence. CoVerifi provides a several channel news validity search, meaning that "fake news" which has fizzled detection by computerized techniques can be detected via client criticism and vice versa. Since several stages are close together, CoVerifi may represent a variety of networks habits. Convexify can also be analyzed to evaluate unused types of content territorially specific stages as a result Identifying the credibility of news Ranging between 50–63%. Others can use Convexify according to their accessible code. counting sending it with paid administrations for more prominent adaptability. Convexifies code can then be associated with a different classification display or a various news Application to interpret distinct types of data. The limitation of this study that It is still not ready for large-scale deployment.

A study by (Pasi, De Grandis, & Viviani, 2020) was found that Internet content that is relevant to users' needs and objectives requires assessing the data's credibility. With the growth of the internet especially social networks and the potential for users to generate material, assessing knowledge authenticity is now a major research issue, with misinformation, fabrications, and other types of deception spreading almost without any basic means of trusted internetworksries. The study by (Pasi, De Grandis, & Viviani, 2020) suggested a system based on several newsrelated requirements that enhances the information's usability. In this paper The CredBank dataset was used to test the suggested model using machine learning algorithms, the question was raised of the validity of knowledge gathering on social networking sites. In the news website (especially Twitter) a multi-criteria decision-making method (MCDM), The SVM classifier was developed using the linear kernel function. The additional classifiers were created by using the default settings of the scikit-learn functions measure AUC, Acc, Prec, Rec, F1, respectively for these classifiers SVM classifier (0.80, 66%, 66%, 99%, 80%), KNN classifier (0.62, 68%, 70%, 87%, 77%), Naive Bayes classifier (0.78, 71%, 71%, 93%, 80%), Decision Tree classifier (0.75,

76%, 89%, 82%, 81%), results of Random Forest (0.87, 79%, 79%, 90%, 84%) respectively. The Python programming language was used to perform the categorization and experimentation stages. Random forest has been given the best accuracy that equal to 79%. data science that has been proposed in recent years by literature that uses aggregation operators and previous domain k as the basis for an evaluation of the credibility of news, was proposed., facilitate the reputation evaluation model to be designed in a versatile manner. Hybridizing the MCDM model with certain feasible and beneficial. Other aggregation operators such as fuzzy integrals and structured approaches to learning the different value factors to be paired through integrity functionality should be examined and assessed.

Scholars of (Roychoudhury & Srivastava, 2020) stated that by using text mining techniques and NLP after extracting the posts of users in social networks that have meaningful text, numbers, repeated words, extracted information from user's contents to help in understanding the user's behavior by developing a adaptable coordinating conspires and Non-availability of needed data either conflicting data for the same user over these frameworks are factors for digital characters, the dataset was collected between May 2018 and July 2018, based on posts, tweets from Facebook and twitter respectively, These models were developed with using the following five classifiers to measure Accuracy, Precision, Recall, F1 Score, AUC respectively: Naïve Bayes (72.8%, 62.4%, 67.1%, 64.7%, 53.1%), Logistic Regression (76.5%, 70.1%, 75.2%, 72.6%,

60.3%), Random Forest (86.4%, 83.2%, 84.1%, 83.6%, 62.2%), SVM (82.5%, 81.2%, 82.2%, 81.7%, 68.1), Neural network (91.2%, 82.5%, 84.5%, 83.5%, 83.4%) The most likely matched Facebook profile was predicted using these classifiers. After constructing the model, test data was retrieved from the training dataset and used to validate it. the paper by (Roychoudhury & Srivastava, 2020) achieved to the highest accuracy nearly 91.2% using neural network between several classifiers. The future study can be Used sentiment analysis and TF–IDF (Abdel-Fattah, Khedr, & Nagm Aldeen, 2017) models to improve the accuracy .
The summarized table 2 presents the previous studies for the pervious section.

Table 2. A summarized table for news credibility in social networking sites using text mining

Ref.	Data sources	Methods and techniques	Results & accuracy	Limitations and Future work
(Fadel, 2020)	Arabic dataset collected fr om Twitter (1162 tweets) in 2019 and covered news credible and non- credible to be assessment.	applied 8 classifiers with 8 features, each classification algorithms employed by using one feature as (correlation with naive based, relief with decision table, info gain with Random trees).	Relief with decision table classifier 81.35%	-comparing between some trusted studies, -using higher dataset -analyzing tweets images
(Mouty & Gazdar, 2019)	It contains 9000 Arabic tweets with their meta-data.	Classification by using more than one algorithm as (Support Vector Machine (SVM), Decision Tree, Random Forest, naïve based) measured the Precision, Recall, F-measure,	Random Forest has the best results (0.776, 0.778, 0.776, 77.81%) respectively	-Using higher size sample -using other social networks sites.

Continued on following page

Table 2. Continued

Ref.	Data sources	Methods and techniques	Results & accuracy	Limitations and Future work
(Kolluri & Murthy, 2021)	Dataset (PHEME) consists of real and fake news from Facebook and twitter	-Developed machine learning model using python. -Using (SVM) classifier, a Logistic Regression classifier, and a Bernoulli Naive Bayes.	Identifying the range of news credibility between 50– 63%.	not suitable to use in a large-scale.
(Pasi, De Grandis, & Viviani, 2020)	CREDBANK datas et	Machine learning algorithm applied python using these classifiers (SVM, KNN, Decision Trees, Naive Bayes, and Random Forests) measuring AUC, Acc, Prec, Rec, F1respectively	Random forest with finest results (0.87, 79%, 79%, 90%, 84%)	aggregation operators as fuzzy integrals.
(Roychoudhury & Srivastava, 2020)	dataset from twitter and Facebook during 2018	Five machine learning classifiers applied (Nave Bayes, Logistic Regression, Random Forest, SVM, and Artificial Neural Network) to measure Accuracy, Precision, Recall, F1 Score, and AUC respectively	neural network cl assifier has the best results (91.2%, 82.5%, 84.5%, 83.5%, 83.4%) respectively	Using sentiment analysis and TF–IDF models to improve the accuracy

STATISTICAL METHODS FOR NEWS CREDIBILITY IN SOCIAL NETWORKS

Because of the rapid spreading of fake news and misinformation last years on social networks, such as during the 2016 US Presidential election, the representation of extraordinary story and news has been dwindling. A study by (Waheeb Yaqub, et al., 2020) stated that reveal of news credibility helped in decreasing of tendency to share fake news, in this study depends on four perspectives: Fact Checkers, News Networks, Public as the measurement of credibility from American's point of view, Artificial Intelligence (AI). The number of participants were 1,512, Out of 18,144 headlines, 18,135 people expressed an intent to share, and 9 people did not respond. They constructed in (Waheeb Yaqub, et al., 2020) a Binomial logistic regression using stepwise approach, then The Chi-square test is used to assess the distinction between different models applying the Likelihood Ratio Test (LRT). The treatment, headline type, political background, internet activity especially social networks, and typical demographic characteristics like gender and age were all included as independent variables. the study has observed that credibility indicators can help fight false news. Their efficiency, even so, depending on the type of indication and the user's personality traits. Their results represented the initial step in a design investigation of credibility indicator content and distribution; fundamental social concerns about how content is distributed and the odds ratio across Fact Checkers and News Networks, that was the most effective index, was 0.457. in the future work it will be used more headline and publishing their results on social networks as a real-world.

A statistical survey by (Bates & Sousa, 2020) aimed to compare between the real news from (individuals and tabloid) and fake(individuals and tabloid) by measuring the news credibility on social networks especially Facebook posts in United Kingdom according to location, age (41% were between 30 and 39), educational level and gender (76% were females), the researchers used methodological scrutiny, the sample collected from UK users on social networks having Facebook accounts in 2019, there were 167 respondents who answered and completed all the questions from 201. according to the survey 60%

of participants uses Facebook more than once through a day. Measuring the Alpha coefficients for two aspects of the credibility scale over four situations then applying a general linear model after testing equality of error variances, achieving a p.05. statistical significance from the sample, the next step is to predict the interaction, the result of their survey in (Bates & Sousa, 2020) illustrated that as every point of author credibility increases the likelihood of people engaging with a post by nearly 17 times assuming the article is genuine news released by a person. If the story is accurate and published by a tabloid, the rise would be slightly lower, 9.3 approximately, the credibility has been measured by using statistic methods, the significant result that the study helped in predicting how can people on social networks will react to a post by comment or reactions, and in their future work it will be Developed other method to measure credibility of users.

A research in (Martin & Hassan, 2020) showed the legitimacy of the networks was analyzed as an indicator for publishing fabricated political news exposure (FNE) reporting via online in 2018; people who use the web through several countries of Arab such as Lebanon, and the United Arab Emirates (N = 4,616). In three nations, regression models that study fake news cases on networks using coefficients clarified significant bulk of variation in this case range between (3percent to 26 percent). However, the prediction that participants' credibility rankings of mass networks would poorly predict FNE was only moderately endorsed, with only two of the five countries showing a connection. The perception that misinformation must be banned online was the greatest significant predictor of FNE. The researchers in (Martin & Hassan, 2020) focused on how individuals in the countries of Arab react for presenting the false political news on social networks, SPSS was used to run multiple linear regression models with each country and using one-way ANOVA and LSD post hoc to compare between results, just in two modern democracies do news credibility tests display a negative association with the perceived prevalence of false news on the internet. Estimations of considerable news accuracy and realness were not irrelevant to fabricated news in the other nations. Support for anti-fake news initiatives was the best indicator of FNE. People that support the news networks could be less suspicious of online information in certain nations, in this study, Results of this study that Lebanon is between (88.1% and7 7.8%), Qatar is between (74.3% and 72.2%), Saudi Arabia is between (85.3% and 84.4%), the United Arab Emirates is between (68.7% and 68.6%), and Tunisia is between (66.6%,75.6%) were the countries with the most inaccurate online news. the future study can be contained five-point questions instead of 4 questions,

The lack of trust in the networks may be linked to a pessimistic perception of uncertain or without references news sources. The Lebanese and Tunisians had the 2nd and 3rd highest Fabricated news scores after the Saudis. accordingly. In the following review, knowledge flows more openly in Tunisia as well as Lebanon than those in the Arab countries. Apart from Saudi Arabia, the two-elements indicators of Fabricated news, the research by (Martin & Hassan, 2020) targeted significance consistency of varied samples and various languages.

The summarized table 3 presents the previous studies for the pervious section.

Table 3. A summarized table for statistical methods for news credibility in social networks

Ref.	Data sources	Methods and techniques	Results & accuracy	Limitations and Future work
(Waheeb Yaqub, et al., 2020)	72 headlines selected from news articles No. of participant=1512.	Binomial logistic regression applied using s tepwise approach and applied Likelihood Ratio Test (LRT) using the Chisquare test.	Fact checkers and news networks effecti veness 0.457	-Using higher data set in the future
(Bates & Sousa, 2020)	Data gathered from a survey (UK respondents who have a Facebook	-Measured alpha coefficient. -applied levene's testing of equality errors of variance. -repeated measured of general linear model for credibility.	-individual authors: 17 tim es credible. -UK	Extend the credibility rating scale from five to seven or
	account) in 2019	-applied 4 logistics regression analysis.	tabloids = 9 tim es.	nine
(Martin & Hassan, 2020)	Data collected from annual Networks Use in the Middle East survey, published by Northwestern University in Qatar (2018) about five Arab countries	-SPSS was used to run multiple linear regression models with each country. -Using one-way ANOVA and LSD post hoc to compare between results. Using (Four-point questions)	Lebanon is having the most inaccurate news among (88.1% and77.8%).	Using 5 point questions.

STATISTICAL MODELS FOR NEWS CREDIBILITY IN SOCIAL NETWORKS

As reported by (Luo, Hancock, & Markowitz, 2020) that the purpose of measuring the news credibility on social networks to distinguish between the fake news and real one, so researchers applied the reality of the situation concept for the credibility of news, enhancing our awareness about how one-sided handling and coverage of networks news and facts can affect detection accuracy when using social networks prompts. they employed in (Luo, Hancock, & Markowitz, 2020) a database of real and fake headlines which have been posted to social networks. The sample size consisted of 30 headlines credible and five fakes for each category. The three types of headlines and the number of participants in each type were health (n=116), politics (n=115) and science (n=106). The total no of participants was 379 in their study, The truth-bias and veracity consequences in the detection accuracy of false and true news were investigated experimentally of TDT's predictions for veracity evaluations of news in social networks, The effectiveness of subject on message credibility was investigated applied a one-way analysis of variance (ANOVA). as well as the truth-bias and veracity impact in detection average accuracy of fake and real headlines has been noted 51%. In their future study they will Use more than one item scale to measure credibility, use more data to measure.

Scholars of (Gong & Eppler, 2021) Investigate the influence of a TV reporter by reporting the news and the extent of credibility of the news sources ratings, as well as whether the gender is male or female, the news environment, and empathetic anxiety mitigated this influence. 352 were participated in this study as a sample from a Survey Sampling International (SSI) in 2017, a statistical approach of analysis of variance (ANOVA), was applied divides observed variance data into multiple elements for use in testing 3 hypotheses. to study the relationship between both the dependent and independent variables, multiple post hoc analysis was applied too. According to the findings of this study (Gong & Eppler, 2021), H3 has the higher credibility source, observers get the highest level of analysis of presenters with an overpowering presence, and that evaluation drops suddenly then increased mix-ups. Once it came to judging brief news segments, viewers could be more accepting of women commentator's 3 conveyance errors than the male commentator, however the two sexual orientations endured similarly when the quantity of slipups expanded for 5. In all news through a day, each impact of transferring accidents on credibility, rating could be predictable. After that, observers with a low level of compassionate consideration had a stricter outlook on secures making mistakes than observers with a significant degree of compassionate concern, Future studies may be including news anchors from various ethnic origins to investigate the influence of ethnicity on trustworthiness judgments in the news broadcasting setting.

A study by (Henke, Leissner, & Möhring, 2019) given the recent buzz about the news networks's lack of credibility, news-casting experts and examinations often recommended that journalists focused on improving the quality of their reporting, data were collected in July 2017 that consists of questions about networks and news in general from a survey, there were 406 respondents in total (46.3% male) as a result, Even more it has been computed and analyzed of the various three dimensions (enjoyment, interest, and perceived vividness) revealed that only the article's perceived vividness was influenced by the manipulations. It also observed significant, minor impact for statistical data ($n^2 = .010$; $p < .05$) and using ANCOVA statistics. a slightly positively bigger main effect for the article's perceived vividness, rely on facts and verification. Using reports on confidence and truthfulness as a guideline, researchers investigate what the use of multiple types of approvals means for the legitimacy and reliability evaluation of papers, accordingly, the learning process from a crowd's point of view. An online observation was carried out as part of the research. to see if the existence of logical sources, factual data, and their interpretation in an online article affected the outcome. The findings indicated that these approvals increased the perceived credibility. Adding factual references, observable facts, and descriptions to a paper may not minimize its apparent satisfaction, but it can increase its obvious uniqueness in the eyes of networks audiences or users. Then, the study by (Henke, Leissner, & Möhring, 2019) has checked for potential impacts on the selection criteria of truth dimension of confidence, but it couldn't find any. The confidence in the consistency of depictions/credibility factor then was checked. Our root factor had a slight but important impact; that Documents with sources were deemed more trustworthy than otherwise without sources. The limitation of their study that the sample is not almost accurate.

The researchers found that social networks sites have been as an organization in (Tandoc Jr, 2018), The study focused on how people using social networks to find news from a variety of sources, such as a news organization (The Straits Times) and a friend, Facebook users were able to recognize between the 2 kinds of news sources more simply than they did with a traditional news agency or a friend, according to the study. The objective is to How do people tell the difference between these two kinds of sources? The study used a two-factor (source: news organization (n=37) vs. Facebook friend(n=35) two-factor (motivation: high vs. low) two-factor (message repetition) mixed experimental design, with source as a between-subjects component and inspiration and message repetition as within-subjects variables. The research by (Tandoc Jr, 2018) is being carried out in Singapore; the respondents were 82 that completed their answers that chosen from a famous Singaporean university. Each respondent read two new articles, one with a high level of motivation and the other with a low level of motivation. Motivation was adjusted depending on the geographic location of the topic under discussion, as in prior experiments. For instance, in the case of one research, high-proximity news stories, or those concerning local events, had a positive impact on one's health. a sample independent was used to 4 hypotheses, a repeated-measures analysis of covariance was used. H1and H3 had the best accuracy of news. The study found that a strong interaction impact across source type and motivation. In their following studies they will discuss How to understand the difference between the user's friends publishing news and news shared by organizations on social networks.

The summarized table 4 presents the previous studies for the pervious section.

Table 4. A summarized table for statistical models for news credibility in social networks

Ref.	Data sources	Methods and techniques	Results & accuracy	Limitations and Future work
(Luo, Hancock, & Markowitz , 2020)	Dataset consists of real and fake news headlines from social networks.	a one-way analysis of variance (ANOVA)	Average accura cy is 51%	Limitation -real data will be more accurate Future study -using a higher dataset.
(Gong & Eppler, 2021)	Data gathered from a Survey Sampling International (SSI) in 2017	-ANOVA applied for testing three hypotheses (H1, H2, H3) -multiple post hoc analysis were applied.	H3 has the higher credibility source.	The news should be including various ethnic origins to investigate the influence of ethnicity on credibility.
(Henke, Leissner, & Möhring, 2019)	Data collected from an online survey about networks and news in 2017	ANCOVA is applied.	perceived vividness is positively related.	Improving the used sample to be more accurate.
(Tandoc Jr, 2018)	Facebook profiles for students studying in a Singapore university and	a sample independent was used to 4 hypotheses. a repeated-measures analysis of covariance was used.	H1 and H3 had the best accuracy of news.	How to understand the difference between the user's friends publishing
	straits news in Singapore			news and news shared by organizations on social networks

TEXT MINING BASED NEWS CREDIBILITY IN SOCIAL NETWORKS

Because of social networks' increasingly assertive presence as a source of news, internet users can no further rely on conventional journalism to verify facts as in (Waruwu, Tandoc Jr,, Duffy, Kim, & Ling, 2020). Rather, they must decide for themselves whether the news is credible or not. Most of the content credibility analyzed have treated news viewers' information assessment as a solely cognitive task, meaning that people can come to accurate conclusions without the help of others. The activity of networks news verification by viewers is re-conceptualized in this paper as a cycle of shared aim of performance. The data used in this study comes from five focus groups (FGDs) in Singapore with a total of 30 social networks users aged between (18-32). The participants were gathered through Singapore's famous university Via a series of focus group discussions, Categories depend on the inductive analysis. the study by (Waruwu, Tandoc Jr,, Duffy, Kim, & Ling, 2020) has explored the social influence, methods, and implications that sustain it. News identification is a practical act of cognitive processing associated with self of participants. To determine news credibility, the tactics revolve around the need for social cues and community engagement. Participants were more likely to double-check social networks reports that affected them or their social connections. Trying to maintain personality and retaining mutual trust were defined as two basic reasons for news verification. The research in (Waruwu, Tandoc Jr,, Duffy, Kim, & Ling, 2020) revealed that news authentication is a strategic activity of information processing that is founded in the respondents' self-categorization, rather than a solitary, cognition-driven, and evidence-based activity. The respondents' willingness (or lack thereof) and techniques for authenticating news were impacted by their membership status in distinct social groups, Future study might investigate how young adult authentication patterns are influenced by digital culture. and may be expanded on their findings to contribute to a more comprehensive understanding of the social sides of news authenticity.

A research by (Liao, Chen, Yang, & Yuan, 2020) purposed to find accurate audience by using text mining techniques so the researchers constructed a topic model recommendations system to make analysis on Facebook posts by arranging these posts according their model system (TMRS) the model has been taken advantage of using marketing and advertising by selecting posts automatically, The Graph API on Facebook was employed in this investigation to gain access to people's information without them having to enter their passwords, this API gathered post data, particularly manage pages and ads management, after acquiring fan page tokens to perform the analysis below, All test data was collected between Mar. 2015 to June 2018. The ad post data for various circumstances is shown. there are 688 posts with delivery data in the training data that will be used to create advertising, to generate advertisements, the test data contains 92 posts for 11 fan pages, this study used 3 scenarios, the first tested all advertising posts, the second three marketing professionals selected ad post wording from the Spirits fan pages. The third To the Makeup/Skincare fan sites, implementing the superior solution from 1 and 2. And by applying 2 two models (LSI and LDA), the third scenario using LSA model were more critical and accurate, consists of mix between the 1st scenario (All of the Wine/Spirits fan pages' advertising post wording), and the 2nd scenario (Three marketing professionals selected ad post wording from Wine/Spirits fan pages).

To apply the third scenario the Makeup/Skincare fan sites, applying the superior solution from the 1st and the 2nd. in the researcher overview in (Liao, Chen, Yang, & Yuan, 2020) because of using limited budgeting in marketing ads and was being effective and sufficient. An automated classifier will be used in the future study, but it may be taking a long time.

A research in (Yahav, Shehory, & Schwartz, 2015) Text mining have picked up incredible force in last few years, with user-generated content getting to be broadly accessible. Comment mining is a real unique filed to study, with assumption examination and assertion mining receiving special attention. Data were gathered from seven different Facebook fan pages text preprocessing is a fundamental phase within the comment mining processing technique, in which each semantic term is assigned a significance that typically rises with its presence in the evaluated content but is compensated by the term's incidence within the area of interest. To calculate these weights, several users use the well-known tf-idf equation using several approaches as adjusted TF-IDF approach, n k-fold cross-validation. This study by (Yahav, Shehory, & Schwartz, 2015) investigated the predisposition raised by creating a connection' talk to consider online comments and suggests a transition. The study by (Yahav, Shehory, & Schwartz, 2015) proposed that substance extricated from discourse is regularly exceedingly related, coming about in reliance structures between perceptions within the consider, in this way presenting a measurable inclination. Overlooking this tendency will result in a weak investigation at best, and a completely incorrect conclusion at worst, the researchers in (Yahav, Shehory, & Schwartz, 2015) suggested an improvement to TF-IDF that adjusts for this bias, and then discussed the results of either the bias and correction with

Facebook fan page data, covering various domains such as news, economics, politics, leisure, shopping, sports and entertainment. Their limitation that the researchers did not use a large dataset

A study by (Sarwani, Sani, & Fakhrin, 2019) Nowadays the internet makes a modern era with modern culture that uses advanced networks. Social networks is one of the well-known advanced networks. Facebook is now a popular social networking site among young people. They are usual to passing on their considerations and expression through social networks. analysis of text mining can be utilized to determine one's identity and personality through social networks with the probabilistic neural network algorithm. data was collected from two Facebook fan pages with a lot of activity, the researchers in (Sarwani, Sani, & Fakhrin, 2019) investigated by correlation and its bias to TF/IDF, so the text can be taken from the status that is on Facebook. In this consider, there are three stages, specifically text processing, weighting, and probabilistic neural systems for deciding classification. Text processing divided into forms, namely: tokenization, stop word, and steaming. The results of this study explained that a weighted value is given to the text processing to each single word by using the Term Inverse Document Frequent (TF / IDF) method, then in the last step the Probabilistic Neural Network classifier is used to declare personalities. this study reached 60% accuracy. In the future it could be used other combined algorithms to enhance the accuracy.

A scholar of (dos Santos, Ramos, & Paraboni, 2020) developed an automated Technique of defining user traits and document behaviors of social networks, especially Facebook, by using machine-learning with natural language processing (NLP). Frameworks of this sort of benefits from the unique link between linguistic knowledge and identity models–like Big Five to give users details. from context and content as input in a non-intrusive as well as low-fetched manner. Despite the fact of presence, a well-established investigate point within the area, several concerns concerning computerized recognition of identity characteristics from text remain open for further study. Data was collected from Facebook about 2.2 million words, this research by (dos Santos, Ramos, & Paraboni, 2020) centered about which constructing psycholinguisticsmotivated frameworks of identity acknowledgment was feasible using multinomial logistic regression, TF-IDF: k-best TF-IDF counts with ANOVA f-value univariate feature selection, TFIDF-weighted averaging word embedding model And Baseline, TF-IDF-weighted averaging was the better face recognition model by result 0.52 While sources of knowledge are not valid for the chosen language under consideration, typical character level formats that still rely on psycholinguistic knowledge rather than collecting such identification aspects from a repository that does not directly forward on this data. These aspects are addressed in this paper through a series of individual tests for identity recognition from Facebook content, posts who is starting comes about ought to help long term advancement of more robust systems of this type.

The summarized table 5 presents the previous studies for the pervious section.

Table 5. A summarized table for text mining based credibility in social networks

Ref.	Data sources	Methods and techniques	Results & accuracy	Limitations and Future work
(Waruwu, Tandoc Jr,, Duffy, Kim, & Ling, 2020)	data collected from 5 focused groups in Singapore with 30 users on social networks.	Categories depend on the inductive analysis.	The respondents' wil lis (or lack thereof) tohare news was linked to their social status.	investigate how young adult authentication patterns are influenced by digital culture.
(Liao, Chen, Yang, & Yuan,	Ad posts from Facebook	-lSA and LDA models applied on three scenarios	The 3rd scenario using LSA mod el is the most accurate	An automated classifier will be used
2020)				
(Yahav, Shehory, & Schwartz, 2015)	Data were gathered from seven different Facebook fan pages.	To calculate tf-idf weight using several approach es as adjusted tf-idf approach, n k-fold cross-validation.	The study has been suggested an improvement to TF-IDF	Their limitation that the researchers did not use a higher dataset
(Sarwani, Sani, & Fakhrin, 2019)	Data collected from 2 Facebook fan pages	using the Term Inverse Document Frequent (TF / IDF) method classifying by using neural network algorithm	60%	Use other combined algorithms to enhance the accuracy
(dos Santos, Ramos, & Paraboni, 2020)	2.2 million words from Facebook	- multinomial logistic regression - TFIDF: k-best TF-IDF counts with ANOVA f-value univariate feature selection - TF-IDF-weighted averaging word embedding model - Baseline	0.52 -TF-IDFweighted averaging (better face recognition)	Use other language

CONCLUSION AND FUTURE WORK

These days, the way of connection between people is really changed and converted to online method. Social networks became the privilege for people to communicate in this era especially because of the covid-19 virus. Using text mining techniques is not only for declaring personality of users but for detecting the accuracy of news and helping organization to take correct decisions. To enhance the quality or service in any company or organization in the future and supporting their chunk in the market so social networks can help in this because of People's opinions (Myneni & Dandamudi, 2019), (WU, SUN, & TAN, 2013). Several researchers have expressed their views on the Text mining techniques used by various models in the most common social networking sites for example Facebook and twitter. As a series of these experiments, text analytics could be broken down to text clusters, classification algorithms, Apriori algorithms extraction and pattern analysis according to text mining applications. text mining is always in continuous development.

While Arabic language is significantly valuable, so the importance of analysis Arabic text using text mining is valuable, it was discovered in the considers the following that Arabic language text was under-represented at social netting sites using text mining analytics. In official transactions, the language of Arabic is used, discussing networks and news discussions. according to a report by (Tarek Kanan, et al., 2019). It is spoken as the main language by almost 380 million people worldwide, and in official transactions and speeches in the social networking sites it is spoken in Arabic A scholar of (Fouadi, El Moubtahij, Lamtougui, & SATORI, 2020) pointed out the importance of Arabic language so the researchers were confronted with many of its speakers and a shortage of Arabic materials, which provided a challenge for their related work.

Text mining techniques examination and looking into Arabic textual data from Facebook is a future work and research. Over and above, the sentimental study of the Arabic text should be considered in future studies. The Arabic language is phenotypical, with written form with free sentence structure, unusual syntax and cut off marks to overcome the persistent misunderstanding between the types which seem to be the same, as a result, the context is essential to remove the uncertainty that exists in similar types of opinion recognition.

REFERENCES

Abdel-Fattah, M. A., Khedr, A. E., & Nagm Aldeen, Y. (2017). An Evaluation Framework for Business Process Modeling Techniques. *International Journal of Computer Science and Information Security, 15*(5), 382–392.

Abed, A., Yuan, J., & Li, L. (2017). A Review of Towered Big-Data Service Model for Biomedical Text-Mining Databases. *International Journal of Advanced Computer Science and Applications, 8*(8), 12.

Andryani, R., Negara, E. S., & Triadi, D. (2019). Social Media Analytics: Data Utilization of Social Media for Research. *Journal of Information Systems and Informatics, 1*(2), 13.

Arianto, R., Leslie Warnars, H. H., Gaol, F. L., & Trisetyarso, A. (2018). Mining Unstructured Data in Social Media for Natural Disaster Management in Indonesia. *2018 Indonesian Association for Pattern Recognition International Conference (INAPR)*, 5.

Axenbeck, J., & Kinne, J. (2020). Web mining for innovation ecosystem mapping: A framework and a large-scale pilot study. *Scientometrics*, 31.

Balaji, P., & S, S. (2019, August). Web 2.0: An Evaluation of Social Media Networking Sites. *International Journal of Innovative Technology and Exploring Engineering, 8*(10), 8.

Bates, N., & Sousa, S. C. (2020). Investigating Users' Perceived Credibility of Real and Fake News Posts in Facebook's News Feed. UK Case Study. In Advances in Artificial Intelligence, Software and Systems Engineering. Springer.

Chen, Q., Peng, Y., & Lu, Z. (2019). BioSentVec: Creating sentence embeddings for biomedical texts. In *2019 IEEE International Conference on Healthcare Informatics* (p. 5). Academic Press.

Dahab, M. Y., Idrees, A. M., Hassan, H. A., & Rafea, A. (2010). Pattern Based Concept Extraction for Arabic Documents. *International Journal of Intelligent Computing and Information Sciences, 10*(2).

De Maio, C., Fenza, G., Gallo, M., Loia, V., & Volpe, A. (2020). Cross-relating heterogeneous Text Streams for Credibility Assessment. *International of Electrial and Electronics Engineers.*

Fadel, M. A. (2020). Evaluating the Credibility of Arabic News in Social Media through the use of Advanced Classifier Algorithms. *International Journal of Advanced Trends in Computer Science and Engineering*, *9*(4), 15.

Fu, Y., Jin, H., & Zhao, Y., & Cao, W. (2019). A novel text mining approach for scholar information extraction from web content in Chinese. *Future Generation Computer Systems*, 35.

Go¨k, A., Waterworth, A., & Shapira, P. (2015). Use of web mining in studying innovation. *Scientometrics*, *102*(1), 19.

Gong, Z. H., & Eppler, J. (2021). Exploring the Impact of Delivery Mistakes, Gender,and Empathic Concern on Source and Message. *Journalism Practice*, 21. doi:10.1080/17512786.2020.1870531

Grimes, S. (2008). *Unstructured data and the 80 percent rule*. CarbridgeBridgepoints.

Gupta, M. K., & Chandra, P. (2020). A comprehensive survey of data mining. *International Jounal of Information Technology (Singapore)*, *12*(4), 15.

Haggag, M. H., Khedr, A. E., & Montasser, H. S. (2015). A Risk-Aware Business Process Management Reference Model and Its Application in an Egyptian University. *International Journal of Computer Science and Engineering Survey*, *6*(2).

Hassan, H. A., Dahab, M. Y., Bahnassy, K., Idrees, A. M., & Gamal, F. (2015). Arabic Documents Classification Method a Step towards Efficient Documents Summarization. *International Journal on Recent and Innovation Trends in Computing and Communication*, *3*(1), 351–359.

Hassan, H. A., Dahab, M. Y., Bahnasy, K., Idrees, A. M., & Gamal, F. (2014). Query answering approach based on document summarization. *International Open Access Journal of Modern Engineering Research*, *4*(12).

Hassan, H. A., & Idrees, A. M. (2010). Sampling technique selection framework for knowledge discovery. In *The 7th International Conference on Informatics and Systems (INFOS)*. IEEE.

Hassan Fouadi, H. E., Fouadi, H., El Moubtahij, H., Lamtougui, H., & Satori, K. (2020). Applications of deep learning in Arabic sentiment analysis. Research perspective. *2020 1st International Conference on Innovative Research in Applied Science, Engineering and Technology, IRASET 2020*.

Helmy, Y., Khedr, A. E., Kolief, S., & Haggag, E. (2019). An Enhanced Business Intelligence Approach for Increasing Customer Satisfaction Using Mining Techniques. *International Journal of Computer Science and Information Security*, *17*(4).

Henke, J., Leissner, L., & Möhring, W. (2019). How can Journalists Promote News Credibility? Effects of Evidences on Trust and Credibility. *Journalism Practice*, *14*(3), 21.

Hossein Hassani, C. B., Hassani, H., Beneki, C., Unger, S., Mazinani, M. T., & Yeganegi, M. R. (2020). Text Mining in Big Data Analytics. *Big Data and Cognitive Computing*, *4*(1), 34.

Huang, J.-Y., & Liu, J.-H. (2019). Using social media mining technology to improve stock price forecast accuracy. *Journal of Forcasting*, 13.

Hung, J.-L., & Zhang, K. (2011). Examining mobile learning trends 2003–2008: A categorical meta-trend analysis using text mining techniques. *Journal of Computing in Higher Education*, *24*(1), 17.

Idrees, A. M., & Ibrahim, A. B. (2015). Enhancing information technology services for ebusiness-the road towards optimization. In *13th International Conference on ICT and Knowledge Engineering (ICT & Knowledge Engineering 2015)* (pp. 72-77). IEEE.

Idrees, A. M., & Ibrahim, M. H. (2018). A Proposed Framework Targeting the Enhancement of Students' Performance in Fayoum University. *International Journal of Scientific and Engineering Research*, *9*(11).

Idrees, A. M., Ibrahim, M. H., & El Seddawy, A. I. (2018). Applying spatial intelligence for decision support systems. *Future Computing and Informatics Journal, 3*, 384-390.

Islam, F., Alam, M. M., Shahadat Hossain, S. M., Motaleb, A., Yeasmin, S., Hasan, M., & Rahman, R. M. (2020). Bengali Fake News Detection. In *Proceedings of 2020 IEEE 10th International Conference on Intelligent Systems* (p. 7). Institute of Electrical and Electronics Engineer Inc.

Joo, S., Lu, K., & Lee, T. (2020). Analysis of content topics, user engagement and library factors in public library social media based on text mining. *Online Information Review*, *44*(1), 21.

Khedr, A., Kholeif, S., & Hessen, S. (2015, March). Adoption of cloud computing framework in higher education to enhance educational process. *International Journal of Innovative Research in Computer Science and Technology*, *3*(3), 150–156.

Khedr, A., Kholeif, S., & Hessen, S. (2015, April). Enhanced Cloud Computing Framework to Improve the Educational Process in Higher Education: A case study of Helwan University in Egypt. *International Journal of Computers and Technology*, *14*(6), 5814–5823.

Khedr, A. E. (2012). Towards Three Dimensional Analyses for Applying E-Learning Evaluation Model: The Case of E-Learning in Helwan University. *IJCSI International Journal of Computer Science Issues*, *9*(4), 161–166.

Khedr, A. E. (2013). Business Intelligence framework to support Chronic Liver Disease Treatment. *International Journal of Computers and Technology*, *4*(2), 307–312.

Khedr, A. E., & El Seddawy, A. I. (2015). A Proposed Data Mining Framework for Higher Education System. *International Journal of Computers and Applications*, *113*(7), 24–31.

Khedr, A. E., & Idrees, A. M. (2017). Adapting Load Balancing Techniques for Improving the Performance of e-Learning Educational Process. *Journal of Computers*, *12*(3), 250–257.

Khedr, A. E., & Kok, J. (2006). Adopting Knowledge Discovery in Databases for Customer Relationship Management in Egyptian Public Banks. *IFIP World Computer Congress, TC 12*, 201-208.

Kim, J.-C., & Chung, K. (2018). Associative Feature Information Extraction Using Text Mining from Health Big Data. *Wireless Personal Communications*, *105*(2), 17.

Kolluri, N. L., & Murthy, D. (2021). CoVerifi: A COVID-19 news verification system. *Online Social Networks and Media*, 13.

Küçük, D., & Can, F. (2020). Stance Detection: A Survey. *Association for Computing Machinery*, *53*(1), 37.

Kumar, G. R., Basha, S. R., & Rao, S. B. (2020, January). A summarization on text mining techniques for information extracting from applications and issues. *Journal of Mechanics of Continua and Mathematical Sciences, 15*(1), 9. doi:10.26782/jmcms.spl.5/2020.01.00026

Lee, S., Nah, S., Chung, D. S., & Kim, J. (2020). Predicting AI News Credibility: Communicative or Social Capital or Both? *Communication Studies, 71*(3), 21.

Liao, C.-H., Chen, L.-X., Yang, J.-C., & Yuan, S.-M. (2020). A Photo Post Recommendation System Based on Topic Model for Improving Facebook Fan Page Engagement. Symmetry, 17(7), 18.

Likhitha, S., Harish, B. S., & Keerthi Kumar, H. M. (2019). A Detailed Survey on Topic Modeling for Document and Short Text Data. *International Journal of Computers and Applications, 178*(39), 9.

Liu, J., Sarkar, M. K., & Chakraborty, G. (2013). Feature-based Sentiment Analysis on Android App Reviews Using SAS® Text Miner and SAS® Sentiment Analysis Studio. *SAS Global Forum, 1*(7), 8.

Liu, Y., Wang, Q., & Huang, Y. (2018, July). Research on expert opinion credibility rating. *International Journal of Innovative, 14*(6), 8.

Luo, M., Hancock, J. T., & Markowitz, D. M. (2020). Credibility Perceptions and News Headlines on Social Media: Effects of Truth-Biasand Endorsement Cues. *Communication Research*, 25.

Martin, J. D., & Hassan, F. (2020). News Media Credibility Ratings and Perceptions of Online Fake News Exposure in Five Countries. *Journalism Studies, 21*(16), 20. doi:10.1080/1461670X.2020.1827970

Mohammad Zoqi Sarwani, D. A., Sarwani, M. Z., Sani, D. A., & Fakhrin, F. C. (2019). Personality Classification through Social Media Using Probabilistic Neural Network Algorithms. *International Journal of Artificial Intelligence & Robotics, 1*(1), 7.

Mohsen, A. M., Hassan, H. A., & Idrees, A. M. (2016). A Proposed Approach for Emotion Lexicon Enrichment. *International Journal of Computer, Electrical, Automation, Control and Information Engineering, 10*(1).

Mohsen, A. M., Hassan, H. A., & Idrees, A. M. (2016). Documents Emotions Classification Model Based on TF IDF Weighting. *International Journal of Computer Electrical Automation Control and Information Engineering, 10*(1).

Mohsen, A. M., Idrees, A. M., & Hassan, H. A. (2019). Emotion Analysis for Opinion Mining From Text: A Comparative Study. *International Journal of e-Collaboration, 15*(1).

Mostafa, A., Khedr, A. E., & Abdo, A. (2017). Advising Approach to Enhance Students' Performance Level in Higher Education Environments. Journal of Computational Science, 13(5), 130–139.

Mouty, R., & Gazdar, A. (2019). The Effect of the Similarity Between the Two Names of Twitter Users on the Credibility of Their Publications. In *Joint 2019 8th International Conference on Informatics, Electronics & Vision (ICIEV) & 3rd International Conference on Imaging, Vision & Pattern Recognition (IVPR)*. IEEE.

Mukerji, N. (2018). What is fake news? *Ergo, 5*(35), 24.

Myneni, M. B., & Dandamudi, R. (2019). Harvesting railway passenger opinions on multi themes by using social graph clustering. *Journal of Rail Transport Planning & Management*, 10.

Namugera, F., Wesonga, R., & Jehopio, P. (2019). Text mining and determinants of sentiments: Twitter social media usage by traditional media houses in Uganda. *Computational Social Networks*, *6*(1), 21.

Nazier, M. M., Khedr, A. E., & Haggag, M. (2013). Business Intelligence and its role to enhance Corporate Performance Management. *International Journal of Management & Information Technology*, *3*(3).

Neo, L. R. (2021). Linking Perceived Political Network Homogeneity with Political Social Media Use via Perceived Social Media News Credibility. *Journal of Information Technology & Politics*.

Othman, M., Hassan, H., Moawad, R., & Idrees, A. M. (2016). Using NLP Approach for Opinion Types Classifier. *Journal of Computers*, *11*(5), 400–410.

Othman, M., Hassan, H., Moawad, R., & Idrees, A. M. (2018). A linguistic approach for opinionated documents summary. *Future Computing and Informatics Journal*, *3*(2), 152158.

Ozbay, F. A., & Alatas, B. (2019). Fake news detection within online social media using supervised artificial intelligence algorithms. *Journal Pre-proof*, 21.

Pasi, G., De Grandis, M., & Viviani, M. (2020). Decision Making over Multiple Criteria to Assess News Credibility in Microblogging Sites. In *IEEE Conference on Fuzzy Systems*. Institute of Electrical and Electronics Engineers Inc.

Pintar, D., Humski, L., & Vranić, D. M. (2019). Analysis of Facebook Interaction as Basis for Synthetic Expanded Social Graph Generation. *IEEE Access, 7*, 15.

Rameshbhai, C. J., & Paulose, J. (2019). Opinion mining on newspaper headlines using SVM and NLP. *Iranian Journal of Electrical and Computer Engineering*, *9*(3), 12.

Roychoudhury, B., & Srivastava, D. K. (2020). Words are important: A textual content based identity resolution scheme across multiple online social networks. *Knowledge-Based Systems*, 17.

Salloum, S. A., Al-Emran, M., Abdel Monem, A., & Shaalan, K. (2017). A Survey of Text Mining in Social Media: Facebook and Twitter Perspectives. *Advances in Science, Technology and Engineering Systems Journal*, *2*(1), 7.

Sayed, M., Salem, R., & Khedr, A. E. (2019). A Survey of Arabic Text Classification Approaches. *International Journal of Computer Applications in Technology*, *95*(3), 236251.

Stieglitz, S., & Dang-Xuan, L. (2012). Social media and political communication: A social media analytics framework. *Social Network Analysis and Mining*, *3*(4), 15.

Sultan, N., Khedr, A. E., Idrees, A. M., & Kholeif, S. (2017). Data Mining Approach for Detecting Key Performance Indicators. *Journal of Artificial Intelligence*, *10*(2), 59–65.

Tandoc, E. C. Jr. (2018). Tell Me Who Your Sources Are. *Journalism Practice*, *13*(2), 14.

Tarek Kanan, O. S.-d., Kanan, T., Sadaqa, O., Aldajeh, A., Alshwabka, H., Al-Dolime, W., . . . Alia, M. A. (2019). A Review of Natural Language Processing and Machine Learning Tools Used to Analyze Arabic Social Media. *2019 IEEE Jordan International Joint Conference on Electrical Engineering and Information Technology (JEEIT)*.

Viviani, M., & Pasi, G. (2017). Credibility in social media: Opinions, news, and health, information—a survey. *Wiley Interdisciplinary Reviews. Data Mining and Knowledge Discovery*, 25.

Waheeb Yaqub, O. K., Yaqub, W., Kakhidze, O., Brockman, M. L., Memon, N., & Patil, S. (2020Effects of Credibility Indicators on social Media News Sharing Intent. In *Conference on Human Factors in Computing Machinery-Proceedings* (p. 14). AMC.

Waruwu, B. K., Tandoc, E. C., Duffy, A., Kim, N., & Ling, R. (2020). Telling lies together? Sharing news as a form of social authentication. *New Media & Society*, 18.

Wesley, R., dos Santos, R. M., dos Santos, W. R., Ramos, R. M., & Paraboni, I. (2020). Computational personality recognition from Facebook text: Psycholinguistic features, words and facets. *New Review of Hypermedia and Multimedia*, *25*(4), 21.

Wu, J., Sun, H., & Tan, Y. (2013). Social media research: A review. *Journal of Systems Science and Systems Engineering*, *23*(3), 26.

Xia, L., Luo, D., Zhang, C., & Wu, Z. (2019). A Survey of Topic Models in Text Classification. In *2019 2nd International Conference on Artificial Intelligence and Big Data*. IEEE.

Xu, K., Wang, F., Wang, H., & Yang, B. (2020, February). Detecting Fake News Over Online Social Media via Domain. *Tsinghua Science and Technology*, *25*(1).

Yahav, I., Shehory, O., & Schwartz, D. (2015). Comments Mining With TF-IDF: The Inherent Bias and Its Removal. *International Transactions on Knowledge and Data Engineering, 14*(8), 14.

Zhang, X., & Ghorbani, A. A. (2019, March). An overview of online fake news: Characterization, detection, and discussion. *Information Processing & Management*, 26.

Zhou, M., Duan, N., Liu, S., & Shum, H.-Y. (2020). Progress in Neural NLP: Modeling, Learning, and Reasoning. Elsevier Ltd.

Chapter 18
Design and Develop a Decision-Making Assistance Model for Agriculture Product Price Prediction:
Deep Learning

Rajeev Kudari
Koneru Lakshmaiah Education Foundation, India

Aggala Naga Jyothi
https://orcid.org/0000-0001-8487-1126
Vignan's Institute of Information Technology, India

Abdul Mannan Mohd
Mahatma Gandhi University, India

Prabha Shreeraj Nair
S. B. Jain Institute of Technology Management and Research, India

ABSTRACT

Most of India's wealth and economy are derived from agriculture. Crop production price forecasting has always been a challenge for farmers. Climatological changes as well as other market variables have resulted in significant losses for farmers. Despite their best efforts, farmers are unable to sell their crops for the prices they want. A decision-assistance model for agricultural product price forecasting is being developed in this project. Farming decisions may be made using this method, which takes into consideration elements like yearly rainfall, WPI, and so on. A year's worth of forecasts are available from the technology. The system employs a machine learning regression approach known as decision tree regression.

DOI: 10.4018/978-1-7998-9640-1.ch018

INTRODUCTION

Agriculture and agricultural goods are the backbone of the Indian economy. For the vast majority of the people, agricultural production is their primary source of income. For agricultural purposes, 60 percent of the country's land is devoted. Our research aims to provide a practical solution to the issue of accurately predicting crop value, in order to provide farmers with predictable revenues (Nelson et al., 2002). A vast variety of agricultural goods may be found on the. There are several variables that affect agricultural product prices and even the same commodity might be priced differently in various marketplaces. At any moment, food production commodities prices might increase or fall, inflicting havoc on the economy. The Decision Tree Regression approach is used in this system to predict crop value from a validated dataset using data from the dataset.

Customers may have more possibilities thanks to an outstanding agricultural prices forecast model. Furthermore, the findings are given in the form of a net appliance that farmers may use with ease. Since farmers may plant their harvests based on future expenditures, they will profit from the work done here to predict the expenditures of horticulture items. In agriculture, there is a standard rate for everything, and these fees are spread out across the full year (Wallenius et al., 2008). In the event that these fees are made known to farmers in due course, this ensures a return on investment (ROI). This graph may be used by horticultural specialists to predict advertising revenues for agricultural producers.

Artificial intelligence (AI) technologies have helped shape global agricultural policy, and that's what this essay sets out to explore in detail. Several publications were uncovered after a search of the main scientific archives was conducted.. Agent-based models, cellular automata, and genetic algorithms are the most often utilized AI approaches, according with data. Another application for these models is to predict agriculture productivity and irrigation and land consumption (Keating et al., 2003). There's little doubt that artificial intelligence (AI) has a critical role to play in formulating agricultural policy.

Agricultural policy, artificial intelligence (AI), and decision-making make up the article's conceptual model, all of which are described specifically in relation to agriculture. For a greater understanding of public policy and its decision-making procedure, as well as how AI has been applied in decision-making procedures are first examined. Governments use many techniques and customs to deal with common issues, such as "the formulation of these policies is founded on societal interests," says. As a consequence, all public policy creation processes have three stages: formulation, implementation, and assessment. Public policy decision-making must be as aggressive as feasible in order to identify the best available implementation option at the formulation stage.

The use of AI technologies in the development of the agricultural sector public policies has shown to be highly advantageous, and their implementations are related in three different ways. When it comes to agricultural public policy, the first question is how certain rural areas will respond to it. An agent-based approach is the AI used to generate this prediction. The second use of artificial neural networks and support vector machines is to assist agricultural public authorities make much better judgments, demonstrating that they are the ideal tools for the task (Lee et al., 002). Using land and water administration as inputs, the third app determines different sorts of agriculture products and examines the results in terms of output units and ecological repercussions. There were cellular automata, fuzzy logic, and Bayesian networks involved. Genetic algorithms, neural networks, and support vector machines have all been found to enhance industrial processes. Public authorities benefit from using AI in agriculture because it helps them make better decisions based on a wide range of criteria and data points.

An agricultural policy model is developed based on the findings of this study will be developed to meet the specific needs of a particular area (Ye et al., 2021).. An agricultural public policy that is competent is intended to result in economic development, decreased disparities, food security, and a reduction in the negative environmental effect of the area in which it is executed.

LITERATURE REVIEW

Organizing Machine learning method for Crop acquiesce forecasting in farming zone

The Random Forests and Decision Tree techniques are used to anticipate the yearly agricultural acquiescence. It utilizes information from Kaggle to help in its learning. Faster results are possible since the approach is simple and does not need much computing power. With unstructured data, it would not work. Consequently, its applicability to diverse kinds of data is not homogeneous.

Crop acquiesce forecast throughout proximal sensing and machine learning techniques

IoT-based technologies are used to collect information from around New Brunswick and Prince Island. ML methods such as linear regression, SVM, elastic-net and k-nearest neighbours (KNN) are utilised once the data has been gathered. Its accuracy is superior than that of the other approaches. It was able to explain how climate, environment, and other external variables impact agriculture production. It was tested on a small sample size. Tests on a much bigger database could improve its reliability.

Expanding harvest cost prediction provision utilizing open information from markets of Taiwan. Autoregressive integrated moving average (ARIMA), partial least square (PLS), artificial neural network (ANN), and PLS coupled with reply exterior technique (RSMPLS) were applied and contrasted. RSMPLS may be used to create a non-linear link between previous expenses (Ye et al., 2021).. In addition, the urbanised services is connected to the intelligent agro administrative platforms, providing a barrier for obtaining previous pricing and projecting future prices. The algorithm has just one variable (prices). More factors, like weather, marketplace locations, and plantation area, will produce greater precise results.

Crops recommendation systems for farming based on demand and market conditions. Classifying and suggesting crops for farmers is done using a Decision Tree algorithms Only data from the past year was included in the dataset in order to improve the relevance of the findings. Text-to-speech translation was made easier using a graphical user interface based on NLP technologies. Neuro linguistic programming techniques in the GUI will be a huge asset for the farming community. The findings will be more accurate since the dataset is constantly updated with the most current year's information. There are too many variables to take into account while deciding on a crop. To get the most accurate results, additional data is needed.

Agricultural products expenses may be predicted using this strategy. Farmers have been helped by a decision-making assistance model in projecting expenses. Farmers will have to go into a portal with their credentials in order to access their accounts under this notion. Commodities names and previous selling prices must be entered by producers. Farmers may make better judgments and anticipate expenses by using this program's usual pricing by a specific yield, which is based on previous prices. By relying on previously recorded prices by customers, they're effectively taking an average; they're relying on inefficient data to forecast crop prices (Lambin E F et al., 2000). As a consequence, this model is not always going to provide you the most accurate and reliable data.

A well-structured investigation into how well the Hadoop platform predicts crop acceptance using a random forest technique. MapReduce is used in conjunction with the Random Forest Algorithm on the Hadoop platform. Effectiveness in dealing with enormous volumes of information will increase the application's scalability. A modular and platform-independent approach may not be possible for agricultural yield forecast difficulties.

METHODOLOGY

Programming languages: Python and flask are the two most crucial ones to have. System models may be defined using Python, a high-level computer language. If you want to make use of these libraries, you'll need to have at least Python 3.x and a working Python 3.5+ environments. Additional components are needed, including flask html-python. Flask is a Python-based web framework that may be used to create dynamic websites (Weaver et al., 2013). None of its own tools or libraries are required to use it.

The algorithm's output may now be seen by anybody with accessibility. We're using the following strategies to develop a website that lets visitors explore the land available, the acreage now being cultivated, the number of producers that use the property, and the many types of crops that have been grown. There are social networking links at the bottom of the page that farmers may use to connect with each other. To acquire answers to their queries, people may phone the helpline numbers or visit the website; there is also a mobile application that users can access. This is a method that makes it simpler for a farmer to make choices by letting him know what is happening now and what will happen eventually.

The present work focuses on supervised learning models, which are types of machine learning. The conceptual replica of the system is determined by the system design in separate structure and several views of the systems. The suggested framework may be used to estimate the prices of crops. This model represents a vast amount of data being gathered and precompiled to eliminate unwanted data such as NULL and absurd values. During the pre - processing stage, we separated the dataset is divided into two sets: one for learning and one for assessment. Training the database to detect the agricultural prices that is included in the database using proper supervised learning techniques. Find crop prices using machine learning algorithms for each new piece of data that has been added to the dataset. The model's efficacy and capacities are determined after data collection using an appropriate machine learning methodology; we employed a number of machine learning techniques, including random forest, polynomial regression and decision trees (Damos, 2015). Accuracy and precision will be determined for the proposed model. This system design incorporates data flow, machine learning techniques, and modules for identifying yield pricing and feature assortments.

Figure 1. Proposed system architecture

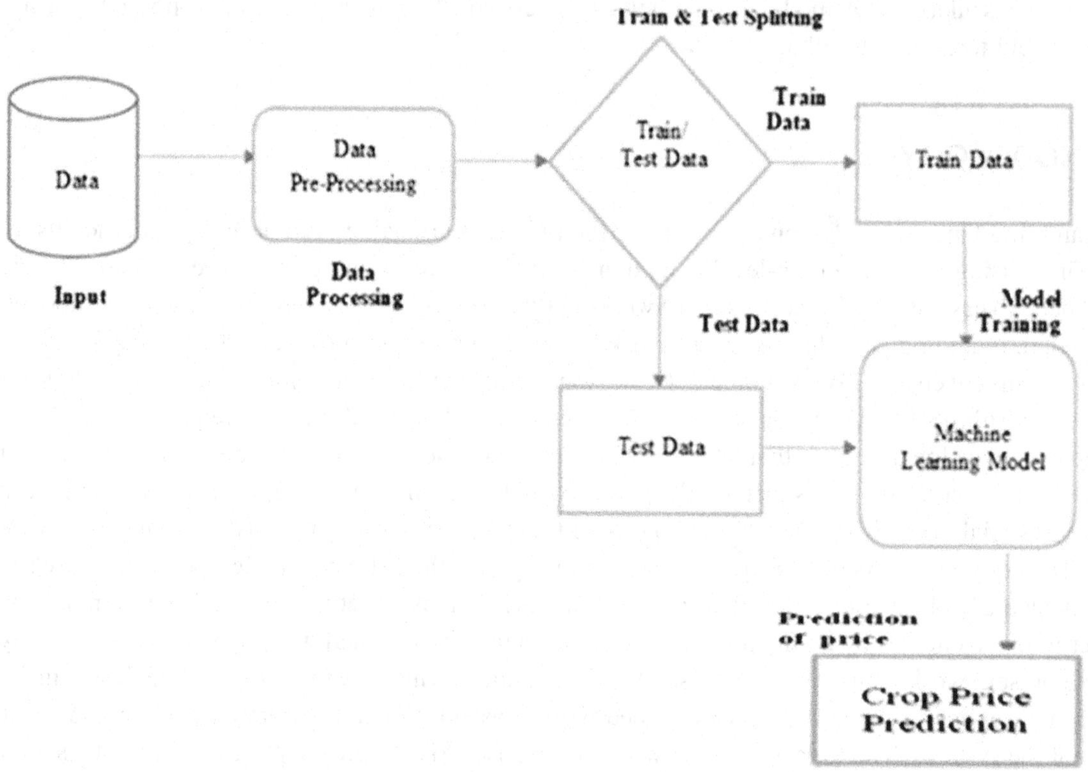

Following modules are an important element of our study shown in Fig 1.

1. Information collection
2. Investigation of data
3. Forecast by means of ML
4. Uses internet-based software

Information collection

For the database, agricultural information was acquired from an open source and utilised. Many databases are available that may be used to collect information. We were able to get rainfall data relevant to each crop.

Investigation of data

It is a sort of exploratory data analysis known as Data Investigation. Data analysis (EDA) is a crucial step that must be accomplished prior to any demonstration when a component collects and analyses data. In order to prevent arousing suspicions, an associate in nursing data study must practically grasp the notion of the data (Han et al., 2017).. Understanding the data's structure, appropriation of qualities, and therefore the distance to special features and linkages within the informational indexes are all made easier with the help of knowledge investigation.

Forecast by Means of ML

The Decision Tree is one of the most extensively used and effective methods for administering learning in classroom. Both relapsing and categorization concerns may be addressed using it, although the latter is the more common use. These tree-structured classifiers have three types of hubs. In the graph, the Root Node is the most important node, since it represents the whole test. It's possible to divide it into nodes with the help of partitioning (Erdem & Keane 1996). There is a distinct difference between a collection's highlights and its choice rules when it comes to the hubs and branches. The Leaf Nodes finally have their say in the matter. For resolving decision-making issues, this computation is quite useful.

Figure 2. Decision tree model

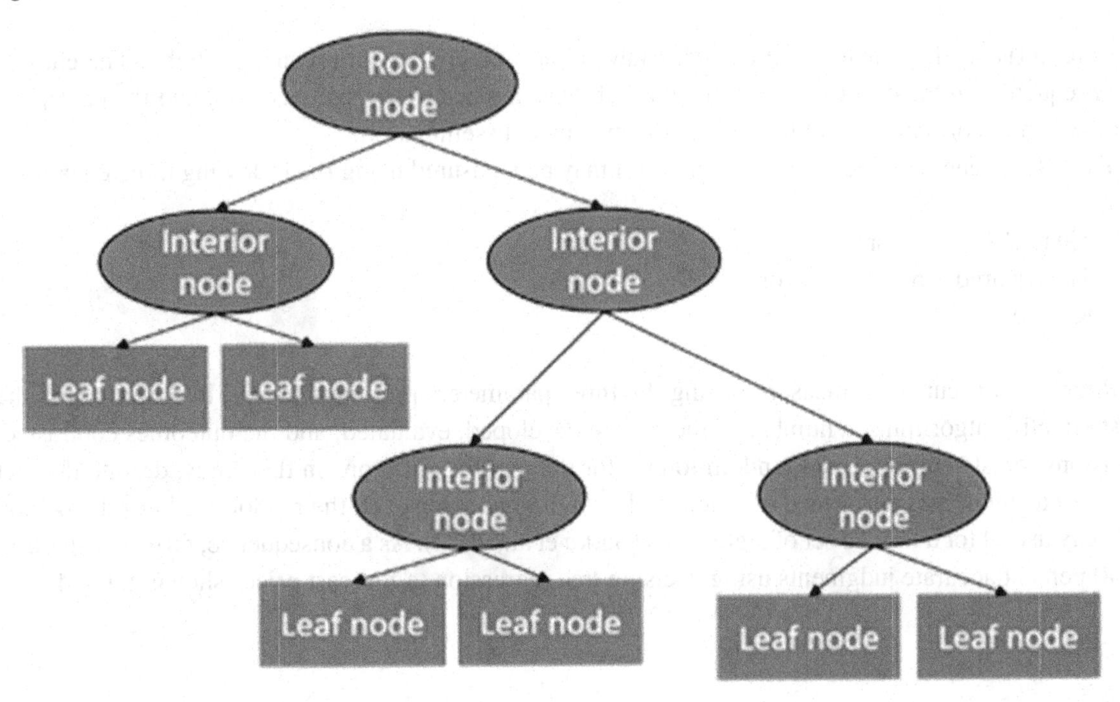

Until it reaches the leaf node, a particular data point negotiates with the whole tree by answering True/False. The final projection is based on an average of the dependant variable's value across each leaf node. After a few rounds, the Tree is able to come up with an accurate value for the data point shown in Fig 2

Implementation Steps for the Algorithm

Step 1: - Create a database including information for the wholesale cost indexes and the rainwater.
Step 2:- The independent variable "x" is represented by all of the columns and rows in database 1.

Step 3:- All rows and columns 2 of database "y," which is the required parameter, should be selected.
Step 4:- decision trees with a high degree of robustness
Step 5:- Be able to predict what the new value will be.
Step 6:- Visualize the final product and verify its correctness before moving on.

Results

It is possible to use the decision tree as an efficient and appropriate complete regression method, even on small datasets. Criteria are used to build the tree (Gonçalves et al., 2022). Consideration is given to one of these characteristics, which is known as "entropy." Classification uses entropy as the primary criterion for evaluating impurities. It comes from:

$$E(t) = -\sum_{i=1}^{c} (i|t) \log_2 p(i|t)$$

Here, p (ilt) is the extent of the tests that have a place to course c for a specific hub t. The entropy is subsequently in the event that all tests at a hub have a place to the same course, and the entropy is maximal on the off chance that we have uniform course dissemination.

The effectiveness of decision tree regression may be measured using the following three equations:

1. Mean absolute error
2. The squared error of the error
3. R2 Score

Regression accuracy is measured using the three parameters presented above. In order to find the most effective algorithms, a number of them were developed, evaluated, and the outcomes contrasted. Decisions forests outperformed random forests for the majority of crops in this study, despite the fact that both technologies performed excellently shown in Fig 3. However, the randomized forests method was only useful for a limited set of algorithms (Sarku et al., 2022). As a consequence, farmers can make 95.40 percent accurate judgments using decision tree regression to forecast prices shown in Fig 4.

Figure 3. Website home page includes land area and crops information

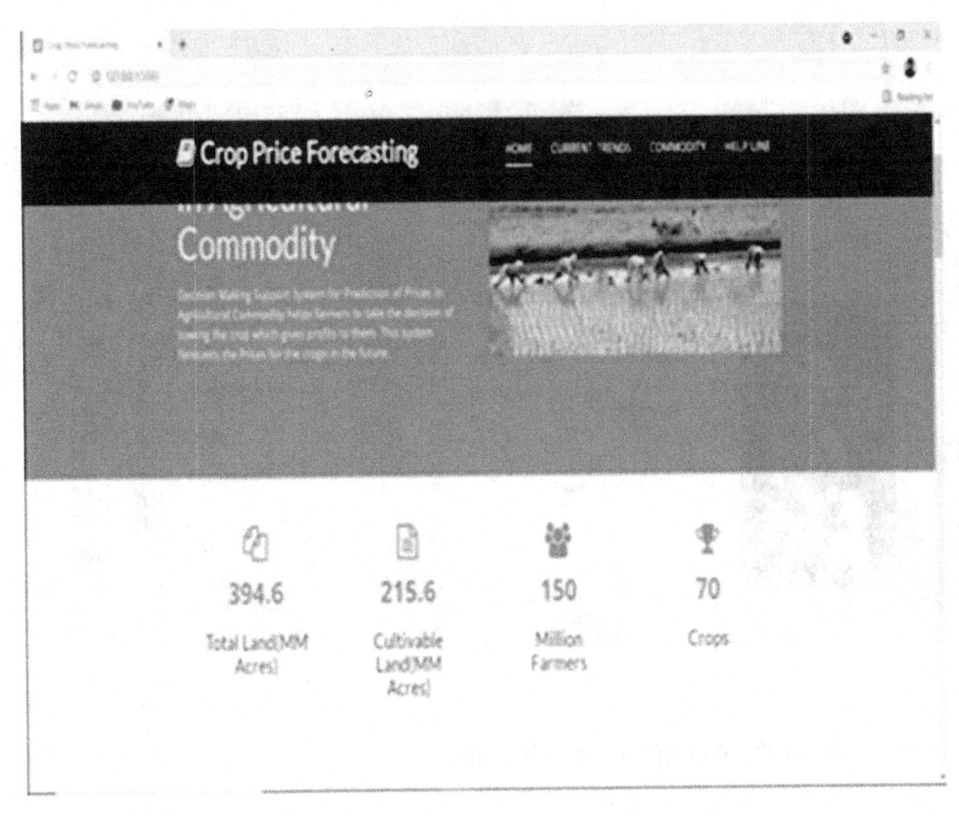

Figure 4. Website displays long term and short term crops for the sake of users

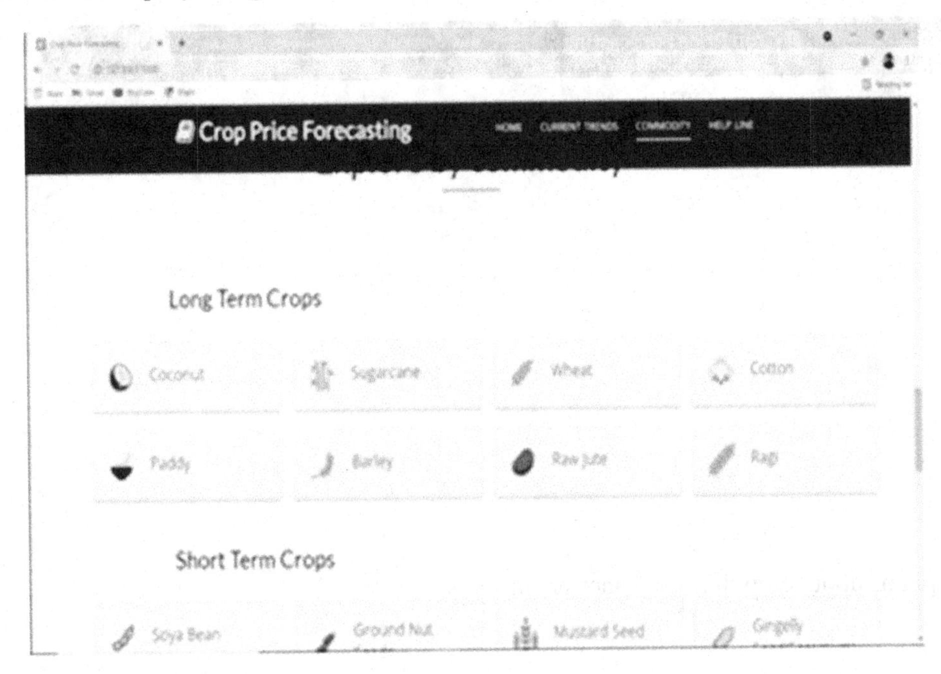

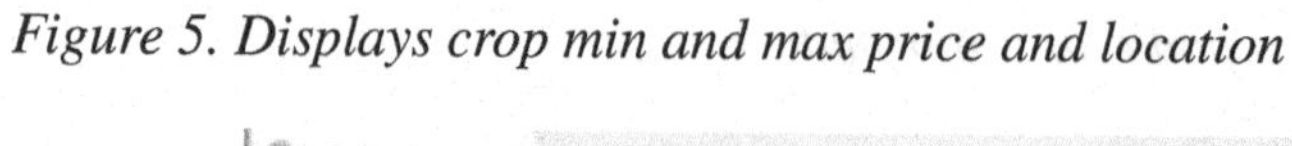

Figure 5. Displays crop min and max price and location

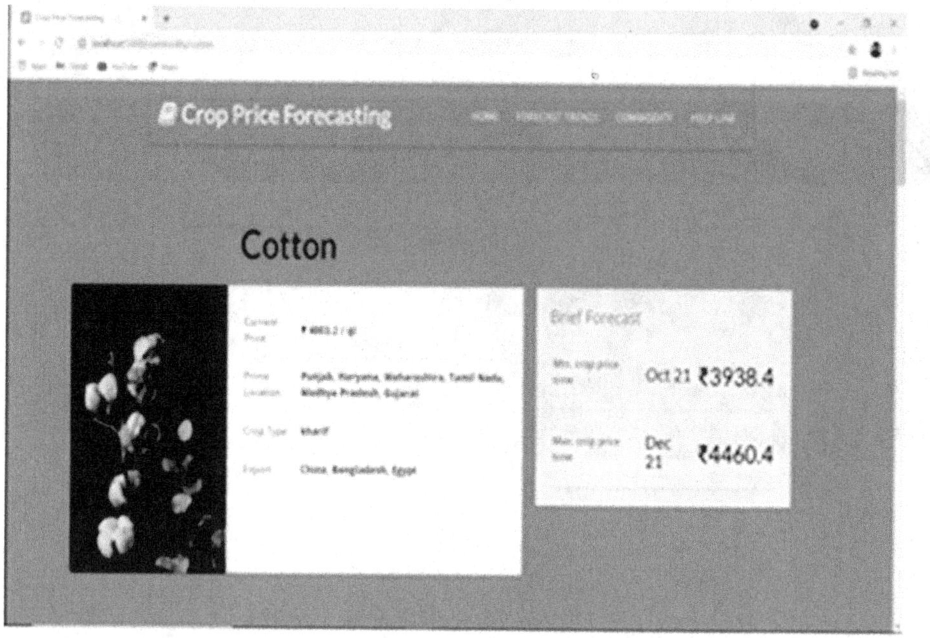

The Fig 5 clearly explains about crop price and location

Figure 6. The crop's predicted tendencies, such as its price over the next year and graphs depicting pricing statistics for the current and the future year

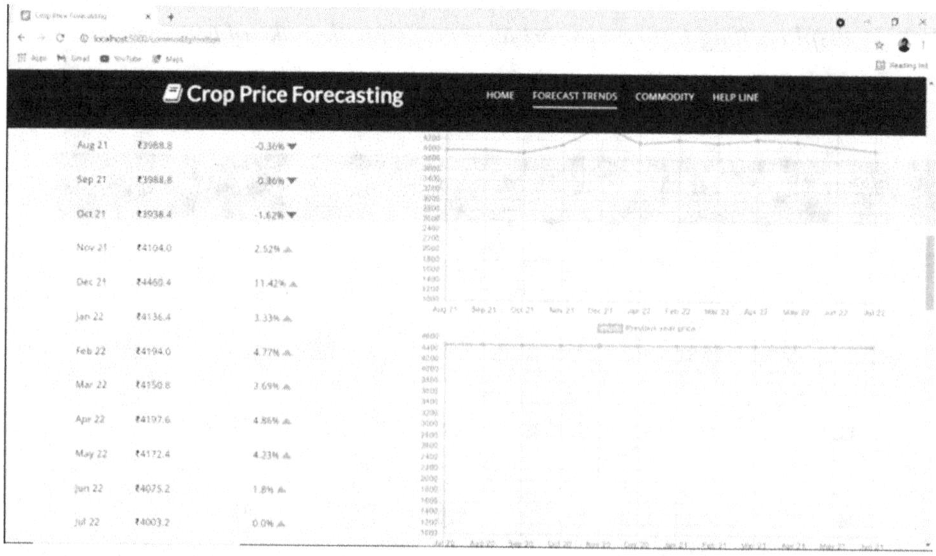

Crop Price Forecasting

HOME FORECAST TRENDS COMMODITY HELP LINE

Aug 21	₹3988.8	-0.36% ▼
Sep 21	₹3988.8	-0.36% ▼
Oct 21	₹3938.4	-1.62% ▼
Nov 21	₹4104.0	2.52% ▲
Dec 21	₹4460.4	11.42% ▲
Jan 22	₹4136.4	3.33% ▲
Feb 22	₹4194.0	4.77% ▲
Mar 22	₹4150.8	3.69% ▲
Apr 22	₹4197.6	4.86% ▲
May 22	₹4172.4	4.23% ▲
Jun 22	₹4075.2	1.8% ▲
Jul 22	₹4003.2	0.0% ▲

The Fig 6 explains about crop disease detaction anaylsis

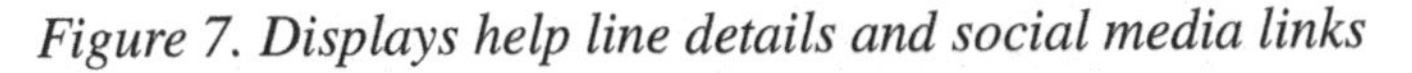

Figure 7. Displays help line details and social media links

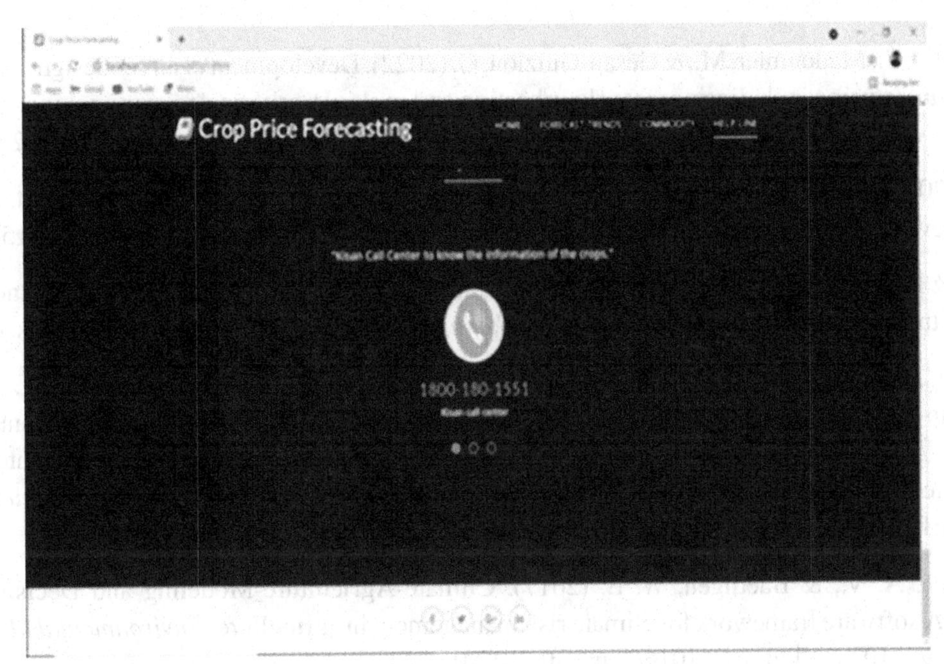

The models don't really ask for any user intervention. The most profitable and least profitable crops are mentioned, and the recent trends are shown. In addition, it provides information on the most significant commodity crops in the country (Azzaro-Pantel et al., 2022). It's possible to look at the crops in terms of how long they'll last, and there's a list of both. After selecting the crop, the farmer may next browse to the commodity, where he or she will be able to read data such as a product's maximum or minimum price per quintal and its site of growth as well as its export destinations. They may see the 12 months forecast next to the graphs that display the current year and the future year's data shown in Fig 7.

CONCLUSION

Using this approach, producers will be able to make better choices about what crops to plant, when to plant them, and even where to put them! Prior to the start of the growing process, our technologies can tell farmers exactly how much their harvest will cost. It is possible to use a number of algorithms for price prediction, such as decision trees and neural networks, and SVM. A decision tree is a crucial part of our strategy. For example, it has been trained on a wide range of kharif and ragi crops (such as rice and barley) and has an extremely high degree of accuracy in its predictions. Using a web-based tool, the researchers predict crop prices and forecasts using efficient machine learning approaches and user-friendly interfaces. Predicting the market price and demand based on the training datasets is possible. This is done by determining the root mean square error (RMSE) for each approach, and then selecting the most accurate system. With this strategy, it is possible for farmers to make better bids for their commodities in order to get a higher price. Farmers benefit from this strategy since it lessens their difficulties and helps them avoid taking their own lives.

REFERENCES

Azzaro-Pantel, C., Madoumier, M., & Gésan-Guiziou, G. (2022). Development of an ecodesign framework for food manufacturing including process flowsheeting and multiple-criteria decision-making: Application to milk evaporation. *Food and Bioproducts Processing*, *131*, 40–59. doi:10.1016/j.fbp.2021.10.003

Damos, P. (2015). Modular structure of web-based decision support systems for integrated pest management. A review. *Agronomy for Sustainable Development*, *35*(4), 1347–1372. doi:10.100713593-015-0319-9

Erdem, T., & Keane, M. P. (1996). Decision-making under uncertainty: Capturing dynamic brand choice processes in turbulent consumer goods markets. *Marketing Science*, *15*(1), 1–20. doi:10.1287/mksc.15.1.1

Gonçalves, C., Honrado, J. P., Cerejeira, J., Sousa, R., Fernandes, P. M., Vaz, A. S., Alves, M., Araújo, M., Carvalho-Santos, C., Fonseca, A., Fraga, H., Gonçalves, J. F., Lomba, A., Pinto, E., Vicente, J. R., & Santos, J. A. (2022). On the development of a regional climate change adaptation plan: Integrating model-assisted projections and stakeholders' perceptions. *The Science of the Total Environment*, *805*, 150320. doi:10.1016/j.scitotenv.2021.150320 PMID:34543791

Han, E., Ines, A. V., & Baethgen, W. E. (2017). Climate-Agriculture-Modeling and Decision Tool (CAMDT): A software framework for climate risk management in agriculture. *Environmental Modelling & Software*, *95*, 102–114. doi:10.1016/j.envsoft.2017.06.024

Keating, B. A., Carberry, P. S., Hammer, G. L., Probert, M. E., Robertson, M. J., Holzworth, D., Huth, N. I., Hargreaves, J. N. G., Meinke, H., Hochman, Z., McLean, G., Verburg, K., Snow, V., Dimes, J. P., Silburn, M., Wang, E., Brown, S., Bristow, K. L., Asseng, S., ... Smith, C. J. (2003). An overview of APSIM, a model designed for farming systems simulation. *European Journal of Agronomy*, *18*(3-4), 267–288. doi:10.1016/S1161-0301(02)00108-9

Lambin, E. F., Rounsevell, M. D., & Geist, H. J. (2000). Are agricultural land-use models able to predict changes in land-use intensity? *Agriculture, Ecosystems & Environment*, *82*(1-3), 321–331. doi:10.1016/S0167-8809(00)00235-8

Lee, B. S., Alexander, M. E., Hawkes, B. C., Lynham, T. J., Stocks, B. J., & Englefield, P. (2002). Information systems in support of wildland fire management decision making in Canada. *Computers and Electronics in Agriculture*, *37*(1-3), 185–198. doi:10.1016/S0168-1699(02)00120-5

Nelson, R. A., Holzworth, D. P., Hammer, G. L., & Hayman, P. T. (2002). Infusing the use of seasonal climate forecasting into crop management practice in North East Australia using discussion support software. *Agricultural Systems*, *74*(3), 393–414. doi:10.1016/S0308-521X(02)00047-1

Sarku, R., Van Slobbe, E., Termeer, K., Kranjac-Berisavljevic, G., & Dewulf, A. (2022). Usability of weather information services for decision-making in farming: Evidence from the Ada East District, Ghana. *Climate Services*, *25*, 100275. doi:10.1016/j.cliser.2021.100275

Wallenius, J., Dyer, J. S., Fishburn, P. C., Steuer, R. E., Zionts, S., & Deb, K. (2008). Multiple criteria decision making, multiattribute utility theory: Recent accomplishments and what lies ahead. *Management Science*, *54*(7), 1336–1349. doi:10.1287/mnsc.1070.0838

Weaver, C. P., Lempert, R. J., Brown, C., Hall, J. A., Revell, D., & Sarewitz, D. (2013). Improving the contribution of climate model information to decision making: The value and demands of robust decision frameworks. *Wiley Interdisciplinary Reviews: Climate Change*, *4*(1), 39–60. doi:10.1002/wcc.202

Ye, K., Piao, Y., Zhao, K., & Cui, X. (2021). A Heterogeneous Graph Enhanced LSTM Network for Hog Price Prediction Using Online Discussion. *Agriculture*, *11*(4), 359. doi:10.3390/agriculture11040359

Chapter 19
Prediction and Prevention of Malicious URL Using ML and LR Techniques for Network Security:
Machine Learning

S. Mythreya
Koneru Lakshmaiah Education Foundation, India

A. Sampath Dakshina Murthy
https://orcid.org/0000-0002-9960-6373
Vignan's Institute of Information Technology, India

K. Saikumar
https://orcid.org/0000-0001-9836-3683
Koneru Lakshmaiah Education Foundation, India

V. Rajesh
Koneru Lakshmaiah Education Foundation, India

ABSTRACT

Understandable URLs are utilized to recognize billions of websites hosted over the present-day internet. Opposition who tries to get illegal admittance to the classified data may use malicious URLs and present them as URLs to users. Such URLs that act as an entry for the unrequested actions are known as malicious URLs. These wicked URLs can cause unethical behavior like theft of confidential and classified data. By using machine learning algorithm SVM, we can detect the malicious URLs. One of the essential features is to permit the benevolent URLs that are demanded by the customer and avoid the malicious URLs. Blacklisting is one of the basic and trivial mechanisms in detecting malicious URLs.

DOI: 10.4018/978-1-7998-9640-1.ch019

INTRODUCTION

A Uniform Resource Locator (URL) naturally termed as web address, reference to a web resource that specify its position on a computer network. URL has exact composition and arrangement. Assailant often alters one or added elements of the structure of the URL to betray used for scattering their malicious URLs. Malicious URLs are the links influence the customers by clicking on an infected URL, users can download ransom are, virus or any type of malware that will compromise user's machine or even their network (Odeh et al., 2021). A malicious URL is just a link that, when clicked, takes the user to a harmful website or page on the internet. As the name says, a malicious URL can do nothing but harm. To accomplish malicious objectives like as advancing a political agenda, stealing confidential information about individuals or businesses, or just making a fast money is the usual motivation for rogue web pages. Fake and authentic websites may both include dangerous links, which should be taken into consideration. A cybercriminal scan may either produce a completely phony and harmful website, or it can create dangerous URLs for legal websites. Drive-by-downloads, phishing, and other forms of social engineering and spam are just a few of the ways malicious URLs are spread. According to statistics, attackers using spreading malicious URLs ranked first among 10 attack techniques (Wanda & Jie 2019). The three major URL scattering methods are malicious URLs, both net URLs and phishing. From the enlargement in the number of malicious URL distributions over the successive days, it is apparent that there is necessitate learning and applying methods to identify and avoid the malicious URLs. To identify and avoid malicious URLs we used machine learning algorithm Logistic Regression. By using this algorithm, it is Easy to detect malicious URL and get rid of malware activities.

Phishing is an online social designing assault that intends to take an individual's computerized personality by imitating a trustworthy substance. The aggressor sends an assault vector, which can be an email, a visit meeting, a blog entry, or whatever else, that contains a connection (URL) to a malevolent site that is utilized to inspire individual data from the people in question. We are especially keen on fostering a framework for URL examination and grouping to forestall phishing attacks. Rather than getting to the site and acquiring highlights from it, URL investigation is interesting to keep the distance between the aggressor and the person in question. It's additionally quicker than the Internet look as far as recovering substance from the objective site and organization level properties, which were used in earlier examinations (Fang et al., 2021). We investigate an assortment of parts of URL examination, remembering execution examination for both adjusted and uneven datasets in both a static and live exploratory setting, just as online versus clump learning. Because of the consistently changing nature of attacks and the design of the present Web pages, detecting malicious Web pages has become a critical responsibility. Attackers use a variety of attack construction tactics. As a result, feature selection and dataset preparation are crucial for detecting fraudulent Web sites. While existing technologies offer a potential answer for detecting fraudulent Web pages, there are still gaps in the detection process. We have introduced a static examination of URL strings for successful recognizable proof of pernicious Web pages in this work. Just the static parts of Web page URLs have been thought of. From benign and malicious URL benchmarks, we collected 79 static properties of URLs and domain names. On our dataset, we tested Support Vector Machine (SVM), AdaBoost, J48, Random Forest (RF), Random Tree (RT), Naive Bayes (NB), Logistic Regression (LR), SGD, & BayesNet batch learning algorithms. For all of the classification models, our experimental research indicates encouraging detection results, with a detection rate of 95 percent to 99 percent and a very low false positive rate (FPR) and false negative rate (FNR).

RELATED WORKS

Studies related to detecting malicious URL was investigated and applied a long ago. And these studies use known / benign urls. A database query will be executed whenever a new URL is accessed. If the accessed URL is blacklisted, then it will be treated as Malicious URL and also a warming will be generated when the URL is found as Malicious or else it will be considered safe when a URL is not blacklisted. And the major drawback of this blacklisting method will be hard to find a novel URL that is not in the specified list (Wang & Wang 2015). The main types of ML (machine learning) algorithms are utilized to identify malicious URLs based on the URLs behavior are SVM. LR, NB, Decision Trees (DT), Online learning etc. In this particular paper Logistic Regression is used. The accuracy will be presented in experimental results. The URLs behavior can be of two types i.e., static and dynamic. The authors presented analyzing and extracting methods on static behavior URLs. Logistic regression can be used to detect the malicious URLs which are static. This machine learning algorithm will predict the malicious URLs and gives warning to user.

This section provides a summary of relevant work in the identification of malicious websites. Recent years have seen an increase in the usage of data mining tools to identify malicious URLs.

A lightweight approach for detecting malicious webpages that combines static analysis & simulation with supervised learning techniques. BINSPECT's experimental evaluation yielded accuracy of over 97 percent and low false signals.

Investigated machine learning classifiers' statistical methods for detecting malicious URLs based on lexical and host-based aspects of URLs. Classifiers achieved 95-99 percent accuracy in their experiments, according to the researchers. Although this research achieves excellent detection accuracy, extracting host-based information takes time, in real-time systems, this might result in a delay.

a low-overhead approach for identifying and preventing JavaScript malware, was proposed by Curtsinger et al. ZOZZLE employs a Bayesian categorization of a hierarchical feature of the JavaScript abstract syntax tree in order to anticipate malware (Basit et al., (2021). In testing 1.2 million benign JavaScript samples, ZOZZLE found a low false-positive rate of 0.0003 percent.

The Ghost in the Browser (2007)

In introduced a summary of the present state of the malware on the web. It is revealed that there are several strategies that are turning web pages into malware infection sectors (Urooj et al., 2022). The major identified features of organizers that are accountable for facilitating browser utilization: third party widgets, advertising, web customer security, and user-contributed substance. Through the analysis, this paper shows how the categories are used to exploit web browsers.

All Your I Frames Points to Us (2008)

The fact that malicious URLs are spreading so far and raising concerns regarding browser's safety by starting drive by downloads. Mavromattes attempted to fill the gaps by giving a comprehensive look from several viewpoints (Al-Sarem et al., 2022). This study uses large data to continuously detect and monitor the websites behavior that are leading to drive by download. In analyzing over 66 million URLs reveals the extent of the problem.

A Virtual Client Honeypot (2009)

In published a paper that identifies malicious attacks by implementing a virtual client Honeypot.

The Nocebo Effect on the Web: An Analysis of Fake Antivirus Distribution (2010)

Proposed a paper in the year 2018 that identifies malware attacks by analyzing and implementing FV Distribution.

Recognition of Drive-by Download Attacks, as well as Malicious JavaScript Code (2010)

In proposed a paper that identifies malware attacks and vulnerable java code.

METHODOLOGIES

Machine-learning technique

Methodology in Machine Learning — this technique is split into two parts: the primary one is the Machine Learning representation, and the subsequent is the Datasets as shown in fig 1 (Abu Al-Haija & Al-Badawi 2022).

Figure 1. Classification of machine learning algorithms

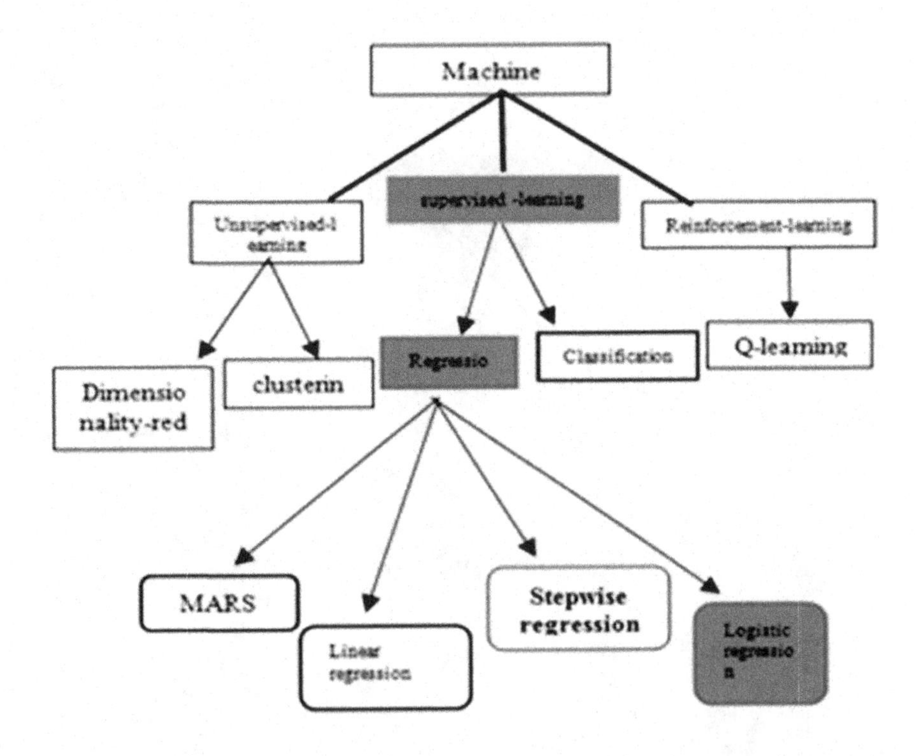

Machine learning model

Machine learning may be a department of artificial intelligence (AI) that permits computers to memorize and make strides without having to be unequivocally modified (Disha & Waheed 2022). Machine learning centers on making computer calculations that can get to information and learn on their ow.

The erudition procedure found by data like direct consideration, or instruction, in order to seek for patterns in the data and draw enhanced inferences about the feature based on the examples provided. The basic goal is to allow computers to learn on their own, with no human intervention or aid, and to control behaviors as a result. Learning under supervision.

In the context of AI and ML, supervised learning is a type of framework that provides mutually input and preferred output data. To provide a learning framework for prospect data processing, input and output data are tagged for categorization (Selim et al., 2021). Although supervised learning models provide a number of advantages over unsupervised learning, they do have some disadvantages.

Logistic regression

The dependent variable in the computed regression approach may be stated as a binary value (0 or 1, true or false, yes or no), meaning that the result might take one of two forms: positive or negative. For example, it can be used when determining the chance of a positive or negative event (Basit et al., 2021). The same method is utilized, however the value of Y changes between 0 and 1. An example of this would be an equation with two variables; each denoted by the letters x1 & x2 (taking value or 1) as shown in fig 2.

Figure 2. Logistic function

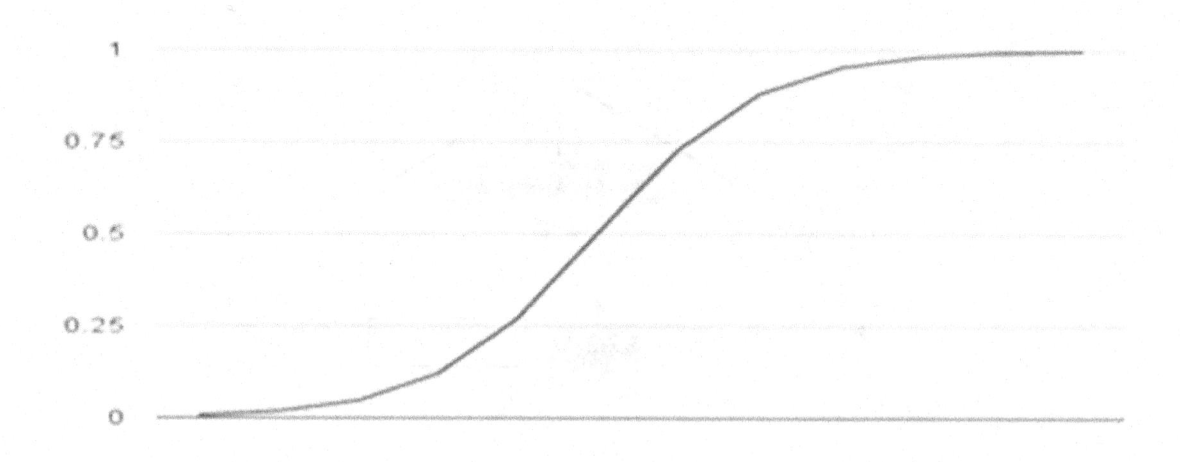

Comparison of Liner and Logistic Regression

The most essential types of regression that are commonly employed are linear and logistic regression. The key distinction among the two is that when the dependent variable is binary, calculated relapse is utilized. Linear regression, on the other hand, is utilized when the dependent variable is persistent, and the relapse line is linear in form

Regression is a method for predicting the worth of a response (needy) variable using one or more forecaster variables, with the variable being numeric (Yeboah-Ofori & Boachie 2019). Linear, multiple, logistic, polynomial, non-parametric, and more types of regression exist.

Implementation

In Machine Learning, the training data set is the actual dataset that is worn to train the representation to execute different tasks. This is the data that the present growth process models learn from a variety of APIs and algorithms in order to teach the machine to work autonomously as shown in fig 3.

Figure 3. Dataset categorization

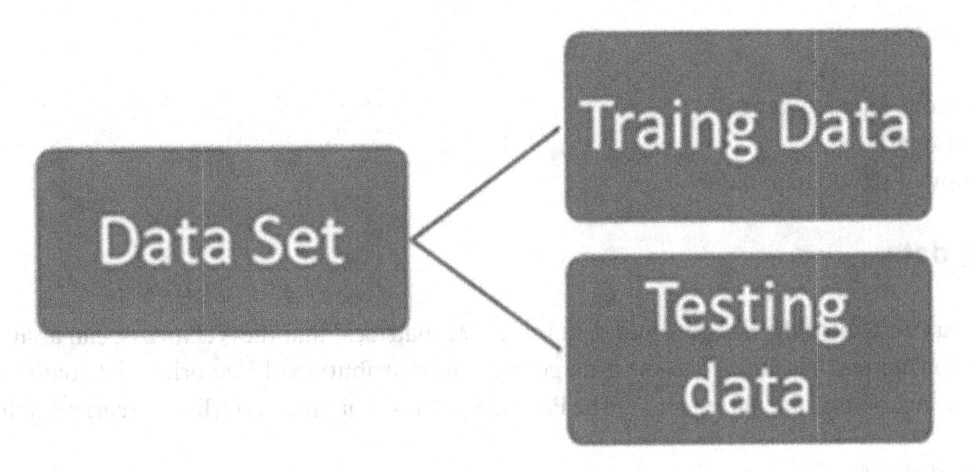

Training dataset

The following are two different kinds of data sets that are used at distinct phases of the development process: Training and Test. The training dataset is the more important of the two, even as the test dataset serves as a seal of endorsement that you don't need to utilize until the end of the project.

Testing dataset

Present data that is often utilized to provide a balanced evaluation of completed finals that fit on the training dataset. Fundamentally certain data is utilized to determine whether or not the model is reacting or performing effectively.

Preparing data

As a result, the URLs differ from our ordinary text documents, and we must devise our possess sanitization process to extract the necessary data from raw URLs. Use the code supplied in the experimental code to build our purification purpose in Python to filter URLs. Here it provides us with the URL data-set values we need to progression and test the representation (Hasan et al., 2019). The data set is divided into two pillars: one for URLs and the added for brand. We used the Tf-idf machine learning text feature removal method as of the sk-learn python package in this case.
Features measured

1. Blacklist doubt
2. Lexical Features

Blacklist

1. Google crawlers and yahoo phish tank have compiled a list of known fraudulent sites.
2. SORBS, URIBL, & SURBL alicious URLs from a number of domain contributors.

Lexical Features

1. Hostname + path tokens
2. URL Length
3. Entropy of the domain name

Reading data

It is necessary to recited the datasets into data frames & matrices, that the Vectorizer can assume. The text extraction approach is defined as the arrangement and distribution of Vectorizer data onto the term-frequency & inverse document frequency. The Pandas Python component is utilised to carry out the work.

Splitting data

We usually divide the data we use into two categories: training & test datasets. The training set contains data that represents a recognized outcome, and the representation learns from it so that it can be universal to new data (Wanda & Jie 2019). In order to assess how well our model predictions perform on this subset, we utilise the test dataset (or subset). To utilize the divide technique, we must first introduce the panda's library.

training set—a subset to train a model.(80%)
test set—a subset to test the trained model.(20%)

Training representation

To train the representation, use the logistic method from the python sci-kit package, which is imported using sklearn model. (Import Logistic Regression from sk learn linear model.) It learns from a train data set. It prints the trained model's score after learning.

Testing model

Input the multiple URLs into the taught representation. It determines whether a URL is good or evil and returns a positive or negative answer as shown in fig 4.

Figure 4. Comparison of logistic regression

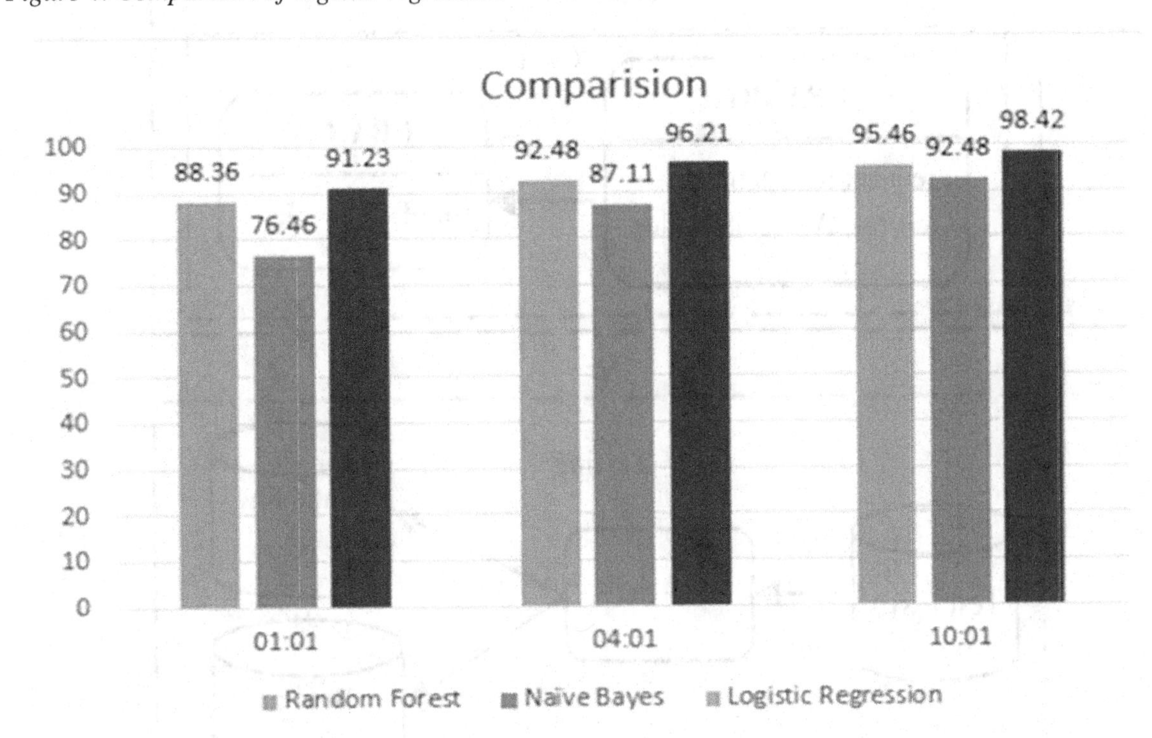

PROPOSED METHODOLOGY

The goal of our suggested method is to examine a URL and identify it as harmful or benign. Fig 5 depicts the architecture of our method. To train and predict the CBA model, we first collect URL and content-based (HTML and JavaScript) information from URLs. Malicious behaviour on a website may be detected with the use of the webpage's attributes (Chiramdasu et al., 2021). Phishing, malware, and drive-by download websites are all taken into consideration in our approach. In the next sections, we'll go through each of our system's modules.

Figure 5. The proposed strategy's architecture

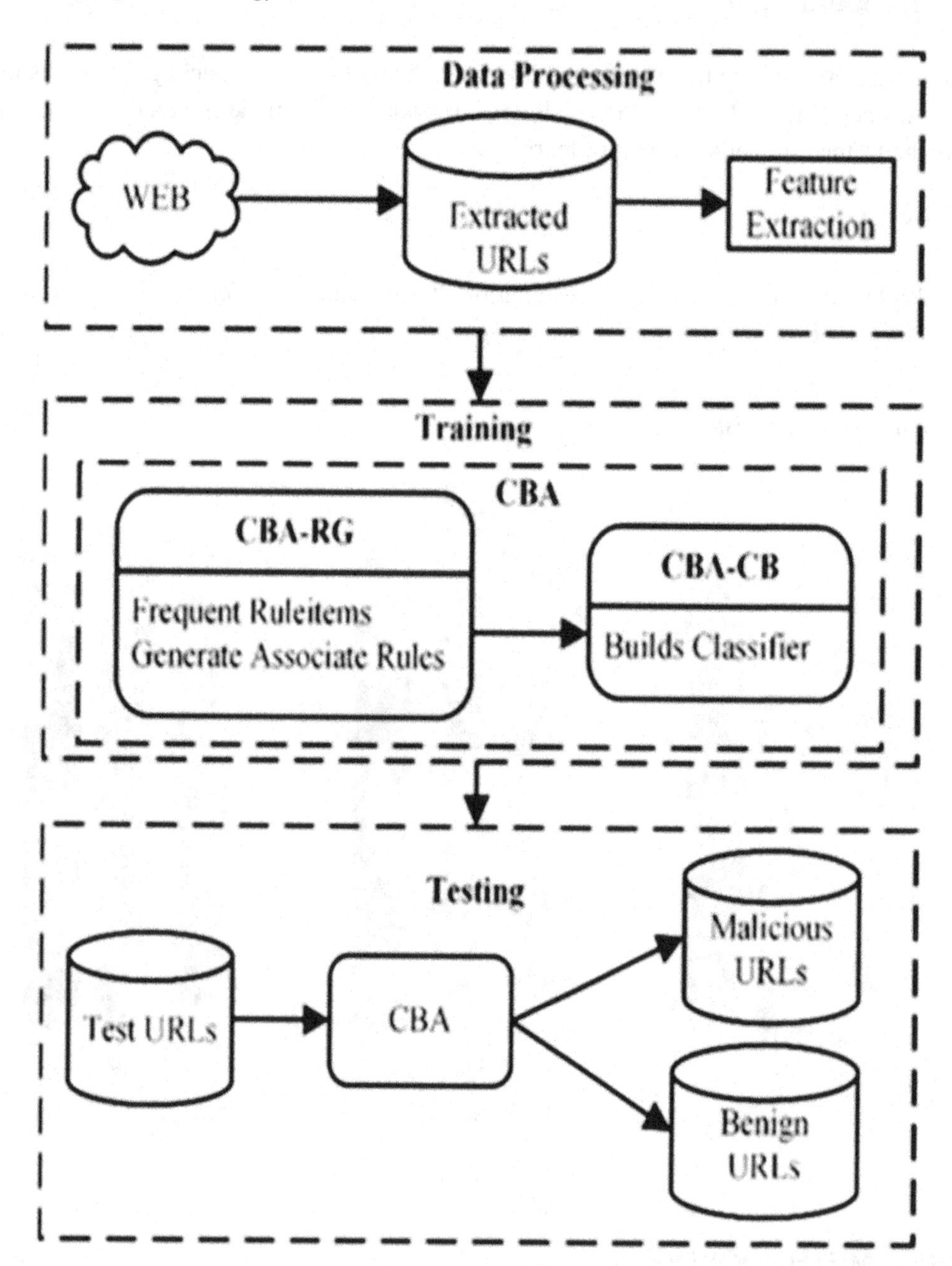

URL features

There are two types of URL-based features: lexical & host-based characteristics. To distinguish among malicious too benign URLs, we extract lexical features (textual attributes) from them. Because the extraction of host-based characteristics such as IP address attributes, WHOIS properties, & geographic properties is time-consuming, we do not include them. The following are the features of URLs:

The higher the entropy of a domain name, the greater the randomness of its URLs' contents, as measured by the unpredictability factor. Entropy is utilised to come up with a random domain name. It is useless to block URLs that utilise domain generation algorithms (DGA) to update their domains often. To create new domains for malware, DGA is a programme that may be used on demand or on the move (Yeboah-Ofori & Boachie 2019). URLs with a high degree of entropy are a sure indicator that malicious activity is taking place. Valid URL entropies may be used to set thresholds for detecting counterfeit URLs. The Shannon entropy formula is used to calculate the entropy of domain names: where H(x) is the Shannon entropy of string x, b is the base of the logarithm applied, and p(x) is the probability mass function (Hasan et.al., 2019).

$$H(x) = -\sum_{i=0}^{n} p(x_i) \log_b p(x_i) \quad (1)$$

In order to find CARs that meet user-defined minimum support (minsup) and confidence (minconf) levels, the CBA-RG employs an a priori technique. The frequency of an itemed in all transactions is measured by support (Wanda & Jie 2019). Confidence is a term used to describe the degree of assurance that can be placed on any given collection of information. The formulae below can be used to calculate support and confidence. Devised an a priori method for discovering Boolean association rules in frequently occurring item sets. All frequent rule items are determined by iterating through the historical extracted characteristics of URLs (training dataset) using CBA-RG, which is then utilised to construct CARs:

$$Support(A) = \frac{support\_count(A)}{total\ number\ of\ transactions} \quad (2)$$

$$Confidence(A \rightarrow B) = \frac{support\_count(A \cup B)}{support\_count(A)} \quad (3)$$

That ratio of transactions containing A to total payments is used in Equation (2) as a measure of the rule object's approval (Chiramdasu et.al., 2021). ruleitems (A ∪ B) supporting count (A) is the volume of coins with ruleitems (A ∪ B) in it, and the volume of transactions with ruleitems in it, support count(A) The faith of a rule is shown in equation (3). Condset ⇒ y is a set of objects, y ∈ Y is a class name, and [condset, y] specifies a rule: [condset (malicious or benign). When it comes to categorising a URL, for example, the following rule item is relevant:

$$\{(A = 1), (B = 1)\} \Rightarrow \{class = benign\} \quad (4)$$

where A and B are characteristics (features). If-then statements are used to interpret rules. If A and B are both 1 in the example above, the URL is considered innocuous.

RESULTS AND DISCUSSION

Using machine learning methodologies, the Model will be sketched to recognize harmful URLs. The two distinct quantities of the process are the machine learning model and datasets.

Case 1

Figure 6. Before entering the URL

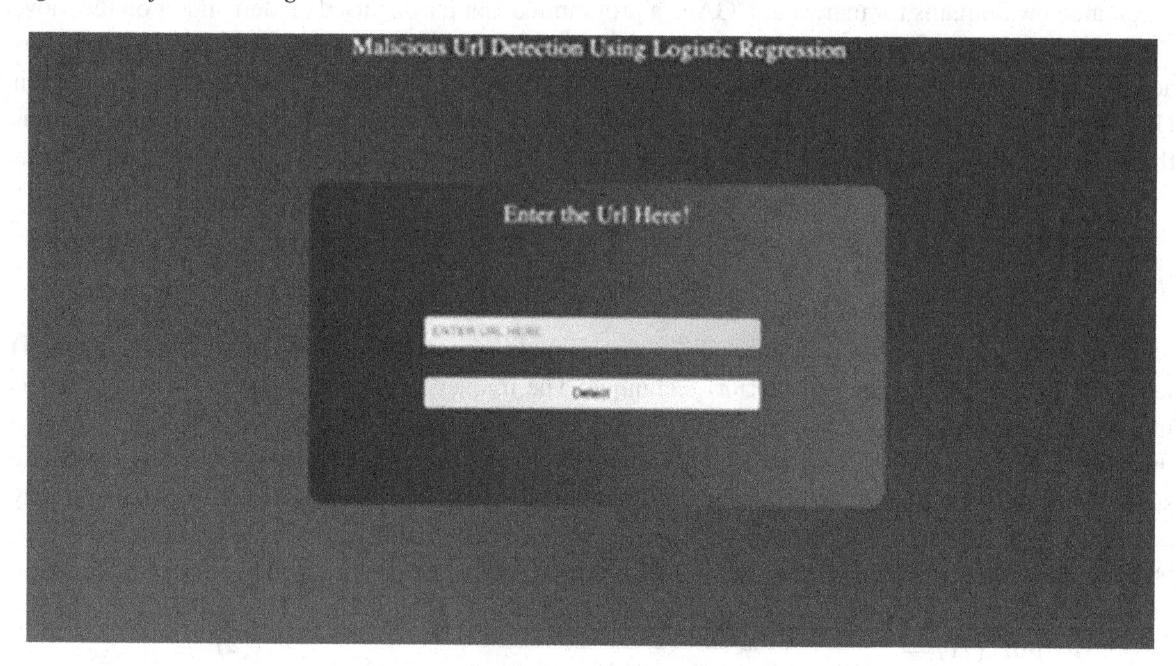

Fig 6 clearly explains about URL entering from LR algorithm, in this using malicious URL detection module can helps the process and providing accurate searching.

Case 2

Figure 7. After entering the URL

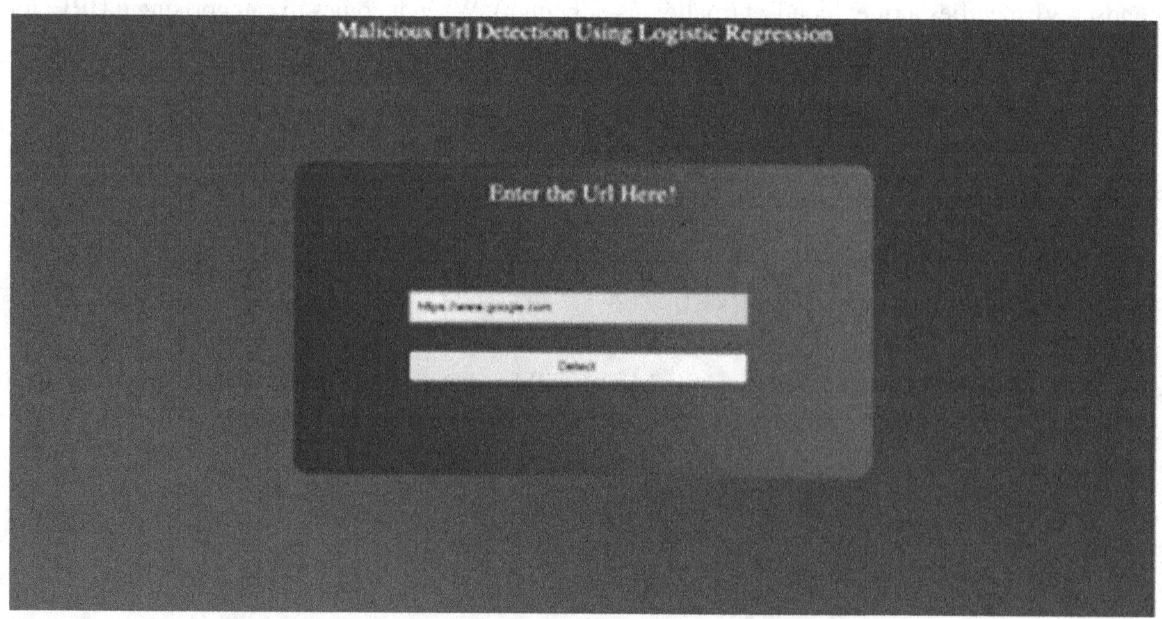

The fig 7 is clearly describing that URL checking by proposed algorithm, using this model virus detection URLs are to be getting blocked.

Case 3

Figure 8. After getting the result

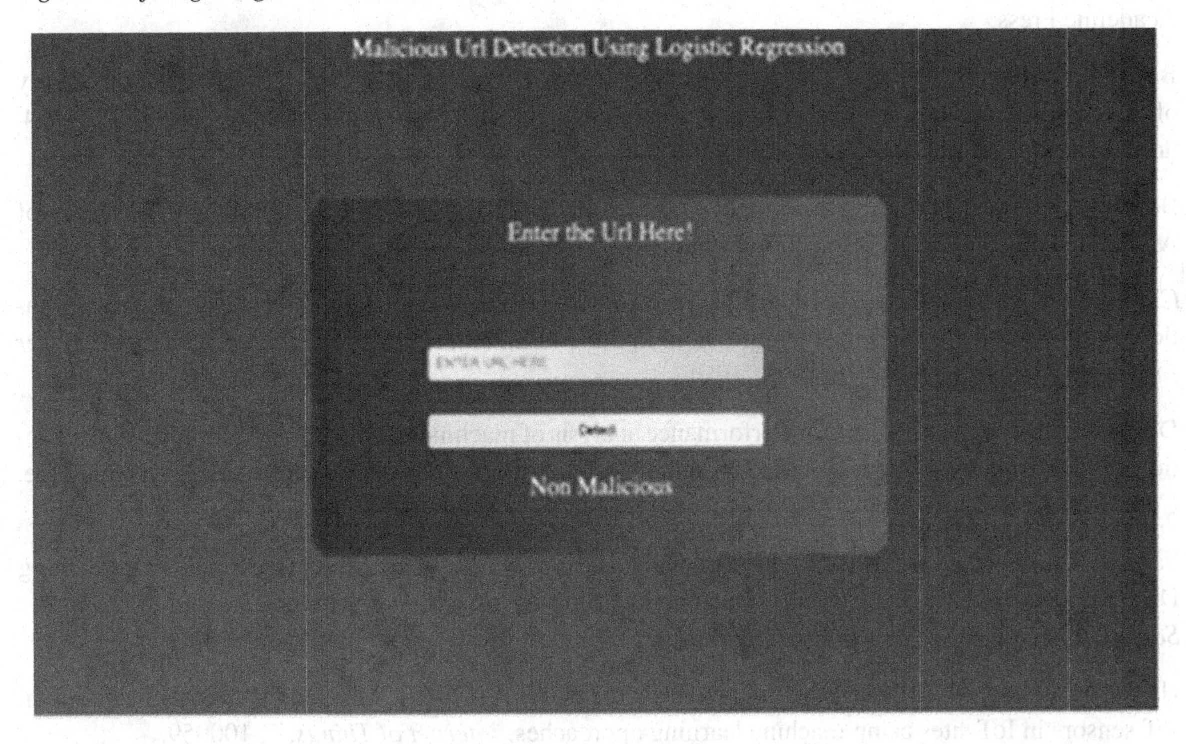

Fig 8 clearly explains URL functioning information, in this malicious, as well as Nonmalicious URLs, are separated and valuable information is placed.

CONCLUSION

Many cyber security and networking models focus on malicious URL detection. The majority of computer attacks begin with a fake webpage being visited. A phishing website may trick a user into disclosing personal information, or a drive-by download can infect a computer with malware. We used a machine learning method called logistic regression to detect phishing URLs, when compared to other algorithms like random forest and nave bays, has the highest learning accuracy. There is a plan in the works to increase training & testing data and identify different levels of accurate, which may subsequently be distributed as online information to all networked devices.

REFERENCES

Abu Al-Haija, Q., & Al-Badawi, A. (2022). Attack-Aware IoT Network Traffic Routing Leveraging Ensemble Learning. *Sensors (Basel)*, *22*(1), 241.

Al-Sarem, M., Saeed, F., Alkhammash, E. H., & Alghamdi, N. S. (2022). *An Aggregated Mutual Information Based Feature Selection with Machine Learning Methods for Enhancing IoT Botnet Attack*. Academic Press.

Basit, A., Zafar, M., Liu, X., Javed, A. R., Jalil, Z., & Kifayat, K. (2021). A comprehensive survey of AI-enabled phishing attacks detection techniques. *Telecommunication Systems*, *76*(1), 139–154. doi:10.100711235-020-00733-2 PMID:33110340

Basit, A., Zafar, M., Liu, X., Javed, A. R., Jalil, Z., & Kifayat, K. (2021). A comprehensive survey of AI-enabled phishing attacks detection techniques. *Telecommunication Systems*, *76*(1), 139–154.

Chiramdasu, R., Srivastava, G., Bhattacharya, S., Reddy, P. K., & Gadekallu, T. R. (2021, August). Malicious url detection using logistic regression. In *2021 IEEE International Conference on Omni-Layer Intelligent Systems (COINS)* (pp. 1-6). IEEE.

Disha, R. A., & Waheed, S. (2022). Performance analysis of machine learning models for intrusion detection system using Gini Impurity-based Weighted Random Forest (GIWRF) feature selection technique. *Cybersecurity*, *5*(1), 1–22.

Fang, L. C., Ayop, Z., Anawar, S., Othman, N. F., Harum, N., & Abdullah, R. S. (2021). URL Phishing Detection System Utilizing Catboost Machine Learning Approach. *International Journal of Computer Science & Network Security*, *21*(9), 297–302.

Hasan, M., Islam, M. M., Zarif, M. I. I., & Hashem, M. M. A. (2019). Attack and anomaly detection in IoT sensors in IoT sites using machine learning approaches. *Internet of Things*, *7*, 100059.

Odeh, A., Keshta, I., & Abdelfattah, E. (2021, January). Machine LearningTechniquesfor Detection of Website Phishing: A Review for Promises and Challenges. In *2021 IEEE 11th Annual Computing and Communication Workshop and Conference (CCWC)* (pp. 813-818). IEEE. 10.1109/CCWC51732.2021.9375997

Selim, G. E., Hemdan, E. E. D., & Shehata, A., & El-Fishawy, N. (2021). An efficient machine learning model for malicious activities recognition in water-based industrial internet of things. *Security and Privacy*, *4*(3), e154.

Urooj, U., Al-rimy, B. A. S., Zainal, A., Ghaleb, F. A., & Rassam, M. A. (2022). Ransomware Detection Using the Dynamic Analysis and Machine Learning: A Survey and Research Directions. *Applied Sciences (Basel, Switzerland)*, *12*(1), 172. doi:10.3390/app12010172

Wanda, P., & Jie, H. J. (2019). URLDeep: Continuous Prediction of Malicious URL with Dynamic Deep Learning in Social Networks. *International Journal of Network Security*, *21*(6), 971–978.

Wanda, P., & Jie, H. J. (2019). URLDeep: Continuous Prediction of Malicious URL with Dynamic Deep Learning in Social Networks. *International Journal of Network Security*, *21*(6), 971–978.

Wang, P., & Wang, Y. S. (2015). Malware behavioural detection and vaccine development by using a support vector model classifier. *Journal of Computer and System Sciences*, *81*(6), 1012–1026. doi:10.1016/j.jcss.2014.12.014

Yeboah-Ofori, A., & Boachie, C. (2019, May). Malware Attack Predictive Analytics in a Cyber Supply Chain Context Using Machine Learning. In *2019 International Conference on Cyber Security and Internet of Things (ICSIoT)* (pp. 66-73). IEEE.

Chapter 20
Security Challenges in Internet of Things

Aiyshwariya Devi R.
Dr. M. G. R. Educational and Research Institute, India

S. Srinidhi
https://orcid.org/0000-0001-8187-5818
Dr. M. G. R. Educational and Research Institute, India

ABSTRACT

A straight explanation of internet of things denotes the standard internet protocol belonging to human-to-entities and entities-to-entities transmission. Similarly, corporeal entities implanted with RFID, sensor, and so forth permits the entity to interconnect. Comparably, if we see IoT security as more important than the practical problem, we need rules and virtuous security systems. There are several problems in internet of things which burst out of the solution (RIFD tag security, cellular security, grid communication security, seclusion guard, and data handling security). In this chapter, the authors analysed several challenges for secure IoT, security concerns in IoT, and which safeguard the IoT assets in devices and data in contradiction of hacking and stealing. The authors give scrutiny to the challenges integrated in WSN and IoT for ecosystem observing.

INTRODUCTION

The Internet of Things indicate that link of extremely diverse network objects and network follows a numeral of transmission, World Technical News such as: Human-to-Human, Human-to-Entities, Entities-to-Entities. Likewise, Internet of Things is a notion that report a prospect where every object attaches to the net and also, identify itself to implement. Internet of Things denotes to the Object, which are known uniquely and uses the organization of internet. IoT has tetrad Main Characteristics which are Sensing, heterogamous Access, Services, and information processing and extra attributes like Security and Privacy. IoT may be termed as machine-to-machine transmission or Cyber physical system, in others Countries. IoT is nearly recognize by RFID, Detector technologies, Cellular Technologies. The situation extends

DOI: 10.4018/978-1-7998-9640-1.ch020

entities toward being sense then regulator faraway crossways the exist web Edifice. Internet is the path way that interconnects the globe for sending, conference gaming, online swapping and so on (Chuah, 2014). Concerning the Security issue, Various Challenges barrier the development of IoT tenders, some reason of addition to gather new technologies like detector grid and portable net, internet will constitute submissive and energetic entities, transmission those entities is must (Agrawal & Das, 2011). Where, IoT able to transmit, the data over the internet without human interplay. Over the creations about IoT original safety problems drive become toward light.

The Period of Internet of Things were brought many ways to interconnect with Internet, like Updating the fridge, and so on. which it's no longer as human, were the machine will do. It will be working without our interference or involvement. IOT has the Internet mode of communication. For Example, Fan will be Receiving the information, regarding should run fast or slow or medium it adjusts accordingly. Whereas, the light also can Communicate whether to turn on or off accordingly to sensors, which are continuous communicate with the Internet (Design Rush, 2020).

Internet of Things were attractive, because easy to use. Mainly, they can control anything from anywhere. Meanwhile, Internet of Things also have with safety standards for some precaution because it's not developed so well. Therefore, not to try to phase any situation. They kept with safety measures. Still, the manufactures, industries are trying to improve the components of Internet of Things. Simply, we can say Internet of Things is not ripened so far, and it's not completely safe.

The Problem may occur through lack of information, Maintenance, Update, Not-Convenient Manufacture Values, Physical Hardening, and so on (Team, 2019).

Whereas many resources started to use Internet of Things because it entrenched device connectivity, it can give solution in the frame of data gathering, tracking, supply chain management, and so on. Need of Internet of Things has improved. So, amount of data from Billions of sensors were ties with Cloud Computing, Deep Learning, Machine Learning, Artificial Intelligence, Big Data, and so on (Banach, 2020).

Returning to 2016, the lack of acquiescence was raised, where manufactures and industries faced a problem has Internet of Things Video Camera are automatically switching on. Which leads to Mirai Malware, Botnet Attack, and so on. Internet of Things has some risks too.

The reminder of the paper is structured as comes after: in Sect 2, we describe about the related works. Sect 3, we discuss about the Internet of Things design. Sect 4, presents the security features in Internet of Things. Sect 5, we discuss regarding disputes for secure Internet of Things. Sect 6, we describe the Security Concerns in Internet of Things. Sect 7, presents the result and discussion. Finally, the work Concluded in Sect 8.

RELATED WORKS

The works of IoT include Xiong li et al. suggested in the theory about believed in safety design for IoT. The Deficiency point based on structure able to assumed as follows: 1) it worried with a living thing, which factors are vital. The Factors which are main data and implement, 2) it is thoughts elderly algorithms, techniques for security, those are not acceptable for IoT, and does not spectacle novel design, 3) the algorithms and techniques, thought in every layer, it's huge to accomplish in the IoT System. This is because restricted power apparatus for instances sensors and RFID, these contemplate while framework based on IoT Schemes.

Arijit U et al., suggested in series to construct safety system for IoT. This series reveal warning and difficulty of stunted security IoT devices. The tools were discussed by the system may be stolen, and it can be noticed by monitor cars or Cameras.

Kiang Z et al., suggested in security construction on assembly traffic for the IoT based. The Series design is concerned in assemblage traffic which communicated over the IoT. Therefore, it is contemplated a motive blend as it can be relevant only for assemblage.

Gang et al., suggested in widespread investigation towards IoT safety problem. It's debate about some characteristics for instance identifying and controlling of sensors outlying. Furthermore, it resists the Denial-of-Service strike to detector node.

Hui S et al., suggested within certification besides entree regulator procedures for the IoT System. Unspecified series absorbed arranged simple and methodical elliptical curve cryptosystem safe key (Suchitra & Vandana, 2016).

In 2017, Lisa Goeke proposed a paper and it was named as Security Challenges of the Internet of Things. The paper consists of how today life to be secure by Internet of Things and to convey the basis of solid to be delivered. They include IoT Component, RFID, WSN, NFC. Lisa Goeke mentioned RFID is low cost and high in Security of Issues. The Hi- Tech of NFC is defenceless to all attacks. To ensure the Data Transmission, NFC delivers an active Counter Measure (Goeke, 2017).

In 2012, Gao et al., planned a protocol aimed at RIFD in IoTs, Random Oracle Method is satisfied and proved (Gao et al., 2012) (Matin, 2008). Tags, Readers, RFID middleware are from RIFD system in Internet of Things, were Secured. Unique EPC is available in every system of the object. Random Oracle Model is applied, to describe the RFID System Model (Alomair et al., 2012).

In 2015, Christian Dancke Tuen proposed a paper and it was named as Security in IOT's Systems. The Paper includes the present of Internet of Things, Low Security Standard causes, in Internet of Things which provided solution to the Problems. The Scheming device of Internet of Things as a guide for the Developers. According to the paper three distinct constraints are used. The paper also has Challenges in IoT, there Solution to be solved (Christian Dancke Tuen, 2015).

In 2009, Pierre de Lease et al., proposed and named as Self-Managed Security Cells. The paper contain about the services, were linked to the key software engineering areas, which includes and the idea of Internet of Things were introduced. The importance has remained put on essential the supplies to safe the resources in the wide area. The Hi- Tech has numerous mechanisms were aimed regarding automation, regionalization, interoperability, and security were Contextualization (de Leusse et al., 2009).

In 2014 Ning Ye et al. wrote a paper titled an efficient authentication and access control scheme for perception layer of Internet of Things. They proposed an efficient authentication and access control method based on general view of the security issues for perception layer of the Internet of Things (IoT). The main focus of the proposed method is establishing session key based on Elliptic Curve Cryptography (ECC) to enhance mutual authentication between the user and sensor nodes, and intermediate processes. The ABC-based authorization technique was adopted for the control policy. Their design focuses on the concept of a base station (BS) which collects data and controls the sensor nodes. [14].

In 2014, Ning Ye et al., proposed a paper and it was called as efficient authentication and access control scheme for perception layer of IoT. They planned a well-organized access control method and authentication grounded on overall opinion of the security matters for perception layer of the IoTs. The chief attention of the projected method is founding meeting main grounded on ECC to improve communal authentication amongst the operator and intermediate processes and Sensor bulges. Based authorization system was accepted aimed at the controller policy. Their plan efforts on the idea of a BS which gathers information and controls the sensor bulges (YE et al., 2014).

In 2012, Renu Aggarwal proposed a paper under the Security which has RDI that is attached in device to connect each other. Anyways, this security will get affected by others (Aggarwal & Das, 2012).

Mattias T. Gebie and Habtamu Abie, in 2017 proposed a paper and named as Risk-Based Adaptive Authentication for IoT. They assumed a method named the Adaptive Verification and risk-based verification. The Adaptive Verification is a refuge prototypical is variations in performing originally by viewing and running the conditions and variations based on opinion to precaution scheme in contradiction of threat overall the internet. In their opinion it is a verification key that endlessly viewing and running the altering atmosphere and accepts its key enthusiastically under by scheme of condition to spoil a structure in contradiction of unidentified hazard. The prototypical usages an original Bayes machine learning procedure to give a proper channel and to verify it, between the rarer device and its gate. Therefore, the proposed paper does not include regarding channel character for calculation and to predicate the prototypical by forecasting dangers by means of the original Bayes classification (Gebrie & Abie, 2017).

In 2015, Ke'ahi Cooper, achieved a Lead Level review the area under the security in Internet of Things and to examine few future procedures for Internet of Things devices. The Paper consist of Three main features, they are Wireless Sensor Network, Radio Frequency Identification and Internet Protocol Version 6. Under the Communication Protocols were examinate and replicate and as an outcome obligation on the disclosure of evidence switched and the validity of the units collaborating was also confirmed throughout the replication. The Paper outcomes to the Security issues, for verification, Elliptic curve digital signature procedure was projected. The Paper given the solution to all verification, privacy, truth, non-repudiation, etc. It has basement and Solution to these issues. According, to the Author review, the solution proved in the proposed paper is more than enough. Anyways, if we do additional research to enhance the new features, so new cryptographic mechanism, that will be more efficient compared to the present mechanism (Ke'ahi Cooper, 2015).

In 2016, Dimitri Jonckers proposed a paper named as Security Mechanism for Internet of Things. In Paper it is mentioned regarding the Gateway Architecture in Smart home setting based on security and discretion. The Paper furthermore discussed regarding the verification, authentication, etc., The Gateway has some policies too and it will specify the needed for the device. Whereas, Communication channel will be implemented outside of the house. Performance Test have been completed; the gateway provides the result according to their performant manner. The suppleness in supportive numerous refuge breadwinners and opportunity to speech facilities consistently infer the gateway which has aid to developers' condition to secure the claims for varied in Internet of Things (Dimitri Jonckers, 2016).

In 2016, Homeland Security Proposed paper and named as Strategic Principles for Securing the Internet of Things. Premeditated principles were implemented to progress the refuse range of manufacture, etc., May be, it will risk the security implementation. So, to secure some incident will be Included as internet connection, devices, etc. Some Principle, can be included as: Design phase incorporate security, security should be enabling, Names and Password can be cracked, creating a new device by the most usable Operating System, which can be very useful and economically possible. The publication given further more information regarding the measures taken by the government and industries further researches based on Security in Internet of Things (Homeland Security, 2016).

Internet of Things is further more developing component. Internet of Things plays a vital role. Its development in architecture and measures under security, is going a wide range. Security in Internet of Things is more needed in many devices. Therefore, industries, government and nations have to concerted in the development of Internet of Things with some consideration for Security. Internet of Things is more feasible and can be used in many parts of world in agriculture, etc. Some issues, technology to that solution we can see in WSN is Limitation of power and storage capacity attacks towards protocol were secure routing protocol in WSN. The solution is Secure Routing Protocols designed especially for WSN. Whereas for RFID Tags the issue is Multiple tags of working scope in one reading, the technology is RFID of conflict collision prevention, and solution is anti-collision algorithm. Same like for Network Layer, Adaptation Layer and Application Layer are mentioned below. Figure 1 represents Security issues and their present projected results in IoT.

Figure 1. Security issues and their present projected results in IoT

	WSN	RFID TAGS	NETWORK LAYER	ADAPTATION LAYER	APPLICATION LAYER
ISSUES	Limitation of power, & storage capacity attacks towards protocol	Multiple tags of working scope in one reading	DDos /Dos Attack	DDOS Attack	DDOS Attack
TECHNOLOGY	Secure Routing protocol in WSN	RFID of Conflict collision prevention	Network Encryption Technologies and Access Control	Supervision Capability	Intrusion detection System
SOLUTION	Secure Routing Protocols designed specifically for WSN.	Anti – Collision Algorithm	Access Control	Information Disclosure Protection	Guard Dog, other Vendors.

IOT DESIGN

The IoT can be suitable of related diverse varied entities direct to the internet, so flexible layered architecture is need. The Figure 2 represents the IoT Security (Sujithra & Padmavathi, 2016) (Sharma & Jinwala, 2015).

Figure 2. IoT Security

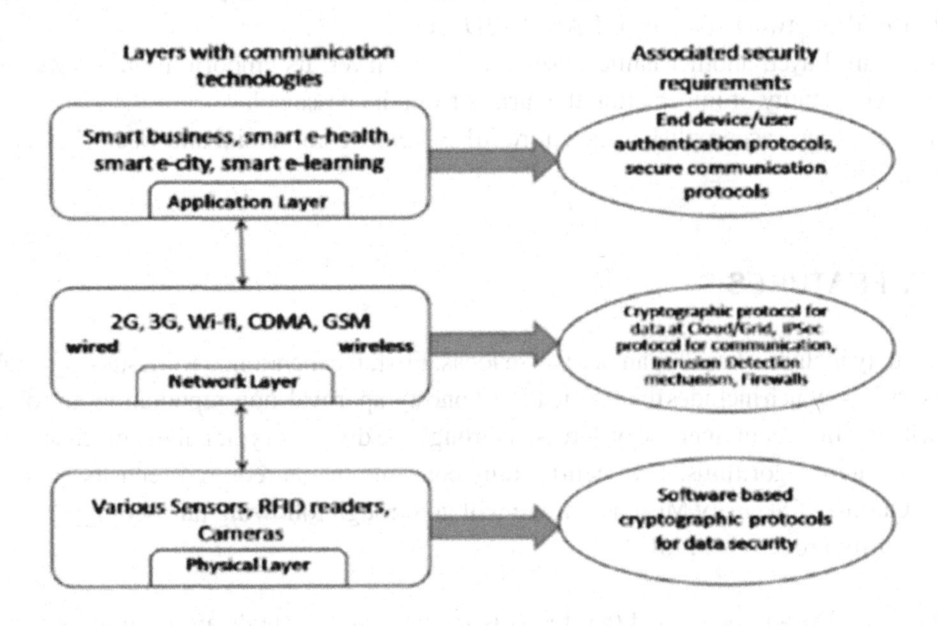

IoT Design has trio Layer namely, Perception Layer, Grid Layer, Application Layer.

- Perception Layer is the Corporeal Sheet, it has sensor to sense and it's gathered information for the atmosphere. It's percept some corporeal framework or recognizes others with intelligent purposes in surrounding. Perception Layer with communication technologies has Various Sensors, RFID Readers, Cameras, Infra-red, QR Code, ZigBee etc... Based on Associated security requirement of Software based cryptographic protocols for data Security.
- Grid Layer is accountable that one way interacting with others smart entities, net devices and attendants. Its characteristics are castoff in conveying and refinement detector data. Grid Layer with communication technologies has 2G, 3G, Wi-fi, CDMA, GSM, etc.... in wired and wireless, Network Communication Software and Physical Components (Server, Network nodes). Based on Associated security requirement of Cryptographic protocol for data at Cloud or Grid, IPSec Protocol for Communication, Intrusion Detection Mechanism, Firewalls.
- Application Layer is accountable towards conveyance request precise server to admin. Intercourse Permits several requests in which IoT contain position, viz clever families, keen fitness, etc.... Application Layer with communication technologies has smart business, smart e-health, smart e-city, smart e-Learning, smart homes, health care etc..... Based on Associated security requirement of End Device / user authentication protocols, secure communication protocols.

Layers explain the main design of the IoT, but it's not sufficient for investigate on IoT, since investigate frequently emphases on finer features of IoT. Therefore, we have few more covered architectures suggested in the poetry. In these, single is pentamerous Layered design, whichever involves the processing and commercial layer. At last, the pentamerous Layer arise from Perception, passage, middle ware, application, and commercial layer. The capacity of the perception, and application layer alike as mention above. So, we will see the function of remaining layers:

- Passage Layer transmit the detector data information from the perception sheet to the middle ware sheet, over the network like tuner, LAN, RFID, 3G.
- Middle Ware Layer, another name arises processing layer. Its emporium, investigate and procedure a huge quantity of information that arises from the passage layer.
- Commercial Layer accomplishes the entire IoT Scheme, commercial and User's Privacy (Sethi & Sarangi, 2017).

SECURITY FEATURES

The phrase Safety include a broad span of diverse ideas. First, it's mentioned from starting establishment of security services, which includes the verification, honesty, approval, non-repudiation, and obtainability. Those security facilities containers be progressed through the diverse cryptanalytic mechanism, like hash function or signature algorithms. The security only concentrates on required security services. Figure 3 represents Outline of Security Mechanism We will go through following phraseology to examine and categorize security properties in IoT:

- The Security Design mentioned that towards scheme fundamentals are been involve in the management of the security relationship in middle of things and course of action, these security interactions are hold throughout the entire cycle.
- The Security Model of a bulge report, that in which way does the security framework, procedure and application are processed the thing. These characteristics involves like process separation, safe packing of keyboarding material, many more. The Thing Security Model indicates the Security Services from Node A, Node B and Configuration Entity.
- Safety bootstrapping indicates an involvement, through which one entity firmly unites the IoT on an assumed position and while period. The Bootstrapping indicates from Security Services to Application in Node A, Application to Application in Configuration Entity, Application to Security Services from Configuration Entity.
- Network Security indicates the apparatus appeal inside the net to check reliable process of IoT. Network Security indicates the Network and L2 from Node A and Node B.
- Application Security warranty only reliable case based on request supervision within IoT can Communicate through respectively, while illegal case mayn't impede. Application Security indicates the Transport to Transport from Node B to Node A (Yu-Chee & Aiello, 2014).

Figure 3. Outline of security mechanism

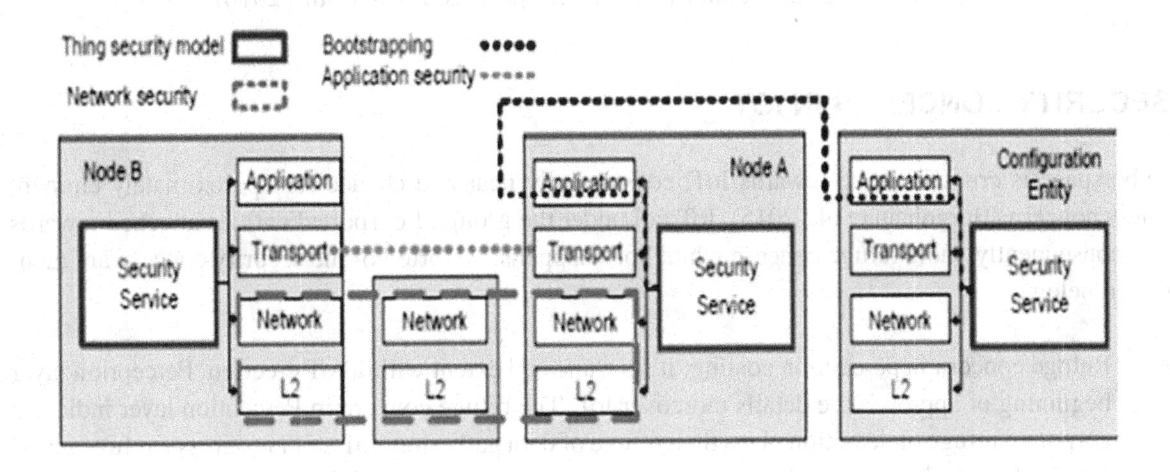

DISPUTES FOR SECURE IOT

The Disputes in the functional in addition practical characteristics based on IoT take place disputed below:

- Constraints and heterogenous Interaction:
  Combine resources constrained nets and the strong net is challenges since the ensue conglomeration of the duo nets obscures protocol design and system operation.
- Bootstrapping of a Refuge Area:
  Generating a refuge dominion since usual about formerly diverse IoT plans is additional crucial process within entire life based on entity and within the IoT grid. Bootstrapping mentioned the procedure through a contraption is related to unity extra, to a grid, or to a scheme. The method its achieved turn over the design: federal or dispersed.
  Within dispersed tactic, a Diffe-Hellman kind of handclasp container permit two squint the approve continuously usual confidential. Widespread, IKEv2, HIP, TLS, DTLS, can perform main interactions then system of safety relations minus accessible relationship to a belief centrality. If not contemplate the assets restriction of entities, license and credential cuffs taken hire towards fixed interconnect competences huge devolved. HIP and Diet HIP don't straight usage license aimed at recognise a congregation, however license hold ability occur based on HIP then identical code of behaviours reason could be castoff for Diet HIP. It's notable, that Diet HIP don't need an entity toward execute cryptological confusions. Therefore, some unimportant execution of Diet HIP instead of confirm credentials except a confusion role is executed through the object.
- Seclusion Guard:
  Knowledge seclusion occur straight considered based on discretion, towards IOT data. Place data of awareness terminal is a Major data resource of objects, and one of the complex details

need to be safe. Furthermore, there are also seclusion matter in information processing, like conduct scrutiny based on data excavating (Sujithra & Padmavathi, 2016).

SECURITY CONCERNS IN IOT

Cyberspace is crucial edifice towards IoT, consequently nearby a chance for approximately eminent safety concern (Borgohain et al., 2015). IoT fall under the group of corporeal entities attached towards net; consequently, more refuge concern whitethorn happens. Assorted of the security concern are mentioned below:

- Refuge concern in perception coating: It's a truncated extent within IoT erection. Perception layer beginning of approach the details moreover IoT. The refuge concern in Perception layer indicates corporeal refuge of detection plans then refuge of data gathering. IoT can't progress a refuge guard scheme besides its risk near hit because of variety, energy restricted, humble and weedy defensive ability of detection node whichever influence the refuge of WSN, RFID. The RFID incorporate refuge difficulties like data outflow, data chasing, tampering. The refuge difficulty endured in perception layer incorporate detention entry bulge, congestion attack, DoS bouts, node replication bouts and frontward bouts (Gou et al., 2013).
  - Refuge concern in Wireless Sensor Networks: WSN is a grid besides junction, this intellect then regulator around surrounding. Similarly authorizes dynamic in middle of peoples or processers in addition nearby atmosphere. WSN incorporate detector knots, actuator knots and likewise. WSN is a group of knots, therefore, there is chance of security concern.
  - Refuge concern in Radio-Frequency Identification technology in IoT: RFID technology is on the whole castoff as RFID labels for computerized interchange of details minus in the least physical participation. The RFID tags are several risk bouts since exterior because of the erroneous security position of the RFID technology (Kasinathan et al., 2013).
  - Refuge concern in corporeal layer: Corporeal layer carryout various characteristics like inflection and demodulation, encoding and decoding, selection in addition to the generation based on carrier frequency, communication and greeting of data (Kasinathan et al., 2013).
- Refuge concern in network layer: Network Layer is responsible for interacting to other's clever objects, grid devices and servers. Its characteristics are castoff in conveying and refinement detector information. IoT reality approximately vulnerable in the net such as illicit entree, honesty, DoS bouts, destruction, virus bout, man-in-the-middle bout, confidentiality, data eavesdropping, and etc.… IoT discern a huge quantity based on plans, therefore various arrangement towards information composed, and the information had an enormous, numerous -cause and heterogeneous features (Gou et al., 2013). The features of the network layer are conquering (Granjal et al., 2015).
- Refuge concern in application layer: Application Layer is accountable for conveyance request exact server to the operator. It Permits several applications in which IoT can be position, for example smart homes, smart fitness, etc. … Appeal of IoT is the outcome of closely combination in middle of interaction technology, processer hi-tech and manufacturing proficient analysed through talented on the way to discover appeal cutting-edge more particular. The refuge concern in application layer incorporates snooping and interfering (Gou et al., 2013). Deposit accomplishes the authority based on circulation organization (Borgohain et al., 2015). A trail grounded DoS bout

were beginning in application layer through restorative a detector node, towards generate a high circulation in the way concerning the sordid position.

Figure 4. Interaction protocol

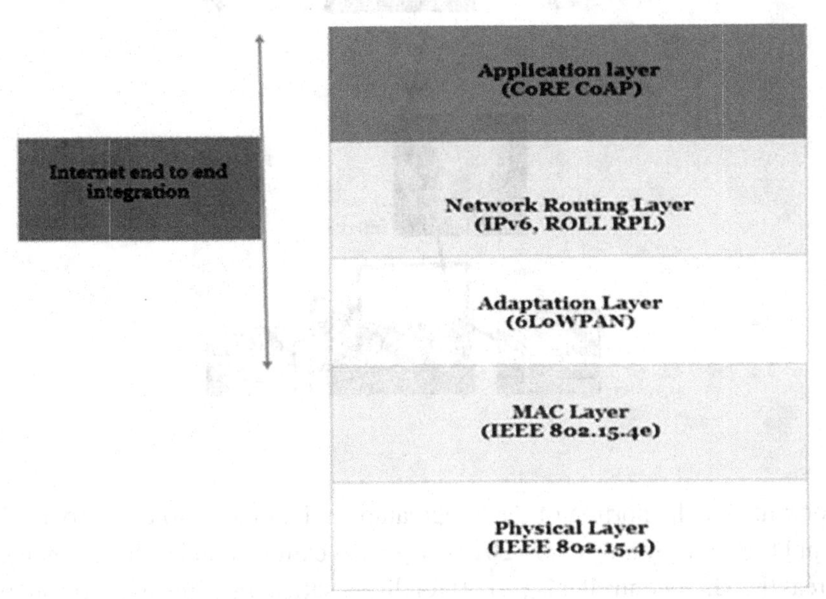

The Figure 4 indicates the Interaction Protocol. The Protocol pillar net interaction with detection devices in IoT. Whereas the Application Layer has CoRE CoAP, Network Routing Layer has IPv6, ROLL RPL, Adaption Layer has 6LoWPAN, MAC Layer has IEEE 802.15.40, Physical Layer has IEEE 802.15.4 towards internet end to end integration (Granjal et al., 2015).

RESULT AND DISCUSSION

We will Scrutiny the challenges were integrated with WSN and IoT for ecosystem observing. The detector node was architecture to perceive the ecosystem framework, viz temperature and humidity were pragmatic by detector SHT75. The detector SHT75 linked to MSP430 Microcontroller (Chen et al., 2013). To Exhibit the data, LCD Components is castoff, and were knot with Microcontroller. The XBeePro stocks the data after dispensation to the earthing scheme. The receiver of XBeePro were linked with SPI peripheral component. The earthing scheme accepts the data since detector node after transmission. The detector node is conveyed the data to the earthing scheme (Sabit et al., 2011). We can review the data from aloofness since of internet capability and software stood castoff is the team spectator. Internet communal the data to the exterior ecosphere. The structural design of detector node component in addition to the planned base station is shown in the figure 5.

Figure 5. Detector node and earthing scheme

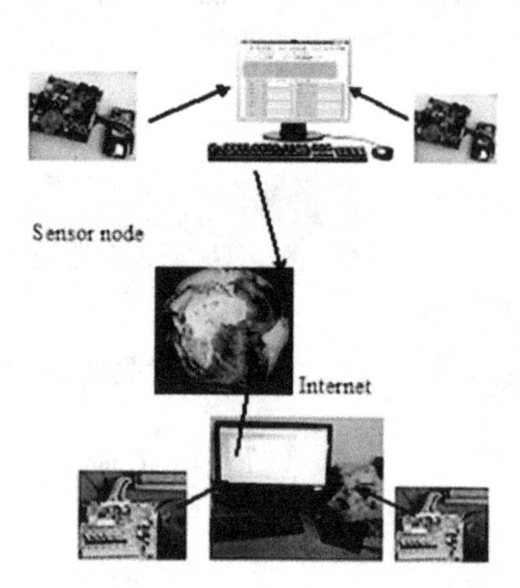

The detector will take the portion of the temperature and humidity to mark zone of the temperature and humidity and transmit (Somov et al., 2013). The data exhibited at earthing scheme is shown in the Figure 6. Whereas the Microcontroller is cast off for dispensation, monitoring, and collaborating the data. In mark area, the constraints are to degree by the detector node. The genuine data is transmitted from the earthing scheme to the detector node before getting the ecological limits (Bhattacharjee et al., 2012).

Figure 6. Exhibition based on temperature and humidity amount produced at the earthing scheme

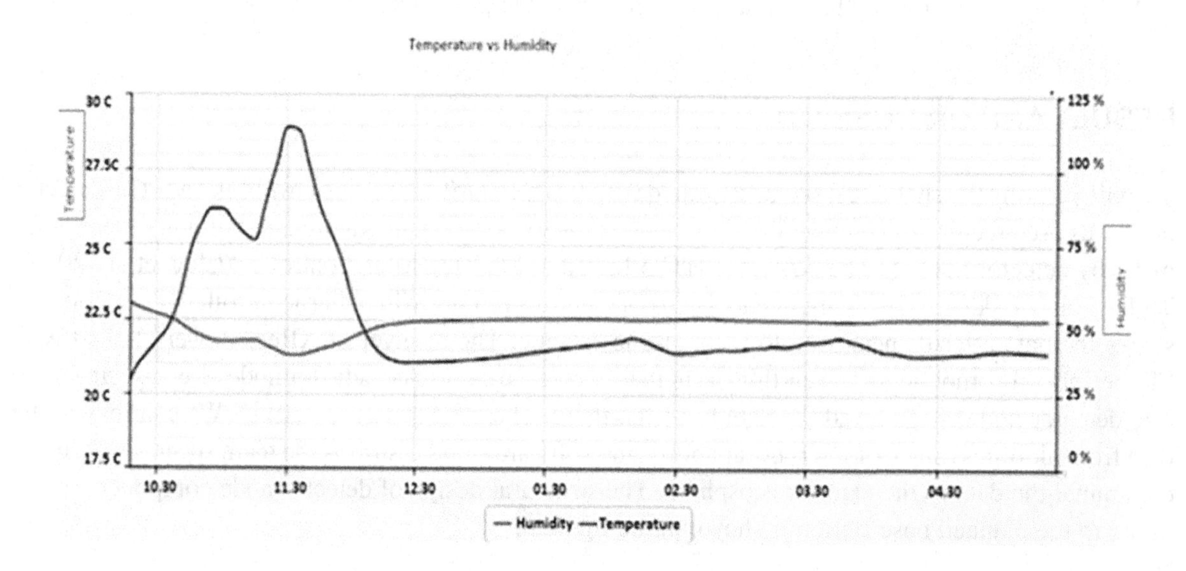

The Exhibition based on Humidity vs Temperature were shown above in Figure 6. Where Temperature and Humidity are in different amount produced at the earthing scheme. Humidity is shown in blue colour amount production and Temperature are shown in brown colour amount production. Humidity is almost moderate and there is a small difference in range of amount produced at the earthing scheme of 10:30 to 11:30 at 22.5c,50%. Temperature is not moderate, there is a huge difference in the range of amount produced at the earthing scheme of 10:30 to 12:30 at 20c,25% to 30c,125%. After, 12:30 range it's almost moderate, there is only small difference.

CONCLUSION

Over-all Expansion based on IoT, a various dissimilar Wireless Interaction hi-tech and Net Construction transpire collection, in addition to the Interaction grid surrounding had enhanced enlarge composite, the straightforward network security concern by all classes of commercial is extra composite and strenuous to crack. IoT Security is high Organization trade, network Security System is accepted later the Interaction System Construction, and Various Composite varied Interaction System may have collision on all-inclusive Security concern owed to its features. IoT makes the interaction in middle of Virtual Creation and Corporeal creation apart from being interconnected to data Security, interaction, nevertheless also incorporate high – priority communal purposes, cerebral acreage guard, discretion on high – priority nationwide elementary trades and communal key amenities. If Security concern aren't directed, it's huge danger on the Application. So, IoT Security concern is leap to incline everywhere, encourage IoT Security.

REFERENCES

Aggarwal, R., & Das, M. L. (2012). RFID security in the context of "internet of things." *Proceedings of the First International Conference on Security of Internet of Things - SecurIT '12*, 51–56. 10.1145/2490428.2490435

Agrawal, S., & Das, M. L. (2011). Internet of things -A paradigm shift of future internet applications. *2011 Nirma University International Conference on Engineering*, 1–7. 10.1109/NUiConE.2011.6153246

Alomair, B., Clark, A., Cuellar, J., & Poovendran, R. (2012). Scalable RFID Systems: A Privacy-Preserving Protocol with Constant-Time Identification. *IEEE Transactions on Parallel and Distributed Systems*, *23*(8), 1536–1550. doi:10.1109/TPDS.2011.290

Banach, Z. (2020, February 19). *The Challenges of Ensuring IoT Security*. Netsparker. https://www.netsparker.com/blog/web-security/the-challenges-of-ensuring-iot-security/

Bhattacharjee, S., Roy, P., Ghosh, S., Misra, S., & Obaidat, M. S. (2012). Wireless sensor network-based fire detection, alarming, monitoring and prevention system for Bord-and-Pillar coal mines. *Journal of Systems and Software*, *85*(3), 571–581. doi:10.1016/j.jss.2011.09.015

Borgohain, T., Kumar, U., & Sanyal, S. (2015). Survey of Security and Privacy Issues of Internet of Things. *International Journal of Advanced Networking Applications*, *5*, 2372–2378. https://arxiv.org/abs/1501.02211

Chen, Z., Shi, Z., & Guo, Q. (2013). Design of wireless sensor network node for carbon monoxide monitoring. *Telecommunication Systems*, *53*(1), 47–53. doi:10.100711235-013-9675-4

Christian Dancke Tuen. (2015, June). *Security in Internet of Things Systems*. Norwegian University of Science and Technology. https://scholar.google.no/citations?view_op=view_citation&hl=en&user=jRh3XQoAAAAJ&alert_preview_top_rm=2&citation_for_view=jRh3XQoAAAAJ:u5HHmVD_uO8C

Chuah, J. W. (2014). The Internet of Things: An overview and new perspectives in systems design. *2014 International Symposium on Integrated Circuits (ISIC)*, 216–219. 10.1109/ISICIR.2014.7029576

de Leusse, P., Periorellis, P., Dimitrakos, T., & Nair, S. K. (2009). Self Managed Security Cell, a Security Model for the Internet of Things and Services. *2009 First International Conference on Advances in Future Internet*, 47–52. 10.1109/AFIN.2009.15

Design Rush. (2020, April 3). *7 IoT Security Issues And How To Protect Your Solution*. https://www.designrush.com/agency/software-development/trends/iot-security-issues

Gao, D., Guo, Y., Cui, J. Q., Hao, H. G., & Shi, H. A. (2012). A Communication Protocol of RFID Systems in Internet of Things. *International Journal of Security and Its Applications*, *06*, 91–102. https://m.earticle.net/Article/A210850

Gebrie, M. T., & Abie, H. (2017). Risk-based adaptive authentication for internet of things in smart home eHealth. *Proceedings of the 11th European Conference on Software Architecture: Companion Proceedings*, 102–108. 10.1145/3129790.3129801

Goeke, L. (2017, May 22). *Theseus: Security Challenges of the Internet of Things*. Thesis - Haaga-Helia ammattikorkeakoulu. https://www.theseus.fi/handle/10024/128420

Gou, Q., Yan, L., Liu, Y., & Li, Y. (2013). Construction and Strategies in IoT Security System. *2013 IEEE International Conference on Green Computing and Communications and IEEE Internet of Things and IEEE Cyber, Physical and Social Computing*, 1129–1132. 10.1109/GreenCom-iThings-CPSCom.2013.195

Granjal, J., Monteiro, E., & Sa Silva, J. (2015). Security for the Internet of Things: A Survey of Existing Protocols and Open Research Issues. *IEEE Communications Surveys and Tutorials*, *17*(3), 1294–1312. doi:10.1109/COMST.2015.2388550

Homeland Security (Ed.). (2016, November). *Strategic Principles for security the internet of things*. US Department of Homeland Security. https://www.dhs.gov/sites/default/files/publications/Strategic_Principles_for_Securing_the_Internet_of_Things-2016-1115-FINAL

Jonckers. (2016). *A security mechanism for internet of things in a smart home context*. Ku Leuven. https://www.scriptiebank.be/sites/default/files/thesis/2016-10/text.pdf

Kasinathan, P., Pastrone, C., Spirito, M. A., & Vinkovits, M. (2013). Denial-of-Service detection in 6LoWPAN based Internet of Things. *2013 IEEE 9th International Conference on Wireless and Mobile Computing, Networking and Communications (WiMob)*, 600–607. 10.1109/WiMOB.2013.6673419

Ke'ahi Cooper. (2015). *Security for the Internet of Things*. KTH Royal Institute of Technology. http://kth.diva-portal.org/smash/record.jsf?pid=diva2%3A848663&dswid=-2959.Degree

Matin, G. (2008). *A Study of the Random Oracle Model*. Cambridge University Press. https://scholar.google.com/scholar_lookup?title=Ph.D.+Thesis&author=G.+Martin&publication_year=2008

Sabit, H., Al-Anbuky, A., & GholamHosseini, H. (2011). Wireless Sensor Network Based Wildfire Hazard Prediction System Modeling. *Procedia Computer Science*, *5*, 106–114. doi:10.1016/j.procs.2011.07.016

Sethi, P., & Sarangi, S. R. (2017). Internet of Things: Architectures, Protocols, and Applications. *Journal of Electrical and Computer Engineering*, *2017*, 1–25. doi:10.1155/2017/9324035

Sharma, D., & Jinwala, D. (2015). Functional Encryption in IoT E-Health Care System. *Information Systems Security*, 345–363. doi:10.1007/978-3-319-26961-0_21

Somov, A., Baranov, A., Spirjakin, D., Spirjakin, A., Sleptsov, V., & Passerone, R. (2013). Deployment and evaluation of a wireless sensor network for methane leak detection. *Sensors and Actuators. A, Physical*, *202*, 217–225. doi:10.1016/j.sna.2012.11.047

Suchitra, C., & Vandana, C. P. (2016). Internet of Things and Security Issues. *International Journal of Computer Science and Mobile Computing*, *5*(1), 133–139. https://www.ijcsmc.com/docs/papers/January2016/V5I1201636.pdf

Sujithra & Padmavathi, G. (2016, February). Internet of things - An Overview. *UGC Sponsored Two Day National Conference on Internet of Things*, 232–239.

Team, I. (2019, September 13). *Biggest IoT Security Issues*. Intellectsoft Blog. https://www.intellectsoft.net/blog/biggest-iot-security-issues/

Tseng & Aiello. (2014). Keynote. *2014 IEEE 7th International Conference on Service-Oriented Computing and Applications*. doi:10.1109/SOCA.2014.63

Ye, N., Zhu, Y., Wang, R., Malekian, R., & Qiao-min, L. (2014). An Efficient Authentication and Access Control Scheme for Perception Layer of Internet of Things. *Applied Mathematics & Information Sciences*, *8*(4), 1617–1624. doi:10.12785/amis/080416

Chapter 21
A Forensic Way to Find Solutions for Security Challenges in Cloudserver Through MapReduce Technique

D. Usha
Dr. M. G. R Educational and Research Institute, India

Reshma Raman
Dr. M. G. R Educational and Research Institute, India

ABSTRACT

Cloud computing is a large and distributed platform repository of user information. But it also extensively serves the security threats in the research aspect. This chapter attempts to find the solution to the security challenges through the MapReduce technique in a forensic way. Four security challenges are included in this chapter: losing the user information during the mapping process for different reasons such as the shutdown of the server, which causes parallel or unrelated services to get interrupted; the velocity of attack, which enables security threats to amplify and spread quickly in the cloud; injecting malicious code; and finally information deletion. MapReduce and dynamic decomposition-based distributed algorithm with the help of Hadoop and JavaBeans in the live forensic method is used to find solution to the problem. MapReduce is a software framework and live forensics is the method attempting to discover, control, and eliminate threats in a live system environment. This chapter uses Hadoop's cloud simulation techniques that can give a live result.

INTRODUCTION

An essential idea behind cloud computing is the placement of the service, and lots of the main points like the hardware or software package on that it's running, area unit for the most part inappropriate to the user. These 'time-sharing' services were for the most part overhauled by the increase of the computer

DOI: 10.4018/978-1-7998-9640-1.ch021

that created owning a laptop rather cheaper than successively by the increase of company information centers wherever firms would store huge amounts of knowledge. MapReduce has emerged because of the preferred computing framework for giant processing because of its easy model and parallel execution. MapReduce and Hadoop concepts are implemented by top companies like Yahoo, Google and Facebook for tremendous applications which include machine learning, bioinformatics and cyber security. Security is the term which safeguards all the information from malicious activities. The intruders or hackers were used as the backdoor or loopholes or vulnerabilities. During this paper ascertain the answer for a few security challenges that area unit within the cloud server that will be with the assistance of the newer technology massive information and MapReduce. Mainly in this paper discuss the challenges that are in a cloud environment and others also. Losing the information during the mapping process such as shut down the computer, the velocity of attacking, infecting malicious code, and finally information deletion. The rest of the paper is organized as follows. In section II, literature review, Section III, we review the existing model and its disadvantages also. Section IV explains the security issues in cloud and Section V presents a proposed system and methodology. The experiment results are discussed and finally, Section VI concludes the paper.

LITERATURE SURVEY

Some methods have been proposed in literature for handling security issues in organizations implementing cloud computing. (Yu,W., Xu,G., Chen,Z., & Moulema,P.,2013) discussed security issues, requirements and challenges that cloud service providers (CSP) need to address in cloud engineering: Security issues describe the problems encountered during implementation of cloud computing(CC). Security standards provide some security templates, which are mandatory for cloud service providers. The Open Visualization Format (OVF) is a standard for creating new business models that help the company to sell a product on premises, on demand, or in a hybrid deployment model. Security management models are designed based on the security standards and best practices.

(Wang,J., Crawl,D., Altintas,I., Tzoumas,K., & Markl,V.,2013) addressed countermeasures (anti-viruses, intrusion detection systems) developed to mitigate well-known security threats. The focus is mainly on anomaly-based approaches which are mostly suited for modern protection tools and not for intrusion detectors. The pattern-based changes (example: from thin client connected to the main frame or powerful workstations connecting to thin clients) are observed, which cause some simultaneous changes in work environment and new problems to security of CC.

(Mackey,G et al.,2008) mentioned CC's features like reduced total cost of ownership, scalability and competitive differentiation. They claim CC also minimizes complexity and provides faster and easier acquisition of services to customers. Virtualization is the technique used to deal with quality of service (QOS). Usage of CC is considered to be unsafe in an organization. For dealing with this type of situation, they investigated a few major security issues with CC and also existing countermeasures to those security challenges. Advantages for implementing CC from a different point of view are also discussed. They also stated that some standards are required in CC for security.

(Rosen,J et al.,2013) dealt with the security risks faced in the CC. They provided empirical evidence on security risks and issues encountered during deployment of service delivery models in an organization. The service models are placed in cloud and the empirical validation was made in order to justify the safety of the environment. Security was the main issue while there were also complications with data protection and data privacy in a continuous manner that affected the market.

The cloud computing helps reduces cost of services and improves business outcomes. But to market this and popularize its use by IT user community, there are many security is absolved. They also mentioned that the cloud services pose an attractive target to cyberattacks and criminal activities as these services have information from many organizations and individuals stored in their repositories. The author performs a survey in cloud computing to find out gaps and security concerns and mentions 5 common types of attacks: Denial of service: In this type of attack the attacker prevents the legitimate user from accessing his resources.

Malicious insider attacks: This type of attack the attacker is an insider. This person can easily gain access to sensitive user information namely: passwords, cryptographic keys, etc.

Cross virtual machine side channel attacks: Is the type of attack in which attacker resides in the same physical hardware as that of the target virtual machine and gains access to his sensitive information,

Phishing attacks: In this type of attacks the attackers ends links to the target user through email or instant messages. These links look as if they were sent by a trusted party but through this links the attacker can gain access to user sensitive information, Attacks targeting shared memory: The shared memory between the user and the attacker is used to perform unwanted, unauthorized actions.

A study by Farhan Bashir Sheikheta includes information regarding vulnerable security threats from 11 articles. The authors tabulated their findings i.e., problem discussed and technique used to solve the problem in their paper. But in the end, they conclude expressing that cloud computing from user perspective is suffering from numerous security threats. This, they say, is the only worth mentioning disadvantage in CC. It also performed a systematic review to identify which security requirements need to be further researched. To find that, the author used an existing model with sub-factors namely: access control, attack/harm detection, non-repudiation, integrity, security auditing, physical protection, privacy and confidentiality, recovery and prosecution to categorize their finding from 55 papers. From this review they Found that non repudiation, physical protection, recovery and prosecution are the least researched in security areas. Integrity, access control and security auditingarethemostpopularareas.Asurprisingfindin-gintheirreviewisthat privacy and confidentiality had been observed only in 7% publications. In addition to security requirements, solutions to these identified challenges were also mentioned.

(Zhang,J et al.,2012) in discusses some key security issues of cloud computing (policy, software and hardware security) and techniques implemented to reduce the risk. The author expresses that usage of CC will increase in near future and more companies will share their information to cloud servers, which could attract large groups of hackers. He also says that in future there are possibilities for interoperability and data lock-in issues, which can be reduced by using open standards from the time of CC adoption.

The author concluded by saying that security is always addressed late while adopting CC and also mentioned that security standards are still missing for CC. If an organization wishes to shift to CC but is reluctant due to lack of proper measures or standards, it can refer to Open Cloud Manifesto which is the largest initiative surrounding open standards. These standards are restrictive and so most companies do not wish to follow the Open Cloud Manifest to standards. The proposed methods address security challenges in CC and solutions to overcome these challenges. The following points can be observed from above related work: In the study performed by Ertaul et al., he mentions that CC is considered unsafe to be used by organizations and he also stated CC requires some standards. This provides a need for further research to ensure security for all those who are using CC applications. Eystein Mathisen concluded in their article that security is always addressed late while adopting CC. He also say that no proper security standards for CC exist.

(Md.Tanzim Khorshedetal.,2006) and Farhan Bashir Sheikhetal both stating that security challenges are still a major hindrance for adopting CC. Md.Tanzim Khorshed et al., have identified some threats to CC and proposed a method for automatic detection of network attacks, but it is still not used in real world. IlianaIankoulova et al., identified few security areas of CC to be less researched and also suggest to use another way of categorization in further studies.

From these studies it can be clearly understood that there are no security standards defined, even after a few researchers trying to formulate them. It can also be understood that even though few organizations and researchers tried to formulate strategies to handle security issues in cloud, there are still many companies that are reluctant to join the group of CC users. Their major concern is still security in cloud computing. This research tries to identify every possible challenge cloud faces and their practice/solution from literature and then pick a challenge that has no proper solutions/practices/models proposed and ask the people working in organizations to suggest a set of guidelines/practice to mitigate/control the challenge. This study will help both organizations and academics to identify the extent of research. It also will help to identify a set of solutions/practices/guidelines to harness the power of CC securely. This study will also include benefits of using a specific technique listed out, which can help organizations to choose a solution that fit their requirements.

EXISTING SYSTEM

In the previous scenario the MapReduce (Akhila,M., Amarnath,V., & Vishnu Murthy,G., 2018).task minimizes large scale of data processing by targeting parallel map (Zhang,J et al.,2012) and reduces tasks. In the olden days hash function was used to divide intermediate data to reduce task but is not structured for traffic because network topology and size of the data linked with each key will not be taken for consideration. In spite of many efforts the performance of MapReduce cannot be made possible.

Disadvantages

The existing MapReduce (Neda Maleki & Amir Masoud Rahmani., 2019) jobs use a hash function and decomposition-based distributed algorithm. One of the disadvantages is that in these the network topology and data size are not taken into consideration which affects the performance of the network. Also, the health of the data nodes is not verified. This imposes a loss of data.

SECURITY ISSUES IN CLOUD

Cloud computing and storage provide users with capabilities to store and process their data in third-party data centers. Organizations use the cloud in a variety of different service models and deployment models (private, public, hybrid, and community). Security concerns associated with cloud computing (Yu,W., Xu,G., Chen,Z., & Moulema,P., 2013) are typically categorized in two ways: as security issues faced by cloud providers and security issues faced by their customers (companies or organizations who host applications or store data on the cloud). The responsibility is shared, however, and is often detailed in a cloud (Matsunaga,A., Tsugawa,M., & Fortes,J., 2008) provider's "shared security responsibility model" or "shared responsibility model". The provider must ensure that their infrastructure is secure and that their clients' data and applications are protected, while the user must take measures to fortify their application and use strong passwords and authentication measures. When an organization elects to store data or host applications on the public cloud, it loses its ability to have physical access to the servers hosting its information. As a result, potentially sensitive data is at risk from insider attacks. According to 2010 Cloud Security Alliance report, insider attacks are one of the top seven biggest threats in cloud computing. Therefore, cloud service providers must ensure that thorough background checks are conducted for employees who have physical access to the servers in the data center. Additionally, data centers are recommended to be frequently monitored for suspicious activity.

In order to conserve resources, cut costs, and maintain efficiency, cloud service providers often store more than one customer's data on the same server. As a result, there is a chance that one user's private data can be viewed by other users (possibly even competitors). To handle such sensitive situations, cloud service providers should ensure proper data isolation and logical storage segregation.

The extensive use of virtualization in implementing cloud infrastructure brings unique security concerns for customers or tenants of a public cloud service. Virtualization alters the relationship between the OS and underlying hardware be it computing, storage or even networking. This introduces an additional layer virtualization that itself must be properly configured, managed and secured. Specific concerns include the potential to compromise the virtualization software, or "hypervisor". While these concerns are largely theoretical, they do exist. For example, a breach in the administrator workstation with the management software of the virtualization software can cause the whole datacenter to go down or be reconfigured to an attacker's liking.

CLOUD SECURITY CONTROLS

Cloud security (Yu,W., Xu,G., Chen,Z., & Moulema,P., 2013) architecture is effective only if the correct defensive implementations are in place. An efficient cloud security architecture should recognize the issues that will arise with security management. The security management addresses these issues with security controls. These controls are put in place to safeguard any weaknesses in the system and reduce the effect of an attack. While there are many types of controls behind a cloud security architecture, they can usually be found in one of the following categories:

Deterrent controls-These controls are intended to reduce attacks on a cloud system. Much like a warning sign on a fence or a property, deterrent controls typically reduce the threat level by informing potential attackers that there will be adverse consequences for them if they proceed. (Some consider them a subset of preventive controls.)

Preventive Controls: Preventive controls strengthen the system against incidents, generally by reducing if not actually eliminating vulnerabilities. Strong authentication of cloud users, for instance, makes it less likely that unauthorized users can access cloud systems, and more likely that cloud users are positively identified.

Detective Controls: Detective controls are intended to detect and react appropriately to any incidents that occur. In the event of an attack, a detective control will signal the preventative or corrective controls to address the issue.[11] System and network security monitoring, including intrusion detection and prevention arrangements, are typically employed to detect attacks on cloud systems and the supporting communications infrastructure.

Corrective Controls: Corrective controls reduce the consequences of an incident, normally by limiting the damage. They come into effect during or after an incident. Restoring system backups in order to rebuild a compromised system is an example of a corrective control.

PROPOSED SYSTEM

Mainly here using not a cloud server it's using the simulation. One of the open-source cloud simulations in the Ubuntu operating system is the Hadoop distributed file system (Pranav,N., Devavarapu Sreenivasarao & Shaik Khasim Saheb, 2019). As per the studies it will be clearly described as the Hadoop is one of the best cloud simulating file system environments. That's why this project has taken Hadoop as the cloud environment and it should have lots of advantages also. This work can't directly be processed in the cloud suddenly so that the project here trying to depicts this project system into a cloud environment and also if this project is successfully implemented then it easier for the implementation in the industrial area because Hadoop is widely used industrial purpose. This work should be divided into three parts that are: Preprocessing, Document clustering, and finally Data Analysis and Visualization. The system architecture depicts in figure Fig:1 below.

In the Preprocessing, three processes are there. That is Encryption and Decryption, Compression, and also Upload Data into HDFS. The data are encrypted and compressed before uploading into HDFS. Hadoop (Nishant Rajput and Nikhil Ganage, 2017) is a real cloud platform which can help the cloud service provider in providing cloud computing services to the cloud services.

In HDFS, the second part will be processed. In that process divides into three that are Cluster formation (Yang, H.C., Dasdan,A., Hsiao,R.L., & Parker,D.S.,2007). Generating Multimode Cluster, and DDDA with Framework Monitoring. In Fig: 2, the system will make the Clustering using MapReduce (Chen.S., & Schlosser.,S.W.,2008) and Forensic Analysis. (Luís Filipe., Cruz Nassif., & Eduardo Raul Hruschka., 2013). Mainly here using Document Clustering with K-Means algorithm and Dynamic Decomposition –based Distributed Algorithm. With that one health care algorithm also added in this algorithm for checking the data node are alive or not. After the clustering process, data will be sent into the final and last part of this project, Data Analysis & Visualization. In this part all the security challenges are mentioned in the first paragraph will be solved by the different modules that are: Intelligent Risk Management, Threat Visualization, Prediction Model, and Testing Model.

Intelligent Risk Management: This system must be backed up by intelligent insight data and also it ensures with the ready data to make analysis simply and rapidly. Handle security threats without delay. This module will reduce the entire delay of time attacking with the help of reducing the time that will know by the client or user for immediate action. And also here figuring out the solution for losing the user information during the mapping process for different reasons such as shutdown server. Whenever the risk is coming through the system that risk will be effectively managed by the system. The system depicted in figure Fig: 3. in our system what you have to do means we already stored lots of data with the help of Bigdata techniques. Even if a hacker will hack the data also we can retrieve the data as it is in the original form without any change.

Threat Visualization: It helps to save the class and strength of the cyber security threats. It can assess a possible attack by evaluating data servers and patterns. In this module, figuring out the intensity of the attack and it will be described as the velocity of attacking. The velocity of attacking means that how security threats amplify and spread quickly in a cloud. When attacks the system by the different types of threats or hacking the system or cluster. The system will know about the thing after only attack successfully executed. That delay time is very high. The delay time reduced by the system with the help of this module. This module will reduce the entire delay of time attacking when they reduced delay time of the attack that will know by the client or user.

Prediction Model: Our system enables experts to build a predictive model that can leave an alert as soon as it sees on the entry in it. This module uses for how the system can send an alert message as soon as possible. The Predictive Model and Testing Model will work together and solve the problem of the infecting malicious code. When the system sends an alert message that intruder enters into the system, it will predicts the intruder will be hacks system or not within a small time duration or later. In this module if the system or data will get any user like a hacker will enter the system or any other type of intruder activities is happening and also we have to alert the message from the system. It will be possible through this module. In Figure Fig: 4 this module predicts hacker activities but also they didn't get any data in that hacker. When the system sends an alert message that intruder enters into the system, it will predicts the intruder will be hacks system or not within a small time duration or later.

Testing Model: This system gives an infrastructure for this business database and process. This module helps keep hackers at bay. It also solves the Malware attack against this system and network to check for exploitable vulnerabilities. And the system will recover the original data even intruder was changed it or not. This will be a solution to the security challenges that the information deletion and also shows in Fig: 5. If there are any types of malware attacks and checks is there any vulnerabilities is for the system that will also be checked through this module. After all the process this work will get the original data without any change. If the hacker can change the data then also this project will get the data in that original form and it is shown in Fig: 6. Finally, this system gives all the contents with the compression also encrypted form. So the user can open secured output using the encrypted key.

ADVANTAGES

The different data nodes assigned to the name nodes are checked periodically in constant time intervals to verify whether they are alive or not. The aggregator which is used to group the data from the data node is to be placed correctly by using an aggregator placement problem. Increasing the speed compared to existing system Faster and efficient.

CONCLUSION

The proposed system tries to solve the security challenges mentioned in the abstract that are losing information during the mapping process for the different reasons, the velocity of attacking, infecting malicious code, information deletion. And here its implement solutions for these problems with the help of the Hadoop cloud environment features and it can be implemented at the industrial level with Hadoop and big data concept also.

REFERENCES

Ahmad, F., Lee, S., Thottethodi, M., & Vijaykumar, T. (2013). Mapreduce with communication overlap. *Journal of Parallel and Distributed Computing*, *73*, 608–620.

Akhila, M., Amarnath, V., & Vishnu Murthy, G. (2018). Applications of MapReduce in Big Data concepts. *International Journal of Research*, *5*, 20.

Chen, S., & Schlosser, S. W. (2008). *MapReduce meet wider varieties of applications*. Intel Research Pittsburgh.

Condie, T., Conway, N., Alvaro, P., Hellerstein, J. M., Gerth, J., Talbot, J., Elmeleegy, K., & Sears, R. (2010). Online aggregation and continuous query support in MapReduce. *Proceedings of the 2010 ACM SIGMOD International Conference on Management of Data*, 1115–1118. 10.1145/1807167.1807295

Cruz Nassif & Hruschka. (2013). Document Clustering for Forensic Analysis: An Approach for Improving Computer Inspection. *IEEE Transactions on Information Forensics and Security*, 8.

Mackey, G., Sehrish, S., Bent, J., Lopez, J., Habib, S., & Wang, J. (2008). Introducing map-reduce to high-end computing. *Petascale Data Storage Workshop*, 1–6.

Maleki & Rahmani. (2019). MapReduce: An infrastructure review and research insights. *The Journal of Supercomputing*, *75*(10), 1–69.

Matsunaga, A., Tsugawa, M., & Fortes, J. (2008). Cloudblast:Combining mapreduce and virtualization on distributed resources for bioinformatics applications. *IEEE Fourth International Conference*, 222–229. doi:10.1145/2465351.2465371

Pranav, Sreenivasarao, & Saheb. (2019). Augmenting the outcomes of Health-Care-Services with Big Data Analytics: A Hadoop based approach. *International Journal of Scientific & Technology Research*, *8*, 12.

Rafique & Khan. (2013). Exploring Static and Live Digital Forensics: Methods, Practices and Tools. *International Journal of Scientific & Engineering Research*, *4*, 10.

Rajput, Ganage, & Thakur. (2017). Review Paper On Hadoop And Map Reduce. *International Journal of Research in Engineering and Technology*, *6*(9).

Rosen, J., Polyzotis, N., Borkar, V., Bu, Y., Carey, M.J., Weimer, M.T., & Ramakrishnan, R. (2013). *Iterative mapreduce for large scale machine learning*. Academic Press.

Venkataraman, S., Bodzsar, E., Roy, I., AuYoung, A., & Schreiber, R. S. (2013). Presto: Distributed machine learning and graph processing with sparse matrices. *Proceedings of the 8th ACM European Conference on Computer Systems,* 197-210.

Wang, J., Crawl, D., Altintas, I., Tzoumas, K., & Markl, V. (2013). Comparison of distributed data-parallelization patterns for bigdata analysis: A bioinformatics case study. *Proceedings of the Fourth International Workshop on Data Intensive Computing in the Clouds*.

Yang, H.C., Dasdan,A., Hsiao, R.L., & Parker, D.S. (2007). MapReduce merge: simplified relational data processing on large clusters. *Proceedings of the 2007 ACM SIGMOD International Conference on Management of Data*, 1029–1040.

Yu, W., Xu, G., Chen, Z., & Moulema, P. (2013). A cloud computing-based architecture for cybersecurity situation awareness. *Communications and Network Security, IEEE Conference*, 488–492.

Zhang, J., Chen, R., Fan, X., Guo, Z., Lin, H., Li, J.Y., Lin, W., Zhou, J., & Zhou, L. (2012). Optimizing data shuffling in dataparallel computation by understanding user-defined functions. *Proceedings of the 7th Symposium on Networked Systems Design and Implementation (NSDI)*.

Chapter 22
The Interaction Between Technologies, Techniques, and People in Higher Education Through Participatory Learning

Vijaya Lakshmi Dara
REVA University, India

Saradha M.
https://orcid.org/0000-0001-5621-0830
REVA University, India

ABSTRACT

Education is an essential factor in the development of a nation. We should make it appropriate according to the changing scenario of the country. Learning is an opportunity to reflect upon the social, economic, cultural, and moral issues faced by human beings. Nowadays most of the institutions are working only for a degree, students and teachers are running after attaining or providing degrees, and not towards knowledge and wisdom. A good teacher can bring the entire world to the classroom. The overall scenario of higher education in India does not match the global quality standards. Education has to develop appropriately according to the time and changing scenarios of the world. It contributes to national development through specialized knowledge and skills. So, higher education has to come out of the static environment and become more dynamic and more futuristic. The solution to all the problems is providing quality education, and teachers are the main ingredients in giving quality education.

INTRODUCTION

Education is an essential factor in the development of our nation. We should make it appropriate according to the changing scenario of the country. Learning is an opportunity to reflect upon the social, economic, cultural, and moral issues facing by human beings. Nowadays most of the institutions are

DOI: 10.4018/978-1-7998-9640-1.ch022

working only for a degree, students and teachers are running after attaining or providing degrees, and not towards knowledge and wisdom. A good teacher can bring the entire world to the classroom. The higher education in our country should strive to match the global quality standards. Education has to develop appropriately according to the time and changing scenarios of the world. It contributes to national development through specialized knowledge and skills. So, higher education has to come out of the static environment and become more dynamic and more futuristic. The solution to all the problems is by providing quality education, and teachers are the main ingredients in giving quality education.

The pedagogy in India was initially started in the 5000 BC, which was then called the Gurukulam or the Gurukula System, where students were taught various subjects and the way of leading their life by the Well-knowledgeable Gurus(Kumar, V et al., 2021). After the independence there were a lot of efforts put forward by the government to spread the importance of education in the country, therefore modern schools were set-up across the country, Education was then made compulsory by the government for all kinds of people up to the age of 14. To make sure that the prevalent mode of education is more than books and include practical implication of the theoretical terms and improve their skill sets. The student will have a practical check over the theoretical terms here rather than passively receiving the information through their books(Nagarajan, Senthil Murugan et al., 2021). The research mainly concentrates on the different knowledge management learning techniques that are adopted by various institutions which contributes to the overall annual performance and help in the overall development of the students.

The methodology of participatory learning originated in the southern hemisphere in the late 1970s. It was used as a method of consultation and communication with rural police of some of the major developing countries Particularly in Africa. It was evolved based on the principle that poor and exploited people can and should be enabled to analyze their, own reality, and to examine their own problems. In essence to it concentrates on empowerment.

The word participatory or participation refers to experiential learning i.e the action of taking part in the activities and projects, the act of sharing in the activities of a group. In the participatory learning approach all I participants are active as they set their goals and work to achieve the goal. Participatory learning has its origin from participatory education. The purpose of participatory learning is to educate, improve skills, build self confidence, creative thinking among students of UG & PG from Management studies in state of Karnataka.

Participatory education has its origin in the early 1960's in the works of Paulo Freire. The components of participatory education can be traced from John Dewey's progressive education, Vygotsky's socio cultural theory, Situated Learning Theory of Lave and Wenger, Constructivism, Collaborative and Cooperative learning. In participatory learning, process the learners share ideas, knowledge, opinion, materials and resources in order to reach a common agreement or to make joint decisions in a transparent way(Muthukumaran V et al., 2021). Thus, learning may be defined as the process of making a new or revised interpretation of the meaning of an experience, which guides subsequent understanding, appreciation and action (J.Menzirow, 1991).

REVIEW OF LITERATURE

The learning and learning environment has a great impact on student's academic performance (Chapman and Adams, 2002). Quality teaching and good learning environment ensures that teachers involve all the children to participate in their learning (Clarke, 2003).

Garrison (2004) examined the transformative potential of blended learning in the context of the challenges facing in higher education. Blended learning gives a consistent explanation with the values of traditional higher education institutions and has the proven potential to enhance both the effectiveness and efficiency of meaningful learning experiences.

Motivation affects cognitive processes, which affects what learners pay attention to and how effectively they process it. For instance, motivated learners often make a concerted effort to truly understand classroom material to learn it meaningfully – and consider how they might use it as their own lives (Ormrod, 2006).

Teachers should also create an active learning environment that enhances students perceived automatically and competence, providing students with choices and opportunities for self- directed learning, and planning learning activities that might increase their feeling of mastery. In fact, intrinsic motivation was shown to be a factor of great importance that can lead to higher perceived learning in the course (Ferreira, Cardo sob and Abrantese, 2011).

Motivation is the important drive thathelps us to carry out activities. Motivation is a hidden feeling we motivated when we feel like doing something and we are able to sustain the effort required during the time required to achieve the objective we set ourselves. Teachers should motivate the students, to mobilize the capabilities and potential of each student for academic success. (Ferreira, Cardosob& Abrantesc, 2011).

Alzbeta Kucharcikova (2016), the aim of the article is to show how it is possible to increase the efficiency and attractiveness of the subject at university using participatory) methods.

Expository approach was assessed by utilizing pre-test post-test experimental design. Purposive stratified sampling was used to select a sample of fifty students. Psycho productive evaluation test items were used at the pretest and post-test phases. The mean gain sores of the pretest and post-test further analyzed using t test statistics resulting to the findings that participatory approach was more effective than expository approach. The reason was that participatory approach was student cantered.

Latha (2013) conducted a study titled 'promoting participatory learning through activity based teaching. This study presented an activity-based approach to promote participatory learning, using informal teaching patterns through simple activities. To motivate the learners and to produce better results by assisting them educationally and morally resulted in creating a conductive classroom atmosphere. It was implemented in a class of forty-five. The participatory learning was found to be motivating and effective. Students were educated to learn by learning how to learn. The study also revealed that peers in the group take up the task of helping their friends. Interaction with peers and faculty creates a better rapport and most of the students work with a sense of commitment.

Omollo, Nyakrura and Mbalamula (2017) conducted a study on 'Application of Participatory teaching and learning approach in Teacher Training Colleges in Tanzania. In this study, the use of participatory approach in teaching and learning in teachers colleges was exposed with the aim of finding out whether tutors apply participatory approach in the classroom in order to improve teaching and learning among triners. Quantitative and qualitative data was collected. The study applied stratified sampling, simple random sampling and purposive sampling procedures to select 96respondents. The study also used semi structured interview, focused group discussions, questionnaires, observation and document review to collect data. The quantitative data were manually analyzed and presented in frequencies, percentages and tables while qualitative data were subjected to content analysis. The study found that tutors use participatory approach in teaching and learning with the questions and answer technique in the classroom. The study found that tutors have the positive attitude towards the use of participatory approach.

Another study by Gal Rubio, Iglesias and Gonzalez(208) 'Evaluation of participatory teaching methods in undergraduate medical students learning along the first academic course' used qualitative and quantitative analysis of over 200medical students. The perception of participatory teaching methodologies and their impact on learning process was studied.

Learning instead of being a passive recipient of knowledge and information. Learner's specific learning needs were given central importance. This led to the development of learner Centered Approach in language teaching (Trikes, 2000)

According to Nunan (1989) in Trikes (2000) learner Centered approach originated from Communicative Language teaching which gives central importance to the student is planning, application and assessment processes students and teachers collaboratively develop the curriculum.

According to Brackenbury (2012) learner centered approach challenges both the teacher and the student. Students should be pro active learner to face complicated problems and develop new ways of performing and thinking teachers face the challenge of giving liberty to students, being careful about the content and uses on the individual and collective needs of students (McCombs and Miller, 2006).

According to (Sarfraz and Akhtar, 2013) learner Centered Approach focuses on specific needs of learners and emphasizes developing specific strategies that will cater to these needs and enhance effective learning. The analysis of Weimer (2002) learner Centered Approach develops a sense of responsibility among students by catering to their needs and involving them in the learning process.

The analysis done by Weimer (2002) and Wilson (2005) teachers should allow students to participate in some decision-making processes related to their learning. Weimer (2002) examined that teacher share their power with students in student-centered approach. However, they insist to transfer their whole power to the students but give them liberty to participate in the decisions that are directly related to their learning. The teacher should share power with students in responsible ways in learners Centered Approach. So, students are motivated to make decisions under the teacher's guidance about the activities, assignments, course policies and assessment criteria(V Vinoth Kumara et al., 2021).

According to examination of Wilson (2005), students should be given liberty to choose material for their class.

According to Nunan (1988), students participate indecision-making processes related to the content of curriculum and teaching methodology. Jurmo's (1989) point of view participation of learners can be done at different levels. The main type of participation can be registering for the course and being physically present in the class. Learner Centered Approach motivates and encourages students to have highest level of participation by taking responsibility of their learning.

Jinal Jani and Girish Tere (2015), the Digital India programme introduced by government of India is important for the development of digital education in the country. Digital India drive is a project initiated by Government of India for Creation of digital empowered society across the country. It will help in mobilizing the capability of information technology across government departments and helps in delivering the different governments and help in delivering the different governments programs and services. Digital India will help in creating job, providing high speed internet and digital locker system and so forth. The main important components of digital India are namely digital infrastructures creation, digital delivering services and resources and digital education.

Shikha Dua et al., (2015). They have discussed the different issues, trends and challenges of digital education in India and suggested the empowering Innovative classroom model for learning. The future trend of digital education includes digitalized classroom, video based learning, and game based learning and so forth. The opinions marked the challenges of digital transformation in education and how to overcome these, and the measures that have to be employed.

Himakshi Goswami (2016). Listed out the different opportunities and challenges of digital India programme in India. This programme was introduced by government of India will help in changing country into a digitally empowered economy. This will guide the government of India to integrate the Government Departments with the people of India. The aim of this programme is to reduce the paper work and help in providing different Government services electronically to citizens for sustainable environment.

Joseph, K. et. al (2017) carried a study on improving critical thinking levels through participatory learning: Analyzing the Nature Of Expression In Makerere University Lecture theatres. The research evaluated the nature of learner's freedom of expression in lecture theatres in Makerere University. Based on Students' voices, interactions are still very restricted and depriving the right to expression. Promotion of free expression is still a distant dream for many students, and learners are not a free will to learn, but the choice that enforced upon, in changing the mindset of learners and teachers in such situation is not adaptive.

Kirti Matliwala (December, 2016), the needs of higher education cannot be met by the Government alone. It needs the support of the Government, the Private providers and perhaps selectively participation of foreign universities. We have to come up with the mindset and take a realistic attitude, taking into consideration the fact that a major revolution is taking place in higher education in the world. We have to take certain steps for improvement of our higher education system.

Younis Ahmad Sheikh, (December, 2018), India is a vastcountry with skilled human resource potential, to utilize this potential properly is the issue which needed to discuss. Opportunities are available but how to get benefits from these opportunities are available but how to get benefits from these opportunities and how to make them accessible to others is the matter of concern. In order to have sustainable rate of growth, there is need to increase the number of institutes and the quality of higher education in India. Indian education sector has to revamp the resources and access with much better investment into this sector, for enabling the students to meet future needs of country and the world.

Henriikka Vartiainen, (2014), the research paper interested in developing international collaboration for future research on design-oriented knowledge creation and participatory leaning in networked communities. This could provide us with interesting opportunities through which to approach global phenomena such as sustainable development as a shared design object for learning and crowd sourcing in an international network of students, teachers, researchers, experts, and intersected others. At the same time, it would provide researchers with interesting opportunities to examine how and in what ways the participants from different backgrounds use and share their own interests, andaffordable tools and technologies, and social environments as resources for learning driven by joint co-development.

Cini C.K (2018) the article "A Study on the Attitude of Students towards Participatory Learning at Graduate Level". In this paper, the investigator tried to understand the meaning and strategies suitable for learning economics at under graduate level. The research found that, all participants are active as they set their own goals and work towards them to achieve the goal. The study also found that, in the learning process the learners share ideas, knowledge, opinion, materials and resources in order to reach a common agreement or to make joint decisions in a transparent way.

Higher education is meantto impart deepest understanding in the minds of students, rather than a relatively superficial grasp that must be acceptable elsewhere in the system. In higher education, the students have to think for critically so as to be able to stand intellectually on their own feet (Barnett, R.1997).

Braxton, J.M., Jones, W.A., Hirschy, A.S., & Hartley III, H.V. (2008). The role of active learning in the college student persistence. New roadmap for Teaching and Learning, 2008 (115), 71-83.

Ebert-May, D., Brewer, C., & Allred, S. (1997). Innovative lectures: Teaching for active learning. Bioscience, 601-607. Innovation in Large Lectures Teaching for Active Learning opens in new window.

Felder, R.M., & Brent, R. (1996). Navigating the bumpy road to student-centered instruction. College teaching, 44(2), 43-47

Lumpkin, A., Achen, R.M., & Dodd, R.K., (2015). Student Perceptions of Active Learning. College Student Journal, 49(1), 121-133. Student Perceptions of Active Learning opens in new window.

Michael, J. (2006). "Where's the evidence that active learning works?" Advances in physiology education 30(4), 159-167. Where's the evidence that active learning works? Opens in new window.

Prince, M. (2004). "Does active learning work? A review of the research." Journal of Engineering Education, Washington, 93,223-232. Does Active Learning Work? A Review of the Research.

Yazedjian, A., & Kolkhorst, B.B. (2007). Implementing small-group activities in large lecture classes. College Teaching 55(4) 164-169. Implementing small-group activities in large lecture classes, open in new window.

The teacher centered traditional approach mainly focuses on teachers as authoritative figures (Mascolo, 2009). Some of the scholars have analysed students as "empty vessels" whose primary role is receiving information that is passed through the teachers, and then later be assessed according to the knowledge that was transmitted to them (Murray & Hourigan, 2008). Therefore, student learning is measured with the help of scored tests results and other types of assessments, projects.

Figure 1.

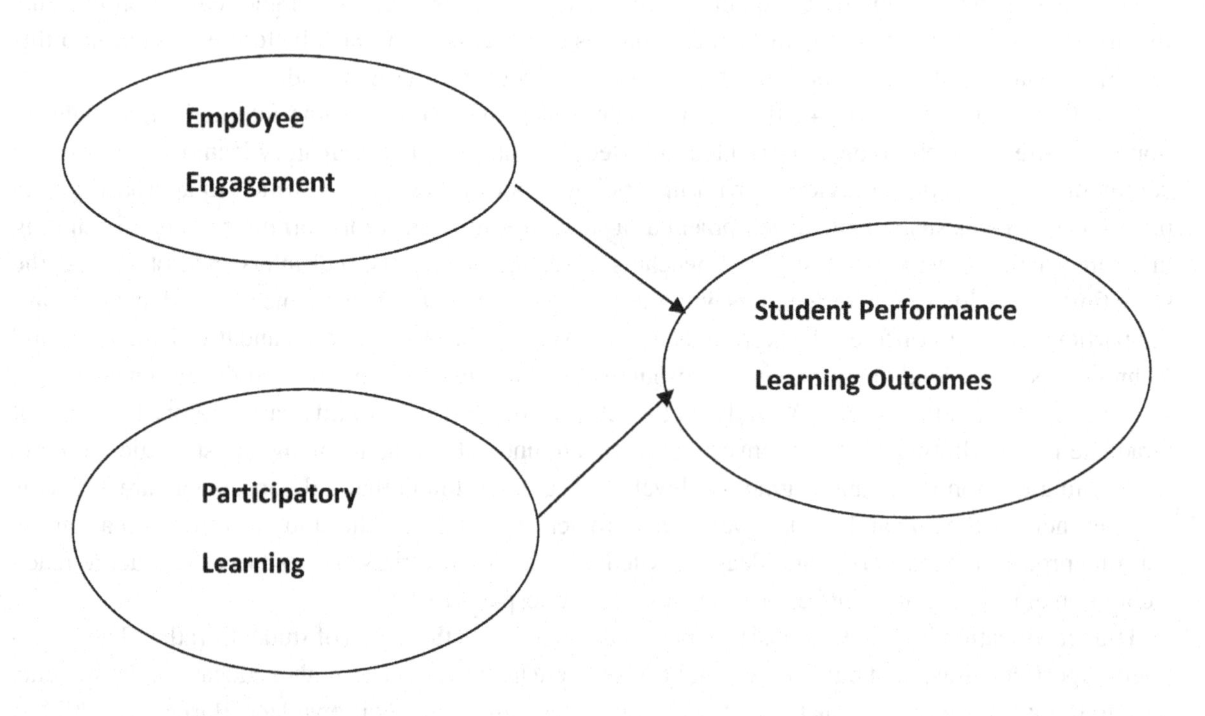

EMERGING TECHNOLOGIES

The technology has different meanings to faculty members engaging in different subjects. Teachers articulate their lectures according to their subjects and use technology according to the requirement. Digital devices are technology driven towards online and blended systems, scientific artifacts, tools, and other facilitating objects (Brown and Sammut, 2012). Technology relates to engineering procedures that assist in the creation of new gadgets. It is now commonly used even in the arena of teaching fraternity who are comfortable using the latest technology for teaching can be termed as those set of individuals who are welcoming the change in the gamut of teaching (Gershon, 2017). These are the teachers who are the working in adopting this new digital teaching pedagogy across the globe. The theoretical framework which can be used to understand the online teaching and learning process is the community of inquiry (CoI) model, which consists of three critical factors: social element, cognitive elements, and teaching element (Garrison et al., 2000). It is the interactions of all three elements of the model that facilitates the educational experience for participants, as illustrated as the ability of participants to which the participants in any particular setting can make meaning through sustained communication" (Garrison et al., 2000). Teaching presence is composed of the design of the educational experience and the creation of sound knowledge to better society (Garrison et al., 2000). "teaching practices during COVID 19", "adoption of technology in higher education during COVID 19", "Learning and teaching pedagogy during COVID 19," "Indian business schools and covid 19", "digital learning during the lockdown", "online teaching during a pandemic." "Education policy during COVID 19" and "physical learning during the shutdown and the combinations of these words. Both the authors performed data management and cleaning. Both the authors screened the articles on title and abstract. Authors viewpoint papers, systematic literature reviews, and studiesand other languages. The selected research papers were not limited and Europe to gain an international view of the topic. Both the authors analyzed the written papers in an ethical way. The key themes identified and discussed include "online teaching in higher education ", and the "shift towards online teaching during COVID 19."

Downloaded the full length articles identified were reviewed the authors maintained to keep highly cited articles out of all downloaded articles for the present study. The authors did not try to rate quality of studies included in this paper. The authors also included finding of some preprint articles and peer-reviewed articles. Most of the articles cited are from renowned publishers' articles. Referred to the articles like Elsevier, Emerald, Sage, Springer, Taylor and Francis, and Wiley. The different database searches identified 100 articles, of which 30 full-text articles were assessed, and eighteen were included in this paper. No relevant articles were returned searching the Global Research Database on technological innovation participatory learning approach in higher education.

The instructing teaching methods have been changed with the data and correspondence innovation (ICT) developments (Konig et al., 2020). For example, ICT has worked with the resources appropriation of understudy driven practices. For example, learning through projects (Law, Pelgrum, and Plomp, 2009) that helped the advancement of deliberate learning (Koh and Chai, 2014), request based learning (Bell et al., 2013) and learning through issue arrangement (walker et al., 2012). Earlier Scientists have introduced solid contentions for ICT as an impetus for a transformation of the showing teaching methods (Beauchamp and Kennewell, 2010). The model recognizes the connection between the information on a workforce about the innovation, instructional method, and content for a solid use of ICT for conveyance in a classroom (Herring et al., 2016). There has been a lot of exploration on the variables that influence the acknowledgment of innovation in instruction. This incorporates the reception of e-learning among the students, e.g., Boateng et al. (2016), Sanchez et al. (2013), Zhou also, Xu (2007) and educators, e.g., Holzmann et al. (2020), Salinaz et al. (2017), Buck enmeyer (2010), Nicolle and Lou (2008), Kotrlik and Redmann(2009). Be that as it may, the COVID 19 pandemic caused a circumstance where in both the instructors and students needed to receive the innovation not by decision but rather as a fundamental prerequisite for the instruction framework's smooth working. The selection accompanied numerous difficulties related to the absence of information about the utilization of innovation by both students and resources, trouble finding and choosing an appropriate stage for online class conveyance, cost of the permit, and issues identified with the foundation inaccessibility of the internet in far off regions. This turns the requirement for research from factors that influence the innovation reception to the elements that would influence the proceeded with utilization of innovation for mixed learning and students advantage. All things being equal, research that can direct the conduct change technique for the two understudies and resources would be required. Also, the substance conveyance and assessment design required a critical update. The vulnerability of occasions set a predicament for the schooling organizations and policy marked about the example of assessment. A need- based methodology was followed at the school level. Some essential class students were elevated straightforwardly to the following class; an online assessment was led for a few higher semester classes and disconnected tests for the individuals who showed up for the board (auxiliary and senior optional) tests. The colleges and business schools significantly received online mode for leading the assessments as the immediate advancement could influence the profession and situation. To the extent advanced education is concerned, it appears to be that the encouraging instructional methods would receive the mixed learning and showing mode for higher viability.

Table 1.

Factors	Formal learning	Informal learning
Pedagogical approach	Teacher-centered	Learning centered
Interaction	With the teacher	Interpersonal
Location	Classroom	Workplace, home, community, etc.
Process	Highly structured	Not structured
Consciousness	Always	Not necessarily
Modality	Explicit	By experience
Knowledge	Vertical	Horizontal
paradigm	Acquisition	Propositional

Advantages of Participatory Learning

- It is a pupil centered method of teaching learning wherein the entire process revolves around the student and is mainly concerned with the vital development of the student
- Develops engagement of students in learning process
- Focuses on students demands
- Helps to improve students overall development and helps them to improve their skills
- Enhances the Grasping power of Individuals
- Teachers with certain research skills and training are required

- Students require in detailed specified explanation
- Students adjusted may not be able to cope up with the participatory methods
- Work intensive

Hurdles for Participatory Learning

Huge classes and heterogeneity are the major hurdles for the success of Participatory Learning.

The problem of huge classes is addressed to a large extent when a variety of topics are brought into the class and all sections of the class are engaged. Huge classes continue to be an impediment to Participatory Learning. Limiting the number of students per class, increasing the number of sections appear as obvious remedies albeit impractical to implement.

Heterogeneity poses a problem particularly in English teaching. At whatever level on teaches, one has a section of the class uninvolved. If one assumes that the class has a poor level of language competence and tries to simplify or use the bilingual method, the students who are good are subjected to boredom and vice-versa.

Participatory Learning Strategies and Effective Teaching Practices

- Learners
- Teacher/Facilitator
- Meaningful Content
- Teaching & Learning Strategy (Team world Planning/ Organizing/ Group Discussion arid active participation)
- Learning Environment
- Evaluation (peer, self and Teacher).

Figure 2.

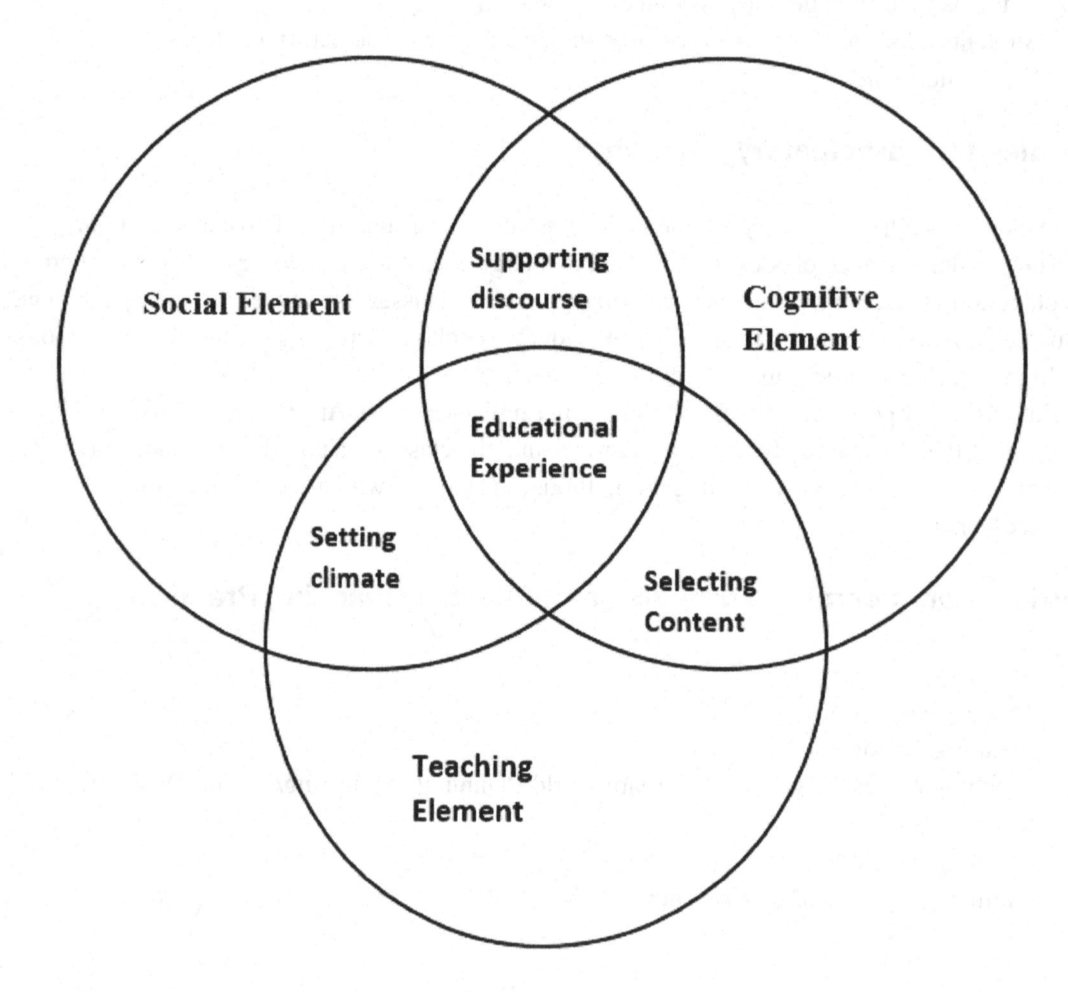

Relationship between Lack of Participatory Learning and Rejection of Employment

- Over 80% Indian engineers are unemployable, lack new age technology skills
- Only 40% of the students do Internship. The people who do don't have the skills to solve problems. Report: Employability assessment company Aspiring Minds.
- Minoti Shrivastava, Key Accounts Manager at Nexus Group says, we are creating learned individuals who lack the skills to survive in a corporate world, in a field work of the industry.
- Article 60% of the engineering graduates un-employed Less than 1% of the engineering students participate in summer internships and just 15% of engineering programmes are offered 3,200, Institutions and accredited by the National Board of Accreditations (NBA) Report: Times of India Education – New Delhi Published on 18/03/2017.
- Seshagiri Rao Tipirneni, Executive Chairman at Sentin group, it is not the dearth of jobs in India which is a problem but lack of skilled persons that baffling the industry and trade.

- According, to the report by https://ww.forbes.com says that schools should concentrate on providing participatory Education to children
- A person who IIT Delhi did not get placed due to no research papers even after having 7.5 CGPA in M.Tech.

Suicidal Attempts by the Youth due to Lack of Skills and Inferiority Complex

Dr Sahar Qazi — All India Institute for Medical Science

- Doctor specialized in the cases of Major Depressive Disorder (MDD) says most of than who were sick and wanted to suicide had more or less symptoms of unemployment and Inferiority complex in them.
- Even few of the other reports from renowned doctors say that people attempt to suicide in the recent years due to Lack of Confidence and Inferiority Complex.

Success Stories of Participatory Learning: Primary Source

Mr. Abhijit. K. Menon, a student from KMWA PU College @Deeksha MahalakshmiPuram, passed put in the year 2014-15 who was working at Raju and Associates" on the basis of his experience gained in Tally ERP during the time of his study in PU-2, which helped him earn his living along with continuing his education during the time of his degree.

Mr. Akhilesh and Mr. Sanjay who similarly are from Gouthama-Siddartha PU College @ Deeksha Rajajinagar, who got passed out in the year 2015-16, immediately got recruited at VI Analytics on the basis of their Tally and Excel learning at their institution, which even helped them, support their families.

TECHNIQUES

Approach Members Constructively: Reward members either verbally or through privilege for taking initiative and for actions of any kind, Everyone needs to know their contributions are appreciated, Even if their comments are not practical, a reply can begin with "That's a good point but what about., or "That's an interesting point, what do others think?"

Activity Profile: Talk to the people and ask their daily activities interview and observe or ask them to write notes.

Assignments (Theoretical and Practical): Ask participants to practice new roles and new skills e.g. ask a different person to act as a chairperson or fill in record book. Set assignments to find out the current market prices for something. As an exercise, work out the likely demand for a product.

Brainstorming: Ask members to think of any ideas that come to mind. All the ideas generated with suspended judgment. In an half hour a group can come up with approximately 250 ideas.

- **Solution to the incident** - As the name itself suggests this method is very similar to case study, The group of students analyzes the situation and further decides what more data needs to be collected regarding the issues

- **Labyrinth** - Under this method at a certain point of time students are given various situations and questions and are asked to solve all of them simultaneously
- **Study visits** - Being one of the major methods in Participatory learning study visits play a major role as they give visual and practical experience to the students.
- **Workshops** - It is a very popular method used to address situations, possibilities, solutions for various problems and the possible outcomes
- **Aquarium** - Just like a traditional aquarium one group of students are placed in a room and arc given a problem, meanwhile another group of students observes them from outside, later upon the finding of solutions they are given are given suggestions by the other students. These kind of methods promote problem solving ability
- **Role plays** - Under this method a group of students are given a particular script to enact. In their script they will be give situations out of which they need to find out the possible outcomes and ways to deal with unforeseen situations.
- **Information collection**: Ask members to collect information on relevant subjects at the local library.
- **Critical Incident**: Use problem situations to analyze advantage and disadvantage and possible solutions to a given situation. Pictures or drawings will help.
- **Describing visual Images**: Choose a photograph or drawing with a clear, relevant message, before displaying the image, ask three volunteers to leave the room. Discuss with the other participants how to describe the picture. Ask person A to return and listen to a description of the image (without seeing it). Let person A tell B and B tell C. Ask C to draw the picture. Discuss. Use this to highlight how messages become distorted when passed from one person to another.

DIRECTIONS FOR PEOPLE IN HIGHER EDUCATION

Education is a very important factor for the development of a country. We should make it appropriate according to the time and changing scenario of the world. Education provides an opportunity to reflect upon the social, economic, cultural, and moral issues facing by a human being. India needs to focus on education for more educated and efficient people to drive our nation. In the world, there are many Indian who well known for their capabilities and skills. To develop India as a digital nation or to become a prosperous partner in global development, India has to strengthen higher education with research and development. This paper is mainly focused on the overall scenario of higher education in India. This paper aims to identify issues and challenges in the field of higher education in India. Finally, the paper concluded here is all stakeholders have to make jointly effort to get solutions of the problems in higher education in India.

Rapid changes of modern world have caused the Higher Education System to face a great variety of challenges. Therefore, training more eager, thoughtful individuals in interdisciplinary fields is required. Thus, research and exploration to figure out useful and effective teaching and learning methods are one of the most important necessities of educational systems (2); Professors have a determining role in training such people in the mentioned field (3). A university is a place where new ideas germinate; roots strike and grow tall and sturdy. It is a unique space, which covers the entire universe of knowledge. It is a place where creative minds converge, interact with each other and construct visions of new realities.

Established notions of truth are challenged in the pursuit of knowledge. To be able to do all this, getting help from experienced teachers can be very useful and effective.

Given the education quality, attention to students' education as a main product that is expected from education quality system is of much greater demand in comparison to the past. There has always been emphasis on equal attention to research and teaching quality and establishing a bond between these two before making any decision; however, studies show that the already given attention to research in universities does not meet the educational quality requirements.

Attention to this task in higher education is considered as a major one, so in their instruction, educators must pay attention to learners and learning approach; along with these two factors, the educators should move forward to attain new teaching approaches. In the traditional system, instruction was teacher-centered, and the students' needs and interests were not considered. This is when students' instruction must change into a method in which their needs are considered and as a result of the mentioned method active behavior change occurs in them (4).Moreover, a large number of graduated students especially bachelor holders do not feel ready enough to work in their related fields (5). Being dissatisfied with the status quo at any academic institution and then making decision to improve it require much research and assistance from the experts and pioneers of that institute. Giving the aforementioned are necessary, especially in present community of Iran; it seems that no qualitative study has ever been carried out in this area drawing on in-depth reports of recognized university faculties; therefore, in the present study the new global student-centered methods are firstly studied and to explore the ideas of experienced university faculties, some class observations and interviews were done. Then, efficient teaching method and its barriers and requirements were investigated because the faculty ideas about teaching method could be itemized just through a qualitative study.

NEW TEACHING METHODS IN HIGHER EDUCATION

Teachers participating in this study believed that teaching and learning in higher education is a shared process, with responsibilities on both student and teacher to contribute to their success. Within this shared process, higher education must engage the students in questioning their preconceived ideas and their models of how the world works, so that they can reach a higher level of understanding. But students are not always equipped with this challenge, nor are all of them driven by a desire to understand and apply knowledge, but all too often aspire merely to survive the course, or to learn only procedurally in order to get the highest possible marks before rapidly moving on to the next subject. The best teaching helps the students to question their preconceptions, and motivates them to learn, by putting them in a situation in which their existing model does not work and in which they come to see themselves as authors of answers, as agents of responsibility for change. That means, the students need to be faced with problems which they think are important. Also, they believed that most of the developed countries are attempting to use new teaching methods, such as student-centered active methods, problem-based and project-based approaches in education.

Figure 3.

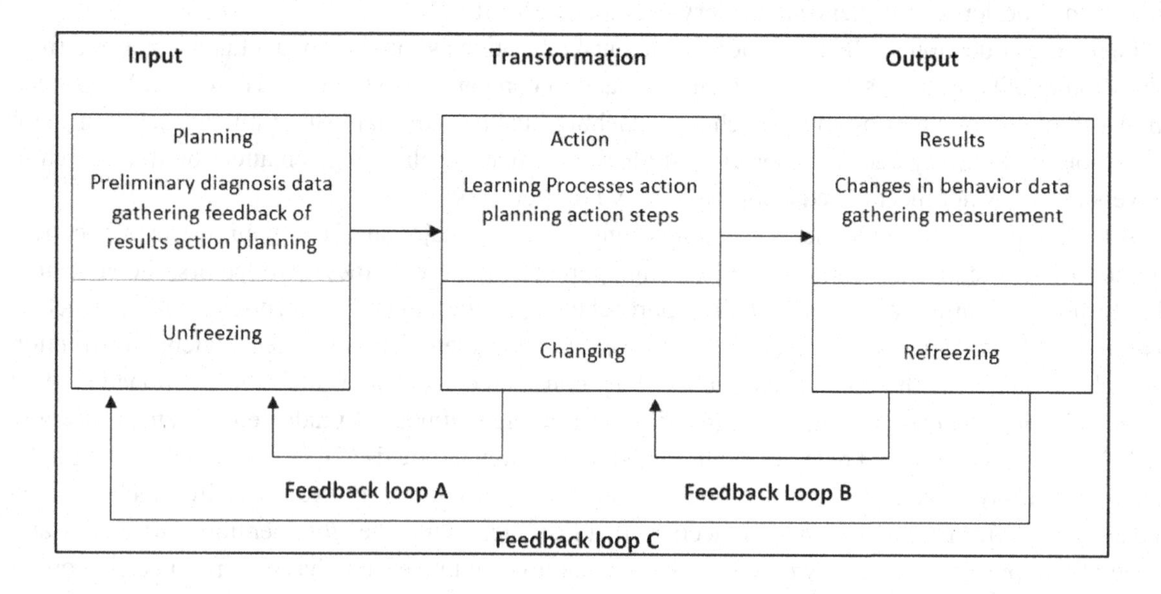

Suggestions Improving the System of Higher Education

There is a need to implement innovative and transformational approach form primary to higher education level to make Indian educational system globally more relevant and competitive. Higher suggestions improving the System of Higher Education:

- Higher educational institutes need to improve quality and reputation.
- There should be a good infrastructure of colleges and universities which may attract the students.
- Government must promote collaboration between Indian higher education institutes and top International institutes and also generates linkage between national research laboratories and research centers of top institutions for better quality and collaborative research.
- There is a need to focus on the graduate students by providing them such courses in which they can achieve excellence, gain deeper knowledge of subject so that they will get jobs after recruitment in the companies which would reduce unnecessary rush to the higher education.
- Universities and colleges in both public private must be away from the political affiliations,
- Favoritism, money making process should be out of education system etc.
- There should be a multidisciplinary approach in higher education so that students' knowledge may not be restricted only up to his own subjects.

Objectives of the Study

- To identify the impact of teacher developing program on the quality of the educational process.
- To develop means and methods to ensure the development of teacher's skills and knowledge, and to identify aesthetics in teaching performance.

- Development of a scientific program containing complete information to assist teacher.
- Activation of teaching aids that helps to develop teacher's performance such as managing of discussions, communication skills, motivation, and simulation exercises, and puzzles.
- Introducing new methods that help teachers developing their teaching performance.

Importance of the **Study**

- According to the researcher's knowledge, there is a lack of studies in the field of improving the quality of the educational process through a program that develop teacher's performance. This encouraged the researcher to conduct this study.
- The researcher is expecting that the results of this study will be of great help to Ministry of Education in the preparation of curricula and teaching methods as well as improving factors affecting the overall quality of the educational process.
- The study will provide teachers with an effective strategy to develop their performance and develop the currently used teaching methods.
- Researcher predicted that this study opens an area for further studies designed to develop different strategies in the field of teaching

Effective Teachers and Teaching

- It is important to distinguish between the related but distinct ideas of teacher quality and teaching quality.
- Teacher quality might be thought of as the bundle of personal traits, skills, and understandings an individual brings to teaching, including dispositions to behave in certain ways. The traits desired of a teacher may vary depending on conceptions of and goals for education; thus, it might be more productive to think of teacher qualities that seem associated with what teachers are expected to be and do.
- Research on teacher effectiveness, based on teacher ratings and student achievement gains, has found the following qualities important:
    - Strong general intelligence and verbal ability that help teachers organize and explain ideas, as well as to observe and think diagnostically.
    - Strong content knowledge – up to a threshold level that relates to what is to be taught.
    - Knowledge of how to teach others in that area (content pedagogy), in particular how to use hands-on learning techniques (e.g. lab work in science and manipulative in mathematics) and how to develop higher order thinking skills.
    - an understanding of learners and their learning and development– including how to assess and scaffold learning, how to support students who have learning differences or difficulties, and how to support the learning of language and content for those who are not already proficient in the language of instruction.
    - Adaptive expertise that allow teachers to make judgments about what is likely to work in a given context in response to students' needs. Although less directly studied, most educators would include this list a set of dispositions to support learning for all students, to teach in a fair and unbiased manner, to be willing and able to adapt instruction to help students succeed, to strive to continue to learn and improve, and to be willing and able to collaborate

with other professionals and parents in the service of individual students and the school as a whole.

CONCLUSION

The needs of higher education cannot be met by the Government alone. It needs the participation of the Government, the private providers and perhaps selectively participation of many universities. We have to free ourselves from the mindset and take a realistic attitude. Taking into consideration the fact that a major revolution is taking place in higher education in the world. We have to take certain steps for improvement of our higher education system.

Learning on the students experience can be seen in the way the students look their teachers and the amount of the enhanced knowledge, skills & ability they perceived / learn in the classroom. From this research, we can find these employee behaviour and attitudes affect students profoundly and to focus on the role of the teacher, which plays an important role in student, learning outcomes so colleges need to find new and innovative ways to support and reward their faculty.

Before beginning of the new topic active or participative learning require the teacher to check what each learner knew about the topic and student life experiences. Students should know the common preconceptions and a possible misconception about the topic encourages meaningful learning. The teachers and principals should make students participate in lessons by giving their experiences concerning the lessons under study.

The efficiency of integrating participatory approach in teaching and learning remains essential to optimize effective learning on the side of students. The participatory learning improves the relationship between students and teachers in a mutual learning process. The students are motivated to be more independent and interdependent. They take their own control of their learning through participatory learning.

At the point when we move towards advanced innovation reception for instructing, a few issues need consideration. To being with, the improvement of a suitable interface for learning and commitment viable with the surviving foundation is required, given the monetary worries of establishments talked about in the initial segments. Second, the endeavors should be directed toward the proceeded with reception of innovation for training.

Third, because of the impediments about the entry level position that empowered the B school understudies to learn in a characteristic workplace, the pathways for powerful experiential discovering that can likewise upgrade the range of abilities and employability of understudies should be decided. Ultimately, strategies to fill the computerized partition for all inclusive learning need prompt consideration.

Participatory approaches of learning are an active approach that encourages students to think themselves and sparks young minds creatively to apply the concepts and gain professional values, knowledge and skills with deeper understanding of the meaning of civic responsibility and prepare themselves for serving the community. Therefore, a teacher has to apply high ethical standards while doing the participative learning.

REFERENCES

Braxton, J.M., Jones, W.A., Hirschy, A.S., & Hartley III, H.V. (2008). The role of active learning in the college student persistence. *New Roadmap for Teaching and Learning, 2008*(115), 71-83.

Ebert-May, D., Brewer, C., & Allred, S. (1997). Innovative lectures: Teaching for active learning. *Bioscience, 47*(9), 601–607. doi:10.2307/1313166

Felder, R. M., & Brent, R. (1996). Navigating the bumpy road to student-centered instruction. *College Teaching, 44*(2), 43–47. doi:10.1080/87567555.1996.9933425

Kucharčíková, A., & Tokarčíková, E. (2016). Use of participatory methods in teaching at the university. *Turkish Online Journal of Science & Technology, 6*(1).

Kumar, V., Niveditha, V. R., Muthukumaran, V., Kumar, S. S., Kumta, S. D., & Murugesan, R. (2021). A Quantum Technology-Based LiFi Security Using Quantum Key Distribution. In Handbook of Research on Innovations and Applications of AI, IoT, and Cognitive Technologies (pp. 104-116). IGI Global.

Kumar, V. V., Raghunath, K. K., Rajesh, N., Venkatesan, M., Joseph, R. B., & Thillaiarasu, N. (2021). Paddy Plant Disease Recognition, Risk Analysis, and Classification Using Deep Convolution Neuro-Fuzzy Network. *Journal of Mobile Multimedia*, 325-348.

Kumar, V. V., Raghunath, K. M., Muthukumaran, V., Joseph, R. B., Beschi, I. S., & Uday, A. K. (2021). Aspect based sentiment analysis and smart classification in uncertain feedback pool. *International Journal of System Assurance Engineering and Management*, 1-11.

Lumpkin, A., Achen, R. M., & Dodd, R. K. (2015). Student Perceptions of Active Learning. *College Student Journal, 49*(1), 121–133.

Mayer, R. E. (2004). Should there be a three-strikes rule against pure discovery learning? *The American Psychologist, 59*(1), 14.

Michael, J. (2006). Where's the evidence that active learning works? *Advances in Physiology Education, 30*(4), 159–167.

Muthukumaran, V., Joseph, R. B., & Uday, A. K. (2021). Intelligent Medical Data Analytics Using Classifiers and Clusters in Machine Learning. In Handbook of Research on Innovations and Applications of AI, IoT, and Cognitive Technologies (pp. 321-335). IGI Global.

Muthukumaran, V., Vinothkumar, V., Joseph, R. B., Munirathanam, M., & Jeyakumar, B. (2021). Improving network security based on trust-aware routing protocols using long short-term memory-queuing segment-routing algorithms. *International Journal of Information Technology Project Management, 12*(4), 47–60. doi:10.4018/IJITPM.2021100105

Nagarajan, S. M., Deverajan, G. G., Chatterjee, P., Alnumay, W., & Muthukumaran, V. (2022). Integration of IoT based routing process for food supply chain management in sustainable smart cities. *Sustainable Cities and Society, 76*, 103448. doi:10.1016/j.scs.2021.103448

Nagarajan, S. M., Muthukumaran, V., Beschi, I. S., & Magesh, S. (2021). Fine Tuning Smart Manufacturing Enterprise Systems: A Perspective of Internet of Things-Based Service-Oriented Architecture. In Handbook of Research on Innovations and Applications of AI, IoT, and Cognitive Technologies (pp. 89-103). IGI Global.

Prince, M. (2004). Does active learning work? A review of the research. *Journal of Engineering Education, Washington*, *93*, 223–232.

Roseth, C. J., Johnson, D. W., & Johnson, R. T. (2008). Promoting early adolescents' achievement and peer relationships: The effects of cooperative, competitive, and individualistic goal structures. *Psychological Bulletin*, *134*(2), 223.

Slavin, R. E. (2014). Cooperative Learning and Academic Achievement: Why Does Groupwork Work? *Anales de Psicología*, *30*(3), 785–791.

Yazedjian, A., & Kolkhorst, B. B. (2007). Implementing small-group activities in large lecture classes. *College Teaching*, *55*(4), 164–169.

Chapter 23
Blockchain-Based Incentive Announcement Network for Communications of Smart Vehicles

Shouryadhar Karamsetty
Asia University, Taiwan

ABSTRACT

The emergence of embedded and connected smart devices, systems, and technologies has given rise to the concept of smart cities in modern metropolises. They've made it possible to connect "anything" to the internet. As a result, the internet of vehicles (IoV) will play a critical role in newly constructed smart cities in the approaching internet of things age. The IoV has the ability to efficiently tackle a variety of traffic and road safety issues in order to prevent fatal crashes. Furthermore, ensuring quick, secure transmission and accurate recording of data is a particular problem in the IoV, particularly in vehicle-to-vehicle (V2V) and vehicle-to-infrastructure (V2I) connections. Furthermore, the authors qualitatively examined the suggested overall system performance and resiliency against popular security assaults. The proposed method solves the primary issues of vehicle-to-x (V2X) communications, such as centralization, lack of privacy, and security, according to computational experiments.

INTRODUCTION

With the help of numerous sensing, networking, and data analysis techniques, automobiles have recently gained more autonomy. Internal and external information about a vehicle can be communicated to base stations or neighboring vehicles via wireless channels, due to onboard sensors. Security is frequently viewed as a critical concern in wireless communication systems due to the open surroundings, especially in automobile networks, which have far more complicated and fast-changing situations. As a result of the lack of appropriate ways, attackers may counterfeit or manipulate important messages, compromising the security and efficiency of vehicle networks. The Internet of Things (IoT) is a network of physical

DOI: 10.4018/978-1-7998-9640-1.ch023

devices, such as home appliances, automobiles, and other goods, that are embedded with network connectivity, actuators, sensors, software, and electronics (Atzei et.al,2017)

. Each device is identified and integrated into the Internet infrastructure using embedded computing systems. Such gadgets can be controlled remotely thanks to the network infrastructure. As a result, computer-based systems are becoming more integrated with the real environment, resulting in cost savings, increased accuracy, and efficiency, as well as less human intervention.

Traditional vehicular ad-hoc networks (VANETs) are being transformed into the Internet of Vehicles (IoV) by the Internet of Things. The Internet of Vehicles (IoV) represents real-time data interaction between vehicles and infrastructures via smart terminal devices, vehicle navigation systems, mobile communication technology, and information platforms that allow for information exchange, sharing of driving instructions, and network system control.

This notion has made information about automobiles and infrastructures easier to collect and share. It also enables data collecting, computation, and exchange in Internet-based systems and other information platforms.

This concept has recently gained traction in the real world. 25 billion things are expected to be connected to the Internet in the near future, with autos accounting for a substantial portion. Intelligent Transportation Systems (ITS) in Japan and Europe have already deployed certain versions of IoV technology, while New Delhi has equipped 55,000 licensed rickshaws with GPS sensors. This concept has sparked a lot of research and commercial interest due to the rapid growth of communication and processing technology. Identity identification, digital signatures, and data encryption are only a few ways that have been extensively researched to meet the first two goals. However, studies focusing on the third goal, data credibility, are significantly lacking. In addition, initiatives to verify identity and maintain data integrity may be futile once the content of a message is compromised.

Smart automobiles are also becoming more connected to the Internet, other adjacent vehicles, and traffic management systems (Bader et.al, 2018). Vehicles are being integrated into the Internet of Things in this fashion (IoT). Despite its undeniable benefits, however, this notion has several drawbacks. Importantly, because of their increased connectivity, smart vehicles are difficult to secure, making them vulnerable to malevolent actors. Furthermore, a sensitive data exchange raises new privacy concerns.

Challenges in IOT

1. **Lack of Privacy**: In most contemporary communication designs, user privacy is not secured. In all the other words, data about the vehicle is shared without the authorization of the owner.
2. **Scalability:** Due to the rapid rise in embedded technologies, the utilization of miniature devices such as actuators and sensors has increased. Meanwhile, the amount of data generated by these devices continues to rise endlessly. As a result, another key IoV difficulty is managing the amount of devices and the data they generate.
3. **Centralization:** Smart car architectures are currently based on centralised, mediated communication mechanisms. All of the vehicles are identified, authenticated, authorized, and connected by central cloud servers. It is unlikely, however, that this model will be scaled.
4. **Threats to Safety:** The number of self-driving features in smart cars is increasing. As a result, a security breach caused by a malfunction caused by malicious software installation can result in car accidents, putting road users at risk.

This paper's reminder is organized as follows. The second section provides an overview of block-chain technologies. Section III introduces the proposed system's detailed designs, which include entities and main procedures. In Section IV, we run a series of tests to ensure that the proposed system is effective and reliable. Finally, Section V brings this paper to a conclusion.

TECHNIQUES IN THE BLOCKCHAIN

The term "block-chain" refers to a set of procedures used in decentralized networks to ensure that all users have access to the same database. Satoshi Nakamoto (Li et.al,2018), first introduced it in order to abstract the essential techniques of the well-known digital currency, Bitcoin. In contrast to typical centralised network structures, block-chain-based networks have no set centre nodes. Every member of the organization has roughly equal power and stores the same block-chain network. Block-chain has been used in a variety of application scenarios due to its high security and reliability, and it is recognized as one of the essential ways to boost global development. Figure.1 shows the block diagram of block chain. A block-chain, as shown in Figure.1, is an ordered collection of blocks, each of which records a specific amount of historical transactions (TXs). Traders generate these TXs, which are disseminated over the entire network. By preserving a digest (i.e., the hash value) of the preceding block, each block is "chained" to the one before it (Shojafar et.al, 2016). As a result, any alteration to a single block would inevitably compromise the chain's integrity. In furthermore, every unit generally uses a nonce, which is the solution to a mathematical problem (Cordeschi et.al, 2015).

Figure 1. Block diagram of block-chain

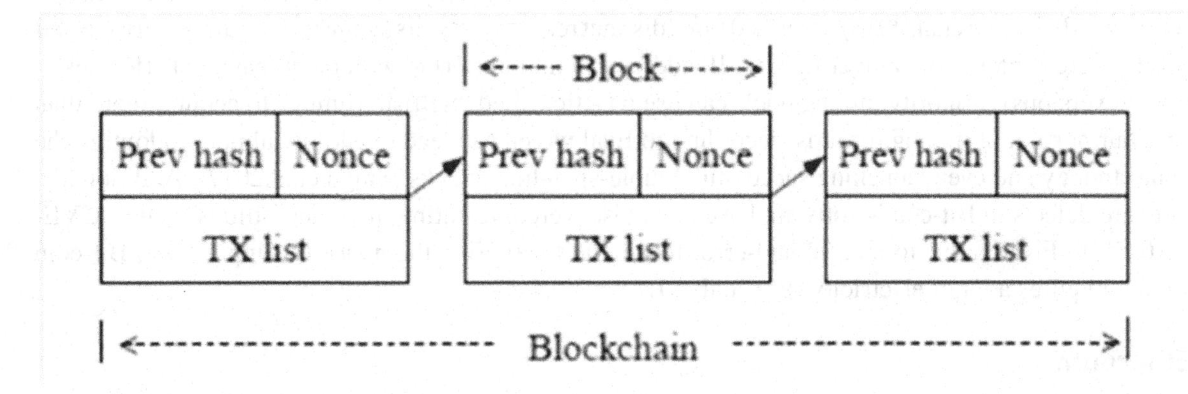

The node that resolves the challenge first, i.e., the miner, is elected as a temporarily centre node and broadcasts its block to others. In recent block-chain-based systems, various miner election systems have been suggested, such as proof-of-stake, proof-of-capacity, and proof-of-work, in which nodes with more processing power, capital, and storage space are more likely to win the election (Aloqaily et.al, 2017).

Figure 2. Architecture of the block-chain-based reputation system

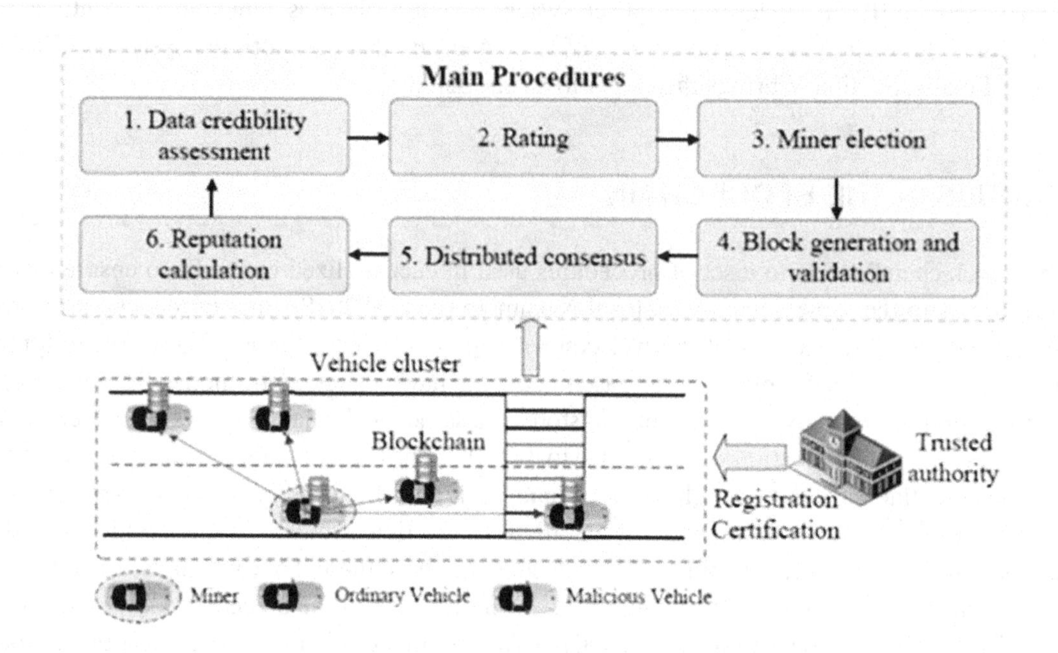

BLOCK CHAIN TECHNIQUE WITH ETHEREUM

Bitcoin's current security is based on the assumption that the majority of nodes in the network will behave correctly in order to retain the system's legitimacy. By tampering with blocks and transaction delivery (AlRidhawi et.al, 2018) described three distinct realistic assaults against Bit-coin's security. Sybil attacks were centered on tampering with IP addresses and interfering with quick payments (for newer Bitcoin versions). Minority mining pools can use a tactic called "Selfish Mining" to acquire more than their fair portion of mining rewards, according to Eyal which can lead to additional pools adopting the same strategy and even launching successful double-spending attacks (Singh et.al,2017). Additionally, software defects in Bit-coin's software have been discovered, resulting in vulnerabilities such as CVE-2010-5139. Finally, due to its trial-and-error technique for finding the nonce to supply PoW, Bit-coin mining requires a lot of electricity (Bai et.al, 2018).

Ethereum

Ethereum is a distributed computing platform based on Block-chain that was created by Vitalik Buterin in 2013 and debuted in 2015 after a successful online crowd sale. Furthermore, it is a programmable Block-chain, allowing users to construct new applications because Ethereum runs the decentralized application programming code (Baralla et.al, 2019). A smart contract is a programming code that allows money, data, and content to be exchanged by generating, implementing, and operating Decentralized Software applications block-chain network(NagaJyothi and Sridevi, 2019).

Ethereum Virtual Machine (EVM)

It is Turing complete software that allows users to deploy and run multiple programs written in various programming languages (such as Solidity) and greatly simplifies the development of Block-chain applications (Jyothi et.al,2020). It is segregated from the rest of the main network because it is a sandbox environment. As a result, each Ethereum network runs its own EVM by following the identical instructions. The Ethereum currency—Ether—is used to pay for computations in the EVM (ETH) (Bartoletti et.al, 2017).

Ether and gas

The transaction sender pays Ether, a volatile Ethereum currency, for executing the code in a contract triggered by a message or a transaction, as well as storage and calculation. In addition, the execution cost or network use cost is indicated in gas. Gas is a fixed unit that describes the amount of compute work required by minors to perform a specific operation. As a result, users utilise ETH to pay for the gas required for the execution (Bogner et.al, 2016).

Transaction

It is a signed data package that contains messages that are sent between Ethereum accounts. The transactions are confirmed through the mining process, which involves producing a signature using the transaction sender's private key. The Ethereum destination address, ETH gasPrice, the transmitted amount and other optional data are all included in the transaction.

Proof-of-Work (PoW)

This is the initial consensus technique used in Block-chain to determine the legitimacy of transactions and assure security. Miners use sophisticated software to solve mathematical problems in PoW. As a result, miners create transaction blocks and add them to Ethereum by connecting them to earlier blocks. Every subsequent block includes the previous block's hash. As a result, the newly added block forms a link in the chain that may be followed until it reaches the first, or genesis, block.

PROPOSED WORK

Overview of the System

The goal of this study is to demonstrate an IoV solution that includes a Real-Time Application (RTA). This approach allows vehicles and other actors in transportation networks to communicate in a secure manner. It tries to overcome constraints like execution time and, as a result, enhances performance. A DISV prototype was created and tested based on the following scenario: if a driver becomes drowsy, the nearest cars should be notified via Block-chain. The proposed solution should primarily consist of three layers, as it is built on an IoT architecture. The physical layer is the perception layer. It is made up of a number of IoT devices with sensors that can recognize and gather information about the surroundings (physical characteristics) as well as detect neighboring smart objects. The perception layer's Android Application for Vehicles (AV) collects and analyses data about the trip, the car, and the driver's behavior. The Android Application for Infrastructure (AP) simulates the role of IoT devices such as radars, traffic lights, roadside electronic signs, and other devices that are integrated into highways.

1. The network layer transports and analyses sensor data as well as connecting sensors to other servers, network devices, and smart things.
2. The Block-chain application and the Central Cloud Server make up the application layer. It provides IoT devices with application-specific services. The Block-chain application, to be more specific, controls communication between cars and other transportation system actors. The Central Cloud Server is in charge of processing and evaluating the data collected as well as managing other actors' invitations.

Figure 3. Block diagram of proposed IOT

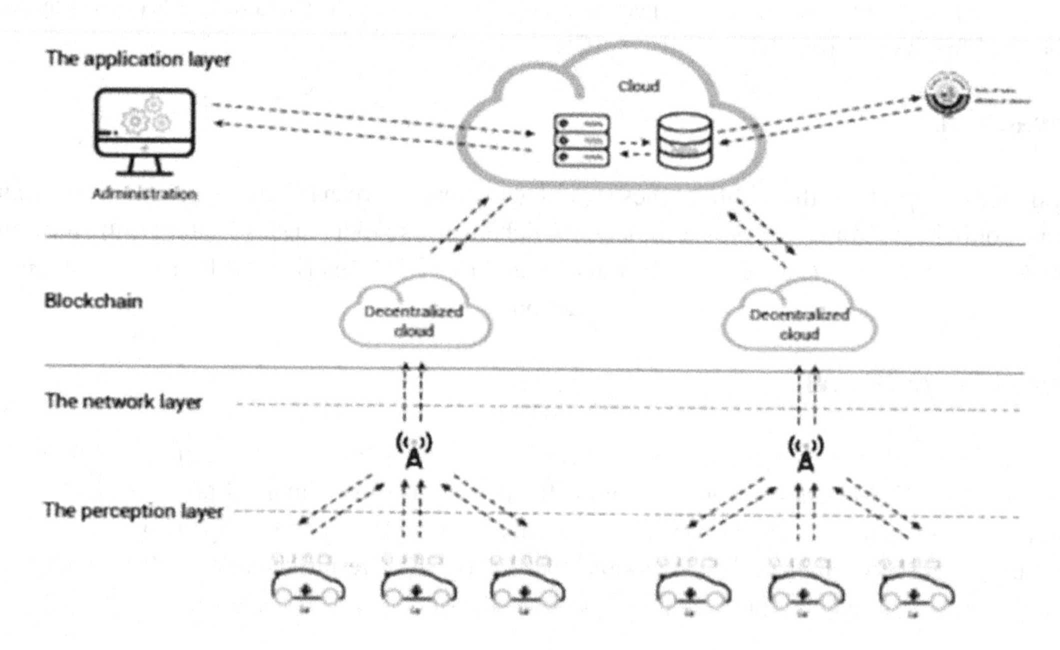

The architecture of the suggested solution is depicted in Figure.3 along with the core workflow, which consists of three main components. The cars first communicate data to a central server. Second, the central server sends an invitation to join to the Block-chain layer based on the data obtained. Finally, the cars can safely share data with other IoV participants in the same area.

The Perception Layer

In order to evaluate probable cases that involve various components, Android applications for vehicles (AV) and infrastructure (AP) have been built, as outlined in the following sections.

Android Application for Vehicle

The Android application AV is made up of two sub-systems. The Vehicle Data Collection System (VDCS) is the initial subsystem, and it collects information on the trip and the vehicle. The next is the Driver Drowsiness Detection system, which collects data on the driver's behavior in order to determine whether or not he is drowsy.

1. VDCS is designed to capture information on the vehicle, such as the make and model, as well as motor statistics such as horsepower, speed, and engine size. Finally, as shown in Figure.4, the system collects trip-related data such as start and end times, distance, and minimum, maximum, and average speed. Every 15 seconds, it is programmed to detect rotational velocity along the Roll, Pitch, and Yaw axes, acceleration, distance, and GPS position.
2. The goal of Driver Sleepiness Detection is to identify driver drowsiness and avoid accidents that may occur as a result. This system is part of the Advanced Driver Assistance System (ADAS), which is an important feature of modern vehicle technology. The purpose of advanced driver assistance systems (ADAS) is to improve safety and provide a pleasurable driving experience.

The Android app contains four pages in total. The first screen allows you to log in with your username and password. Following authentication, the user can either start a new trip or go to the second page to see information about the previous five trips. If the user selects a new journey, the app will begin recording and showing all of the data stated in the preceding section. The captured data will then be sent to the cloud server via the web service. The driver's face will be captured and displayed on the fourth page via the front camera.

Figure 4. The 4 main pages of the AV

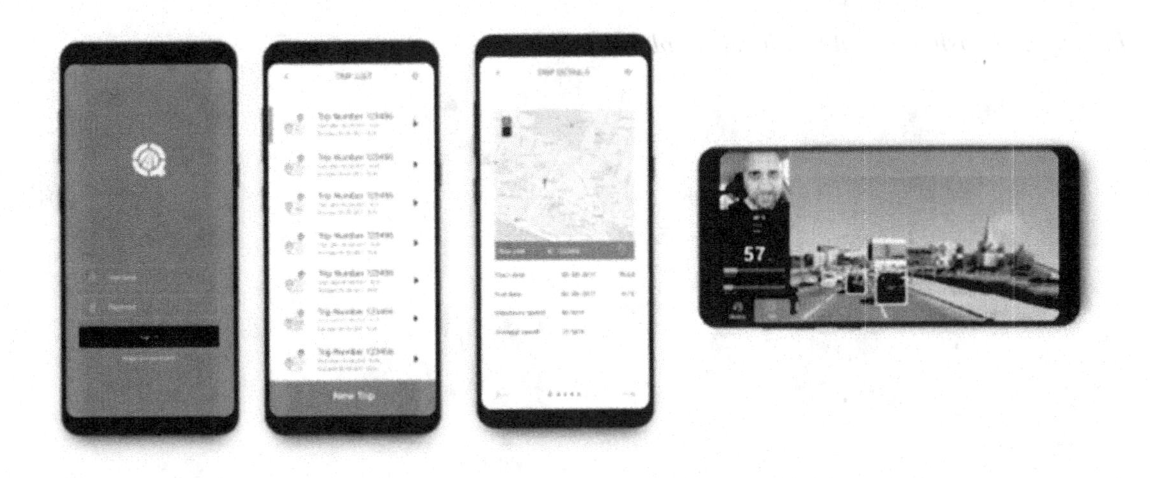

Android Application for Infrastructure

The goal of this program is to imitate the role of IoT devices such as radars, traffic signals, roadside electronic signs, and other devices that are integrated into roads. Traffic jams, car speed, and weather conditions are just a few of the Android app's extra features that can be added to the perception layer.

Application Layer and Network Layer

The network layer connects the servers and sends and processes sensor data. To deliver data to the server, the programmed can use Wi-Fi or mobile internet (3G/3G+/4G).

The hybrid system is used in this data gathering procedure to collect and store data locally before sending it to the server. When the Internet connection is poor or unreliable, this strategy has proven to be quite successful for data collection.

The application layer consists of two main components: a central cloud server and a communication system based on a Block-chain Network.

Cloud server

The end-user receives application-specific services from the central cloud. Before exposing the data to the end-user, it transmits it to the web services for processing and analysis. The web service is an application layer component that facilitates communication between various IoT solution components such as the web site, database server, IoT devices, and embedded systems. Microsoft's Windows Communication Foundation is used to create the REST Architecture and JSON message format online service. It includes information regarding crashes from the General Directorate of Traffic at the Ministry of Interior, as well as road conditions and any other relevant data from other authorities, in addition to data collected from devices. The data is accessible to the end user via a website that provides direct access to online services.

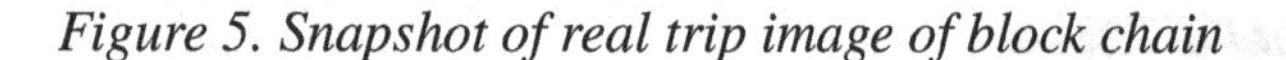

Figure 5. Snapshot of real trip image of block chain

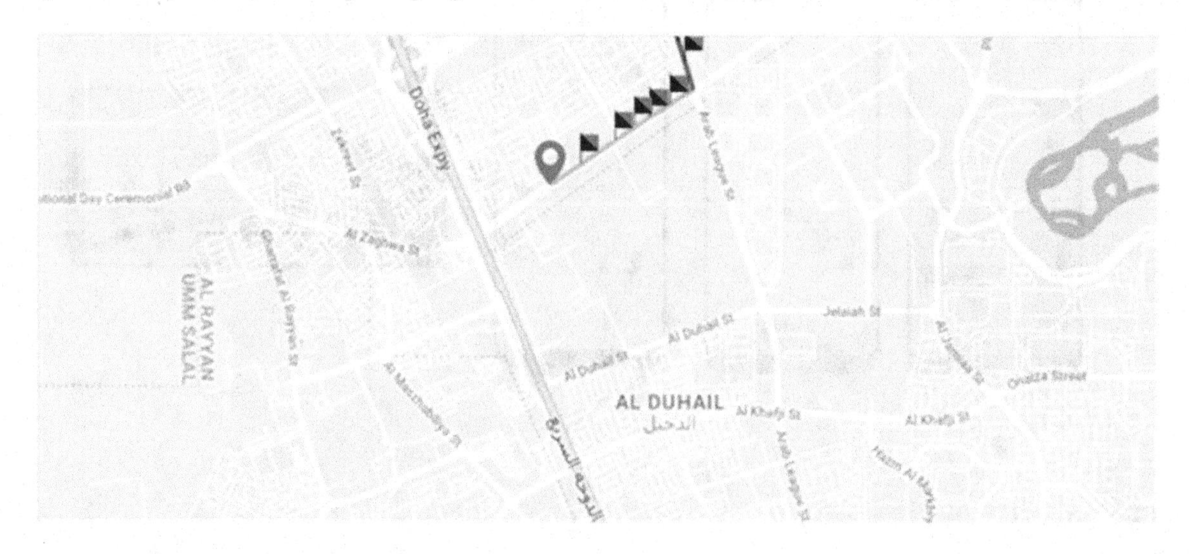

RESULT ANALYSIS AND DISCUSSION

Different approaches can be used to evaluate the performance of software solutions. It's very important to assess the precise features required for the solution's smooth operation. As illustrated, the suggested solution's primary attributes include execution time, costs, availability, integrity, immutability, and security. All of these features must be of the greatest quality in order for the solution to run without a hitch. As a result, this study will focus on these characteristics in order to evaluate the solution's overall effectiveness.

The Ethereum network's Testnet was utilized to deploy the smart contract prototype. The expenses of creating and executing a smart contract are examined in this section. The following values were utilized, which were valid in January 2020: 1 ETH $161.92 US and 1 gas 1 wei (0.000000001 ETH). The minimum gas value for a transaction was set at 1 wei, and the average gas value at the time of investigation was around 0.006845 Ethereum.
1 Gas = 0.006845 Ethereum (ETH)
Gas Price = 6,138,887 Gwei

Execution time

Execution time is one of the most important metrics for evaluating transportation management systems like DISV. In fact, even tiny delays in sending or receiving signals can cause major system failures. Because the mining process relies on complicated problem solving, it is critical to ensure the timely inclusion of each message to the smart contract for the framework for communication between cars and all other actors in the transportation system based on block-chain technology to function properly. Because the suggested prototype is a real-time application, the execution time is critical. The call times for each function of the Android application are measured in computational tests. A server with a configuration of 64 GB RAM and a Core i7-000 was utilized to evaluate the performance of the Ethereum private Block-chain proposed solution.

Figure 6. Execution time of different decentralization of IOT

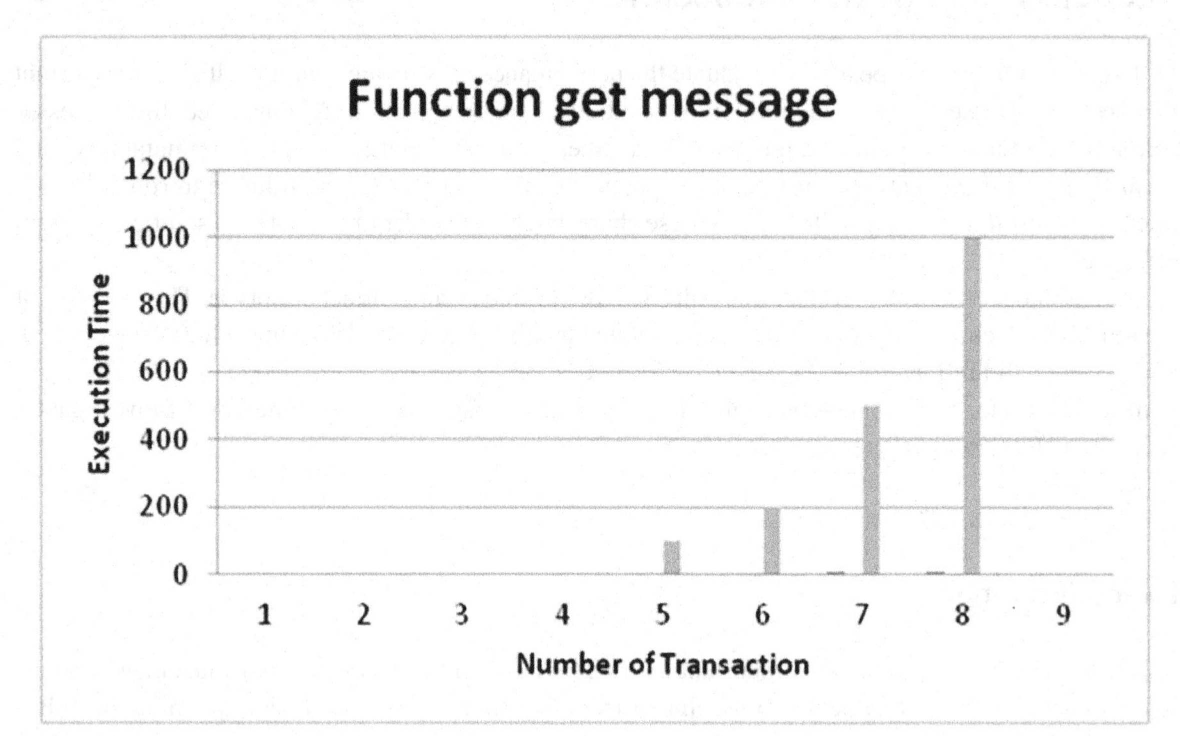

Memory and Power Consumption

Because DISV leverages Blockchain for IoT, it's vital to assess power and memory usage, as IoT devices typically have very low power and processing capacities. In the demo, a Huawei P8 Lite with 2 GB RAM, a Li-Po 2500 mAh battery, and a Hisilicon Kirin 620 Processor was used for compute experiments. Figure.6 shows that the memory usage of the proposed Android solution is substantially lower than that of other commercial apps like the Facebook app (134 MB), WhatsApp (106), and Skype (233 MB). In terms of energy consumption, the proposed solution uses an average of 23.43 mAh, which is comparable to Skype and Facebook, which use 21.66 mAh and 18.56 mAh, respectively, as shown in Figure.7.

Figure 7. Memory consumption of DISV with other applications

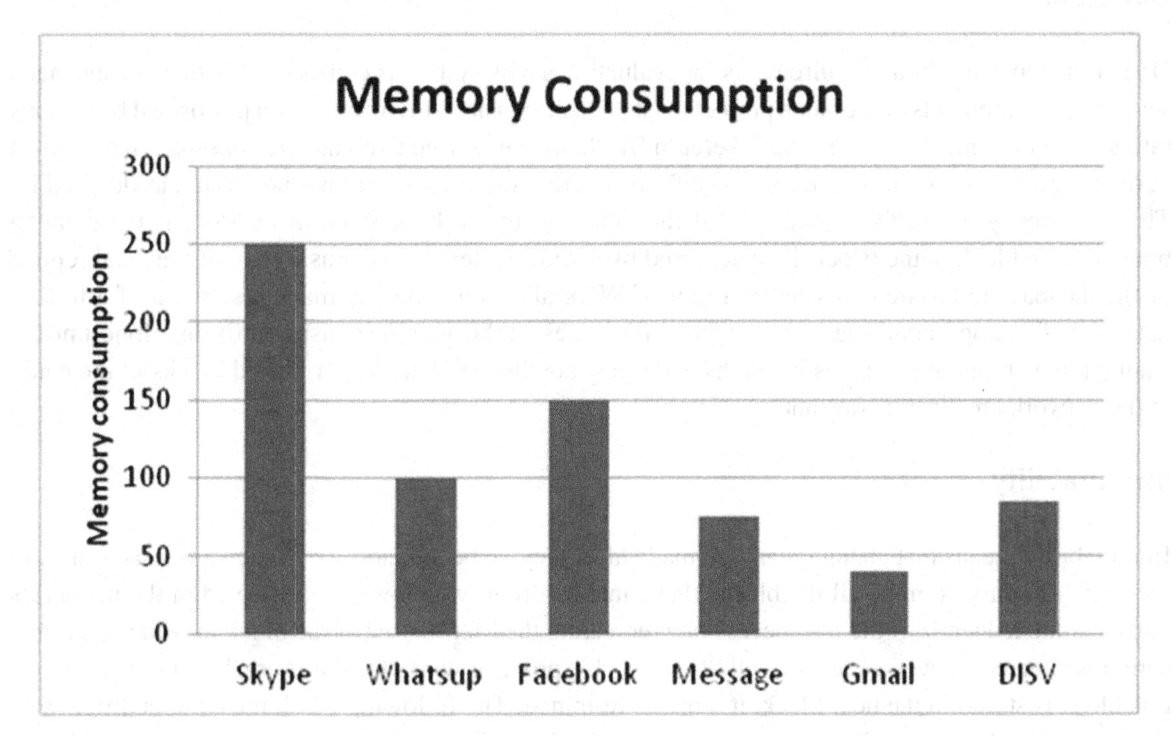

Availability

The availability of transportation management systems like DISV is another important feature. More specifically, even a short temporary system breakdown is likely to cause traffic congestion and crashes. The term "availability" refers to a system's ability to be accessed at any time. A system shutdown (off-line) can be caused by a variety of circumstances, ranging from planned maintenance downtime to unexpected failure. Because the Blockchain is decentralised and strong, assaults such as denial-of-service (DoS) attacks, which exclusively target nodes, are prevented because the central party cannot be a single point of failure. The Blockchain's dissemination, however, is not comprehensive. Mining power is usually restricted to miners that live in close proximity to one another. As a result, they can be isolated by using the internet infrastructure to hijack some border gateway protocol (BGP) prefixes with a routing attack.

Integrity

Another important attribute of systems that share sensitive data among users is integrity. As a result, the suggested software solution's data integrity property must be evaluated. The quality and reliability of data throughout its life cycle is referred to as data integrity. It is intertwined with the idea of data security. Uncorrupted data is full, and it does not alter in terms of its status. Maintaining data consistency throughout its life cycle is critical for assuring security.

Consistency

One of the most important requirements for evaluating a new system is consistency, which means that a series of measurements of the same project must provide consistent findings when performed by various raters using the same approach. The Ethereum Blockchain was used to create the consensus mechanism in the suggested solution. As a result, specific reconciliation methods are not necessary, as detailed in. The consistency mechanism assumes that the branch with the highest Proof-of-Work is the genuine branch. Each block in the Blockchain accepted by a node preserves the consistency of the local replica of the database to ensure consistency. Proof-of-Work allows for the automatic resolution of a fork in the event of a temporary disagreement among the nodes on the genuine consistent truths. Honest nodes cannot adjust to inconsistencies in chains under any conditions. The deeper buried blocks in the chain in the network are always constant.

Immutability

Immutability means that changes can't be made after they've been made. To change a transaction from the past, you must re-mine all the blocks that came before it, which will be reflected in the network's copies of the ledger. It would also necessitate recreating the Merkle tree of the block in which the transaction is situated, as well as redoing all of the block's proof of work. Furthermore, because the hash of this block is stored in the next block, it must be re-mined. The following block must be modified with the new "prior block hash," resulting in a new block hash. In some situations, such a hash might not match the established difficulty level, necessitating re-mining of the block. In actuality, re-mining will have to wait until the chain's final block.

Security

The top online vulnerabilities identified by the Open Web Application Security Project (OWASP) Foundation were used to evaluate the security of websites, web services, and the DISV system's central server. OWASP is a non-profit organization whose purpose is to provide security professionals and developers with practical and fair knowledge on application security. It mostly focuses on critical web app vulnerabilities. The vulnerability assessment analysis identified 10 of the most common assaults. as well as the security measures required to counteract these attacks. The system incorporates the recommended requirement.

CONCLUSION

This research developed a new Decentralized IoT vehicle communication solution (DISV). It consists of three basic layers that investigate the use of Block-chain for communication in the IoV. On the Ethereum Testnet, a smart contract prototype has been implemented. To see if the Block-chain is an efficient and secure platform for IoV communications, this study looked at numerous features of the solution, including availability, integrity, and security. The findings revealed that DISV may be used as a real-time application as well as a solution to the primary difficulties of Vehicle-to-X (V2X) communications, such as security, centralization, and privacy. It can also make data interchange and collaboration between cars, infrastructure, and other intelligent transportation system actors easier. Furthermore, DISV might be viewed as a key component of Advanced Driver Assistance Systems (ADAS), which could improve transportation safety and mobility.

REFERENCES

Al Ridhawi, I., Aloqaily, M., Kantarci, B., Jararweh, Y., & Mouftah, H. T. (2018). A continuous diversified vehicular cloud service availability framework for smart cities. *Computer Networks*, *145*, 207–218.

Aloqaily, M., Al Ridhawi, I., Kantraci, B., & Mouftah, H. T. (2017, October). Vehicle as a resource for continuous service availability in smart cities. In *2017 IEEE 28th annual international symposium on personal, indoor, and mobile radio communications (PIMRC)* (pp. 1-6). IEEE.

Atzei, N., Bartoletti, M., & Cimoli, T. (2017, April). A survey of attacks on ethereum smart contracts (sok). In *International conference on principles of security and trust* (pp. 164-186). Springer. 10.1007/978-3-662-54455-6_8

Bader, L., Bürger, J. C., Matzutt, R., & Wehrle, K. (2018, December). *Smart contract-based car insurance policies. In 2018 IEEE Globecom workshops (GC wkshps).* IEEE.

Bai, X., Cheng, Z., Duan, Z., & Hu, K. (2018, February). Formal modeling and verification of smart contracts. In *Proceedings of the 2018 7th international conference on software and computer applications* (pp. 322-326). Academic Press.

Baralla, G., Pinna, A., & Corrias, G. (2019, May). Ensure traceability in European food supply chain by using a blockchain system. In *2019 IEEE/ACM 2nd International Workshop on Emerging Trends in Software Engineering for Blockchain (WETSEB)* (pp. 40-47). IEEE.

Bartoletti, M., & Pompianu, L. (2017, April). An empirical analysis of smart contracts: platforms, applications, and design patterns. In *International conference on financial cryptography and data security* (pp. 494-509). Springer.

Bogner, A., Chanson, M., & Meeuw, A. (2016, November). A decentralised sharing app running a smart contract on the ethereum blockchain. In *Proceedings of the 6th International Conference on the Internet of Things* (pp. 177-178). Academic Press.

Cordeschi, N., Amendola, D., Shojafar, M., & Baccarelli, E. (2015). Distributed and adaptive resource management in cloud-assisted cognitive radio vehicular networks with hard reliability guarantees. *Vehicular Communications*, *2*(1), 1–12. doi:10.1016/j.vehcom.2014.08.004

Jyothi, G. N., Sanapala, K., & Vijayalakshmi, A. (2020). ASIC implementation of distributed arithmetic based FIR filter using RNS for high speed DSP systems. *International Journal of Speech Technology*, 1–6.

Lei, A., Cruickshank, H., Cao, Y., Asuquo, P., Ogah, C. P. A., & Sun, Z. (2017). Blockchain-based dynamic key management for heterogeneous intelligent transportation systems. *IEEE Internet of Things Journal*, *4*(6), 1832–1843. doi:10.1109/JIOT.2017.2740569

Li, L., Liu, J., Cheng, L., Qiu, S., Wang, W., Zhang, X., & Zhang, Z. (2018). Creditcoin: A privacy-preserving blockchain-based incentive announcement network for communications of smart vehicles. *IEEE Transactions on Intelligent Transportation Systems, 19*(7), 2204–2220. doi:10.1109/TITS.2017.2777990

NagaJyothi, G., & Sridevi, S. (2019). High speed and low area decision feed-back equalizer with novel memory less distributed arithmetic filter. *Multimedia Tools and Applications, 78*(23), 32679–32693.

Shojafar, M., Cordeschi, N., & Baccarelli, E. (2016). Energy-efficient adaptive resource management for real-time vehicular cloud services. *IEEE Transactions on Cloud computing, 7*(1), 196-209.

Singh, M., & Kim, S. (2017). *Blockchain based intelligent vehicle data sharing framework.* arXiv pre-print arXiv:1708.09721.

Chapter 24
The Categorization of Development Boards to Implement the Embedded Systems and Internet of Things With Cloud Database for Volcano Monitoring Drones

Aswin Kumer S. V.
Koneru Lakshmaiah Education Foundation, India

Ayeesha Nasreen M.
https://orcid.org/0000-0003-4771-7319
RMD Engineering College, India

Jayalakshmi S.
QIS College of Engineering and Technology, India

Venkatasubramanian K.
QIS College of Engineering and Technology, India

Lakshmi Bharath Gogu
Srinivasa Institute of Technology and Science, India

ABSTRACT

Instead of sending human beings into volcanoes, drone-bot is used to measure the live lava temperature, and it alerts the ground station to protect people near the surroundings. The thermocouple is used as a temperature sensor. It can measure a wide range of higher temperatures, and it can be interfaced with the TTGo T-Call development board to process and send the temperature data to the ground station through GSM as short message service (SMS). Also the ESP-32 CAM is interfaced with that development board to capture the snapshot of the mountain if the temperature is high and the same snap is shared to the ground station through Wi-Fi. The GPS module is also interfaced with the development board to know the location of the volcano.

DOI: 10.4018/978-1-7998-9640-1.ch024

INTRODUCTION

The Embedded system can be implemented by using the set of sensors and bunch of actuators connected with the Microcontroller or a Development Board with microcontroller like Arduino, ESP8266, ESP32, TTGo-T-Call and Raspberry Pi, etc., and the development board is programmed by using some programming languages like C Programming, C++ Programming, Python Programming, Micro Python programming, etc., to perform specific task. If the embedded system is connected with internet, to monitor and control the actuators based on the sensor data from the remote location with high data security, then it is called Internet of Things. There are many stages in Internet of Things which are to be discussed in this chapter.

Embedded systems

The bunch of sensors and actuators connected with a microcontroller or a development board which can be operated and controlled through a software or firmware to perform the specific task is called an embedded system (Ma & Jiao, 2020). The basic structure of an Embedded system is shown in figure 1. There are variety of microcontrollers available to implement an embedded system (Martínez-Rodríguez, Valle, Brox, & Sánchez-Solano, 2020). They are 8051 microcontroller and its variants, PIC microcontroller and its variants and ARM microcontroller and its variants (Arredondo-Velázquez, Diaz-Carmona, Barranco-Gutiérrez, & Torres-Huitzil, 2020). These are the standard microcontrollers widely used in the industries to perform the particular tasks (Kwak & Lee, 2020). The assembly language programming and Embedded 'C' programming is used for programming these microcontrollers (Zhang, Seo, Donyanavard, Dutt, & Kurdahi, 2021). These microcontrollers never produce noises when handling the inputs and outputs because of its reliability(Muthukumaran V et al., 2018).

The high reliability microcontrollers are always used in the industries for good efficiency. These microcontrollers can be used with the development boards to implement the tasks easily. The other development boards are Arduino and its models which uses AVR family microcontrollers, ESP8266, Node MCU, ESP 12, ESP 32, TTGo-T-Call and its models, and Raspberry Pi and its different versions. These development boards can be programmed by using Embedded 'C++' through Arduino Integrated Development Environment (IDE), Micro python IDE, and Thonny python IDE. The Raspberry pi is the only development board which is having the operating system to perform the tasks(Muthukumaran V et al., 2021).

Figure 1. The basic structure of an embedded system

INTERNET OF THINGS (IOT)

As mentioned earlier, if the embedded system is connected with internet, to monitor and control the actuators based on the sensor data from the remote location with high data security, then it is called Internet of Things. The hardware used to implement the IoT, the Arduino development board along with ESP8266 can be used or the MKR series of Arduino development board can be used. Otherwise, the development board like Node MCU, ESP 12, ESP 32, TTGo-T-Call and its models, and Raspberry Pi and its different versions can be used which supports the Wi-Fi connectivity to implement the IoT. The IoT can be basically classified into two parts. They are, Industrial IoT and Commercial IoT (Pathak, Deb, Mukherjee, & Misra, 2021). The Application Oriented Classification Structure of an IoT is shown in the figure 2.

Figure 2. The application oriented classification structure of an IoT

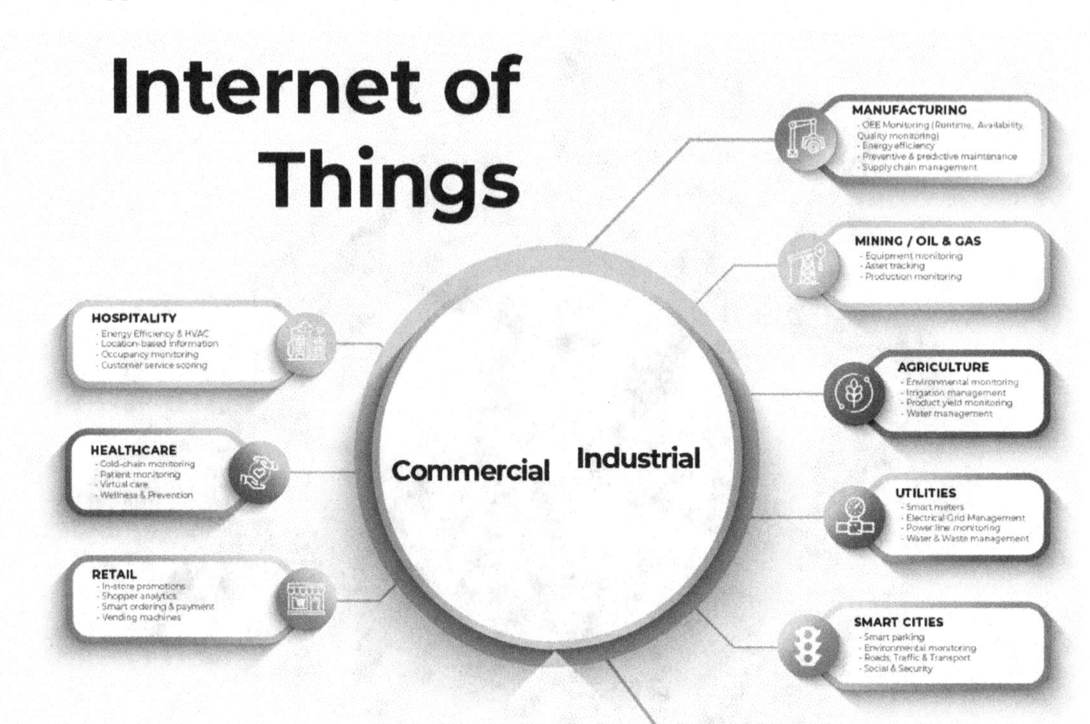

There are various stages in IoT, and each stage will be unique in nature to process the data. The IoT includes sensors, actuators, microcontroller, internet connectivity, cloud database, Big data storage, analytics, and summarization of the data (Khan, Rao, & Camtepe, 2021). The sequence stages in IoT are shown in figure 3.

Figure 3. The sequence stages of an IoT

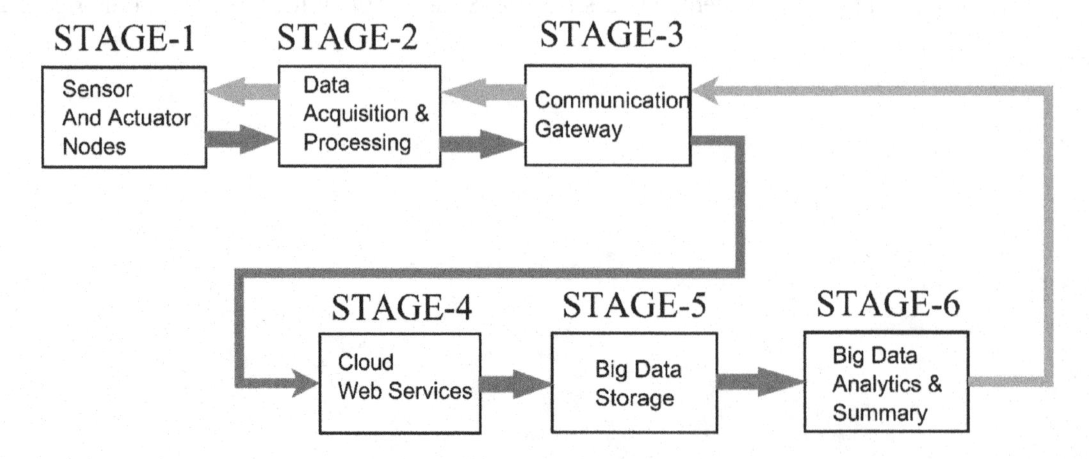

Sensor and Actuator Nodes

The sensors and actuators are playing important roles in an embedded system. The sensors are used for measuring the value of the parameter and the actuators are used to react based on the sensor data. The sensors like (Digital Humidity and Temperature) DHT 11, Infra-Red (IR), Passive Infra-Red (PIR), Ultrasonic sensor (SR-04), Rain sensor and Soil Moisture sensor are used to measure the specific parameters, which are connected with the microcontroller directly or with development board based on the application specific tasks. Some sensors like Barometric pressure sensor, Accelerometer, magnetometer can be connected through I2C protocol and SPI protocol. Some sensors will produce analog output and some sensors will provides both analog and digital outputs(Kumar, V et al., 2021). The actuators like servo motor (SG-90), Dc motor, Relay, opto-isolator, and Stepper motor are used to operate the devices based on the commands received from the microcontroller (Meneghello, Calore, Zucchetto, Polese, & Zanella, 2019). The Pulse Width Modulation (PWM) pins are required to connect some actuators like SG-90, etc(Nagarajan, S. M et al., 2022).

Data Acquisition and Processing

The data acquisition and processing can be possible by using a microcontroller or a development board. The board helps to acquire data from the sensors and process those data based on the program developed by the user and give commands to the actuators to act according to the application. So, there is a bidirectional communication between the module Sensor and actuator nodes and another module data acquisition and processing. The development board may be the combined module of Arduino and ESP 8266 or NODEMCU or ESP 32 or Raspberry pi. Some sensors can be connected directly to the board, but some sensors can be connected using I2C Protocol. So, the availability of protocol should be verified in the development board. Some sensor will work on serial interface protocol, then the software serial can be used to connect that type of sensors. The actuator needs PWM pins to control the signals. The ADC is required in the board to convert some sensor data. The development board should be chosen with that it supports all communication protocols like serial protocol, I2C protocol and SPI Protocol. And, it should have PWM pins, Analog to Digital Converter (ADC), etc. otherwise, the external ADC can be used to implement such applications (Neshenko, Bou-Harb, Crichigno, Kaddoum, & Ghani, 2019). The board should support Wi-Fi connectivity or GPRS connectivity for internet connection.

Communication gateway

The Hyper Text Transfer Protocol (HTTP) and Message Queue Telemetry Transport (MQTT) protocol are the communication gateways which can be accessed through anyone of the following like GPRS, Bluetooth, Li-Fi, Zigbee, LPWAN and Wi-Fi to post the data into cloud webservices through internet connectivity. The IPV4 protocol is widely used in the network layer of the open system interconnection. Based on the requirements and applications, the IPV6 can also be used. The cloud should be accessed to post the data is the aim for using this communication gateway (Swamy & Kota, 2020). The device to device communication can be implemented and the device to cloud communication is established by this gateway(Velliangiri, S et al., 2021)

Cloud web services

The Application Programming Interface (API) gateway should be used to access the cloud web services. The Representational State Transfer (REST) is one of the API which helps to use the cloud web services in the IoT platform. The basic layer to access the cloud web services are HTTP and the above layer to access the cloud web services are REST API. The job of this REST API is to create the users in the network, connect the users in the network by establishing the connection and delete the unnecessary users. The data exchange formats used in the REST API are JSON and XML. The HTTP will request the REST API to GET, POST, PUT and DELETE the data and resources as a handshaking signal. The browser using HTTP client will request the REST API server and get responses to transfer the resources and data (Sandhu, 2022).

Big data storage

The data from the cloud will be huge if it is a real time frequent usage data, then the NO SQL server like Mongo DB is used to store the big data. The other servers like Cassandra, HBASE can also be used for storing the data. This storage database accepts only JSON type of files and XML files. Before sending a file, the file type should be specified. The stored files can be accessed at any time with proper authentication. Every NOSQL database has the different procedure to store the files and the method which is suitable to the application should choose for better access (Hababeh, Gharaibeh, Nofal, & Khalil, 2019).

Big Data Analytics and Summary

The predictive modelling of data and the statistical modelling of the huge data should be performed in this stage. The Tableau, Spark, Cassandra and the ZOHO are some of the tools used to analyse and visualize the big data. Based on this analysis the AI system learns and predicts the things to be implemented in the IoT and all other applications.

LITERATURE SURVEY

For data communication the drone follows the protocols described in the IEEE standards for the encryption and transmission of payloads (Alwateer, Loke, & Fernando, 2019). The camera is also can be installed along with the drone for capturing the scene information for specific and multiple object detection (Wang, et al., 2019). There are multiple types of drones are available which can be identified and used for the different applications (Kim, et al., 2019). The data captured in the drone will be stored and monitored in the cloud database and the drone navigation and charging station of drones are also setup for those implementations(Nagarajan, S. M et al., 2021).

For continuous operation of drone the charging nodes can be setup will long last to provide the power for drones (Park, Kim, Lee, Joo, & Kim, 2021). The drone has to find out the exact location to alert cloud server about the volcano mountain point which can be possible by installing location aware network module(Kim & Moon, 2019). The autonomous vehicle information is available based on the programming module programmed in the drones for navigation purpose which can be easy to track the information and location of the volcanoes (Shi, et al., 2019). The restricted areas, possible flying areas and the safety areas to fly are pre-programmed in the drone for smooth implementation of the monitoring system (Li, Wen, Tian, Li, & Wang, 2019).

METHODOLOGY

ESP32 TT Go T Call

The IoT can be implemented using different ways, in that, choosing the development board is very important at the hardware side(Ezhilmaran, D., & Muthukumaran, V, 2017). Because, the development board should support UART protocol for serial data communication interface, Software serial protocol, I2C protocol, SPI protocol and it should have enough number of PWM pins, Analog (Analog to Digital) input pins, Digital to Analog Converter pins and also it should access the internet through Wi-Fi protocol (Ishigami & Sugiyama, 2020). The ESP 32 TT Go T Call development board has all these features and it is suitable for implementing the IoT applications which is shown in Figure 4 (Kaplan, et al., 2021). This board has MAC address, which is used to connect many similar boards as a Topological Network to gather the sensor data without internet connection is possible (Gómez-Lagos, Candia-Véjar, & Encina, 2021).

Raspberry Pi

The Linux operating system namely Raspbian operating system should be installed in the micro-SD card which should be inserted in the raspberry pi development board(Manikandan, G et al., 2021). The Raspbian operating system has inbuilt python and Thonny python IDE for programming in the Raspberry pi. There is no other development board except Raspberry pi have the operating system with built-in IDE. So, the Raspberry pi development board is a step ahead than other development boards(Manikandan, G et al., 2020). But it is not having Analog input pins and it needs separate Analog to Digital Converter chip to handle analog inputs. The GPIO (General Purpose Input Output) available in the Raspberry pi development board will handle and tolerate up to 3.3v DC, these are the major limitation of this development board. The MQTT broker can be installed in this board to implement the IoT based applications.

IMPLEMENTATION

The drone has to find the route for travelling and delivering message and capturing data from particular location includes many obstacles to reach the destination which can be solved (Alsharoa, Ghazzai, Kadri, & Kamal, 2020). The services can be implemented using the drones which are helpful for scripting and crowdsourcing (Shi, Chang, Yang, Wu, & Wu, 2020). The drones can be controlled by the user as they want in all different environments (Salehi & Hossain, 2021). Those drones will communicate with the signal station using some protocols which supports the improvement in performance and efficiency. With the help of these drone behaviour, the projects can be implemented for finding and troubleshooting the resource breakage in the system (Yazdinejad, et al., 2021). To establish the connection between the drones and signal station, different radio networks are used for better assistance(Linda, G. M et al., 2021).

The Drone-Bot is used to detect and monitor the volcanoes in the surrounding mountains with the help of TTGO T-CALL development Board. The board has ESP 32 Microcontroller with Wi-Fi enabled protocol, and built-in SIM 800 module which is used to include the GSM and access the GPRS(Kumar, D et al., 2021). In some terrain areas the GSM signal is not available because it is restricted in the reserved areas. In absence of GSM, the EPROM in the development board will store the data and return back to the node which is having either Wi-Fi signal connectivity or GSM signal with GPRS connectivity to post the data in the cloud database (Taha & Shoufan, 2019). The ground station and the battery charging nodes are setup to establish the connection with the Drone-Bot. The SIM 28 GPS module is interfaced with the development board to find out the exact location of the volcano spot. The ESP 32 CAM is also interfaced with the development board to capture the images of the volcano mountain if the temperature exceeds the threshold point (Sun, et al., 2020). To measure the temperature, the thermocouple sensor is used which measures the temperature in the range of -27' C to +3000' C.

Figure 4. Block diagram of the proposed method

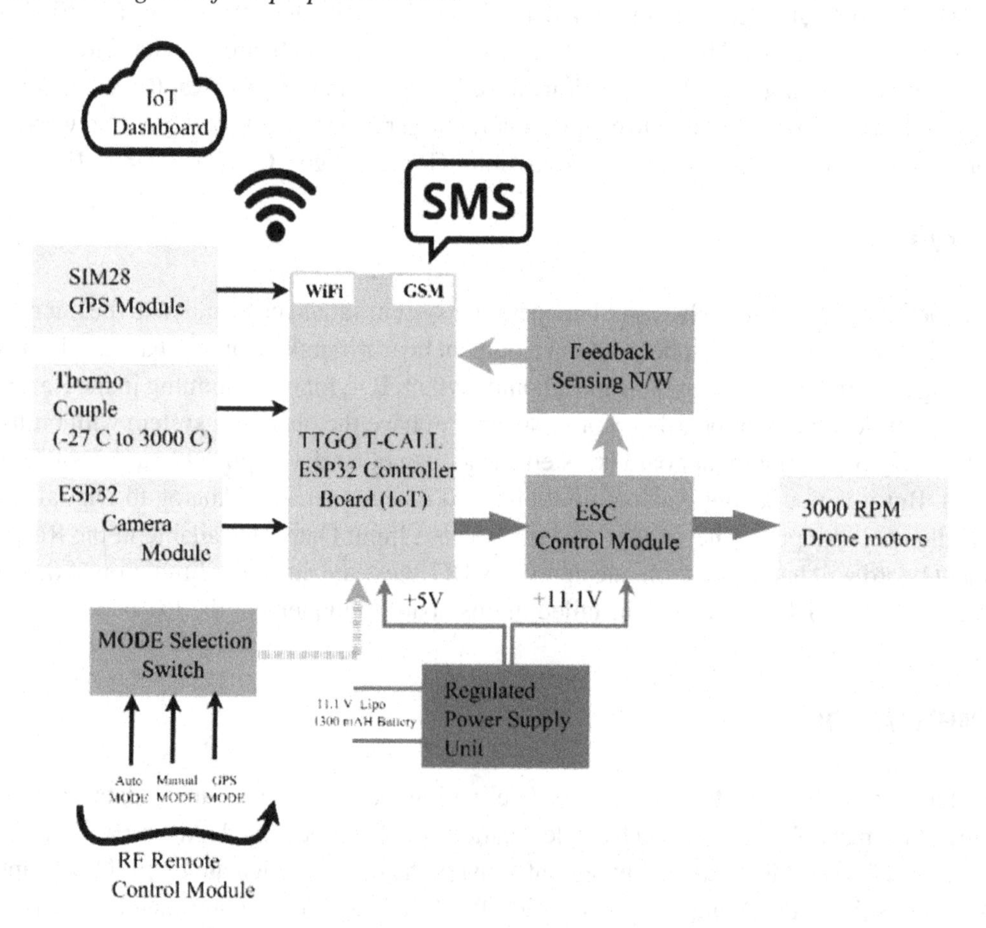

Another important module in this Drone-Bot is Electronic Speed Control (ESC) Modules which are used to drive the Brushless DC (BLDC) motors especially for the Drones. It rotates 3000 Rotations Per Minute (RPM) which helps to fly the drone with proper speed(Sadhasivam, J et al., 2021). The 11.1 V Lipo 1300mAH Battery provides the power supply to the Development board as well as the ESC Control Modules. The ESC control module is also connected and controlled by the Development board (Sun,

Ansari, & Fierro, 2020). The battery charging nodes are kept in the Drone navigating path to provide the continuous battery power supply to the Drone-Bot.

Figure 5. Circuit diagram of the proposed method

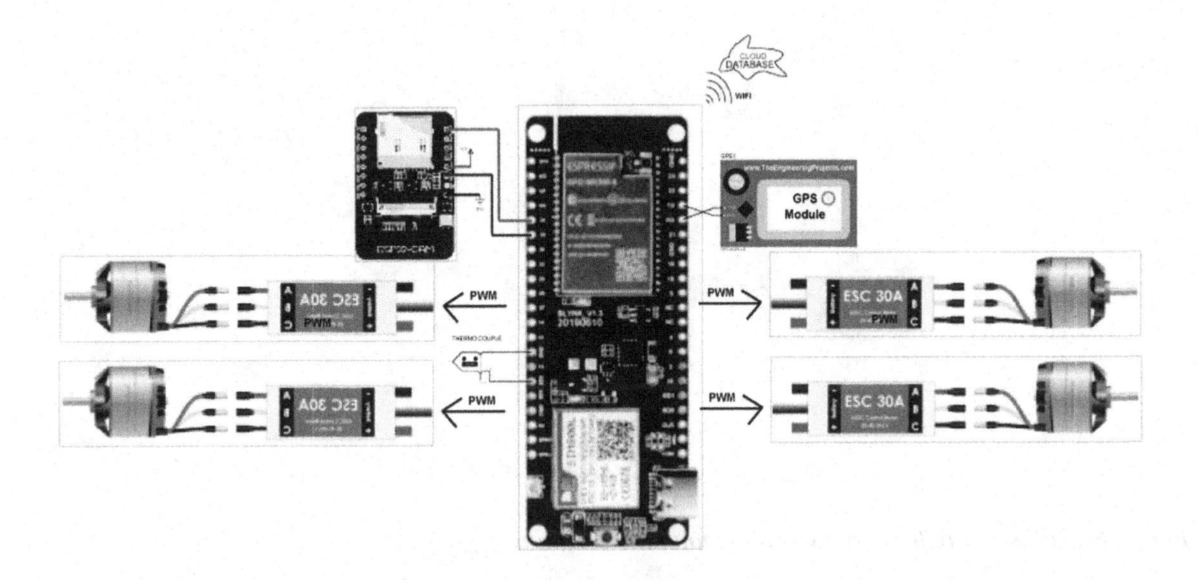

RESULTS AND DISCUSSIONS

The Temperature from the Thermocouple which is downscaled to 100:1 ratio updated time to time in the cloud database which is shown in Fig.6.

Figure 6. The plot of temperature from Thermocouple (Downscaled 100:1)

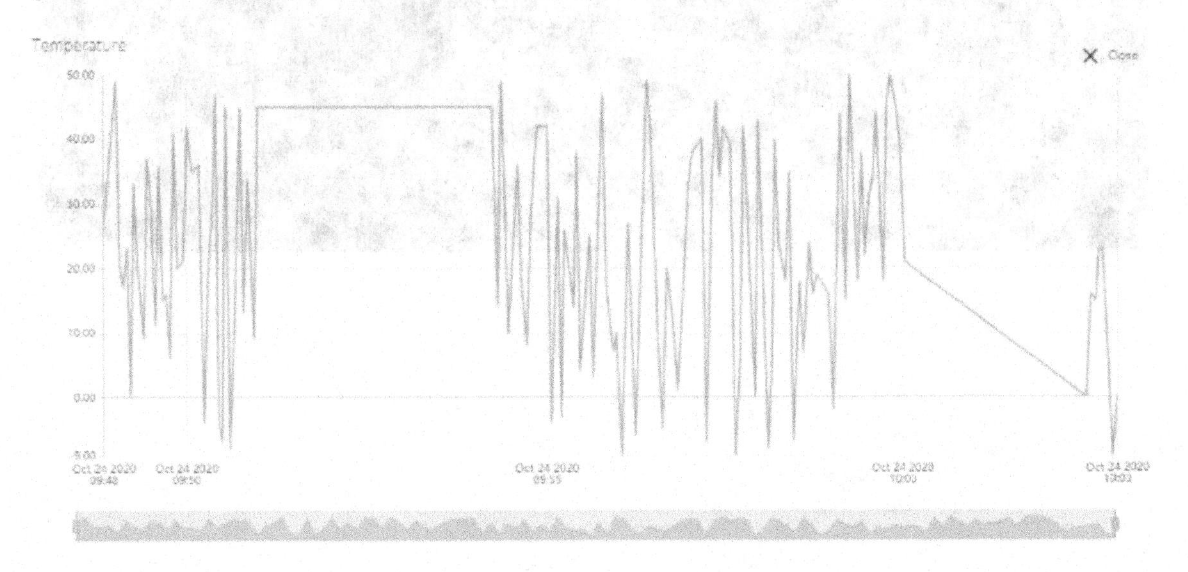

Figure 7. The temperature from Thermocouple (Downscaled 100:1)

Figure 8. GPS sharing the exact volcano mountain location

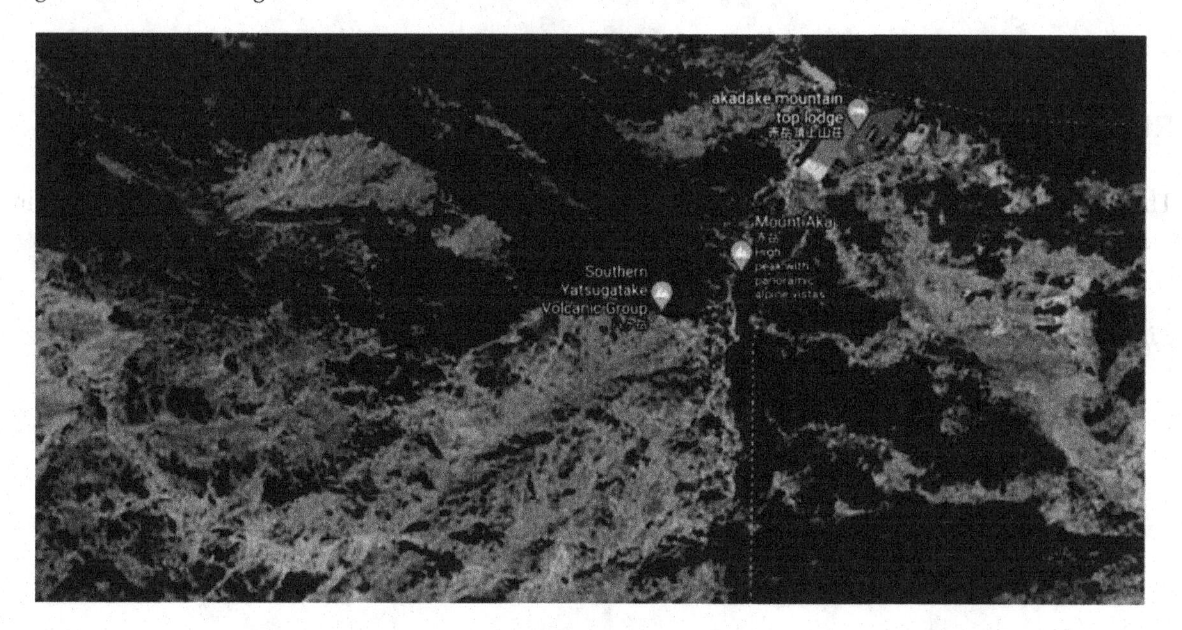

Figure 9. Drone-Bot

Thus, the Drone-Bot is implemented to monitor the volcano mountain status time to time with the help of TTGO T-CALL development board, ESP 32 Ai Thinker CAM board, SIM 28 GPS module, Thermocouple Sensor and the drone module includes the Lipo Battery, ESC control Module and BLDC motors to drive the propellor of the drone.

CONCLUSION

The categorization of Development boards to Implement the Embedded Systems and Internet of Things with the Cloud database were discussed in this chapter and selecting the particular board is based on the applications. The MQTT protocol can be implemented using Raspberry Pi and the ESP 32 is used to gather many analog data from sensors and it can drive more actuators through the PWM pins, if analog output is necessary for any device, which can be supplied by the ESP 32. These two major boards are used as combined to implement the IoT and Robotic applications

REFERENCES

Alsharoa, A., Ghazzai, H., Kadri, A., & Kamal, A. E. (2019). Spatial and temporal management of cellular HetNets with multiple solar powered drones. *IEEE Transactions on Mobile Computing*, *19*(4), 954–968.

Arredondo-Velazquez, M., Diaz-Carmona, J., Barranco-Gutierrez, A. I., & Torres-Huitzil, C. (2020). Review of prominent strategies for mapping CNNs onto embedded systems. *IEEE Latin America Transactions*, *18*(05), 971–982. doi:10.1109/TLA.2020.9082927

Ezhilmaran, D., & Muthukumaran, V. (2017). Authenticated group key agreement protocol based on twist conjugacy problem in near-rings. *Wuhan University Journal of Natural Sciences*, *22*(6), 472–476. doi:10.100711859-017-1275-9

Jayasuruthi, L., Shalini, A., & Kumar, V. V. (2018). Application of rough set theory in data mining market analysis using rough sets data explorer. *Journal of Computational and Theoretical Nanoscience*, *15*(6-7), 2126–2130.

Kim, S., & Moon, I. (2018). Traveling salesman problem with a drone station. *IEEE Transactions on Systems, Man, and Cybernetics. Systems*, *49*(1), 42–52.

Kumar, D., Swathi, P., Jahangir, A., Sah, N. K., & Vinothkumar, V. (2021). Intelligent Speech Processing Technique for Suspicious Voice Call Identification Using Adaptive Machine Learning Approach. In Handbook of Research on Innovations and Applications of AI, IoT, and Cognitive Technologies (pp. 372-380). IGI Global.

Kumar, V., Niveditha, V. R., Muthukumaran, V., Kumar, S. S., Kumta, S. D., & Murugesan, R. (2021). A Quantum Technology-Based LiFi Security Using Quantum Key Distribution. In Handbook of Research on Innovations and Applications of AI, IoT, and Cognitive Technologies (pp. 104-116). IGI Global.

Kumar, V. V., Raghunath, K. K., Rajesh, N., Venkatesan, M., Joseph, R. B., & Thillaiarasu, N. (2021). Paddy Plant Disease Recognition, Risk Analysis, and Classification Using Deep Convolution Neuro-Fuzzy Network. *Journal of Mobile Multimedia*, 325-348.

Kumar, V. V., Raghunath, K. M., Muthukumaran, V., Joseph, R. B., Beschi, I. S., & Uday, A. K. (2021). Aspect based sentiment analysis and smart classification in uncertain feedback pool. *International Journal of System Assurance Engineering and Management*, 1-11.

Li, T., Wen, B., Tian, Y., Li, Z., & Wang, S. (2018). Numerical simulation and experimental analysis of small drone rotor blade polarimetry based on RCS and micro-Doppler signature. *IEEE Antennas and Wireless Propagation Letters*, *18*(1), 187–191.

Ma, D., & Jiao, X. (2019). WoMA: An Input-Based Learning Model to Predict Dynamic Workload of Embedded Applications. *IEEE Embedded Systems Letters*, *12*(3), 74–77. doi:10.1109/LES.2019.2957487

Manikandan, G., Perumal, R., & Muthukumaran, V. (2021). Secure data sharing based on proxy re-encryption for internet of vehicles using seminearring. *Journal of Computational and Theoretical Nanoscience*, *18*(1-2), 516–521.

Martínez-Rodríguez, M. C., del Valle, S. S., Brox, P., & Sánchez-Solano, S. (2020). Hardware Implementation of Authenticated Ciphers for Embedded Systems. *IEEE Latin America Transactions*, *18*(09), 1581–1591. doi:10.1109/TLA.2020.9381800

Meneghello, F., Calore, M., Zucchetto, D., Polese, M., & Zanella, A. (2019). IoT: Internet of threats? A survey of practical security vulnerabilities in real IoT devices. *IEEE Internet of Things Journal*, *6*(5), 8182–8201. doi:10.1109/JIOT.2019.2935189

Muthukumaran, V., Ezhilmaran, D., & Anjaneyulu, G. S. G. N. (2018). Efficient Authentication Scheme Based on the Twisted Near-Ring Root Extraction Problem. In *Advances in Algebra and Analysis* (pp. 37–42). Birkhäuser. doi:10.1007/978-3-030-01120-8_5

Muthukumaran, V., Joseph, R. B., & Uday, A. K. (2021). Intelligent Medical Data Analytics Using Classifiers and Clusters in Machine Learning. In Handbook of Research on Innovations and Applications of AI, IoT, and Cognitive Technologies (pp. 321-335). IGI Global.

Muthukumaran, V., Vinothkumar, V., Joseph, R. B., Munirathanam, M., & Jeyakumar, B. (2021). Improving network security based on trust-aware routing protocols using long short-term memory-queuing segment-routing algorithms. *International Journal of Information Technology Project Management, 12*(4), 47–60.

Nagarajan, S. M., Muthukumaran, V., Beschi, I. S., & Magesh, S. (2021). Fine Tuning Smart Manufacturing Enterprise Systems: A Perspective of Internet of Things-Based Service-Oriented Architecture. In Handbook of Research on Innovations and Applications of AI, IoT, and Cognitive Technologies (pp. 89-103). IGI Global.

Neshenko, N., Bou-Harb, E., Crichigno, J., Kaddoum, G., & Ghani, N. (2019). Demystifying IoT security: An exhaustive survey on IoT vulnerabilities and a first empirical look on Internet-scale IoT exploitations. *IEEE Communications Surveys and Tutorials, 21*(3), 2702–2733. doi:10.1109/COMST.2019.2910750

Park, S., Kim, H. T., Lee, S., Joo, H., & Kim, H. (2021). Survey on Anti-Drone Systems: Components, Designs, and Challenges. *IEEE Access: Practical Innovations, Open Solutions, 9*, 42635–42659.

Pathak, N., Deb, P. K., Mukherjee, A., & Misra, S. (2021). IoT-to-the-Rescue: A Survey of IoT Solutions for COVID-19-like Pandemics. *IEEE Internet of Things Journal.*

Sadhasivam, J., Arun, M., Deepa, R., Muthukumaran, V., Kumar, R. L., & Kumar, R. P. (2021, July). Forex exchange using big data analytics. *Journal of Physics: Conference Series, 1964*(4), 042060. doi:10.1088/1742-6596/1964/4/042060

Sadhasivam, J., Muthukumaran, V., Raja, J. T., Vinothkumar, V., Deepa, R., & Nivedita, V. (2021, July). Applying data mining technique to predict trends in air pollution in Mumbai. [). IOP Publishing.]. *Journal of Physics: Conference Series, 1964*(4), 042055.

Salehi, M., & Hossain, E. (2020). Handover Rate and Sojourn Time Analysis in Mobile Drone-Assisted Cellular Networks. *IEEE Wireless Communications Letters, 10*(2), 392–395.

Sun, H., Yang, J., Shen, J., Liang, D., Ning-Zhong, L., & Zhou, H. (2020). TIB-Net: Drone Detection Network With Tiny Iterative Backbone. *IEEE Access: Practical Innovations, Open Solutions, 8*, 130697–130707.

Sun, X., Ansari, N., & Fierro, R. (2019). Jointly optimized 3D drone mounted base station deployment and user association in drone assisted mobile access networks. *IEEE Transactions on Vehicular Technology, 69*(2), 2195–2203.

Swamy, S. N., & Kota, S. R. (2020). An empirical study on system level aspects of Internet of Things (IoT). *IEEE Access: Practical Innovations, Open Solutions*, *8*, 188082–188134. doi:10.1109/ACCESS.2020.3029847

Taha, B., & Shoufan, A. (2019). Machine learning-based drone detection and classification: State-of-the-art in research. *IEEE Access: Practical Innovations, Open Solutions*, *7*, 138669–138682.

Velliangiri, S., Karthikeyan, P., & Vinoth Kumar, V. (2021). Detection of distributed denial of service attack in cloud computing using the optimization-based deep networks. *Journal of Experimental & Theoretical Artificial Intelligence*, *33*(3), 405–424.

Vinoth Kumar, V., Ramamoorthy, S., Dhilip Kumar, V., Prabu, M., & Balajee, J. M. (2021). Design and Evaluation of Wi-Fi Offloading Mechanism in Heterogeneous Network. International Journal of e-Collaboration, 17(1).

Wang, D., Hu, P., Du, J., Zhou, P., Deng, T., & Hu, M. (2019). Routing and scheduling for hybrid truck-drone collaborative parcel delivery with independent and truck-carried drones. *IEEE Internet of Things Journal*, *6*(6), 10483–10495.

Chapter 25
Autonomous Robotic Technology and Conveyance for Supply Chain Management Using 5G Standards

Hariprasath Manoharan
Panimalar Institute of Technology, India

Pravin R. Kshirsagar
G. H. Raisoni College of Engineering, India

Radha Raman Chandan
Shambhunath Institute of Engineering and Technology, India

Kalpana V.
Vel Tech Rangarajan Dr. Sagunthala R&D Institute of Science and Technology, India

Ashim Bora
Kampur College, India

Abhay Chaturvedi
GLA University, India

ABSTRACT

The process of incorporating robotic technology and autonomous vehicles are increasing in all applications where for all real-time application developments time and energy can be saved for every single movement transfer as compared to human classifications. Thus, considering the advantage of autonomous process without any presence of an individual, the supply chain management can be designed using robotic technology. The robotic technology provides an informal route where all goods can be transported to different places within a short span of time, and any false identification in transfer of goods can also be easily identified. To drive the autonomous vehicle towards correct location, a precise protocol is chosen, which is termed common industrial protocol (CIP) where proper solutions can be achieved for all control applications using time synchronization model. Further, the data monitoring process is trailed using an online contrivance which is termed as internet router (IR) where short distance can be identified using corresponding addressing scheme.

DOI: 10.4018/978-1-7998-9640-1.ch025

CHARACTERISTICS OF SUPPLY CHAIN MANAGEMENT

In topical eons there is a high demand in delivery of goods from one location to another where time consumed for delivery process is much higher. In some cases when the goods are transferred from different parts of the world it is much difficult to trace it in the presence of humans. Also there is a high probability that expected goods will be much different from delivered ones and sometimes the goods will not reach the receiver properly. To overcome the aforementioned drawbacks the process of supply chain management is introduced in all trade sectors. The introduction of supply chain management process manages the movement of goods from commercial to location subdivisions. In addition to delivery process the raw materials are chosen and they are manufactured from different locations and the methodical products are produced within the expected period of time. Thus the process of conception, distribution and return process is handled by supply chain management process to reduce the cost of implementation. Further supply chain management can be subdivided to five separate stages as follows,

- Stage 1: A simple plan for development of products
- Stage 2: Analysis of materials for expansion
- Stage 3: Production and adeptness form
- Stage 4: Approach of distribution
- Stage 5: Transportation of unwanted products

For all the aforementioned stage a step-by-step approach is needed and it varies for all distinct manufacturing industries. For this five stage process the possible problems are also identified at initial stage and a grievance contrivance has been created in case of any disenchantment to the consumer. Thus, in recent days a cloud monitoring process is created to achieve collaboration between different networks. While designing the five stage process more amount of data will be generated and it will be examined by data experts and in this case optimization process in required to minimize inactivity of all developed products. The major reason for such optimization process is to maximize the value of consumers and to accomplish a sustainable growth in delivery of products using physical and information system flows. The implementation of supply chain management process is shown in Figure 1.

Figure 1. Phases of supply chain management using robotic technology

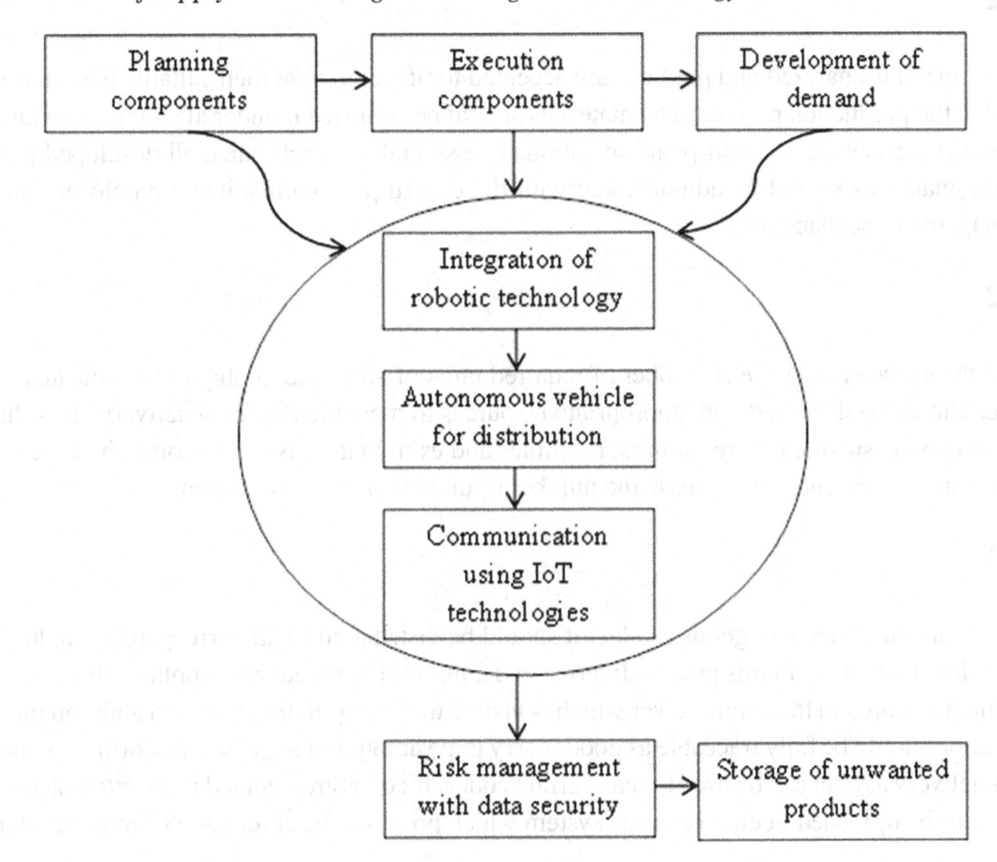

Stage 1

The first stage in all application development for supply chain management process is to develop a unique strategy for development of products where the following plans should be implemented.

- Organization of resource
- Identification of various demand
- Plan for demand production
- Setup strategies and
- Strategy for auctions setup

During development of products it is necessary to update the demand of products that are relevant to all functionalities. If the demand is not reorganized then creation of products will consume more time and as a result it will result in more transactions of product failure.

Stage 2

Once the demand is analyzed and products are accepted for development then suitable materials should be loaded in the production process. The materials should be significant under all environmental conditions where the existence of multi-plant operations is essential to synchronize all developed products. Further the materials should be administered with the design plan with suitable employees and high level management mechanisms.

Stage 3

In stage 3 the capacity plan (i.e.,) number of required units of all products should be indicated and the employees should be delivered with appropriate resources to meet the expected delivery units. In some cases there may be sudden failure of units, machines and even materials can become shortage where a sub-plan should be executed to increase the number of units at parallel conditions.

Stage 4

After development of demand goods product it should be distributed to all corresponding industries or to the individual directive. In this process Internet of Things (IoT) procedure is applied where location is tracked and it is stored in the central sever which is visible to corresponding users. In addition the model of distribution should be fully traceable as goods carry large amount of significant information therefore in case of delivery a robot can be used to scan certain codes. If codes are scanned in an automated manner then data will be uploaded in cloud storage system which provides exact location of moving vehicles.

Stage 5

The last stage in supply chain management process is to store all the unwanted materials at appropriate places and in case if the products that are delivered to each individual is not matching with the ordered ones then it should not be unexploited. Thus a separate location has to be chosen where the robotic technology will be helpful for packing all unwanted materials and if needed in future the data can be gathered and located.

IMPORTANCE OF ROBOTICS IN SCM

In current generation systems it is necessary to implement robots in all business related communications where the data is transferred at desired locations. The advantage of robotic technology in supply chain management process is that cost of industry deployment is reduced and stability can be achieved inside the industrial process. In addition all errors during the production process will be reduced as compared to human faults because sorting time of an individual is reduced completely. In line with above concern it is illustrious that the entire supply chain management process is not interchanged with robotic system as only instructions are given and adjacent exertion will be completed only using individual persons. However achieving adeptness model of robotic technology in supply chain management is a huge task and it is solved by integration of human systems as any dangerous elucidations will be avoided completely.

The process of automation in robotics can be achieved using the following parameters for preventing the hazardous developments.

1. Attaining nontoxic development
2. Effective command process

Attaining nontoxic development

In the process of nontoxic development it is possible to overcome all the harmful effect that is produced by outdoor environments (i.e.) robotic technology can be integrated in high safety working environment and it will not cause any external disturbance to other individuals. Even for large scale operations robots are preferred and all corresponding tasks are accomplished in an individual manner. Moreover in recent time the effect of contagion infections are much higher which in turn reduces the life time of humans due to bit process. But if robots are implemented then all employees can be retained away from different work station plants and even if any repetitive process is present then it can be processed by storing it in memory processor that is present in robots. For example the cause COVID-19 has provided great effect on humans but at the mean time if robots are implemented then all gesture activities by humans can be performed at remote location and monitoring of instructions can be processed. This process provides a great improvement in all real life conditions and the risk management process will be reduced completely. Figure 2 deliberates the solutions on achieving developments through safety process.

Figure 2. Detachment of nontoxic solutions using robotics

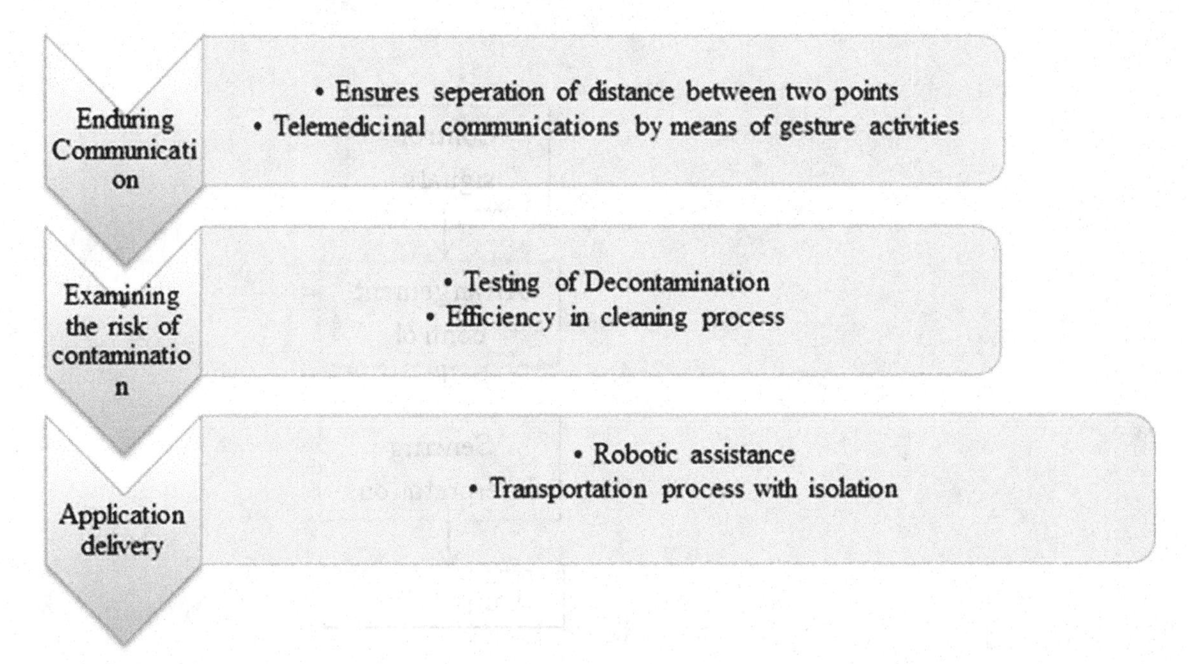

Effective command process

The most important prospect during the production process is to provide commands to robotic systems where all functions will work in an effective way only by using a Graphical Interface Unit (GUI). The main task of GUI is to control a robot by observing the metaphors that are displayed in an implemented device where the robots can able to capture the images easily and after apprehending it the users can able to receive the provide corresponding commands. In case if the process of image extraction is not clear then an alternate mode using a sensing device can be created. However in the process of measuring device a decision making mechanism is essential which is designed using valid input ranges to improve the accuracy of command execution.

In the alternate process a speech modulating device can be integrated with sensors where commands can be generated with vocal sound. In this robots are given incessant training with microphone based systems which makes the robot to understand the commands without any mistake. The abovementioned process not only reduces the error in the command process but also transfers the data in a secured way to corresponding robots. Figure 3 shows the effective command process to robots with interaction from humans.

Figure 3. Effective commands with human interaction

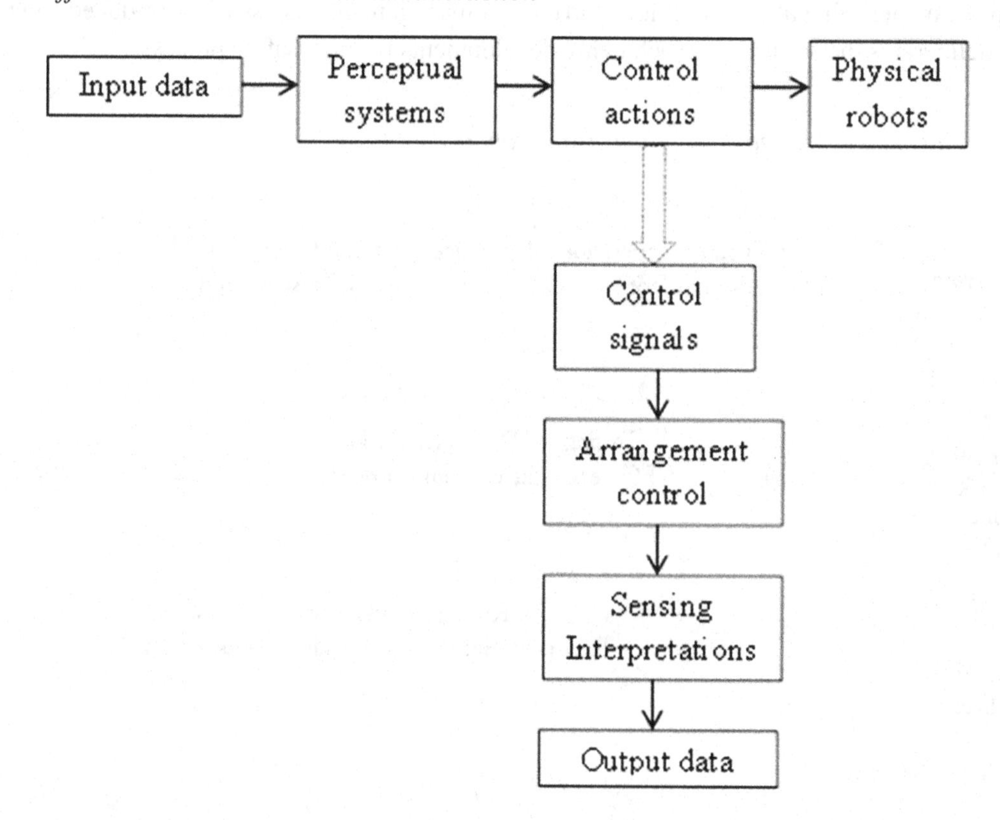

CONSERVATIVE CHALLENGES OF SCM: A SURVEY

There are many issues and challenges that exist in previous generation networks that are implemented specifically for SCM and investigators have also observed several measures that are taken for successful automatic process. In (Chettri, L & Bera, R, 2020) a survey on many periodicals has been prepared by implementing fifth generation systems with IoT. It is reported that the presence of Multiple Input Multiple Output (MIMO) antennas are required as a precise communicating technology in wireless medium. But the presence of antennas is indicated as old generation systems and new signal transmission technique should be identified. In addition to antenna systems it is also necessary to implement a middleware IoT process for effective data automation process (Da Cruz, M. A. A et al., 2021). The middleware IoT should represent various protocols that are present in application layer such as Hyper Text Transfer Protocol (HTTP) etc. Nevertheless the process have defined the need of fifth generation systems as the domain is moving towards the wireless transmission systems with high speed data transmission approach.

It has been contended that the application layer protocols will function effectively in the presence of machine learning algorithm (Donta, P. K et al., 2021) where dynamic mode of operation can be adapted for converting normal process to an intelligent process. In the abovementioned procedure data handling mechanism has been processed with IoT without any intervention from human systems. However the objective is to make the human process to give in voice commands in an effective manner thus satisfying the constraints in SCM. When a human interaction is present then an authentic model has to be designed which it is established with IoT gateway nodes (Taher, B. H et al., 2021). The process has not involved any algorithm for routing the data in shortest path but provides high security in terms of data transfer approach. Many layers and their operations are examined with big data and outcomes have been perceived by interacting with outdoor environment (Lombardi, M et al., 2021). In this case of examination IoT proves to be much effective in support with different layer protocols. The seven layers of interconnection model forms the basis of network security and it is well established using Internet Protocol Version 6 (IPV6) (Triantafyllou, A et al., 2018). If the users are demanding for conversion of SCM to an automated process then classification mechanism should be provided to overcome the parametric challenges such as scalability, transparency in operation etc.

The primary identified problem during data transmission in SCM is delay that is present in existing networks (Aljubayri, M et al., 2021) and to overcome such difficulties a resourceful routing technique is established. Even though some resources are highly sustainable more amount of unconventionalities are present which makes the process to be difficult for sustaining against global environment cases. Therefore a business layer has been added above application layer and in this additional five layer architecture an object management process has been integrated. But the effect of business layer has not been analyzed which is considered as major drawback in implementation process. Further with application and business layer it is predictable that transport layer provides handshaking for improving the number of bytes to varying range of 15 percentage. Though the rise is percentage is demonstrated using battery life time the varying characteristics of transport layer requires a handshaking mechanism before every data transfer to different users. If IoT is operated for SCM then management process requires an inter operation at data link layer in the presence of heterogeneous device (Sharma, G et al., 2018). Thus interoperation is completely satisfied in IoT at data link layer where high speed data transfer has been realized in real time applications and such high speed process provides great advantage for SCM implementation.

The process of IoT has been specifically designed for industrial applications and they are termed as Industrial Internet of Things (IIoT) where all smart manufacturing and production process can be estimated with data acquisition protocols (Amjad, A et al., 2021). Since day-to-day activities are varying it is essential to implement IIoT in all autonomous SCM process and this will circumvent single point of failure which is represented in terms of balancing the load and demand. Still the process of IIoT is applied only for data processing techniques and in the operation of SCM the products will be created thus there is a necessity to reduce the human intrusion. A discrete security solutions has been examined in the implementation of IoT development process (Chmiel, M et al., 2021) where SCM has been managed with unique codes during delivery stage. This process is similar to that is present in all commercial platforms but a real time monitoring is possible with traditional data. But there are many issues that are related to the security process in the network layer of IoT such as poor management of goods, difficulty in management identification etc. For all aforementioned issues the identified limitations and circumstances have been made highly secured using a separate observer layer (Burhan, M et al., 2018). In the enactment process the observer layer uses a 128 bit authentication key that enables the SCM users to check whether the information is protected from several virus platforms. This process of information checking is enabled in all real time applications of SCM thus providing high trust from blocking all intruders. In all the mechanism it is disputed that only 7 layers are highly essential for automation process but in reality for converting SCM to an automatic process 12 layers are needed in addition to physical object association (Bouaouad, A. E et al., 2020). Apart form 7 layers the following layers such as interconnection, access, middleware and cloud layers form the basis of automation process and virtually 32 architectures can be implemented for SCM for providing support to five stage process.

There are several vulnerabilities that are present in the moderation process of SCM as many industries have employed humans for productivity and management purpose. In this type of conversion the intrusion detection process is the first step to be designed where the obstacles should be strengthened with high security in the authentication layer. Conversely in addition to automation with IoT all the human interface system should be converted to an outdoor environment which can be used only for giving commands and the autonomous production and management should be sustained with robotic technology as robots can able to work according to the set of instructions and in final stage the data processing can be added with IoT by following high security procedures (Sierra Marín, S. D et al., 2021).

EFFECT OF 5G STANDARDS IN SCM

In all application of robotic technology a signal transmission standard should be defined predominantly in SCM the manufacturing and delivery procedure cannot be operated without the presence of standard signaling techniques. Therefore, in SCM process for fast transmission of signals, data is needed and it can be achieved using fifth generation networks as the speed of signal transmission is higher as compared to other network systems. For proper communication in Industry 4.0 the communication segment should provide unified network signal and as compared to other standards the one that is specially designed for industrial process should provide response at quick time. Moreover the response time of the robots with 5G standards must be ten times greater than other non-industrial system procedures. Some of the proficient characteristics of 5G in SCM are as follows,

- Radical communications
- Low expectancy
- Immense appliance communication

Radical communications

In radical communication process, an extra security is needed thus satisfying the reliability requirements with robotic technology. Since the robots are operated in wireless standard the reliable speed must be equivalent to wired technology where the down rate of five percentage should be reduced in further cases. In addition if any robots need a surgery then a telemedicine operation can be performed in a secured manner as the information stored in robots have to be conserved in saving application platforms.

Low expectancy

As compared to 4G standards the implementation process of 5G in robotic technology provides a low latency rate of 1 milliseconds and this process of critical time is measured for both manufacturing and delivery procedures in SCM. Further if the manufacturing latency is reduced then the demand cannot be satisfied at any moment thus latency must be maintained at low rate and this expectancy process have to achieved in case of designing autonomous delivery vehicles.

Immense appliance communication

5G standards are having the potential to provide device-to-device communication where the procedure is much similar to human interaction process with the help of a central tower.

Figure 4. Application of 5G standard in SCM

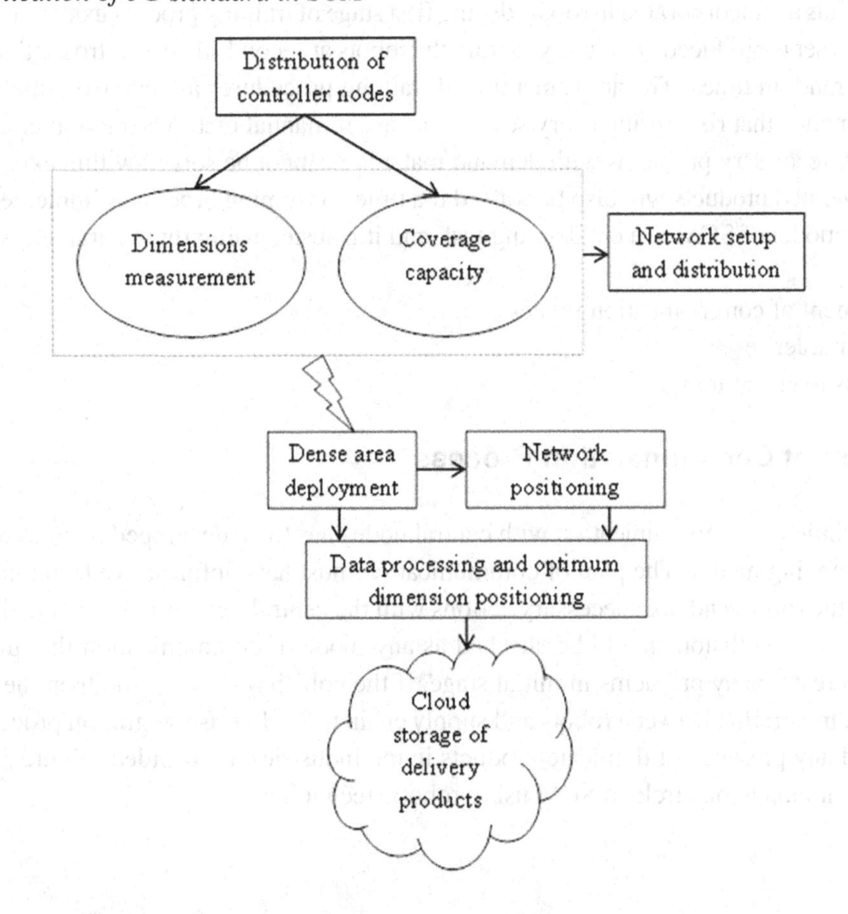

In case of immense appliance communication an ecosystem will be created only if 5G standards are combined with wireless fidelity technologies. The process of immense communication is guaranteed only if SCM is connected to active states where all robots will operate in duly reliable boundaries.

PARAMETRIC MEASUREMENT DIFFICULTIES

The purpose of parametric testing model for SCM applications using robotic technology and 5G standards is to verify whether the operations are performed in a streamlined manner. If the robotic technology is introduced then many industries will emanate onward to modernize the resources and supply of materials where all clients and consumers can ensure aggravation based environmental process. The aggravation model can adapt to any complex problems and enhanced transportation process is guaranteed with proper storage of raw materials at proper locations. In addition to the parametric testing models the process of automation in SCM can also solve the following problems that are faced by Industries. The primary problems such as increase in expenditure, increase in fuel consumption rate due to absence of autonomous driving vehicles can be avoided. Further uprights in multiple supply chain process is one of the uninterrupted problem that occurs in SCM and this type of multiple SCM with third party clients needs to be operated where it must be delivered with short span of time. Thus human interferences are converted to stationary process and only commands are given at outdoor steps. Also any difficulties that arise to robots can be immediately solved by individuals through paramedical procedures.

Moreover all minor complications in the backend such as congestion in connected ports, inappropriate matching of trade cost in backyards and capacity problems in loading of trucks is also solved as the storing commands are incorporated in robots during first stage of training process using previous data set. Even a new data set is produced at industry to train the robots at second stage as introduction of materials will change at random times. The abovementioned training procedures are carried out after a trial test to meet the demands that rise during every season. In case if manual methods are implemented in SCM all the primary, secondary problems with demand matching cannot be solved within expected time and delivery of generated products will also be wasted if a time consuming process is implemented. Usually the parametric model of SCM is a challenging task and it is tested using three parameters as follows,

- Arrangement of communication process
- Time of indolence
- Adeptness to ecosystem

Arrangement of Communication Process

In SCM the technique of communication with central nodes has to be developed in an effective way using strategic planning model. The plan of communication must have informative technologies that can able to control the robots and take necessary actions with the central operator. After calculating the first path the presence of collision should be checked as any mode of communication that interrupts with joint data will create many problems in initial stage. If the collision is not found then the process continues to create integration between robots and supply chain tools. In this integration process all manual procedures and any presence of duplicate products in the industries are avoided. Figure 5 provides the structure of communication circle in SCM using robotic technology.

Figure 5. Organization plan of communication in SCM with robots

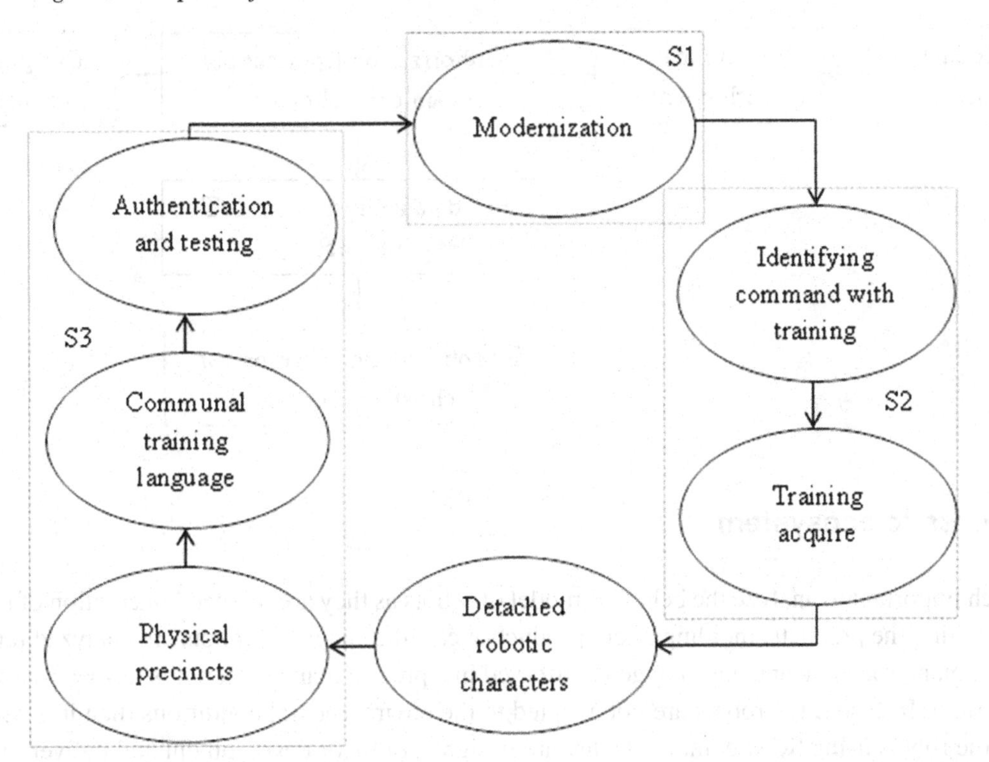

Time of indolence

Once the organization plan is designed then time for executing it using robotic technology should be examined and the reaction time for all commands in this case should be less than 1 milliseconds. This process of time measurement is usually carried out to avoid all emergency case in both production and delivery process as the client will usually check the time of two stages. Since most of the industries are communicating over electronic mail as their expectation and design process should be monitored with quick reply procedure. This expectation will be captured using robots and at initial stage with the help of storage requirements checking process will be completed. If any products are identified with same design then an immediate reply will be sent to the client and if any new product requirements are designed then the corresponding model will be communicated to the consumer. This procedure can also be processed by humans but the time of quick communication will be misplaced. The procedure of time saving process in SCM is deliberated in Figure 6.

Figure 6. Integration time using robots

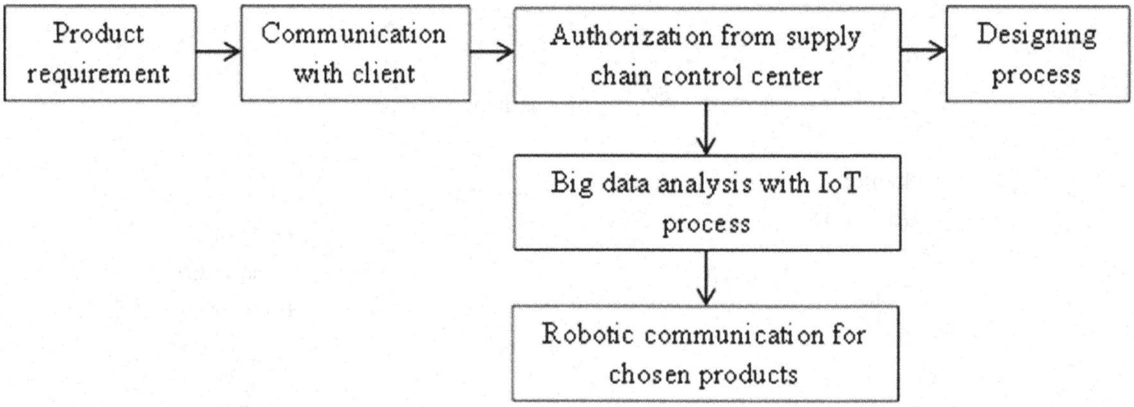

Product requirement → Communication with client → Authorization from supply chain control center → Designing process
Authorization from supply chain control center ↓ Big data analysis with IoT process ↓ Robotic communication for chosen products

Adeptness to ecosystem

It is much important to analyze the behavior model of robots as they are involved in creation of products such as cutting the products, molding of cut products etc. If the change in robots are analyzed for updating the human procedure and any change is observed in a precise manner then it is termed as adeptness to ecosystem. In case if the robots are not adapted to the environmental conditions then it is necessary to train the robots using SCM commands that are designed only for development and delivery process. The process of ecosystem represents suboptimal utilization of available resources, lack of visibility and identification of ineffective process. If all the above listed parameters are modified then the process of SCM using robotics will be adapted to all future technologies.

CONCLUSION

In near future all the application developments will be based on robotic technology as all production process needs to be completed in a quick time. Also the current industries are converting their central procedures to automatic machines where all human workers are eluded form several dangerous process which in turn enhances the life time of humans. The automatic process which includes the presence of robots is also protected in the risk process and even if affected paramedical amenities are provided immediately by experts. Moreover in SCM process the first anticipation by client is the time required for production and delivery of goods this is turn can only be satisfied in the presence of robots. Moreover the entire replacement works cannot be completed without the presence of humans as commands for production and delivery will be provided only by humans to ensure appropriate procedures in SCM management process. Further, 5G standards can be incorporated in the SCM technology which provides a great advantage to connect multiple devices on the same network. This kind of integration in automatic process enables the users to connect in real time to ensure the quality improvement in production fields. Furthermore the client and the industries can be on the same platforms for ensuring the accuracy of designed model with enhanced security features without any external attacks.

REFERENCES

Aljubayri, M., Peng, T., & Shikh-Bahaei, M. (2021). Reduce delay of multipath TCP in IoT networks. *Wireless Networks, 27*(6), 4189–4198. doi:10.100711276-021-02701-3

Amjad, A., Azam, F., Anwar, M. W., & Butt, W. H. (2021). A Systematic Review on the Data Interoperability of Application Layer Protocols in Industrial IoT. *IEEE Access: Practical Innovations, Open Solutions, 9*, 96528–96545. doi:10.1109/ACCESS.2021.3094763

Bouaouad, A. E., Cherradi, A., Assoul, S., & Souissi, N. (2020). The key layers of IoT architecture. *Proceedings of 2020 5th International Conference on Cloud Computing and Artificial Intelligence: Technologies and Applications, CloudTech 2020*, 20–23. 10.1109/CloudTech49835.2020.9365919

Burhan, M., Rehman, R. A., Khan, B., & Kim, B. S. (2018). IoT elements, layered architectures and security issues: A comprehensive survey. *Sensors (Switzerland), 18*(9), 2796. Advance online publication. doi:10.339018092796 PMID:30149582

Chettri, L., & Bera, R. (2020). A Comprehensive Survey on Internet of Things (IoT) Toward 5G Wireless Systems. *IEEE Internet of Things Journal, 7*(1), 16–32. doi:10.1109/JIOT.2019.2948888

Chmiel, M., Korona, M., Kozioł, F., Szczypiorski, K., & Rawski, M. (2021). Discussion on iot security recommendations against the state-of-the-art solutions. *Electronics (Switzerland), 10*(15), 1814. Advance online publication. doi:10.3390/electronics10151814

Da Cruz, M. A. A., Rodrigues, J. J. P. C., Lorenz, P., Korotaev, V. V., & De Albuquerque, V. H. C. (2021). In.IoT - A New Middleware for Internet of Things. *IEEE Internet of Things Journal, 8*(10), 7902–7911. doi:10.1109/JIOT.2020.3041699

Donta, P. K., Srirama, S. N., Amgoth, T., & Rao Annavarapu, C. S. (2021). *Survey on recent advances in IoT application layer protocols and machine learning scope for research directions*. Digital Communications and Networks. doi:10.1016/j.dcan.2021.10.004

Lombardi, M., Pascale, F., & Santaniello, D. (2021). Internet of things: A general overview between architectures, protocols and applications. *Information (Switzerland), 12*(2), 1–21. doi:10.3390/info12020087

Sharma, G., Pandey, N., Hussain, I., & Kathri, S. K. (2018). Design of framework and analysis of Internet of things at data link layer. *2nd International Conference on Telecommunication and Networks, TEL-NET 2017*, 1–4. 10.1109/TEL-NET.2017.8343520

Sierra Marín, S. D., Gomez-Vargas, D., Céspedes, N., Múnera, M., Roberti, F., Barria, P., Ramamoorthy, S., Becker, M., Carelli, R., & Cifuentes, C. A. (2021). Expectations and Perceptions of Healthcare Professionals for Robot Deployment in Hospital Environments During the COVID-19 Pandemic. *Frontiers in Robotics and AI, 8*(June), 1–15. doi:10.3389/frobt.2021.612746 PMID:34150856

Taher, B. H., Liu, H., Abedi, F., Lu, H., Yassin, A. A., & Mohammed, A. J. (2021). A Secure and Lightweight Three-Factor Remote User Authentication Protocol for Future IoT Applications. *Journal of Sensors, 2021*, 1–18. Advance online publication. doi:10.1155/2021/8871204

Triantafyllou, A., Sarigiannidis, P., & Lagkas, T. D. (2018). Network protocols, schemes, and mechanisms for internet of things (IoT): Features, open challenges, and trends. *Wireless Communications and Mobile Computing, 2018*, 1–24. Advance online publication. doi:10.1155/2018/5349894

Chapter 26
E-Collaboration for Management Information Systems Using Deep Learning Technique

Rajalakshmi V.
REVA University, India

Muthukumaran V.
https://orcid.org/0000-0002-3393-5596
REVA University, India

Satheesh Kumar S.
https://orcid.org/0000-0002-2635-4777
REVA University, India

Manjula Sanjay Koti
Dayananda Sagar Academy of Technology and Management, India

Vinothkumar V.
Jain University, India

Thillaiarasu N.
https://orcid.org/0000-0002-7930-0748
REVA University, India

ABSTRACT

Universities are currently confronted with changing student needs, a competitive labour market, and a fast-paced environment. The advancement of communication technology has enabled us to address these issues. Collaboration advances are critical to the current learning process because they train students to work in groups on tasks. In this chapter, the authors present a thorough foundation for an e-collaboration platform that was established during the successful implementation of an e-collaboration solution at the management information systems. The solution makes use of cutting-edge web portal technology and a digital asset management system to create a uniform, centralised platform for system users to collaborate, communicate, and exchange information.

DOI: 10.4018/978-1-7998-9640-1.ch026

INTRODUCTION

Since the beginning of computer use, there has been a need to encourage user cooperation to facilitate everyday chores, communication, work, and training. When computer networking became available, this requirement grew much more pressing. The widespread adoption of computer networks, the Internet, and the World Wide Web are just a few of the factors that have hastened the development of applications, technologies, standards, and systems that facilitate communication and e-collaboration (Muthukumaran V et al., 2021). These technologies, together with the widespread adoption of the Internet, prompted application designers to reconsider how they could employ Information and Communication Technologies (ICT) to assist groups of people. This has had an impact on the design and delivery of e-collaboration services, which enable geographically dispersed users in companies and/or organisations to communicate and collaborate to learn (Computer Supported Collaborative Learning-CSCL) or work (Computer Supported Collaborative Work-CSCW). There are numerous tools, protocols, and technologies available today that can be utilised to construct collaborative systems and applications that meet the criteria and special demands of end-users (Muthukumaran V et al., 2018). E-collaboration is a hot area in research, with a slew of experts contributing to various elements. The large topic's breadth, which includes not only technology but also social and psychological aspects, is the key cause for this major research activity. As a result, different people have different ideas about what e-collaboration is. More specifically, we may state that e-collaboration has been described in a variety of ways in the past, with the number of definitions increasing in recent years. The next section defines the key words in this field (Kumar, V et al., 2021).

Teams have evolved to include new types of contact and collaboration as new technologies, notably ICTs, have been developed. This group could be referred to as a virtual team. A virtual team, like any other team, is defined by (Lipnack and Stamps, 1997) as a group of people who engage through interdependent tasks driven by a common goal (Nagarajan, S. M et al., 2022). A virtual team, unlike traditional teams, collaborates beyond place, time, and organisational boundaries, with linkages strengthened by communication technologies. Virtual team members can work and cooperate in order to communicate with one another. Collaboration and collaboration are closely related concepts that are frequently interchanged. Working together on a common task or process is referred to as collaboration. Cooperation is when two or more people work together to achieve a common objective or advantage(Velliangiri, S et al., 2021). According to (Biuck-Aghai, 2004), we may better comprehend the distinction between collaboration and cooperation by looking at their antonyms: collaboration's antonym is "working alone," whereas cooperation's antonym is "competition." As a result, we believe that collaboration is a better phrase to characterise the pattern of interaction among virtual team members. According to the foregoing, we can consider Biuck Aghai's wide and descriptive word of virtual collaboration, which is defined as collaboration that is conducted without face-to-face interaction and is enabled by technology(Nagarajan, S. M et al., 2022). used a similar definition, stating that e-collaboration is "collaboration among individuals engaged in a common task using electronic technologies." This broad definition considers e-collaboration to be a word that encompasses more than only computer mediated communication (CMC) or computer-assisted collaborative work (CSCW). CSCW is a computer-assisted coordinated activity that involves a group of people working together. As a result, it should be apparent that CSCW is a broad word that encompasses knowledge of how people collaborate in groups, as well as the enabling technologies of computer networking and related hardware, software, services, and methodologies(Ezhilmaran, D., & Muthukumaran, V, 2017).

Prior to the emergence of CSCW, the bulk of computer systems were founded on the incorrect assumption that people work alone and that there is no purpose to employ systems that support collaboration(Manikandan, G et al., 2021). E-collaboration, according to Kock and Nosek should be a broad term because it encompasses a wide range of electronic technologies that aren't (strictly speaking) computers but can be used to enable collaboration among individuals working on a similar task. E-collaboration can take place without any CMC or CSCW, according to this definition(Manikandan, G et al., 2020).

This definition may be acceptable to us. However, we may witness a trend nowadays in which communication devices are (in a wide sense) computers, whether personal computers (PCs), mobile phones, embedded systems, or portable devices(Linda, G. M et al., 2021). In addition, most e-collaboration cases include computers and computer networks, and the current trend in tele-communication networks is to use all-IP networks(Kumar, D et al., 2021). As a result, the focus of this Work is on e-collaboration, which is described as collaboration between individuals or members of virtual teams engaged in a common activity utilising ICT without face-to-face interaction(Munkvold, B. E., and Zigurs, I. 2005; Bhasin, N. K., and Rajesh, A.,2021; Papadimitriou, S. T., and Papadakis, S., 2021; Benali, M et al., 2021; Osmani, M. W et al., 2020).

Groupware is software that allows several users at different workstations to collaborate on a single project(Sadhasivam, J et al., 2021). Groupware is software that emphasises numerous user experiences by coordinating and arranging things so that users may "see" each other while avoiding conflicts(Jayasuruthi, L et al., 2018). The primary premise that groupware makes distinguishes it from other software: groupware makes the user aware that he or she is a member of a group, whereas the bulk of other software strives to hide and shield users from each other (Das, K., and Das, S., 2022; Bhasin, N. K., and Rajesh, A.,2021; Jayasuruthi, L et al., 2018).

E-collaboration is a very complicated topic, as evidenced by the definitions offered in the preceding section, and there is a clear need to shape e-interaction to avoid chaos and failure in virtual teams. Not only technological, but also social and psychological concerns play a role in developing e-interaction to enable e-collaboration. This section discusses the issues, relevant concepts, and useful architectures, systems, protocols, and standards for the development and support of e-collaboration.

Collaboration appears to have several facets when it comes to information technology. To guarantee that the appropriate technologies are used to build and develop groupware systems that can effectively support e-Collaboration, it is vital to understand the distinctions in human interactions. Conversational engagement, transactional interaction, and collaborative interaction are the three basic methods in which humans interact.

A conversational interaction is an information exchange between one or more individuals with the primary goal of discovery or relationship building. For most verbal encounters, communication technology such as phones, instant messaging, and e-mail suffices. Transactional interaction entails the exchange of transaction entities, each of which has the primary role of altering the connection between the parties. Transactional systems that manage state and commit records for persistent storage are the most effective at handling transactional interactions.

The major role of the participants' relationship in collaborative interactions is to change a cooperation entity. The collaborative entity is currently in a state of flux. The evolution of a concept, the production of a design, and the fulfilment of a common objective are all examples. As a result, true collaboration solutions enable many people to contribute to a single product. Collaboration technologies include record or document management, threaded discussions, audit histories, and other systems for capturing the activities of many into a managed content environment.

In order to meet the demand for collaboration and collaborative work, e-Collaboration research focuses on the following key issues: group awareness, multi-user interfaces, concurrency control, group communication and coordination, shared information space, and support for a heterogeneous, open environment that integrates existing single-user applications. According to Biuck-Aghai, there are two major difficulties to e-collaboration:
What is the best way to learn how to collaborate virtually?
How can you tell what's going on (or has gone on) during a virtual collaboration?

The elements of collaborative software suggested by Cerovsek and Turk in order to allow sharing potentially address these difficulties. The following are some of these characteristics:

1. Information sharing entails the exchange of many forms of data that must be interpreted by humans.
2. Apart from information generation, knowledge organization/storage, and knowledge application, knowledge sharing is one of the processes in the knowledge management system.
3. Application sharing, which entails the sharing of code or the availability of applications: Code, components, apps, services, and computing are examples of sharing categories.
4. Workspace sharing provides a virtual place (shared workspace) for employees' work (like in an office) and may include the sharing of multiple levels of sharing.
5. Sharing of resources. All of the types of sharing outlined above, as well as the sharing of other resources including computational resources, processor time, and equipment, are examples of resource sharing.

e-Collaboration systems are frequently classified according to the time/location matrix, which distinguishes between systems that operate at the same time (synchronous) and those that operate at different times (asynchronous), as well as between systems that operate at the same time (face-to-face) and those that operate at different times (asynchronous) (distributed). Poltrock (2002) proposed a new category of collaboration systems based on the time-interaction criterion, i.e. synchronous and asynchronous. Groupware can be split into three groups based on the level of collaboration, according to this classification.

Communication technologies allow people to exchange messages, files, data, or documents to one another, facilitating information sharing. Audio/video conferencing, the telephone, textual chat, instant messaging, and broadcast video are all examples of synchronous tools. E-mail, voice mail, and fax are examples of asynchronous tools in this category.

Collaboration/Conferencing software that allows users to share information in a more engaging manner. Whiteboards, application sharing, meeting facilitation tools, and collaborative virtual environments are examples of synchronous tools in this category. Document management tools, threaded discussions, hypertext, and collaborative workspaces are examples of asynchronous tools in this category.

Tools for collaborative management (coordination) that make it easier to organise and manage group activities. Floor control and session management are examples of synchronous technologies in this area. Workflow management, case management, project management, and calendar and scheduling applications are examples of asynchronous tools in this area.

DEEP LEARNING

Deep learning is a branch of machine learning and a subset of Artificial Intelligence (AI) in which a neural network is used to mimic a "human brain" with a set of input and output values that are interconnected through the hidden layer (s). Deep learning does not necessitate any specific programming to process vast amounts of data with high processing power. Deep learning is defined as "a type of machine learning that learns to represent the world as a layered hierarchy of concepts, with each concept defined in reference to simpler concepts and more abstract representations computed in terms of less abstract representations in Figure 1."

Figure 1. The evolution of Artificial Intelligence

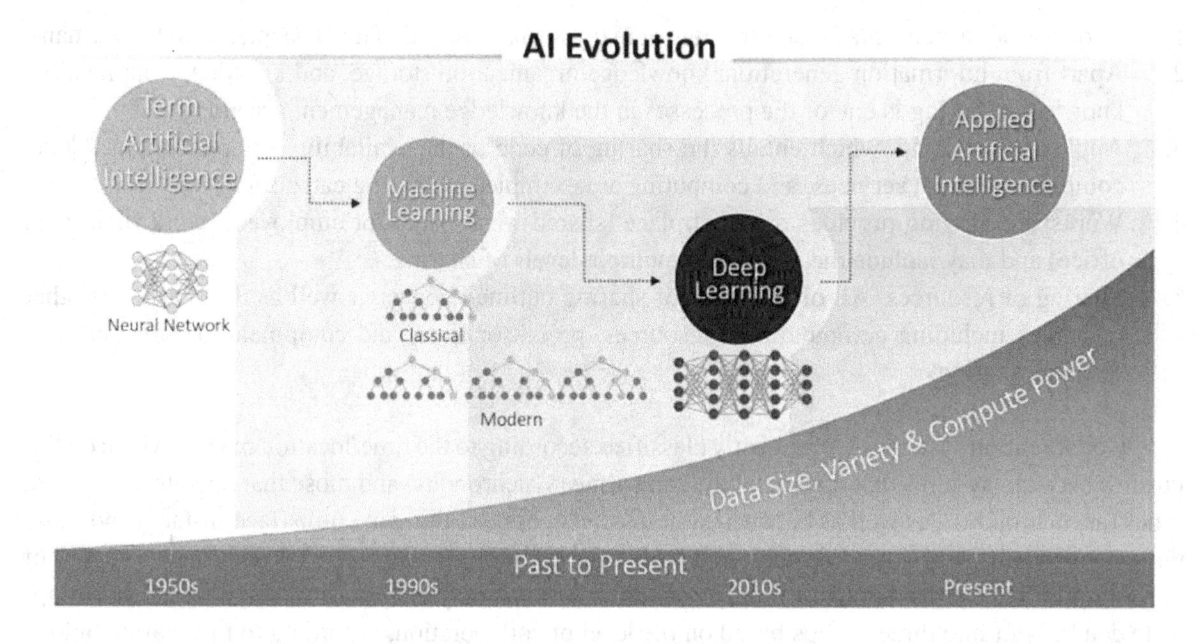

Artificial Neural Networks (ANN) are used to represent billions of neurons in the human brain. Neurons are represented as nodes or neurons. Deep learning is "self-taught and unsupervised feature learning consisting of numerous intermediate hidden layers that process information and from it like the human brain," according to Andrew Ng, and is mostly used in brain simulations to fulfil the following goals.

- Develop significantly more effective and user-friendly algorithms.
- Using growing amounts of data with the availability of fast enough computers to create major advances in Artificial Intelligence and Machine Learning
- To achieve more scalability by building larger neural networks and training them with more data to achieve better performance by feeding more data; to perform automatic feature extraction through feature learning on raw data with more computation to train them; to perform automatic feature extraction through feature learning on raw data with more computation to train them in Figure 2.

Figure 2. The timeline of machine learning methods in security and privacy

	Pre-1990s	1990s	2000s	2010s
SPAM DETECTION	**1978:** First spam email	Spam continues to worsen due to growth in email **1996:** First spam blockers	**2002:** Machine learning methods first proposed for spam detection **2003:** First attempts to regulate spam in the United States	Machine learning spam detection widely embedded in email services Emergence of deep learning-based classifiers
INTRUSION DETECTION	**1980:** First intrusion detection systems **1986:** Anomaly detection systems combine expert rules and statistical analysis	**Early 1990s:** Neural networks for anomaly detection first proposed **1999:** DARPA creates datasets to study intrusion detection systems	Machine learning further studied as a possible tool for misuse-based and anomaly-based intrusion detection	**Late 2010s:** Emergence of large-scale, cloud-based intrusion detection systems Deep learning studied for intrusion detection
MALWARE DETECTION	**Early 1980s:** First viruses found "in the wild" **Late 1980s:** First antivirus companies founded	**Early 1990s:** First polymorphic viruses **1996:** IBM begins studying machine learning for malware detection	**Early 2000s:** First metamorphic viruses Wide number of traditional machine learning methods studied to detect malware	Rise of "next-gen" antivirus detection Emergence of ML-focused antivirus companies

Deep learning techniques

Deep learning methods appear to have outperformed classical methods such as Support Vector Machine (SVM), Naive Bayes, Random Forests, k-means clustering, logistic regression models, and others in terms of findings and performance, according to recent studies. “Traditional machine learning” refers to a wide range of old-school approaches that have been a part of machine learning for decades, at least until “Deep learning” entered the scene in the last 5 to 10 years. Deep learning algorithms are divided into three groups based on their functionality explained in Figure.3.

1. Supervised Learning
2. Unsupervised Learning
3. Semi-supervised Learning
4. Reinforcement Learning

N-Reparents Nodes

Figure 3. Deep learning examples architecture

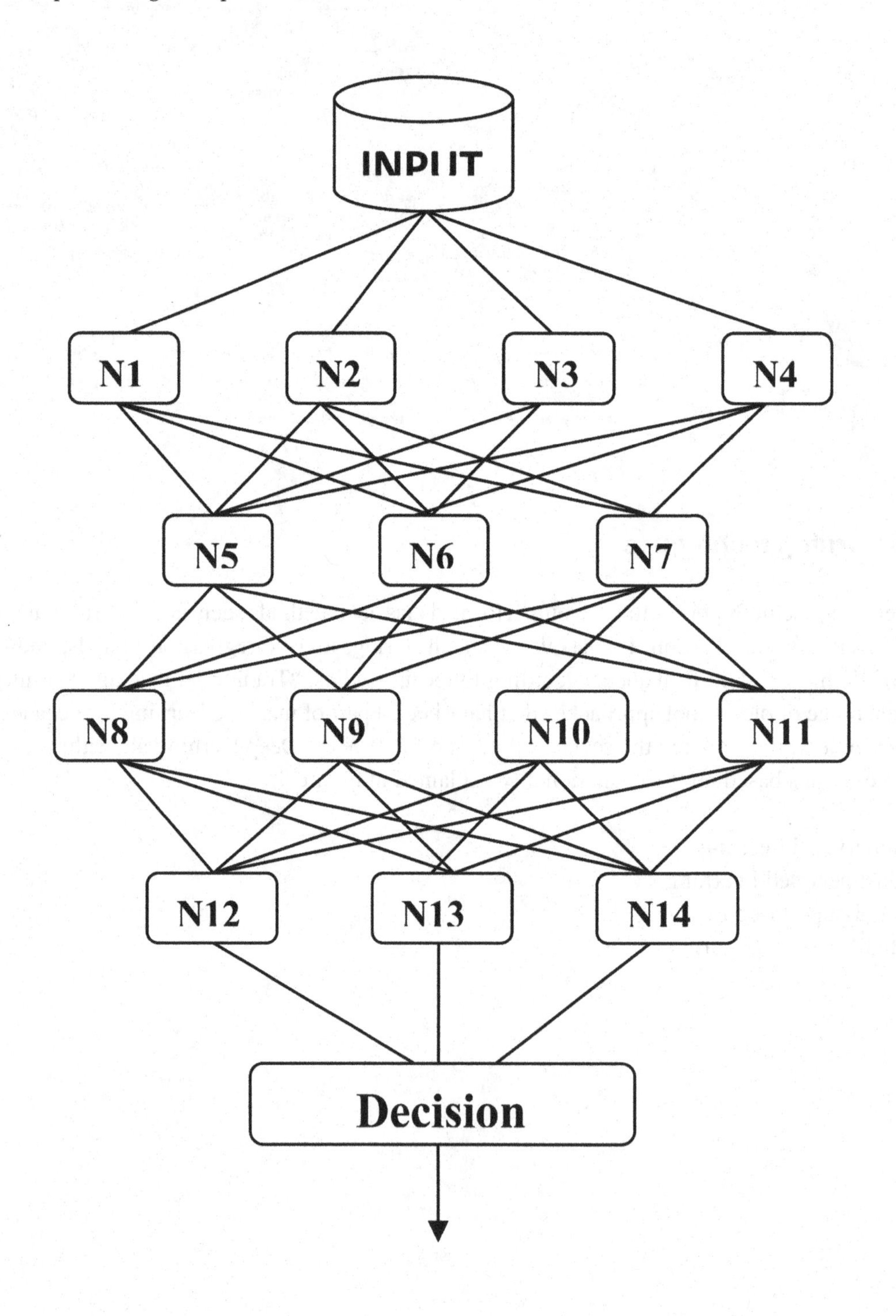

Supervised learning

A model is trained using these datasets to predict output with greater accuracy to categorise fresh inputs from the labelled data sets. By simply comparing network traffic to a known profile, a well-labelled data collection on recent attacks could aid in the detection of future attacks. When mining data, supervised learning can be utilised for classification and regression purposes. An algorithm is used in classification to group data into specific groups depending on its characteristics. For example, consider a data collection with apple and orange characteristics that reliably predicts the output value, i.e. either orange or apple, based on the input values.

Based on multiple data sets, regression employs an algorithm to find and comprehend the "relationship between dependent and independent variables." These methods are extremely useful when it comes to forecasting numerical values from such data sources. For instance, estimating a company's sales expectations. The most extensively used regression techniques are linear regression, logistic regression, and polynomial regression.

Unsupervised learning

Unsupervised learning is a technique for uncovering hidden patterns in unlabelled data sets using an algorithm that analyses and clusters the data without requiring any human participation. The following tasks are accomplished using this method.

Clustering

It is mostly used for grouping in market segmentation and image compression applications. It's a data mining technique that groups unlabelled data into groups based on similarities or differences. The K-means clustering algorithm is an example of this strategy, which forms groups with similar data points from an unlabelled data set, with the k-value algorithm assisting in describing the size and granularity of those groups.

Association

This method is mostly used in market basket analysis and recommendation engines to construct relationships between variables in an unlabelled data set by employing a set of rules or association rules. When buying a laptop online, for example, the page may propose buying a laptop bag and other relevant accessories, or it may make suggestions with a statement such as "Customers who bought this item also bought."

DEEP LEARNING IN SECURITY AND PRIVACY

Many deep learning algorithms have been widely employed in industries such as marketing, finance, and sales throughout the last few decades. Early in 2010, it was difficult to find research papers aimed at safeguarding products and enterprises from malware and hacker attacks. However, major enterprise technology giants such as Google, Microsoft, Salesforce, and Facebook have incorporated deep learning-based algorithms into their products and services to assure data protection and privacy. Google recently included such algorithms to catch 'spam' emails that its Gmail filter had missed. TensorFlow was used to detect spam emails that were generally difficult to detect, such as messages from freshly created domains, image-based emails, and communications with concealed embedded content, among other things.

As technical innovations such as data science and cloud computing become more prevalent, protecting the security and privacy of data on these platforms has become a difficult challenge that necessitates the complete focus of academics and specialists. Deep learning algorithms' accuracy is primarily determined on the quality of labelled data sets, which are not freely available on any platform.

Deep learning helps to ensure security in the following ways:

1. Constant network traffic monitoring identifies dangers on a network.
2. Keep individuals safe when they're online in a "bad area."
3. Provide end-point malware detection by detecting unknown malware by matching known malware behaviour and attributes.
4. Ensure IP reputation analysis, location-based analysis, and suspicious login behaviour in the cloud to protect data.
5. Detect malware in encrypted traffic by identifying malicious patterns in order to uncover hidden risks.

Deep learning techniques have recently outperformed traditional and rule-based machine learning methods in the fields of malware identification and network intrusion detection.

Many research studies were conducted between 1990 and 2010 to secure network data and information sharing security utilising classis machine learning techniques, but the results showed that such algorithms were ineffective in identifying security and privacy related assaults. In fact, despite the presence of more capable tools to detect them, risks to both the government and other commercial industries have grown exponentially. Importantly, malicious attackers have sufficient resources and knowledge of advanced persistent threats, yet cyber security specialists are struggling to keep up with the ever-increasing demand for their services.

1. The proposed system or algorithm must have a clear image of the sorts of attacks to be identified, so that a researcher can create a more particular system, avoiding or reducing misclassifications in the results.
2. If strong relationships between features and attacks of interest are not established while designing an algorithm or implementing a systematic approach, the findings may contain major faults.

E-COLLABORATION SYSTEMS FOR RESEARCH AND TECHNOLOGY IN DEEP LEARNING

In Non-Profit Educational Organisations, the Need for E-Collaboration is Critical

The requirement that people's knowledge and skill levels be always up to date is one of the biggest challenges in today's knowledge-based Internet economy. The greater an organization's ability to develop and distribute information, the better its chances in a highly competitive market. Continuous skill improvement should be ingrained in an organization's culture. However, we must remember that organisational culture is based on the experiences of its members prior to their joining. The habit of always upgrading one's talents should begin at least in university.

As a result, the new economy places a new and difficult task on universities: preparing their alumni for continuous improvement in collaborative groups, as collaboration has been shown to be more efficient than individual work by introducing synergy effects, encouraging creativity, promoting trade-offs, and, finally, being enjoyable. Computer mediated cooperation, group support systems, groupware, cooperative work, digital asset management, and other activities that bring two or more people together to be more creative by using electronic tools are examples.

E-collaboration in non-profit educational organisations reduces expenses in a wide range of domains, which is currently regarded as a non-trivial assertion. However, certain investments are required to accomplish this cost decrease. The initial investment is substantial. However, with a successful implementation of the solution, these investments will pay out many times over. E-collaboration transforms an organisation into an e-organization, in which members communicate at least partially in virtual space. Furthermore, e-collaboration at universities considerably improves the learning process. We feel it is true because of our many years of expertise with e-mail communication and the use of web technology in education, therefore we decided to expand present e-communication into a full e-collaboration solution.

It makes functionalities available to educational institutions that were previously unavailable. Wider availability, instant messaging functionality, and simplicity of interaction are just a few examples. Infrastructure must be built to communicate. For a user, this is not always an easy task. As a result, we believe that implementing communication structure into the system is the greatest answer to this problem. The above characteristics establish and justify the necessity for e-collaboration and, as a result, improve the traditional educational process, which has remained largely unchanged for millennia. With the advancement of information systems and network technology, we can now transform traditional cooperation into e-collaboration, enhancing current interactions between teachers and students as well as between students.

Server for the Network

The foundation for an e-collaboration solution should be built on a network server that can access the organization's internal knowledge resources (presumably file servers, data warehouses, databases, and document repositories). As a result, such a server needs handle at least protocol as well as other internal protocols (such as IPX/SPX or NetBIOS). However, we prefer to store corporate information in a single location (we intentionally ignore security concerns here), in a single format (or with standard metadata descriptions), and in the same style (by utilising the same network protocol). At the same time, in order to respond to user requests, the network server should be able to connect to external knowledge and information resources.

Directory service

Most existing data should be reused, and new data placed into the system should be made available to others. As a result, the e-collaboration solution should store all, or almost all, of the user's data in well-known structures like standard directory services (Novell Directory Services, Active Directory, or Lightweight Directory Access Protocol, to name a few). This should include information about user groups, user privileges, available services, and so on, in addition to keeping users' contact and configuration information.

Messaging server

Communication software is one of the most crucial cooperation aids, especially in a relatively low scattered context. This includes both email and more complex newsgroup servers. All major operating systems support such complicated systems.

Instant messaging server

It is frequently necessary for one person to contact another in order to discuss a topic, exchange ideas, or train each other. In such a case, an instant messaging server can be useful. This allows for the creation of a one-on-one chat room, which is frequently sound and video improved. When someone on the user's private list is online, the standard instant messaging service notifies her. Unfortunately, messaging systems lack a consistent protocol, and there are numerous competing instant messaging systems.

Digital Asset Management System

Digital files are stored in a central repository. This should also serve as a central storage location for any e-collaboration solutions. There are a variety of competitive technologies to choose from, including relational databases, data warehouses, and specialist software solutions. Performance, flexibility, and open standards compatibility are the most important characteristics to consider while selecting the right DAM.

Web portal

The use of web portal technology can be used to establish a uniform and easily accessible user environment. E-mail, a search engine, forums, news, and other services are available on a conventional web portal. As a result, the authors believe that the portal should be used for e-collaboration as well.

Plan that is Meticulous and Detailed

Three server machines were used to build the platform: a directory services server, a portal server, and a DNS server, all of which were hosted on the same Internet domain: kie.ae.poznan.pl. The DC and portal operate on Windows 2000 Server, whereas DNS is based on Sun's SunOS. The portal server makes use of Microsoft's SharePoint Portal Server and the Web Storage System's document repository. Sun's computer is a principal name server for the entire Internet domain, and DC is a primary Windows Domain Controller and Exchange server.

The platform was built on three server machines: a directory services server, a portal server, and a DNS server, all of which were hosted on the same IP address: kie.ae.poznan.pl. Windows 2000 Server is used for the DC and portal, whereas Sun's SunOS is used for DNS. Microsoft's SharePoint Portal Server and the Web Storage System document repository are used by the portal server. Sun's system serves as a major name server for the entire Internet domain, as well as a primary Windows Domain Controller and Exchange server.

- Direct access to Microsoft Exchange and Outlook Web Access is accessible through the site.
- Registration for classes.
- Implementing a procedure for writing essays.
- Control of the essay deadline.
- Management of a student group.
- Group management for teachers and lecturers.
- Controlling the structure of the document repository

The system facilitates e-collaboration by providing storage and a point of contact for remote users. It promotes collaboration in the creation of added value through publication capabilities. During the implementation of the system, for example, a temporary website was set up to support and coordinate the operations of several project groups. The system is also used to coordinate departmental projects and create documents. The departmental platform was used to write this paper. Every new version with comments appeared on the system, thus the authors didn't have to meet face to face. The check-in – check-out feature aided in maintaining a single document version (at no time had two authors worked on a document without knowing of it). If a new version of a document appeared on the plat form, writers received a subscription notification.

CONCLUSION

In today's world, universities must successfully deploy e-collaboration platforms. It not only saves money, but it also equips graduates to compete successfully in the job market. It also helps pupils prepare for lifetime learning, which appears to be the key to future success. We believe that while designing an e-collaboration deployment, three aspects of the project should be considered: corporate culture, a meticulous and precise plan, and technology. Because there are few e-collaboration technologies on the market that give a complete solution to universities' needs, unique implementations have a lot of promise.

REFERENCES

Benali, M., Ghomari, A. R., Zemmouchi-Ghomari, L., & Lazar, M. (2021). Flexible Crowd-Driven Decision Making in Times of Crisis. In *E-Collaboration Technologies and Strategies for Competitive Advantage Amid Challenging Times* (pp. 1–27). IGI Global.

Bhasin, N. K., & Rajesh, A. (2021). Impact of E-Collaboration Between Indian Banks and Fintech Companies for Digital Banking and New Emerging Technologies. *International Journal of e-Collaboration*, *17*(1), 15–35. doi:10.4018/IJeC.2021010102

Bhasin, N. K., & Rajesh, A. (2021). Impact of E-Collaboration Between Indian Banks and Fintech Companies for Digital Banking and New Emerging Technologies. *International Journal of e-Collaboration*, *17*(1), 15–35.

Biuk-Aghai, R. P., & Simoff, S. J. (2004, November). Patterns of virtual collaboration in online collaboration systems. In *Proceedings of the IASTED International Conference on Knowledge Sharing and Collaborative Engineering* (pp. 22-24). Academic Press.

Das, K., & Das, S. (2022). Energy-Efficient Cloud-Integrated Sensor Network Model Based on Data Forecasting Through ARIMA. *International Journal of e-Collaboration*, *18*(1), 1–17. doi:10.4018/IJeC.290292

Dhiman, G., Kumar, V. V., Kaur, A., & Sharma, A. (2021). DON: Deep Learning and Optimization-Based Framework for Detection of Novel Coronavirus Disease Using X-ray Images. *Interdisciplinary Sciences, Computational Life Sciences*, 1–13.

Ezhilmaran, D., & Muthukumaran, V. (2017). Authenticated group key agreement protocol based on twist conjugacy problem in near-rings. *Wuhan University Journal of Natural Sciences*, *22*(6), 472–476. doi:10.100711859-017-1275-9

Jayasuruthi, L., Shalini, A., & Kumar, V. V. (2018). Application of rough set theory in data mining market analysis using rough sets data explorer. *Journal of Computational and Theoretical Nanoscience*, *15*(6-7), 2126–2130.

Kumar, V., Niveditha, V. R., Muthukumaran, V., Kumar, S. S., Kumta, S. D., & Murugesan, R. (2021). A Quantum Technology-Based LiFi Security Using Quantum Key Distribution. In Handbook of Research on Innovations and Applications of AI, IoT, and Cognitive Technologies (pp. 104-116). IGI Global.

Kumar, V. V., Raghunath, K. K., Rajesh, N., Venkatesan, M., Joseph, R. B., & Thillaiarasu, N. (2021). Paddy Plant Disease Recognition, Risk Analysis, and Classification Using Deep Convolution Neuro-Fuzzy Network. *Journal of Mobile Multimedia*, 325-348.

Kumar, V. V., Raghunath, K. M., Muthukumaran, V., Joseph, R. B., Beschi, I. S., & Uday, A. K. (2021). Aspect based sentiment analysis and smart classification in uncertain feedback pool. *International Journal of System Assurance Engineering and Management*, 1-11.

Linda, G. M., Lakshmi, N. S. R., Murugan, N. S., Mahapatra, R. P., Muthukumaran, V., & Sivaram, M. (2021). Intelligent recognition system for viewpoint variations on gait and speech using CNN-CapsNet. *International Journal of Intelligent Computing and Cybernetics*.

Lipnack, J., & Stamps, J. (1997). *Reaching across space, time, and organizations with technology*. John Wiley & Sons, Inc.

Manikandan, G., Perumal, R., & Muthukumaran, V. (2020, November). A novel and secure authentication scheme for the Internet of Things over algebraic structure. In AIP Conference Proceedings: Vol. 2277. *No. 1* (p. 060001). AIP Publishing LLC. doi:10.1063/5.0025330

Manikandan, G., Perumal, R., & Muthukumaran, V. (2021). Secure data sharing based on proxy re-encryption for internet of vehicles using seminearring. *Journal of Computational and Theoretical Nanoscience*, *18*(1-2), 516–521.

Munkvold, B. E., & Zigurs, I. (2005). Integration of e-collaboration technologies: Research opportunities and challenges. *International Journal of e-Collaboration*, *1*(2), 1–24. doi:10.4018/jec.2005040101

Muthukumaran, V., Ezhilmaran, D., & Anjaneyulu, G. S. G. N. (2018). Efficient Authentication Scheme Based on the Twisted Near-Ring Root Extraction Problem. In *Advances in Algebra and Analysis* (pp. 37–42). Birkhäuser. doi:10.1007/978-3-030-01120-8_5

Muthukumaran, V., Joseph, R. B., & Uday, A. K. (2021). Intelligent Medical Data Analytics Using Classifiers and Clusters in Machine Learning. In Handbook of Research on Innovations and Applications of AI, IoT, and Cognitive Technologies (pp. 321-335). IGI Global.

Muthukumaran, V., Vinothkumar, V., Joseph, R. B., Munirathanam, M., & Jeyakumar, B. (2021). Improving network security based on trust-aware routing protocols using long short-term memory-queuing segment-routing algorithms. *International Journal of Information Technology Project Management*, *12*(4), 47–60. doi:10.4018/IJITPM.2021100105

Nagarajan, S. M., Deverajan, G. G., Chatterjee, P., Alnumay, W., & Muthukumaran, V. (2022). Integration of IoT based routing process for food supply chain management in sustainable smart cities. *Sustainable Cities and Society*, *76*, 103448.

Nagarajan, S. M., Muthukumaran, V., Beschi, I. S., & Magesh, S. (2021). Fine Tuning Smart Manufacturing Enterprise Systems: A Perspective of Internet of Things-Based Service-Oriented Architecture. In Handbook of Research on Innovations and Applications of AI, IoT, and Cognitive Technologies (pp. 89-103). IGI Global.

Osmani, M. W., El Haddadeh, R., Hindi, N., & Weerakkody, V. (2020). The Role of Co-Innovation Platform and E-Collaboration ICTs in Facilitating Entrepreneurial Ventures. *International Journal of E-Entrepreneurship and Innovation*, *10*(2), 62–75.

Papadimitriou, S. T., & Papadakis, S. (2021). E-Collaboration in Educational Organizations: Opportunities and Challenges in Virtual Learning Environments and Learning Spaces. *Collaborative Convergence and Virtual Teamwork for Organizational Transformation*, 120-146.

Velliangiri, S., Karthikeyan, P., & Vinoth Kumar, V. (2021). Detection of distributed denial of service attack in cloud computing using the optimization-based deep networks. *Journal of Experimental & Theoretical Artificial Intelligence*, *33*(3), 405–424. doi:10.1080/0952813X.2020.1744196

Compilation of References

Abdel-Fattah, M. A., Khedr, A. E., & Nagm Aldeen, Y. (2017). An Evaluation Framework for Business Process Modeling Techniques. *International Journal of Computer Science and Information Security, 15*(5), 382–392.

Abed, A., Yuan, J., & Li, L. (2017). A Review of Towered Big-Data Service Model for Biomedical Text-Mining Databases. *International Journal of Advanced Computer Science and Applications, 8*(8), 12.

Abu Al-Haija, Q., & Al-Badawi, A. (2022). Attack-Aware IoT Network Traffic Routing Leveraging Ensemble Learning. *Sensors (Basel), 22*(1), 241.

Adil, A. H., Mulla, M., Mulla, M., & Ramakrishnappa, S. (2020). *Usage of Social Media among Undergraduate University Students*. https://www.researchgate.net/publication/347399224_Usage_of_Social_Media_among_Undergraduate_University_Students

Admiraal, W., Huizenga, J., Akkerman, S., & Ten Dam, G. (2011). The concept of flow in collaborative game-based learning. *Computers in Human Behavior, 27*(3), 1185–1194. doi:10.1016/j.chb.2010.12.013

Affleck, J., & Kvan, T. (2008). A virtual community as the context for discursive interpretation: A role in cultural heritage engagement. *International Journal of Heritage Studies, 14*(3), 268–280. doi:10.1080/13527250801953751

Aggarwal, C. C., & Zhai, C. (2012). A survey of text clustering algorithms. In *Mining text data* (pp. 77–128). Springer.

Aggarwal, R., & Das, M. L. (2012). RFID security in the context of "internet of things." *Proceedings of the First International Conference on Security of Internet of Things - SecurIT '12*, 51–56. 10.1145/2490428.2490435

Agrawal, S., & Das, M. L. (2011). Internet of things -A paradigm shift of future internet applications. *2011 Nirma University International Conference on Engineering*, 1–7. 10.1109/NUiConE.2011.6153246

Ahammad, S. H., Rajesh, V., Venkatesh, K. N., Nagaraju, P., Rao, P. R., & Inthiyaz, S. (n.d.). Liver segmentation using abdominal CT scanning to detect liver disease area. *International Journal of Emerging Trends in Engineering Research, 7*(11), 664-669.

Ahammad, S. H., Rajesh, V., Hanumatsai, N., Venumadhav, A., Sasank, N. S. S., Gupta, K. B., & Inithiyaz, S. (2019). MRI image training and finding acute spine injury with the help of hemorrhagic and non hemorrhagic rope wounds method. *IJPHRD, 10*(7), 404. doi:10.5958/0976-5506.2019.01603.6

Ahammad, S. H., Rajesh, V., Neetha, A., Sai Jeesmitha, B., & Srikanth, A. (2019). Automatic segmentation of spinal cord diffusion MR images for disease location finding. *Indonesian Journal of Electrical Engineering and Computer Science, 15*(3), 1313–1321. doi:10.11591/ijeecs.v15.i3.pp1313-1321

Ahammad, S. H., Rajesh, V., & Rahman, M. Z. U. (2019). Fast and accurate feature extraction-based segmentation framework for spinal cord injury severity classification. *IEEE Access: Practical Innovations, Open Solutions*, *7*, 46092–46103. doi:10.1109/ACCESS.2019.2909583

Ahammad, S. H., Rajesh, V., Rahman, M. Z. U., & Lay-Ekuakille, A. (2020). A hybrid CNN-based segmentation and boosting classifier for real time sensor spinal cord injury data. *IEEE Sensors Journal*, *20*(17), 10092–10101. doi:10.1109/JSEN.2020.2992879

Ahammad, S. K., & Rajesh, V. (2018). Image processing based segmentation techniques for spinal cord in MRI. *Indian Journal of Public Health Research & Development*, *9*(6), 317. doi:10.5958/0976-5506.2018.00571.5

Ahmad, F., Lee, S., Thottethodi, M., & Vijaykumar, T. (2013). Mapreduce with communication overlap. *Journal of Parallel and Distributed Computing*, *73*, 608–620.

Ahmad, T., Alvi, A., & Ittefaq, M. (2019, July-September). The Use of Social Media on Political Participation Among University Students: An Analysis of Survey Results from Rural Pakistan. *SAGE Open*, *9*(3), 1–9. doi:10.1177/2158244019864484

Ajit, P. (2016). Prediction of employee turnover in organizations using machine learning algorithms. *Algorithms*, *4*(5), C5.

Akhila, M., Amarnath, V., & Vishnu Murthy, G. (2018). Applications of MapReduce in Big Data concepts. *International Journal of Research*, *5*, 20.

Akhtar, N., & Mian, A. (2018). Threat of adversarial attacks on deep learning in computer vision: A survey. *IEEE Access: Practical Innovations, Open Solutions*, *6*, 14410–14430.

Akyildiz, I., & Stuntebeck, E. (2006). Wireless underground sensor networks: Research challenges. *Ad Hoc Networks*, *4*(1), 669–686. doi:10.1016/j.adhoc.2006.04.003

Akyildiz, I., Su, W., Sankarasubramaniam, Y., & Cayirci, E. (2002). Wireless sensor networks: A survey. *Computer Networks-Journal Elsevier*, *38*(4), 393–422. doi:10.1016/S1389-1286(01)00302-4

Akyildiz, I., & Vuran, M. (2010). *Wireless Sensor Networks. Wiley.*

Al Ridhawi, I., Aloqaily, M., Kantarci, B., Jararweh, Y., & Mouftah, H. T. (2018). A continuous diversified vehicular cloud service availability framework for smart cities. *Computer Networks*, *145*, 207–218.

Alalwan, A. A. (2018). Investigating the impact of social media advertising features on customer purchase intention. *International Journal of Information Management*, *42*, 65–77. doi:10.1016/j.ijinfomgt.2018.06.001

Al-Aufi, A., & Fulton, C. (2015a). Impact of social networking tools on scholarly communication: A cross-institutional study. *The Electronic Library*, *33*(2), 224–241. doi:10.1108/EL-05-2013-0093

Alduayj, S. S., & Rajpoot, K. (2018, November). Predicting employee attrition using machine learning. In *2018 International Conference on Innovations in Information Technology (IIT)* (pp. 93-98). IEEE.

Alhabash, S., & Ma, M. (2017). A Tale of Four Platforms: Motivations and Uses of Facebook, Twitter, Instagram, and Snapchat Among College Students? *Social Medial + Society*, 1-2.

Alhadidi, D., Mohammed, N., Fung, B. C., & Debbabi, M. (2012, July). Secure distributed framework for achieving ε-differential privacy. In *International Symposium on Privacy Enhancing Technologies Symposium* (pp. 120-139). Springer. 10.1007/978-3-642-31680-7_7

Al-Hunaiyyan, A., Alhajri, R. A., & Al-Sharhan, S. (2016). Perceptions and challenges of mobile learning in Kuwait. *Journal of King Saud University-Computer and Information Sciences.*

Al-Jenaibi, B. (2015). Current issues about public relations professionals: Challenges and potentials of PR in UAE organisations. *Middle East Journal of Management*, *2*(4), 330–351. doi:10.1504/MEJM.2015.073568

Al-Jenaibi, B. (2015). The New Electronic Government: Are the federal authorities ready to use e-government? *International Journal of Knowledge Society Research*, *6*(3), 45–74. doi:10.4018/IJKSR.2015070104

Al-Jenaibi, B. (2016). Upgrading Society with Smart Government: The use of smart services among Federal Offices of the UAE *International Journal of Information Systems and Social Change*, *7*(4), 20–51. doi:10.4018/IJISSC.2016100102

Al-Jenaibi, B. (2017). The impact of dubbed serials on students in the UAE. *International Journal of Arab Culture Management and Sustainable Development*, *3*(1), 41–66. doi:10.1504/IJACMSD.2017.10007172

Aljubayri, M., Peng, T., & Shikh-Bahaei, M. (2021). Reduce delay of multipath TCP in IoT networks. *Wireless Networks*, *27*(6), 4189–4198. doi:10.100711276-021-02701-3

Al-Kandari, A. A., Al-Sumait, F. Y., & Al-Hunaiyyan, A. (2017). Looking perfect: Instagram use in a Kuwaiti cultural context. *Journal of International and Intercultural Communication*, *10*(4), 273–290. doi:10.1080/17513057.2017.1281430

Alnjadat, R., Hmaidi, M. M., Samha, T. E., Kilani, M. M., & Hasswan, A. M. (2019). Gender variations in social media usage and academic performance among the students of University of Sharjah. *Journal of Taibah University Medical Sciences*, *14*(4), 390–394. doi:10.1016/j.jtumed.2019.05.002 PMID:31488973

Alomair, B., Clark, A., Cuellar, J., & Poovendran, R. (2012). Scalable RFID Systems: A Privacy-Preserving Protocol with Constant-Time Identification. *IEEE Transactions on Parallel and Distributed Systems*, *23*(8), 1536–1550. doi:10.1109/TPDS.2011.290

Aloqaily, M., Al Ridhawi, I., Kantraci, B., & Mouftah, H. T. (2017, October). Vehicle as a resource for continuous service availability in smart cities. In *2017 IEEE 28th annual international symposium on personal, indoor, and mobile radio communications (PIMRC)* (pp. 1-6). IEEE.

Al-Rahmi, W. M., Othman, M. S., & Yusuf, L. M. (2015). The role of social media for collaborative learning to improve academic performance of students and researchers in Malaysian higher education. *International Review of Research in Open and Distance Learning*, *16*(4), 177–204. doi:10.19173/irrodl.v16i4.2326

Alsalem, F. (2019). Why do they Post? Motivations and uses of Snapchat, Instagram and Twitter Among KUWAIT college students. *Media Watch*, *10*(3). Advance online publication. doi:10.15655/mw/2019/v10i3/49699

Al-Sarem, M., Saeed, F., Alkhammash, E. H., & Alghamdi, N. S. (2022). *An Aggregated Mutual Information Based Feature Selection with Machine Learning Methods for Enhancing IoT Botnet Attack*. Academic Press.

Alsharoa, A., Ghazzai, H., Kadri, A., & Kamal, A. E. (2019). Spatial and temporal management of cellular HetNets with multiple solar powered drones. *IEEE Transactions on Mobile Computing*, *19*(4), 954–968.

Alt, D. (2018). Students' Wellbeing, Fear of Missing out, and Social Media Engagement for Leisure in Higher Education Learning Environments. *Current Psychology (New Brunswick, N.J.)*, *37*(1), 128–138. doi:10.100712144-016-9496-1

Alwagait, E., Shahzad, B., & Alim, S. (2015, October). Impact of social media usage on students' academic performance in Saudi Arabia. *Computers in Human Behavior*, *51*, 1092–1097. doi:10.1016/j.chb.2014.09.028

Al-Zubi, N. (2015). Concealed weapon detection using X-Ray images. *International Journal of Electrical, Electronics and Data Communications*, *3*, 27–30.

Amjad, A., Azam, F., Anwar, M. W., & Butt, W. H. (2021). A Systematic Review on the Data Interoperability of Application Layer Protocols in Industrial IoT. *IEEE Access: Practical Innovations, Open Solutions*, *9*, 96528–96545. doi:10.1109/ACCESS.2021.3094763

Anderson, T., Rourke, L., Garrison, D. R., & Archer, W. (2001). Assessing teaching presence in a computer conferencing context. *Journal of Asynchronous Learning Networks*, *5*(2), 1–17. doi:10.24059/olj.v5i2.1875

Andreas, K., & Michael, H. (2010). Users of the world, unite! The challenges and opportunities of social media. *Business Horizons*, *53*, 61.

Andrejevic, M. (2014). 'Free lunch' in the digital era: Organization is the new content. In *The Audience Commodity in a Digital Age: Revisiting a Critical Theory of Commercial Media* (pp. 193–206). Peter Lang Publishing.

Andryani, R., Negara, E. S., & Triadi, D. (2019). Social Media Analytics: Data Utilization of Social Media for Research. *Journal of Information Systems and Informatics*, *1*(2), 13.

Anim, P. A., Asiedu, F. O., Adams, M., Achempong, G., & Boakye, E. (2019). "Mind the gap": To succeed in marketing politics, think of social media innovation. *Journal of Consumer Marketing*, *36*(6), 806–817. doi:10.1108/JCM-10-2017-2409

Anitha, G., Sunitha Devi, P., Vidhya Sri, J., & Priyanka, D. (2020). Face Recognition Based Attendance System Using Mtcnn and Facenet. *Zeichen Journal.*, *6*(1), 189–195.

Aone, C., Okurowski, M. E., & Gorlinsky, J. (1998, August). Trainable, scalable summarization using robust NLP and machine learning. In *Proceedings of the 17th international conference on Computational linguistics-Volume 1* (pp. 62-66). Association for Computational Linguistics.

Aria, M., & Cuccurullo, C. (2017). Bibliometrix: An R-tool for comprehensive science mapping analysis. *Journal of Informetrics*, *11*(4), 959–975. doi:10.1016/j.joi.2017.08.007

Arianto, R., Leslie Warnars, H. H., Gaol, F. L., & Trisetyarso, A. (2018). Mining Unstructured Data in Social Media for Natural Disaster Management in Indonesia. *2018 Indonesian Association for Pattern Recognition International Conference (INAPR)*, 5.

Ariel, Y., & Avidar, R. (2014). Information, Interactivity, and Social Media. *Atlantic Journal of Communication*, *23*(1), 19–30. doi:10.1080/15456870.2015.972404

Aronson, E., Timothy, D. W., & Akert, R. M. (2010). *Social psychology*. Prentice Hall.

Arredondo-Velazquez, M., Diaz-Carmona, J., Barranco-Gutierrez, A. I., & Torres-Huitzil, C. (2020). Review of prominent strategies for mapping CNNs onto embedded systems. *IEEE Latin America Transactions*, *18*(05), 971–982. doi:10.1109/TLA.2020.9082927

Athira Krishnan, S. K. (2014). A survey on image segmentation and feature extraction methods for acute myelogenous leukemia detection in blood microscopic images. *International Journal of Computer Science and Information Technologies*, *5*(6), 7877-7879.

Atzei, N., Bartoletti, M., & Cimoli, T. (2017, April). A survey of attacks on ethereum smart contracts (sok). In *International conference on principles of security and trust* (pp. 164-186). Springer. 10.1007/978-3-662-54455-6_8

Au-Yong-Oliveira, M., Pesqueira, A., Sousa, M. J., Dal Mas, F., & Soliman, M. (2021). The Potential of Big Data Research in HealthCare for Medical Doctors' Learning. *Journal of Medical Systems*, *45*(1), 1–14. doi:10.100710916-020-01691-7 PMID:33409620

Axenbeck, J., & Kinne, J. (2020). Web mining for innovation ecosystem mapping: A framework and a large-scale pilot study. *Scientometrics*, 31.

Azzaro-Pantel, C., Madoumier, M., & Gésan-Guiziou, G. (2022). Development of an ecodesign framework for food manufacturing including process flowsheeting and multiple-criteria decision-making: Application to milk evaporation. *Food and Bioproducts Processing*, *131*, 40–59. doi:10.1016/j.fbp.2021.10.003

Bader, L., Bürger, J. C., Matzutt, R., & Wehrle, K. (2018, December). *Smart contract-based car insurance policies. In 2018 IEEE Globecom workshops (GC wkshps)*. IEEE.

Bagozzi, R. P., & Dholakia, U. M. (2002). Intentional social action in virtual communities. *Journal of Interactive Marketing*, *16*(2), 2–21. doi:10.1002/dir.10006

Bagozzi, R. P., & Lee, K. H. (2002). Multiple routes for social influence: The role of compliance, internalization, and social identity. *Social Psychology Quarterly*, *65*(3), 226–247. doi:10.2307/3090121

Bai, X., Cheng, Z., Duan, Z., & Hu, K. (2018, February). Formal modeling and verification of smart contracts. In *Proceedings of the 2018 7th international conference on software and computer applications* (pp. 322-326). Academic Press.

Baishya, D., & Maheshwari, S. (2020). Whatsapp groups in academic context: Exploring the academic uses of whatsapp groups among the students. *Contemporary Educational Technology*, *11*(1), 31–46. doi:10.30935/cet.641765

Balaji, P., & S, S. (2019, August). Web 2.0: An Evaluation of Social Media Networking Sites. *International Journal of Innovative Technology and Exploring Engineering*, *8*(10), 8.

Banach, Z. (2020, February 19). *The Challenges of Ensuring IoT Security*. Netsparker. https://www.netsparker.com/blog/web-security/the-challenges-of-ensuring-iot-security/

Bandura, A. (1986). *Social foundations of thought and action: A social cognitive theory*. Prentice-Hall, Inc.

Barak, M. (2018). Are digital natives open to change? Examining flexible thinking and resistance to change. *Computers & Education*, *121*, 115–123. doi:10.1016/j.compedu.2018.01.016

Baralla, G., Pinna, A., & Corrias, G. (2019, May). Ensure traceability in European food supply chain by using a blockchain system. In *2019 IEEE/ACM 2nd International Workshop on Emerging Trends in Software Engineering for Blockchain (WETSEB)* (pp. 40-47). IEEE.

Barthe, G., Köpf, B., Olmedo, F., & Zanella Beguelin, S. (2012, January). Probabilistic relational reasoning for differential privacy. In *Proceedings of the 39th annual ACM SIGPLAN-SIGACT symposium on Principles of programming languages* (pp. 97-110). 10.1145/2103656.2103670

Bartoletti, M., & Pompianu, L. (2017, April). An empirical analysis of smart contracts: platforms, applications, and design patterns. In *International conference on financial cryptography and data security* (pp. 494-509). Springer.

Barzilay, R., & Lee, L. (2003, May). Learning to paraphrase: an unsupervised approach using multiple-sequence alignment. In *Proceedings of the 2003 Conference of the North American Chapter of the Association for Computational Linguistics on Human Language Technology-Volume 1* (pp. 16-23). Association for Computational Linguistics. 10.3115/1073445.1073448

Basheer, S., Anbarasi, M., Sakshi, D. G., & Kumar, V. V. (2020). Efficient text summarization method for blind people using text mining techniques. *International Journal of Speech Technology*, *23*(4), 713–725.

Basit, A., Zafar, M., Liu, X., Javed, A. R., Jalil, Z., & Kifayat, K. (2021). A comprehensive survey of AI-enabled phishing attacks detection techniques. *Telecommunication Systems*, *76*(1), 139–154. doi:10.100711235-020-00733-2 PMID:33110340

Bates, N., & Sousa, S. C. (2020). Investigating Users' Perceived Credibility of Real and Fake News Posts in Facebook's News Feed. UK Case Study. In Advances in Artificial Intelligence, Software and Systems Engineering. Springer.

Battré, D., Ewen, S., Hueske, F., Kao, O., Markl, V., & Warneke, D. (2010, June). Nephele/pacts: a programming model and execution framework for web-scale analytical processing. In *Proceedings of the 1st ACM symposium on Cloud computing* (pp. 119-130). ACM.

Baxendale, P. B. (1958). Machine-made index for technical literature—An experiment. *IBM Journal of Research and Development*, *2*(4), 354–361. doi:10.1147/rd.24.0354

Bay, H., Ess, A., Neubeck, A., & Van Gool, L. (2006). 3D from Line Segments in Two Poorly-Textured, Uncalibrated Images. *Proceedings - Third International Symposium on 3D Data Processing, Visualization, and Transmission,* 496-503.

Benali, M., Ghomari, A. R., Zemmouchi-Ghomari, L., & Lazar, M. (2021). Flexible Crowd-Driven Decision Making in Times of Crisis. In *E-Collaboration Technologies and Strategies for Competitive Advantage Amid Challenging Times* (pp. 1–27). IGI Global.

Benson, V., Morgan, S., & Filippaios, F. (2014). Social career management: Social media and employability skills gap. *Computers in Human Behavior*, *30*, 519–525. doi:10.1016/j.chb.2013.06.015

Berwind, K., Bornschlegl, M. X., Kaufmann, M., Engel, F., Fuchs, M., Heutelbeck, D., & Hemmje, M. (2017). Hadoop Ecosystem Tools and Algorithms. In NoSQL: Database for Storage and Retrieval of Data in Cloud (pp. 177-198). Chapman and Hall/CRC.

Bhasin, N. K., & Rajesh, A. (2021). Impact of E-Collaboration Between Indian Banks and Fintech Companies for Digital Banking and New Emerging Technologies. *International Journal of e-Collaboration*, *17*(1), 15–35. doi:10.4018/IJeC.2021010102

Bhat, A., Rustagi, S., Purwaha, S. R., & Singhal, S. (2020). Deep-learning based group-photo Attendance System using One Shot Learning. *2020 International Conference on Electronics and Sustainable Communication Systems (ICESC)*. 10.1109/ICESC48915.2020.9155755

Bhattacharjee, S., Roy, P., Ghosh, S., Misra, S., & Obaidat, M. S. (2012). Wireless sensor network-based fire detection, alarming, monitoring and prevention system for Bord-and-Pillar coal mines. *Journal of Systems and Software*, *85*(3), 571–581. doi:10.1016/j.jss.2011.09.015

Bilbao-Osorio, B., Dutta, S., & Lanvin, B. (2013). *The global information technology report 2013.* World Economic Forum.

Biuk-Aghai, R. P., & Simoff, S. J. (2004, November). Patterns of virtual collaboration in online collaboration systems. In *Proceedings of the IASTED International Conference on Knowledge Sharing and Collaborative Engineering* (pp. 22-24). Academic Press.

Blau, P. M. (1964). *Exchange and Power in Social Life. Wiley.*

Blok, M., van Ingen, E., de Boer, A. H., & Slootman, M. (2020). The use of information and communication technologies by older people with cognitive impairments: From barriers to benefits. *Computers in Human Behavior*, *104*(106173), 1–9. doi:10.1016/j.chb.2019.106173

Bogner, A., Chanson, M., & Meeuw, A. (2016, November). A decentralised sharing app running a smart contract on the ethereum blockchain. In *Proceedings of the 6th International Conference on the Internet of Things* (pp. 177-178). Academic Press.

Borgohain, T., Kumar, U., & Sanyal, S. (2015). Survey of Security and Privacy Issues of Internet of Things. *International Journal of Advanced Networking Applications*, *5*, 2372–2378. https://arxiv.org/abs/1501.02211

Borkar, N. R., & Kuwelkar, S. (2017). Real-time implementation of face recognition system. *International Conference on Computing Methodologies and Communication (ICCMC)*, *1*(1), 249 – 255.

Bouaouad, A. E., Cherradi, A., Assoul, S., & Souissi, N. (2020). The key layers of IoT architecture. *Proceedings of 2020 5th International Conference on Cloud Computing and Artificial Intelligence: Technologies and Applications, CloudTech 2020*, 20–23. 10.1109/CloudTech49835.2020.9365919

Bowers, J. (2017, August 13). *Social Media for Business: A Marketer's Guide*. Retrieved from Business New Daily: https://www.businessnewsdaily.com/7832-social-media-for-business.html

Boyd, D., & Crawford, K. (2012). Critical questions for big data: Provocations for a cultural, technological, and scholarly phenomenon. *Information Communication and Society*, *15*(5), 662–679. doi:10.1080/1369118X.2012.678878

Boyd, & Ellison, N. B. (2007). Social Network Sites: Definition, History, and Scholarship. *Journal of Computer-Mediated Communication*, *13*(1), 210–230. doi:10.1111/j.1083-6101.2007.00393.x

Boyle, K., & Johnson, T. J. (2010). MySpace is your space? Examining self-presentation of MySpace users. *Computers in Human Behavior*, *26*(6), 1392–1399. doi:10.1016/j.chb.2010.04.015

Brandow, R., Mitze, K., & Rau, L. F. (1995). Automatic condensation of electronic publications by sentence selection. *Information Processing & Management*, *31*(5), 675–685. doi:10.1016/0306-4573(95)00052-I

Brandtzaeg, P. B., Heim, J., & Kaare, B. H. (2010). Bridging and bonding in social network sites–investigating family-based capital. *International Journal of Web Based Communities*, *6*(3), 231–253. doi:10.1504/IJWBC.2010.033750

Braxton, J.M., Jones, W.A., Hirschy, A.S., & Hartley III, H.V. (2008). The role of active learning in the college student persistence. *New Roadmap for Teaching and Learning, 2008*(115), 71-83.

Brečko, B., & Ferrari, A. (2016). The Digital Competence Framework for Consumers. Joint Research Centre Science for Policy Report; EUR 28133 EN. doi:10.2791/838886

Burhan, M., Rehman, R. A., Khan, B., & Kim, B. S. (2018). IoT elements, layered architectures and security issues: A comprehensive survey. *Sensors (Switzerland)*, *18*(9), 2796. Advance online publication. doi:10.339018092796 PMID:30149582

Burt, P., & Adelson, E. (1983). A multiresolution spline with application to image mosaics. *ACM Transactions on Graphics*, *2*, 217–236.

Cacheiro-Gonzalez, M. L., Medina-Rivilla, A., Dominguez-Garrido, M. C., & Medina-Dominguez, M. (2019). The learning platform in distance higher education: Student's perceptions. *Turkish Online Journal of Distance Education*, *20*(1), 71–95. doi:10.17718/tojde.522387

Calderon-Garrido, D., Leon-Gomez, A., & Gil-Fernandez, R. (2019). the Use of the Social Networks Between the Students of Teacher'S Degree in an Environment Exclusively Online. *Vivat Academia*, *147*, 23–39.

Callan, V. J., & Johnston, M. A. (2020). Influences upon social media adoption and changes to training delivery in vocational education institutions. *Journal of Vocational Education and Training*, *00*(00), 1–26. doi:10.1080/13636820.2020.1821754

Campos, N., Nogal, M., Caliz, C., & Juan, A. A. (2020). Simulation-based education involving online and on-campus models in different European universities. *International Journal of Educational Technology in Higher Education*, *17*(1), 8. Advance online publication. doi:10.118641239-020-0181-y

Caplan, S. E., & Turner, J. S. (2007). Bringing theory to research on computer-mediated comforting communication. *Computers in Human Behavior*, *23*(2), 985–998. doi:10.1016/j.chb.2005.08.003

Casilli, A. A. (2017). Global digital culturel digital labor studies go global: Toward a digital decolonial turn. *International Journal of Communication*, *11*, 21.

Castellanos-Reyes, D. (2020). 20 Years of the Community of Inquiry Framework. *TechTrends*, *64*(4), 557–560. doi:10.100711528-020-00491-7

Charmaz, K. (2006). *A Practical Guide through Qualitative Analysis*. Sage.

Chatap, N., & Shibu, S. (2014). Analysis of blood samples for counting leukemia cells using Support vector machine and nearest neighbour. *IOSR Journal of Computer Engineering, 16*(5), 79-87.

Chaudhary, D., Rawat, A., Maurya, D., Patel, A., & Shukla, D. S. (2020). Face Recognition Based Attendance System. *International Journal of Engineering Applied Sciences and Technology.*, *5*(1), 485–487. doi:10.33564/IJEAST.2020.v05i01.085

Chen, Q., Peng, Y., & Lu, Z. (2019). BioSentVec: Creating sentence embeddings for biomedical texts. In *2019 IEEE International Conference on Healthcare Informatics* (p. 5). Academic Press.

Cheng, E. C. K. (2012). Knowledge strategies for enhancing school learning capacity. *International Journal of Educational Management*, *26*(6), 577–592. doi:10.1108/09513541211251406

Cheng, E. C. K. (2013). Applying knowledge management for school strategic planning. *KEDI Journal of Educational Policy*, *10*(2), 339–356.

Cheng, E. C. K. (2019). Knowledge management strategies for sustaining Lesson Study. *International Journal for Lesson and Learning Studies*, *9*(2), 167–178. doi:10.1108/IJLLS-10-2019-0070

Cheng, E. C. K., & Chu, C. K. W. (2018). A Normative Knowledge Management Model for School Development. *International Journal of Learning and Teaching*, *4*(1), 76–82. Advance online publication. doi:10.18178/ijlt.4.1.76-82

Cheng, E. C. K., Wu, S. W., & Hu, J. (2017). Knowledge management implementation in the school context: Case studies on knowledge leadership, storytelling, and taxonomy. *Educational Research for Policy and Practice*, *16*(2), 177–188. doi:10.100710671-016-9200-0

Chen, H. M., Lee, S., Rao, R. M., Slamani, M. A., & Varshney, P. K. (2005). Imaging for concealed weapon detection: A tutorial overview of development in imaging sensors and processing. *IEEE Signal Processing Magazine*, *22*(2), 52–61. doi:10.1109/MSP.2005.1406480

Chen, R., Fung, B. C., Desai, B. C., & Sossou, N. M. (2012, August). Differentially private transit data publication: a case study on the montreal transportation system. In *Proceedings of the 18th ACM SIGKDD international conference on Knowledge discovery and data mining* (pp. 213-221). 10.1145/2339530.2339564

Chen, S., & Schlosser, S. W. (2008). *MapReduce meet wider varieties of applications*. Intel Research Pittsburgh.

Chen, S., Schreurs, L., Pabian, S., & Vandenbosch, L. (2019). Daredevils on social media: A comprehensive approach toward risky selfie behavior among adolescents. *New Media & Society*, *21*(11-12), 2443–2462. doi:10.1177/1461444819850112

Chen, Z., Shi, Z., & Guo, Q. (2013). Design of wireless sensor network node for carbon monoxide monitoring. *Telecommunication Systems*, *53*(1), 47–53. doi:10.100711235-013-9675-4

Chettri, L., & Bera, R. (2020). A Comprehensive Survey on Internet of Things (IoT) Toward 5G Wireless Systems. *IEEE Internet of Things Journal*, *7*(1), 16–32. doi:10.1109/JIOT.2019.2948888

Cheung, C. M. K., Chiu, P.-Y., & Lee, M. K. O. (2011). Online social networks: Why do students use Facebook? *Computers in Human Behavior*, *27*(4), 1337–1343. doi:10.1016/j.chb.2010.07.028

Cheung, C. M. K., & Lee, M. K. O. (2009). Understanding the sustainability of a virtual community: Model development and empirical test. *Journal of Information Science*, *35*(3), 279–298. doi:10.1177/0165551508099088

Cheung, C. M. K., & Lee, M. K. O. (2010). A theoretical model of intentional social action in online social networks. *Decision Support Systems*, *49*(1), 24–30. doi:10.1016/j.dss.2009.12.006

Chia, H. P., & Pritchard, A. (2014). Using a virtual learning community (VLC) to facilitate a cross-national science research collaboration between secondary school students. *Computers & Education*, *79*, 1–15. doi:10.1016/j.compedu.2014.07.005

Chiramdasu, R., Srivastava, G., Bhattacharya, S., Reddy, P. K., & Gadekallu, T. R. (2021, August). Malicious url detection using logistic regression. In *2021 IEEE International Conference on Omni-Layer Intelligent Systems (COINS)* (pp. 1-6). IEEE.

Chiu, C.-M., Cheng, H.-L., Huang, H.-Y., & Chen, C.-F. (2013). Exploring individuals' subjective well-being and loyalty towards social network sites from the perspective of network externalities: The Facebook case. *International Journal of Information Management*, *33*(3), 539–552. doi:10.1016/j.ijinfomgt.2013.01.007

Chmiel, M., Korona, M., Kozioł, F., Szczypiorski, K., & Rawski, M. (2021). Discussion on iot security recommendations against the state-of-the-art solutions. *Electronics (Switzerland)*, *10*(15), 1814. Advance online publication. doi:10.3390/electronics10151814

Christensen, S. P. (2018). *Social Media Use and Its Impact on Relationships and Emotions*. https://scholarsarchive.byu.edu/cgi/viewcontent.cgi?article=7927&context=etd

Christian Dancke Tuen. (2015, June). *Security in Internet of Things Systems*. Norwegian University of Science and Technology. https://scholar.google.no/citations?view_op=view_citation&hl=en&user=jRh3XQoAAAAJ&alert_preview_top_rm=2&citation_for_view=jRh3XQoAAAAJ:u5HHmVD_uO8C

Christofides, E., Muise, A., & Desmarais, S. (2009). Information disclosure and control on Facebook:Are they two sides of the same coin or two different processes? *Cyberpsychology & Behavior*, *12*(3), 341–345. doi:10.1089/cpb.2008.0226 PMID:19250020

Chuah, J. W. (2014). The Internet of Things: An overview and new perspectives in systems design. *2014 International Symposium on Integrated Circuits (ISIC)*, 216–219. 10.1109/ISICIR.2014.7029576

Chun; Seo; Kim. (2003). Image retrieval using bdip and bvlc moments. *IEEE Transactions on Circuits and Systems for Video Technology*, *13*(9), 951–957. doi:10.1109/TCSVT.2003.816507

Chun, Y. D., Kim, N. C., & Jang, I. H. (2008). Content-based image retrieval using multiresolution color and texture features. *IEEE Transactions on Multimedia*, *10*(6), 1073–1084. doi:10.1109/TMM.2008.2001357

Cleveland-Innes, M., & Campbell, P. (2012). Emotional presence, learning, and the online learning environment. *International Review of Research in Open and Distance Learning*, *13*(4), 269–292. doi:10.19173/irrodl.v13i4.1234

Conde & Dominguez. (2018). Scaling the chord and hellinger distances in the range [0, 1]: An option to consider. *Journal of Asia-Pacific Biodiversity*, •••, 76–83.

Condie, T., Conway, N., Alvaro, P., Hellerstein, J. M., Gerth, J., Talbot, J., Elmeleegy, K., & Sears, R. (2010). Online aggregation and continuous query support in MapReduce. *Proceedings of the 2010 ACM SIGMOD International Conference on Management of Data*, 1115–1118. 10.1145/1807167.1807295

Cordeschi, N., Amendola, D., Shojafar, M., & Baccarelli, E. (2015). Distributed and adaptive resource management in cloud-assisted cognitive radio vehicular networks with hard reliability guarantees. *Vehicular Communications*, *2*(1), 1–12. doi:10.1016/j.vehcom.2014.08.004

Cosenz, F., & Bivona, E. (2020). Fostering growth patterns of SMEs through business model innovation. A tailored dynamic business modelling approach. *Journal of Business Research.*

Coursaris, C. K., & Van Osch, W. (2016). A Cognitive-Affective Model of Perceived User Satisfaction (CAMPUS): The complementary effects and interdependence of usability and aesthetics in IS design. *Information & Management*, *53*(2), 252–264. doi:10.1016/j.im.2015.10.003

Crilley, R., & Gillespie, M. (2019). What to do about social media? Politics, populism and journalism. *Journalism*, *20*(1), 173–176. doi:10.1177/1464884918807344

Cruz Nassif & Hruschka. (2013). Document Clustering for Forensic Analysis: An Approach for Improving Computer Inspection. *IEEE Transactions on Information Forensics and Security*, •••, 8.

Da Cruz, M. A. A., Rodrigues, J. J. P. C., Lorenz, P., Korotaev, V. V., & De Albuquerque, V. H. C. (2021). In.IoT - A New Middleware for Internet of Things. *IEEE Internet of Things Journal*, *8*(10), 7902–7911. doi:10.1109/JIOT.2020.3041699

Dabbagh, N., & Kitsantas, A. (2012). Personal Learning Environments, social media, and self-regulated learning: A natural formula for connecting formal and informal learning. *Internet and Higher Education*, *15*(1), 3–8. doi:10.1016/j.iheduc.2011.06.002

Dada, E. G., Bassi, J. S., Chiroma, H., Adetunmbi, A. O., & Ajibuwa, O. E. (2019). Machine learning for email spam filtering: Review, approaches and open research problems. *Heliyon*, *5*(6), e01802.

Dahab, M. Y., Idrees, A. M., Hassan, H. A., & Rafea, A. (2010). Pattern Based Concept Extraction for Arabic Documents. *International Journal of Intelligent Computing and Information Sciences*, *10*(2).

DailyMail. (2014). *University students spend six hours a day on Facebook, YouTube and sending texts—even during lectures*. http://www.dailymail.co.uk/news/ article-2664782/University-students-spend-SIX-HOURS-day-Facebook-YouTube-sending-texts-lectures.html

Damale, R. C., & Pathak, B. V. (2018). Face Recognition Based Attendance System Using Machine Learning Algorithms. *2018 Second International Conference on Intelligent Computing and Control Systems (ICICCS)*. 10.1109/ICCONS.2018.8662938

Damos, P. (2015). Modular structure of web-based decision support systems for integrated pest management. A review. *Agronomy for Sustainable Development*, *35*(4), 1347–1372. doi:10.100713593-015-0319-9

Dash, S., Shakyawar, S. K., Sharma, M., & Kaushik, S. (2019). Big data in healthcare: Management, analysis and future prospects. *Journal of Big Data*, *6*(1), 1–25. doi:10.118640537-019-0217-0

Das, K., & Das, S. (2022). Energy-Efficient Cloud-Integrated Sensor Network Model Based on Data Forecasting Through ARIMA. *International Journal of e-Collaboration, 18*(1), 1–17. doi:10.4018/IJeC.290292

Datta, Joshi, Li, & Wang. (2008). Image retrieval: Ideas, influences and trends of the new age. *ACM Computing Surveys, 40*, 5.

de Leusse, P., Periorellis, P., Dimitrakos, T., & Nair, S. K. (2009). Self Managed Security Cell, a Security Model for the Internet of Things and Services. *2009 First International Conference on Advances in Future Internet*, 47–52. 10.1109/AFIN.2009.15

De Maio, C., Fenza, G., Gallo, M., Loia, V., & Volpe, A. (2020). Cross-relating heterogeneous Text Streams for Credibility Assessment. *International of Electrial and Electronics Engineers.*

Deaves, A., Grant, E., Trainor, K., & Jarvis, K. (2019). Students' perceptions of the educational value of twitter: A mixed-methods investigation. *Research in Learning Technology, 27*(0). Advance online publication. doi:10.25304/rlt.v27.2139

Deepak Kumar, B. (2011). Evaluation of the Medical Records System in an Upcoming Teaching Hospital—A Project for Improvisation. *Journal of Medical Systems.*

Demchenko, Y., Grosso, P., De Laat, C., & Membrey, P. (2013, May). Addressing big data issues in scientific data infrastructure. In *2013 International conference on collaboration technologies and systems (CTS)* (pp. 48-55). IEEE.

Dennis, E. E., Martin, J. D., Wood, R., & Saeed, M. (2016). Media Use in the Middle East. Doha: Mideastmedia.org.

Design Rush. (2020, April 3). *7 IoT Security Issues And How To Protect Your Solution.* https://www.designrush.com/agency/software-development/trends/iot-security-issues

Devaraju, V. K., & Rao, S. S. (n.d.). A Real and Accurate Vegetable Seeds Classification Using Image Analysis and Fuzzy Technique. *Turkish Journal of Physiotherapy and Rehabilitation, 32*, 2.

Dewey, J. (1991). *The Child and the Curriculum*. University of Chicago Press. (Original work published 1902)

Dhiman, G., Kumar, V. V., Kaur, A., & Sharma, A. (2021). DON: Deep Learning and Optimization-Based Framework for Detection of Novel Coronavirus Disease Using X-ray Images. *Interdisciplinary Sciences, Computational Life Sciences*, 1–13.

Dholakia, U. M., Bagozzi, R. P., & Pearo, L. K. (2004). A social influence model of consumer participation in network- and small- group-based virtual communities. *International Journal of Research in Marketing, 21*(3), 241–263. doi:10.1016/j.ijresmar.2003.12.004

Digital Marketing. (2015). *What are data brokers–and what is your data worth?* Retrieved from https: //www.webfx.com/blog/general/what-are-data-brokers-and-what-is-your-data-worth-infographic/

Dinakaran, M., & Vijayarajan, V. (2013). Feature based image retrieval using fused sifts and surf features. *Research Gate, 10*, 2500–2506.

Ding, Y., & Jiang, J. (2014). Extracting interest tags from twitter user biographies. Information Retrieval Technology, 268-279.

Disha, R. A., & Waheed, S. (2022). Performance analysis of machine learning models for intrusion detection system using Gini Impurity-based Weighted Random Forest (GIWRF) feature selection technique. *Cybersecurity, 5*(1), 1–22.

Domizi, D. (2013, January 10). Microblogging To Foster Connections And Community in a Weekly Graduate Seminar Course. *TechTrends, 57*(1), 43–45. doi:10.100711528-012-0630-0

Dommett, E. J. (2019). Understanding student use of twitter and online forums in higher education. In Education and Information Technologies (Vol. 24, Issue 1, pp. 325–343). doi:10.100710639-018-9776-5

Donta, P. K., Srirama, S. N., Amgoth, T., & Rao Annavarapu, C. S. (2021). *Survey on recent advances in IoT application layer protocols and machine learning scope for research directions*. Digital Communications and Networks. doi:10.1016/j.dcan.2021.10.004

Doulkeridis, C., & Nørvåg, K. (2014). A survey of large-scale analytical query processing in MapReduce. *The VLDB Journal, 23*(3), 355–380.

Duncan, F. (2016, February 2). *So long social media: The kids are opting out of the online public sphere*. The Conversation. Retrieved from https://theconversation.com/so-long-social-media-the-kidsare-opting-out-of-the-online-public-square-53274

Dwork, C. (2008, April). Differential privacy: A survey of results. In *International conference on theory and applications of models of computation* (pp. 1-19). Springer.

Dwyer, C., Hiltz, S., & Passerini, K. (2007). Trust and privacy concern within social networking sites: A comparison of Facebook and MySpace. *AMCIS 2007 proceedings*, 339.

Ebert-May, D., Brewer, C., & Allred, S. (1997). Innovative lectures: Teaching for active learning. *Bioscience, 47*(9), 601–607. doi:10.2307/1313166

Edmundson, H. P. (1969). New methods in automatic extracting. *Journal of the Association for Computing Machinery, 16*(2), 264–285. doi:10.1145/321510.321519

Elantheraiyan, P., & Shankarkumar, S. (2019). A Research on Impact of Social Media on College Students in Chennai District. *International Journal of Innovative Technology and Exploring Engineering, 8*(11S), 675–679. doi:10.35940/ijitee.K1114.09811S19

Erdem, T., & Keane, M. P. (1996). Decision-making under uncertainty: Capturing dynamic brand choice processes in turbulent consumer goods markets. *Marketing Science, 15*(1), 1–20. doi:10.1287/mksc.15.1.1

Erkan, G., & Radev, D. R. (2004). Lexpagerank: Prestige in multi-document text summarization. In *Proceedings of the 2004 Conference on Empirical Methods in Natural Language Processing* (pp. 365-371). Academic Press.

European Commission. (2014). *Commission urges governments to embrace potential of big data*. Retrieved from European Commission press release. Retrieved from http: //europa.eu/rapid/press-release_IP-14-769_en.htm

Ezhilmaran, D., & Muthukumaran, V. (2017). Authenticated group key agreement protocol based on twist conjugacy problem in near-rings. *Wuhan University Journal of Natural Sciences, 22*(6), 472–476. doi:10.100711859-017-1275-9

Fadel, M. A. (2020). Evaluating the Credibility of Arabic News in Social Media through the use of Advanced Classifier Algorithms. *International Journal of Advanced Trends in Computer Science and Engineering, 9*(4), 15.

Fahn, C. S., & Wang, C. H. (2011). *Real-time Multi-Face Recognition and Tracking Techniques Used for the Interaction between Humans and Robots*. Reviews, Refinements and New Ideas in Face Recognition. doi:10.5772/19589

Fallucchi, F., Coladangelo, M., Giuliano, R., & William De Luca, E. (2020). Predicting employee attrition using machine learning techniques. *Computers, 9*(4), 86. doi:10.3390/computers9040086

Fang, L. C., Ayop, Z., Anawar, S., Othman, N. F., Harum, N., & Abdullah, R. S. (2021). URL Phishing Detection System Utilizing Catboost Machine Learning Approach. *International Journal of Computer Science & Network Security, 21*(9), 297–302.

Feigenbaum, L. (2014). *Turning big data into smart data.* Retrieved from https: //tdwi.org/Articles/2014/07/08/Turning-Big-Data-into-Smart-Data-1.aspx?Page=1

Felder, R. M., & Brent, R. (1996). Navigating the bumpy road to student-centered instruction. *College Teaching, 44*(2), 43–47. doi:10.1080/87567555.1996.9933425

Feng, C., Adnan, M., Ahmad, A., Ullah, A., & Khan, H. U. (2020). Towards energy-efficient framework for IoT big data healthcare solutions. *Scientific Programming, 2020*, 2020. doi:10.1155/2020/7063681

Feng, D. (2019). Interdiscursivity, social media and marketized university discourse: A genre analysis of universities' recruitment posts on WeChat. *Journal of Pragmatics, 143*, 121–134. doi:10.1016/j.pragma.2019.02.007

Fernández, A., del Río, S., López, V., Bawakid, A., del Jesus, M. J., Benítez, J. M., & Herrera, F. (2014). Big Data with Cloud Computing: An insight on the computing environment, MapReduce, and programming frameworks. *Wiley Interdisciplinary Reviews. Data Mining and Knowledge Discovery, 4*(5), 380–409.

Firouzi, B. B., Sadeghi, M. S., & Niknam, T. (2010). A new hybrid algorithm based on PSO, SA, and K-means for cluster analysis. *International Journal of Innovative Computing, Information, & Control, 6*(7), 3177–3192.

Freeman, D. M. (2010, May). Converting pairing-based cryptosystems from composite-order groups to prime-order groups. In *Annual International Conference on the Theory and Applications of Cryptographic Techniques* (pp. 44-61). Springer. 10.1007/978-3-642-13190-5_3

Freitas, D., & Kouroupetroglou, G. (2008). Speech technologies for blind and low vision persons. *Technology and Disability, 20*(2), 135–156. doi:10.3233/TAD-2008-20208

Fuchs, C. (2010). Labor in informational capitalism and on the Internet. *The Information Society, 26*(3), 179–196. doi:10.1080/01972241003712215

Fu, Y., Jin, H., & Zhao, Y., & Cao, W. (2019). A novel text mining approach for scholar information extraction from web content in Chinese. *Future Generation Computer Systems*, 35.

Gambardella, A., & McGahan, A. M. (2010). Business-model innovation: General purpose technologies and their implications for industry structure. *Long Range Planning, 43*(2-3), 262–271. doi:10.1016/j.lrp.2009.07.009

Gao, D., Guo, Y., Cui, J. Q., Hao, H. G., & Shi, H. A. (2012). A Communication Protocol of RFID Systems in Internet of Things. *International Journal of Security and Its Applications, 06*, 91–102. https://m.earticle.net/Article/A210850

Garcia, T., & Wang, T. (2013). Analysis of Big Data Technologies and Method - Query Large Web Public RDF Datasets on Amazon Cloud Using Hadoop and Open Source Parsers. *Semantic Computing (ICSC), IEEE Seventh International Conference.* http://architects.dzone.com/

Garcia, E., Moizer, J., Wilkins, S., & Haddoud, M. Y. (2019). Student learning in higher education through blogging in the classroom. *Computers & Education, 136*, 61–74. doi:10.1016/j.compedu.2019.03.011

Garrison, D. R. (2011). Communities of Inquiry in Online Learning. Encyclopedia of Distance Learning, 352–355. doi:10.4018/978-1-60566-198-8.ch052

Garrison, D. R. (2017). Other Presences? *The Community of Inquiry.* http://www.thecommunityofinquiry.org/editorial7

Garrison, D., Anderson, T., & Archer, W. (2000). Critical Inquiry in a Text-Based Environment: Computer Conferencing in Higher Education. *The Internet and Higher Education*, 2(2–3), 87–105. doi:10.1016/S1096-7516(00)00016-6

Garrison, R., Anderson, T., & Archer, W. (2002). Critical Inquiry in a Text-Based Environment. *The Internet and Higher Education*, 2(2), 87–105. http://dergipark.gov.tr/saufenbilder/issue/20673/220600

Gattim, N. K., Pallerla, S. R., & Bojja, P. (2019). Plant Leaf Disease Detection Using SVM Technique. *International Journal of Emerging Trends in Engineering Research, 7*(11), 634–637. doi:10.30534/ijeter/2019/367112019

Gayathri, S., Swapna, B., Kamalahasan, M., & Balavinoth, S. (2020). Retire away Essential Accuracy for Darkness Discovery and Elimination. *Test Engineering and Management, 83*, 2411–2417.

Gazit, T., Aharony, N., & Amichai-Hamburger, Y. (2019). Tell me who you are and I will tell you which SNS you use: SNSs participation. *Online Information Review, 44*(1), 139–161. doi:10.1108/OIR-03-2019-0076

Gebrie, M. T., & Abie, H. (2017). Risk-based adaptive authentication for internet of things in smart home eHealth. *Proceedings of the 11th European Conference on Software Architecture: Companion Proceedings*, 102–108. 10.1145/3129790.3129801

Geeraerts, K., Vanhoof, J., & Van den Bossche, P. (2016). Teachers' perceptions of intergenerational knowledge flows. *Teaching and Teacher Education, 56*, 150–161. doi:10.1016/j.tate.2016.01.024

Ghazal, A., Rabl, T., Hu, M., Raab, F., Poess, M., Crolotte, A., & Jacobsen, H. A. (2013, June). Bigbench: Towards an industry standard benchmark for big data analytics. In *Proceedings of the 2013 ACM SIGMOD international conference on Management of data* (pp. 1197-1208). ACM.

Giannikas, C. (2019). Facebook in tertiary education: The impact of social media in e-learning. *Journal of University Teaching & Learning Practice, 17*(1), 2020. https://ro.uow.edu.au/jutlpAvailableat:https://ro.uow.edu.au/jutlp/vol17/iss1/3

Gillick, D., & Favre, B. (2009, June). A scalable global model for summarization. In *Proceedings of the Workshop on Integer Linear Programming for Natural Langauge Processing* (pp. 10-18). Association for Computational Linguistics. 10.3115/1611638.1611640

Go¨k, A., Waterworth, A., & Shapira, P. (2015). Use of web mining in studying innovation. *Scientometrics, 102*(1), 19.

Goeke, L. (2017, May 22). *Theseus: Security Challenges of the Internet of Things*. Thesis - Haaga-Helia ammattikorkeakoulu. https://www.theseus.fi/handle/10024/128420

Gonçalves, C., Honrado, J. P., Cerejeira, J., Sousa, R., Fernandes, P. M., Vaz, A. S., Alves, M., Araújo, M., Carvalho-Santos, C., Fonseca, A., Fraga, H., Gonçalves, J. F., Lomba, A., Pinto, E., Vicente, J. R., & Santos, J. A. (2022). On the development of a regional climate change adaptation plan: Integrating model-assisted projections and stakeholders' perceptions. *The Science of the Total Environment, 805*, 150320. doi:10.1016/j.scitotenv.2021.150320 PMID:34543791

Gong, Z. H., & Eppler, J. (2021). Exploring the Impact of Delivery Mistakes, Gender,and Empathic Concern on Source and Message. *Journalism Practice*, 21. doi:10.1080/17512786.2020.1870531

Gou, Q., Yan, L., Liu, Y., & Li, Y. (2013). Construction and Strategies in IoT Security System. *2013 IEEE International Conference on Green Computing and Communications and IEEE Internet of Things and IEEE Cyber, Physical and Social Computing*, 1129–1132. 10.1109/GreenCom-iThings-CPSCom.2013.195

Granjal, J., Monteiro, E., & Sa Silva, J. (2015). Security for the Internet of Things: A Survey of Existing Protocols and Open Research Issues. *IEEE Communications Surveys and Tutorials, 17*(3), 1294–1312. doi:10.1109/COMST.2015.2388550

Greenhow, C., & Lewin, C. (2016). Social media and education: Reconceptualizing the boundaries of formal and informal learning. *Learning, Media and Technology, 41*(1), 6–30. doi:10.1080/17439884.2015.1064954

Grimes, S. (2008). *Unstructured data and the 80 percent rule*. CarbridgeBridgepoints.

Groves, R. M., Fowler, F. J., Couper, M. P., Lepkowski, J. M., Singer, E., & Tourangeau, R. (2009). *Survey Methodology*. John Wiley & Sons.

Gulnar, B., Balcı, S., & Cakır, V. (2010). Motivations of Facebook, Youtube and similar web sites users. *Bilig, 54*, 161–184.

Gupta, M. K., & Chandra, P. (2020). A comprehensive survey of data mining. *International Jounal of Information Technology (Singapore), 12*(4), 15.

Haggag, M. H., Khedr, A. E., & Montasser, H. S. (2015). A Risk-Aware Business Process Management Reference Model and Its Application in an Egyptian University. *International Journal of Computer Science and Engineering Survey, 6*(2).

Haig, M. (2019). *Notes on a nervous planet.* Penguin.

Hallikainen, P. (2015). *Why People Use Social Media Platforms: Exploring the Motivations and Consequences of Use.* Academic Press.

Han, E., Ines, A. V., & Baethgen, W. E. (2017). Climate-Agriculture-Modeling and Decision Tool (CAMDT): A software framework for climate risk management in agriculture. *Environmental Modelling & Software, 95*, 102–114. doi:10.1016/j.envsoft.2017.06.024

Hao, C., & Ying, Q. (2011). Research of Cloud Computing Based on the Hadoop Platform. *Computational and Information Sciences (ICCIS), 2011 International Conference.*

Hargittai, E. (2007). Whose Space? Differences Among Users and Non-Users of Social Network Sites. *Journal of Computer-Mediated Communication, 13*(1), 276–297. doi:10.1111/j.1083-6101.2007.00396.x

Harris, T. (2010). *Cloud Computing- An Overview, Whitepaper.* Torry Harris Business Solutions.

Hasan, M., Islam, M. M., Zarif, M. I. I., & Hashem, M. M. A. (2019). Attack and anomaly detection in IoT sensors in IoT sites using machine learning approaches. *Internet of Things, 7*, 100059.

Hassan Fouadi, H. E., Fouadi, H., El Moubtahij, H., Lamtougui, H., & Satori, K. (2020). Applications of deep learning in Arabic sentiment analysis. Research perspective. *2020 1st International Conference on Innovative Research in Applied Science, Engineering and Technology, IRASET 2020.*

Hassan, H. A., & Idrees, A. M. (2010). Sampling technique selection framework for knowledge discovery. In *The 7th International Conference on Informatics and Systems (INFOS).* IEEE.

Hassan, H. A., Dahab, M. Y., Bahnasy, K., Idrees, A. M., & Gamal, F. (2014). Query answering approach based on document summarization. *International Open Access Journal of Modern Engineering Research, 4*(12).

Hassan, H. A., Dahab, M. Y., Bahnassy, K., Idrees, A. M., & Gamal, F. (2015). Arabic Documents Classification Method a Step towards Efficient Documents Summarization. *International Journal on Recent and Innovation Trends in Computing and Communication, 3*(1), 351–359.

Hawi, N. S., & Samaha, M. (2017). The Relations Among Social Media Addiction, Self-Esteem, and Life Satisfaction in University Students. *Social Science Computer Review, 35*(5), 576–586. doi:10.1177/0894439316660340

Heider, A., Gerken, M., van Dinther, N., & Hülsbeck, M. (2020). Business model innovation through dynamic capabilities in small and medium enterprises–evidence from the German Mittelstand. *Journal of Business Research.*

Heller, M. (2020). What is face recognition? AI for big brother. *InfoWorld, 1*(1), 1–10.

Helmy, Y., Khedr, A. E., Kolief, S., & Haggag, E. (2019). An Enhanced Business Intelligence Approach for Increasing Customer Satisfaction Using Mining Techniques. *International Journal of Computer Science and Information Security, 17*(4).

Henke, J., Leissner, L., & Möhring, W. (2019). How can Journalists Promote News Credibility? Effects of Evidences on Trust and Credibility. *Journalism Practice, 14*(3), 21.

Holmes, B. (2013). School Teachers' Continuous Professional Development in an Online Learning Community: Lessons from a case study of an eTwinning Learning Event. *European Journal of Education, 48*(1), 97–112. doi:10.1111/ejed.12015

Homeland Security (Ed.). (2016, November). *Strategic Principles for security the internet of things.* US Department of Homeland Security. https://www.dhs.gov/sites/default/files/publications/Strategic_Principles_for_Securing_the_Internet_of_Things-2016-1115-FINAL

Hong, A., Xiao, W., & Ge, J. (2021, May). Big Data Analysis System Based on Cloudera Distribution Hadoop. In *2021 7th IEEE Intl Conference on Big Data Security on Cloud (BigDataSecurity), IEEE Intl Conference on High Performance and Smart Computing,(HPSC) and IEEE Intl Conference on Intelligent Data and Security (IDS)* (pp. 169-173). IEEE.

Hong, S., & Nadler, D. (2012). Which candidates do the public discuss online in an election campaign?: The use of social media by 2012 presidential candidates and its impact on candidate salience. *Government Information Quarterly, 29*(4), 455–461. doi:10.1016/j.giq.2012.06.004

Hong, Y., Vaidya, J., Lu, H., Karras, P., & Goel, S. (2014). Collaborative search log sanitization: Toward differential privacy and boosted utility. *IEEE Transactions on Dependable and Secure Computing, 12*(5), 504–518. doi:10.1109/TDSC.2014.2369034

Hossein Hassani, C. B., Hassani, H., Beneki, C., Unger, S., Mazinani, M. T., & Yeganegi, M. R. (2020). Text Mining in Big Data Analytics. *Big Data and Cognitive Computing, 4*(1), 34.

Hrastinski, S. (2008). What is online learner participation? A literature review. *Computers & Education, 51*(4), 1755–1765. doi:10.1016/j.compedu.2008.05.005

Huang, J.-Y., & Liu, J.-H. (2019). Using social media mining technology to improve stock price forecast accuracy. *Journal of Forcasting*, 13.

Huang, Ravi Kumar, Mandar, Zhu, & Ramin. (1997). Image indexing using color correlograms. *Computer Vision and Pattern Recognition, Proceedings. IEEE Computer Society Conference*, 762-768.

Hung, J.-L., & Zhang, K. (2011). Examining mobile learning trends 2003–2008: A categorical meta-trend analysis using text mining techniques. *Journal of Computing in Higher Education, 24*(1), 17.

Hussein, N. (2016). Multisensor of thermal and visual images to detect concealed weapon using harmony search image fusion approach. *Pattern Recognition Letters.*

Idrees, A. M., Ibrahim, M. H., & El Seddawy, A. I. (2018). Applying spatial intelligence for decision support systems. *Future Computing and Informatics Journal, 3*, 384-390.

Idrees, A. M., & Ibrahim, A. B. (2015). Enhancing information technology services for ebusiness-the road towards optimization. In *13th International Conference on ICT and Knowledge Engineering (ICT & Knowledge Engineering 2015)* (pp. 72-77). IEEE.

Idrees, A. M., & Ibrahim, M. H. (2018). A Proposed Framework Targeting the Enhancement of Students' Performance in Fayoum University. *International Journal of Scientific and Engineering Research, 9*(11).

Ifinedo, P. (2016). Applying uses and gratifications theory and social influence processes to understand students' pervasive adoption of social networking sites: Perspectives from the Americas. *International Journal of Information Management, 36*(2), 192–206. doi:10.1016/j.ijinfomgt.2015.11.007

Ihm, J. (2019). Communicating without nonprofit organizations on nonprofits' social media: Stakeholders' autonomous networks and three types of organizational ties. *New Media & Society, 21*(11-12), 2648–2670. doi:10.1177/1461444819854806

Imamverdiyev, Y., & Abdullayeva, F. (2018). Deep learning method for denial of service attack detection based on restricted Boltzmann machine. *Big Data, 6*(2), 159–169. doi:10.1089/big.2018.0023 PMID:29924649

Inamorato dos Santos, A., Punie, Y., Castaño-Muñoz, J. (2016). *Opening up Education: A Support Framework for Higher Education Institutions*. JRC Science for Policy Report, EUR 27938 EN. doi:10.2791/293408

Instagram. (2016). *Our story: A quick walk through our history as a company*. Retrieved from https://www.instagram.com/ press/?hl=en

Inthiyaz, S., Prasad, M. V. D., Lakshmi, R. U. S., Sai, N. B. S., Kumar, P. P., & Ahammad, S. H. (2019). Agriculture Based Plant Leaf Health Assessment Tool: A Deep Learning Perspective. *International Journal of Emerging Trends in Engineering Research, 7*(11), 690–694. doi:10.30534/ijeter/2019/457112019

Islam, F., Alam, M. M., Shahadat Hossain, S. M., Motaleb, A., Yeasmin, S., Hasan, M., & Rahman, R. M. (2020). Bengali Fake News Detection. In *Proceedings of 2020 IEEE 10th International Conference on Intelligent Systems* (p. 7). Institute of Electrical and Electronics Engineer Inc.

Iwamoto, D., & Chun, H. (2020). The emotional impact of social media in higher education. *International Journal of Higher Education, 9*(2), 239–247. doi:10.5430/ijhe.v9n2p239

Izuagbe, R., Ifijeh, G., Izuagbe-Roland, E. I., Olawoyin, O. R., & Ogiamien, L. O. (2019). Determinants of perceived usefulness of social media in university libraries: Subjective norm, image and voluntariness as indicators. *Journal of Academic Librarianship, 45*(4), 394–405. doi:10.1016/j.acalib.2019.03.006

Jacob, Toft, & Pedersen. (2011). Study group surf: Feature detection description. *Research Gate, 4*, 52-56.

Jasso-Medrano, J. L., & López-Rosales, F. (2018). Measuring the relationship between social media use and addictive behavior and depression and suicide ideation among university students. *Computers in Human Behavior, 87*, 183–191. doi:10.1016/j.chb.2018.05.003

Jayasuruthi, L., Shalini, A., & Kumar, V. V. (2018). Application of rough set theory in data mining market analysis using rough sets data explorer. *Journal of Computational and Theoretical Nanoscience, 15*(6-7), 2126–2130.

Jhaver, M., Gupta, Y., & Mishra, A. K. (2019, November). Employee Turnover Prediction System. In *2019 4th International Conference on Information Systems and Computer Networks (ISCON)* (pp. 391-394). IEEE.

Jiang, D., Ooi, B. C., Shi, L., & Wu, S. (2010). The performance of mapreduce: An in-depth study. *Proceedings of the VLDB Endowment International Conference on Very Large Data Bases, 3*(1-2), 472–483.

Jiang, W., & Clifton, C. (2006). A secure distributed framework for achieving k-anonymity. *The VLDB Journal, 15*(4), 316–333. doi:10.100700778-006-0008-z

Jonckers. (2016). *A security mechanism for internet of things in a smart home context*. Ku Leuven. https://www.scriptiebank.be/sites/default/files/thesis/2016-10/text.pdf

Joo, S., Lu, K., & Lee, T. (2020). Analysis of content topics, user engagement and library factors in public library social media based on text mining. *Online Information Review, 44*(1), 21.

Juergensen, J., & Leckfor, C. (2018). Stop Pushing Me Away: Relative Level of Facebook Addiction Is Associated With Implicit Approach Motivation for Facebook Stimuli. *Psychological Reports, 122*(6), 2012–2025. doi:10.1177/0033294118798624 PMID:30189800

Jyothi, G. N., Sanapala, K., & Vijayalakshmi, A. (2020). ASIC implementation of distributed arithmetic based FIR filter using RNS for high speed DSP systems. *International Journal of Speech Technology*, 1–6.

Kaiser, B. (2019). *Targeted: The Cambridge Analytica Whistleblower's inside story of how big data, Trump, and Facebook broke Democracy and how it can happen again.* HarperCollins.

Kamath, D. R. S., Jamsandekar, D. S. S., & Naik, D. P. G. (2019). Machine Learning Approach for Employee Attrition Analysis. *Int. J. Trend Sci. Res. Dev.*, 62-67.

Kamboj, S., Kumar, V., & Rahman, Z. (2017). Social media usage and firm performance: The mediating role of social capital. *Social Network Analysis and Mining, 7*(1:51), 1-14.

Kamboj, S., Sarmah, B., Gupta, S., & Dwivedi, Y. (2018). Examining branding co-creation in brand communities on social media: Applying the paradigm of Stimulus-Organism-Response. *International Journal of Information Management, 39*, 169–185. doi:10.1016/j.ijinfomgt.2017.12.001

Kaminskiene, L., Žydžiunaite, V., Jurgile, V., & Ponomarenko, T. (2020). Co-creation of learning: A concept analysis. *European Journal of Contemporary Education, 9*(2), 337–349. doi:10.13187/ejced.2020.2.337

Kampylis, P., Punie, Y., & Devine, J. (2015). Promoting Effective Digital-Age Learning: A European Framework for Digitally-Competent Educational Organisations, EUR 27599 EN. *Publications Office of the European Union., JRC98209.* Advance online publication. doi:10.2791/54070

Kang, Y. S., Min, J., Kim, J., & Lee, H. (2013). Roles of alternative and self-oriented perspectives in the context of the continued use of social network sites. *International Journal of Information Management, 33*(3), 496–511. doi:10.1016/j.ijinfomgt.2012.12.004

Kaplan, M., & Haenlein, M. (2010). "Users of the world, unite! The challenges and opportunities of social media"(PDF). *Business Horizons, 53*(1), 61. doi:10.1016/j.bushor.2009.09.003

Karim, F., Oyewande, A., Abdalla, L. F., Chaudhry Ehsanullah, R., & Khan, S. (2020). Social media use and its connection to mental Health: A systematic review. *Cureus.* Advance online publication. doi:10.7759/cureus.8627 PMID:32685296

Karthikeyan, T., Sekaran, K., Ranjith, D., Vinoth Kumar, V., & Balajee, J. M. (2019). Personalized Content Extraction and Text Classification Using Effective Web Scraping Techniques. *International Journal of Web Portals, 11*(2), 41–52. doi:10.4018/IJWP.2019070103

Kasinathan, P., Pastrone, C., Spirito, M. A., & Vinkovits, M. (2013). Denial-of-Service detection in 6LoWPAN based Internet of Things. *2013 IEEE 9th International Conference on Wireless and Mobile Computing, Networking and Communications (WiMob)*, 600–607. 10.1109/WiMOB.2013.6673419

Katal, A., Wazid, M., & Goudar, R. H. (2013, August). Big data: issues, challenges, tools and good practices. In *2013 Sixth international conference on contemporary computing (IC3)* (pp. 404-409). IEEE.

Katz, E., Blumler, J. G., & Gurevitch, M. (1974). Utilization of mass communication by the individual. In J. Blumler & E. Katz (Eds.), *The uses of mass communications: current perspectives on gratifications research* (pp. 19–32). Sage.

Kaur, A., & Kaur, L. (2016). Concealed weapon detection from images using SIFT and SURF. *Online International Conference on Green Engineering and Technologies (IC-GET)*, 1-8. 10.1109/GET.2016.7916679

Ke'ahi Cooper. (2015). *Security for the Internet of Things.* KTH Royal Institute of Technology. http://kth.diva-portal.org/smash/record.jsf?pid=diva2%3A848663&dswid=-2959.Degree

Keating, B. A., Carberry, P. S., Hammer, G. L., Probert, M. E., Robertson, M. J., Holzworth, D., Huth, N. I., Hargreaves, J. N. G., Meinke, H., Hochman, Z., McLean, G., Verburg, K., Snow, V., Dimes, J. P., Silburn, M., Wang, E., Brown, S., Bristow, K. L., Asseng, S., ... Smith, C. J. (2003). An overview of APSIM, a model designed for farming systems simulation. *European Journal of Agronomy, 18*(3-4), 267–288. doi:10.1016/S1161-0301(02)00108-9

Kelly, N., & Antonio, A. (2016). Teacher peer support in social network sites. *Teaching and Teacher Education, 56*, 138–149. doi:10.1016/j.tate.2016.02.007

Kelman, H. (1958). Compliance, identification, and internalization: Three processes of attitude change. *The Journal of Conflict Resolution, 1*(1), 51–60. doi:10.1177/002200275800200106

Kelman, H. C. (1974). Social influence and linkages between the individual and the social system: further thoughts on the processes of compliance, identification, and internalization. In J. T. Tedeschi (Ed.), *Perspectives on social power* (pp. 125–171). Aldine.

Kennedy, E., & Laurillard, D. (2019). The potential of MOOCs for large-scale teacher professional development in contexts of mass displacement. *London Review of Education, 17*(2), 141–158. doi:10.18546/LRE.17.2.04

Khatun, A., Fazlul Haque, A. K. M., Ahmed, S., & Rahman, M. M. (2015). Design and implementation of iris recognition based attendance management system. *2015 International Conference on Electrical Engineering and Information Communication Technology (ICEEICT)*. 10.1109/ICEEICT.2015.7307458

Khedr, A. E. (2012). Towards Three Dimensional Analyses for Applying E-Learning Evaluation Model: The Case of E-Learning in Helwan University. *IJCSI International Journal of Computer Science Issues, 9*(4), 161–166.

Khedr, A. E. (2013). Business Intelligence framework to support Chronic Liver Disease Treatment. *International Journal of Computers and Technology, 4*(2), 307–312.

Khedr, A. E., & El Seddawy, A. I. (2015). A Proposed Data Mining Framework for Higher Education System. *International Journal of Computers and Applications, 113*(7), 24–31.

Khedr, A. E., & Idrees, A. M. (2017). Adapting Load Balancing Techniques for Improving the Performance of e-Learning Educational Process. *Journal of Computers, 12*(3), 250–257.

Khedr, A. E., & Kok, J. (2006). Adopting Knowledge Discovery in Databases for Customer Relationship Management in Egyptian Public Banks. *IFIP World Computer Congress, TC 12*, 201-208.

Khedr, A., Kholeif, S., & Hessen, S. (2015, April). Enhanced Cloud Computing Framework to Improve the Educational Process in Higher Education: A case study of Helwan University in Egypt. *International Journal of Computers and Technology, 14*(6), 5814–5823.

Khedr, A., Kholeif, S., & Hessen, S. (2015, March). Adoption of cloud computing framework in higher education to enhance educational process. *International Journal of Innovative Research in Computer Science and Technology, 3*(3), 150–156.

Kietzmann, J. H., Hermkens, K., McCarthy, I. P., & Silvestre, B. S. (2011). Social media? Get serious! Understanding the functional building blocks of social media. *Business Horizons, 54*(3), 241–251. doi:10.1016/j.bushor.2011.01.005

Killian, S., Lanon, J., Murray, L., Avram, G., Giralt, M., & O'Riordan, S. (2019). Social Media for Social Good: Student engagement for the SDGs. *International Journal of Management Education, 17*(100307), 1–12. doi:10.1016/j.ijme.2019.100307

Kim, G., Park, S.-B., & Oh, J. (2010). An examination of factors influencing consumer adoption of short message service (SMS). *Computer Human Behavior*, 1152–1161.

Kim, J.-C., & Chung, K. (2018). Associative Feature Information Extraction Using Text Mining from Health Big Data. *Wireless Personal Communications, 105*(2), 17.

Kim, S., & Moon, I. (2018). Traveling salesman problem with a drone station. *IEEE Transactions on Systems, Man, and Cybernetics. Systems*, *49*(1), 42–52.

King, D. (2015). Why Use Social Media? *Library Technology Reports*, *51*(1), 6–9.

Kitchin, R. (2013). Big data and human geography opportunities, challenges and risks. *Dialogues in Human Geography*, *3*(3), 262–267. doi:10.1177/2043820613513388

Koc, M., & Gulyagci, S. (2013). Facebook addiction among Turkish college students: The role of psychological health, demographic, and usage characteristics. *Cyberpsychology, Behavior, and Social Networking*, *16*(4), 279–284. doi:10.1089/cyber.2012.0249 PMID:23286695

Kolluri, N. L., & Murthy, D. (2021). CoVerifi: A COVID-19 news verification system. *Online Social Networks and Media*, 13.

Köse, Ö. B., & Doğan, A. (2019). The Relationship between Social Media Addiction and Self-Esteem among Turkish University Students. *Turkish Green Crescent Society.*, *6*(1), 175–190.

Kosyakova, I. V., Zhilyunov, N. Y., & Astashev, Y. V. (2020). Prospects for the Integration of Environmental Innovation Management on the Platform of Information and Communication Technologies. *Advances in Intelligent Systems and Computing.*, *908*, 345–355. doi:10.1007/978-3-030-11367-4_34

Kouser, R. R., Manikandan, T., & Kumar, V. V. (2018). Heart disease prediction system using artificial neural network, radial basis function and case based reasoning. *Journal of Computational and Theoretical Nanoscience*, *15*(9-10), 2810–2817.

Kreijns, K., Van Acker, F., & Van Buuren, H. (2014). Community of Inquiry: Social presence revisited. *E-Learning and Digital Media*, *11*(1), 5–18. Advance online publication. doi:10.2304/elea.2014.11.1.5

Krishen, A. S., Berezan, O., Agarwal, S., & Kachroo, P. (2019). Social media networking satisfaction in the US and Vietnam: Content versus connection. *Journal of Business Research*, *101*, 93–103. doi:10.1016/j.jbusres.2019.03.046

Kucharčíková, A., & Tokarčíková, E. (2016). Use of participatory methods in teaching at the university. *Turkish Online Journal of Science & Technology*, *6*(1).

Küçük, D., & Can, F. (2020). Stance Detection: A Survey. *Association for Computing Machinery*, *53*(1), 37.

Kukulska-Hulme, A. (2012). How should the higher education workforce adapt to advancements in technology for teaching and learning? *Internet and Higher Education*, *15*(4), 247–254. doi:10.1016/j.iheduc.2011.12.002

Kulkarni, Y., Warhade, K. K., & Bahekar, S. (2014). Primary nutrients determination in the soil using UV spectroscopy. *International Journal of Emerging Engineering Research and Technology*, *2*(1), 198–204.

Kumar, D., Swathi, P., Jahangir, A., Sah, N. K., & Vinothkumar, V. (2021). Intelligent Speech Processing Technique for Suspicious Voice Call Identification Using Adaptive Machine Learning Approach. In Handbook of Research on Innovations and Applications of AI, IoT, and Cognitive Technologies (pp. 372-380). IGI Global.

Kumar, V. V., Raghunath, K. K., Rajesh, N., Venkatesan, M., Joseph, R. B., & Thillaiarasu, N. (2021). Paddy Plant Disease Recognition, Risk Analysis, and Classification Using Deep Convolution Neuro-Fuzzy Network. *Journal of Mobile Multimedia*, 325-348.

Kumar, V. V., Raghunath, K. M., Muthukumaran, V., Joseph, R. B., Beschi, I. S., & Uday, A. K. (2021). Aspect based sentiment analysis and smart classification in uncertain feedback pool. *International Journal of System Assurance Engineering and Management*, 1-11.

Kumar, V., Niveditha, V. R., Muthukumaran, V., Kumar, S. S., Kumta, S. D., & Murugesan, R. (2021). A Quantum Technology-Based LiFi Security Using Quantum Key Distribution. In Handbook of Research on Innovations and Applications of AI, IoT, and Cognitive Technologies (pp. 104-116). IGI Global.

Kumar, D. S., Kumar, C. S., Ragamayi, S., Kumar, P. S., Saikumar, K., & Ahammad, S. H. (2020). A test architecture design for SoCs using atam method. *Iranian Journal of Electrical and Computer Engineering*, *10*(1), 719.

Kumar, G. R., Basha, S. R., & Rao, S. B. (2020, January). A summarization on text mining techniques for information extracting from applications and issues. *Journal of Mechanics of Continua and Mathematical Sciences*, *15*(1), 9. doi:10.26782/jmcms.spl.5/2020.01.00026

Kumar, M. S., Inthiyaz, S., Vamsi, C. K., Ahammad, S. H., Sai Lakshmi, K., Venu Gopal, P., & Bala Raghavendra, A. (2019). Power optimization using dual sram circuit. *International Journal of Innovative Technology and Exploring Engineering*, *8*(8), 1032–1036.

Kumar, V. V., Ramamoorthy, S., Kumar, V. D., Prabu, M., & Balajee, J. M. (2021). Design and Evaluation of Wi-Fi Offloading Mechanism in Heterogeneous Networks. *International Journal of e-Collaboration*, *17*(1), 60–70. doi:10.4018/IJeC.2021010104

Ku, Y. C., Chu, T. H., & Tseng, C. H. (2013). Gratifications for using CMC technologies: A comparison among SNS, IM, and e-mail. *Computers in Human Behavior*, *29*(1), 226–234. doi:10.1016/j.chb.2012.08.009

Ku, Y.-C., Chen, R., & Zhang, H. (2013). Why do users continue using social networking sites? An exploratory study of members in the United States and Taiwan. *Information & Management*, *50*(7), 571–581. doi:10.1016/j.im.2013.07.011

Laffey, J., Lin, G. Y., & Lin, Y. (2006). Assessing social ability in online learning environments. *Journal of Interactive Learning Research*, *17*(2), 163–177.

Lallier, E., & Farooq, M. (2000). A real time pixel-level based image fusion via adaptive weight averaging. *Proceedings of the Third International Conference on Information Fusion, 2*, WEC3/3-WEC313.

Lambin, E. F., Rounsevell, M. D., & Geist, H. J. (2000). Are agricultural land-use models able to predict changes in land-use intensity? *Agriculture, Ecosystems & Environment*, *82*(1-3), 321–331. doi:10.1016/S0167-8809(00)00235-8

Lambton-Howard, D., Kiaer, J., & Kharrufa, A. (2020). 'Social media is their space': Student and teacher use and perception of features of social media in language education. *Behaviour & Information Technology*, *0*(0), 1–16. doi:10.1080/0144929X.2020.1774653

Lang, N. (2015). Why teens are leaving Facebook: It's 'meaningless.' *The Washington Post*. Retrieved from https://www.washingtonpost.com/news/the-intersect/wp/2015/02/21/whyteens-are-leaving-facebook-its-meaningless/

Laurillard, D., Kennedy, E., Charlton, P., Wild, J., & Dimakopoulos, D. (2018). Using technology to develop teachers as designers of TEL: Evaluating the learning designer. *British Journal of Educational Technology*, *49*(6), 1044–1058. doi:10.1111/bjet.12697

Lavanya, B. L. (2017). A Study on Employee Attrition: Inevitable yet Manageable. *International Journal of Business and Management Invention*, *6*(9), 38–50.

Lebedko, M. (2014). Globalization, Networking and Intercultural Communication. *Intercultural Communication Studies*, *23*(1), 28–41.

Lee, B. S., Alexander, M. E., Hawkes, B. C., Lynham, T. J., Stocks, B. J., & Englefield, P. (2002). Information systems in support of wildland fire management decision making in Canada. *Computers and Electronics in Agriculture*, *37*(1-3), 185–198. doi:10.1016/S0168-1699(02)00120-5

Lee, I. (2017). Big data: Dimensions, evolution, impacts, and challenges. *Business Horizons, 60*(3), 293–303. doi:10.1016/j.bushor.2017.01.004

Lee, S., Nah, S., Chung, D. S., & Kim, J. (2020). Predicting AI News Credibility: Communicative or Social Capital or Both? *Communication Studies, 71*(3), 21.

Lei, A., Cruickshank, H., Cao, Y., Asuquo, P., Ogah, C. P. A., & Sun, Z. (2017). Blockchain-based dynamic key management for heterogeneous intelligent transportation systems. *IEEE Internet of Things Journal, 4*(6), 1832–1843. doi:10.1109/JIOT.2017.2740569

Lenard, T. M., & Rubin, P. H. (2013). The big data revolution: Privacy considerations. Technology Policy Institute.

Leong, L.-Y., Hew, T.-S., Ooi, K.-B., Lee, V.-H., & Hew, J.-J. (2019). A hybrid SEM-neural network analysis of social media addiction. *Expert Systems with Applications, 133*, 296–316. doi:10.1016/j.eswa.2019.05.024

Leung, L. (2013). Generational Differences in Content Generation in Social Media: The Roles of the Gratifications Sought and of Narcissism. *Computers in Human Behavior, 29*(3), 997–1006. doi:10.1016/j.chb.2012.12.028

Leung, L., & Wei, R. (2000). More than just talk on the move: A use-and-gratification study of the cellular phone. *Journalism & Mass Communication Quarterly, 77*(2), 308–320. doi:10.1177/107769900007700206

Li, C., Qian, X., & Liu, Y. (2013). Using supervised bigram-based ILP for extractive summarization. In *Proceedings of the 51st Annual Meeting of the Association for Computational Linguistics (*Volume 1*: Long Papers)* (pp. 1004-1013). Academic Press.

Liao, C.-H., Chen, L.-X., Yang, J.-C., & Yuan, S.-M. (2020). A Photo Post Recommendation System Based on Topic Model for Improving Facebook Fan Page Engagement. Symmetry, 17(7), 18.

Librizzi, J. (2017). The Haunted Animal: Peirce's Community of Inquiry and the Formation of the Self. *All Theses & Dissertations.* 317.

Likhitha, S., Harish, B. S., & Keerthi Kumar, H. M. (2019). A Detailed Survey on Topic Modeling for Document and Short Text Data. *International Journal of Computers and Applications, 178*(39), 9.

Li, L., Liu, J., Cheng, L., Qiu, S., Wang, W., Zhang, X., & Zhang, Z. (2018). Creditcoin: A privacy-preserving blockchain-based incentive announcement network for communications of smart vehicles. *IEEE Transactions on Intelligent Transportation Systems, 19*(7), 2204–2220. doi:10.1109/TITS.2017.2777990

Lim, T. S., Sim, S. C., & Mansor, M. M. (2009). RFID based attendance system. *2009 IEEE Symposium on Industrial Electronics & Applications.* 10.1109/ISIEA.2009.5356360

Lin, H., & Bilmes, J. (2010, June). Multi-document summarization via budgeted maximization of submodular functions. In *Human Language Technologies: The 2010 Annual Conference of the North American Chapter of the Association for Computational Linguistics* (pp. 912-920). Academic Press.

Lin, C. Y., & Hovy, E. (2003). Automatic evaluation of summaries using n-gram co-occurrence statistics. In *Proceedings of the 2003 Human Language Technology Conference of the North American Chapter of the Association for Computational Linguistics* (pp. 150-157). 10.3115/1073445.1073465

Linda, G. M., Lakshmi, N. S. R., Murugan, N. S., Mahapatra, R. P., Muthukumaran, V., & Sivaram, M. (2021). Intelligent recognition system for viewpoint variations on gait and speech using CNN-CapsNet. *International Journal of Intelligent Computing and Cybernetics.*

Lin, F., Lin, S., & Huang, T. (2008). Knowledge sharing and creation in a teachers' professional virtual community. *Computers & Education*, *50*(3), 742–756. doi:10.1016/j.compedu.2006.07.009

Lipnack, J., & Stamps, J. (1997). *Reaching across space, time, and organizations with technology*. John Wiley & Sons, Inc.

Li, T., Wen, B., Tian, Y., Li, Z., & Wang, S. (2018). Numerical simulation and experimental analysis of small drone rotor blade polarimetry based on RCS and micro-Doppler signature. *IEEE Antennas and Wireless Propagation Letters*, *18*(1), 187–191.

Litjens, G., Sánchez, C. I., Timofeeva, N., Hermsen, M., Nagtegaal, I., Kovacs, I., ... Van Der Laak, J. (2016). Deep learning as a tool for increased accuracy and efficiency of histopathological diagnosis. *Scientific Reports*, *6*(1), 1–11.

Liu, J., Sarkar, M. K., & Chakraborty, G. (2013). Feature-based Sentiment Analysis on Android App Reviews Using SAS® Text Miner and SAS® Sentiment Analysis Studio. *SAS Global Forum, 1*(7), 8.

Liu, Y., Wang, Q., & Huang, Y. (2018, July). Research on expert opinion credibility rating. *International Journal of Innovative, 14*(6), 8.

Liu, Q., Li, P., Zhao, W., Cai, W., Yu, S., & Leung, V. C. (2018). A survey on security threats and defensive techniques of machine learning: A data driven view. *IEEE Access: Practical Innovations, Open Solutions*, *6*, 12103–12117.

Loc, N. P., Tong, D. H., Thao, V. T. T., Han, N. N., Tram, T. H., Thoa, D. T., & Co, L. V. (2019). Students' social networking: Current status and impact. *International Journal of Scientific and Technology Research*, *8*(12), 3602–3605.

Lombardi, M., Pascale, F., & Santaniello, D. (2021). Internet of things: A general overview between architectures, protocols and applications. *Information (Switzerland)*, *12*(2), 1–21. doi:10.3390/info12020087

Lumpkin, A., Achen, R. M., & Dodd, R. K. (2015). Student Perceptions of Active Learning. *College Student Journal*, *49*(1), 121–133.

Luo, M. M., Chea, S., & Chen, J. S. (2011). Web-based information service adoption: A comparison of the motivational model and the uses and gratifications theory. *Decision Support Systems*, *51*(1), 21–30. doi:10.1016/j.dss.2010.11.015

Luo, M., Hancock, J. T., & Markowitz, D. M. (2020). Credibility Perceptions and News Headlines on Social Media: Effects of Truth-Biasand Endorsement Cues. *Communication Research*, 25.

Luo, T., Freeman, C., & Stefaniak, J. (2020). "Like, comment, and share"—professional development through social media in higher education: A systematic review. *Educational Technology Research and Development*, *68*(4), 1659–1683. doi:10.100711423-020-09790-5

Mackey, G., Sehrish, S., Bent, J., Lopez, J., Habib, S., & Wang, J. (2008). Introducing map-reduce to high-end computing. *Petascale Data Storage Workshop*, 1–6.

Ma, D., & Jiao, X. (2019). WoMA: An Input-Based Learning Model to Predict Dynamic Workload of Embedded Applications. *IEEE Embedded Systems Letters*, *12*(3), 74–77. doi:10.1109/LES.2019.2957487

Madge, C., Meek, J., Wellens, J., & Hooley, T. (2009). Facebook, social integration and informal learning at university: "It is more for socialising and talking to friends about work than for actually doing work. *Learning, Media and Technology*, *34*(2), 141–155. doi:10.1080/17439880902923606

Magretta, J. (2002). Why business models matter. *Harvard Business Review*, *80*(5), 86–92. PMID:12024761

Maithili, K., Vinothkumar, V., & Latha, P. (2018). Analyzing the Security Mechanisms to Prevent Unauthorized Access in Cloud and Network Security. *Journal of Computational and Theoretical Nanoscience*, *15*(6), 2059–2063. doi:10.1166/jctn.2018.7407

Majeski, R. A., Stover, M., & Valais, T. (2018). The Community of Inquiry and Emotional Presence. *Adult Learning*, *29*(2), 53–61. doi:10.1177/1045159518758696

Maleki & Rahmani. (2019). MapReduce: An infrastructure review and research insights. *The Journal of Supercomputing*, *75*(10), 1–69.

Malhotra, Y., & Galletta, D. F. (1999). Extending the technology acceptance model to account for social influence: theoretical bases and empirical validation. *Proceedings of the 32nd Hawaii international conference on system sciences*, 1–11. 10.1109/HICSS.1999.772658

Manca, S., & Ranieri, M. (2017). Implications of social network sites for teaching and learning. Where we are and where we want to go. *Education and Information Technologies*, *22*(2), 605–622. doi:10.100710639-015-9429-x

Manikandan, G., Perumal, R., & Muthukumaran, V. (2020, November). A novel and secure authentication scheme for the Internet of Things over algebraic structure. In. AIP Conference Proceedings: Vol. 2277. *No. 1* (p. 060001). AIP Publishing LLC. doi:10.1063/5.0025330

Manikandan, G., Perumal, R., & Muthukumaran, V. (2021). Secure data sharing based on proxy re-encryption for internet of vehicles using seminearring. *Journal of Computational and Theoretical Nanoscience*, *18*(1-2), 516–521.

Martínez-Rodríguez, M. C., del Valle, S. S., Brox, P., & Sánchez-Solano, S. (2020). Hardware Implementation of Authenticated Ciphers for Embedded Systems. *IEEE Latin America Transactions*, *18*(09), 1581–1591. doi:10.1109/TLA.2020.9381800

Martin, J. D., & Hassan, F. (2020). News Media Credibility Ratings and Perceptions of Online Fake News Exposure in Five Countries. *Journalism Studies*, *21*(16), 20. doi:10.1080/1461670X.2020.1827970

Matin, G. (2008). *A Study of the Random Oracle Model*. Cambridge University Press. https://scholar.google.com/scholar_lookup?title=Ph.D.+Thesis&author=G.+Martin&publication_year=2008

Matsunaga, A., Tsugawa, M., & Fortes, J. (2008). Cloudblast:Combining mapreduce and virtualization on distributed resources for bioinformatics applications. *IEEE Fourth International Conference*, 222–229. doi:10.1145/2465351.2465371

Matthews, C. (2014, January 15). Facebook: More than 11 million young people have fled Facebook since 2011. *Time Magazine*. Retrieved from http://business.time.com/2014/01/ 15/more-than-11-million-young-people-have-fled-facebooksince-2011/

Matzat, U., & Vrieling, E. M. (2016). Self-regulated learning and social media – a 'natural alliance'? Evidence on students' self-regulation of learning, social media use, and student–teacher relationship. *Learning, Media and Technology*, *41*(1), 73–99. doi:10.1080/17439884.2015.1064953

Mayer, R. E. (2004). Should there be a three-strikes rule against pure discovery learning? *The American Psychologist*, *59*(1), 14.

Mayer-Schönberger, V., & Cukier, K. (2013). *Big data: A revolution that will transform how we live, work, and think.* Houghton Mifflin Harcourt.

McClure, C., & Seock, Y.-K. (2020). The role of involvement: Investigating the effect of brand's social media pages on consumer purchase intention. *Journal of Retailing and Consumer Services*, *53*(101975), 1–8. doi:10.1016/j.jretconser.2019.101975

McLeod, S. (2014). *The Interview Method.* Retrieved on April 17, 2014 fromhttp://www.simplypsychology.org/interviews.html

McQuail, D. (2010). *Mass communication theory: An introduction.* Sage Publications.

Meneghello, F., Calore, M., Zucchetto, D., Polese, M., & Zanella, A. (2019). IoT: Internet of threats? A survey of practical security vulnerabilities in real IoT devices. *IEEE Internet of Things Journal, 6*(5), 8182–8201. doi:10.1109/JIOT.2019.2935189

Michael, J. (2006). Where's the evidence that active learning works? *Advances in Physiology Education, 30*(4), 159–167.

Michikyan, M., Subrahmanyam, K., & Dennis, J. (2015). Facebook use and academic performance among college students: A mixed-methods study with a multi-ethnic sample. *Computers in Human Behavior, 45*, 265–272. doi:10.1016/j.chb.2014.12.033

Michos, K., & Hernández-Leo, D. (2020). CIDA: A collective inquiry framework to study and support teachers as designers in technological environments. *Computers & Education, 143*, 103679. doi:10.1016/j.compedu.2019.103679

Miller, M. F. (2019). Why Hate the Internet? Contemporary Fiction, Digital Culture, and the Politics of Social Media. *Arizona Quarterly: A Journal of American Literature, Culture, and Theory, 75*(3), 59–85.

Mills, N. (2011). Situated Learning through Social Networking Communities: The Development of Joint Enterprise, Mutual Engagement, and a Shared Repertoire. *Journal, 28*(2), 345–368. doi:10.11139/cj.28.2.345-368

Moghavvemi, S., Paramanathan, T., Rahin, N., & Sharabati, M. (2017). Student ' s perceptions towards using e-learning via Facebook Sedigheh Moghavvemi, Tanuosha Paramanathan, Nurliana Md Rahin &. *Behaviour & Information Technology, 0*(0), 1–20. doi:10.1080/0144929X.2017.1347201

Mohammad Zoqi Sarwani, D. A., Sarwani, M. Z., Sani, D. A., & Fakhrin, F. C. (2019). Personality Classification through Social Media Using Probabilistic Neural Network Algorithms. *International Journal of Artificial Intelligence & Robotics, 1*(1), 7.

Mohammad, M. N., Kumari, C. U., Murthy, A. S. D., Jagan, B. O. L., & Saikumar, K. (2021). Implementation of online and offline product selection system using FCNN deep learning: Product analysis. *Materials Today: Proceedings, 45*, 2171–2178.

Mohapatra, S., Patra, D., & Satpathy, S. (2012). Leukemia diagnosis using color-based clustering and unsupervised blood microscopic image segmentation. *International Journal of Computer Information Systems and Industrial Management Applications*, 477–485.

Mohapatra, S., Patra, D., & Satpathi, S. (2010, December). Image analysis of blood microscopic images for acute leukemia detection. In *2010 International Conference on Industrial Electronics, Control and Robotics* (pp. 215-219). IEEE. 10.1109/IECR.2010.5720171

Mohsen, A. M., Hassan, H. A., & Idrees, A. M. (2016). A Proposed Approach for Emotion Lexicon Enrichment. *International Journal of Computer, Electrical, Automation, Control and Information Engineering, 10*(1).

Mohsen, A. M., Hassan, H. A., & Idrees, A. M. (2016). Documents Emotions Classification Model Based on TF IDF Weighting. *International Journal of Computer Electrical Automation Control and Information Engineering, 10*(1).

Mohsen, A. M., Idrees, A. M., & Hassan, H. A. (2019). Emotion Analysis for Opinion Mining From Text: A Comparative Study. *International Journal of e-Collaboration, 15*(1).

Mostafa, A., Khedr, A. E., & Abdo, A. (2017). Advising Approach to Enhance Students' Performance Level in Higher Education Environments. Journal of Computational Science, 13(5), 130–139.

Mouty, R., & Gazdar, A. (2019). The Effect of the Similarity Between the Two Names of Twitter Users on the Credibility of Their Publications. In *Joint 2019 8th International Conference on Informatics, Electronics & Vision (ICIEV) & 3rd International Conference on Imaging, Vision & Pattern Recognition (IVPR)*. IEEE.

Mukerji, N. (2018). What is fake news? *Ergo, 5*(35), 24.

Mukherjee, A., Datta, J., Jorapur, R., Singhvi, R., Haloi, S., & Akram, W. (2012). Shared disk big data analytics with Apache Hadoop. *High Performance Computing (HiPC), 19th International Conference.*

Mulyono, H., & Suryoputro, G. (2020). The use of social media platform to promote authentic learning environment in higher education setting. *Science for Education Today, 10*(2), 105–123. doi:10.15293/2658-6762.2002.07

Munkvold, B. E., & Zigurs, I. (2005). Integration of e-collaboration technologies: Research opportunities and challenges. *International Journal of e-Collaboration, 1*(2), 1–24. doi:10.4018/jec.2005040101

Muthukumaran, V., Joseph, R. B., & Uday, A. K. (2021). Intelligent Medical Data Analytics Using Classifiers and Clusters in Machine Learning. In Handbook of Research on Innovations and Applications of AI, IoT, and Cognitive Technologies (pp. 321-335). IGI Global.

Muthukumaran, V., Ezhilmaran, D., & Anjaneyulu, G. S. G. N. (2018). Efficient Authentication Scheme Based on the Twisted Near-Ring Root Extraction Problem. In *Advances in Algebra and Analysis* (pp. 37–42). Birkhäuser. doi:10.1007/978-3-030-01120-8_5

Muthukumaran, V., Kumar, V. V., Joseph, R. B., Munirathanam, M., & Jeyakumar, B. (2021). Improving Network Security Based on Trust-Aware Routing Protocols Using Long Short-Term Memory-Queuing Segment-Routing Algorithms. *International Journal of Information Technology Project Management, 12*(4), 47–60. doi:10.4018/IJITPM.2021100105

Muthukumaran, V., Vinothkumar, V., Joseph, R. B., Munirathanam, M., & Jeyakumar, B. (2021). Improving network security based on trust-aware routing protocols using long short-term memory-queuing segment-routing algorithms. *International Journal of Information Technology Project Management, 12*(4), 47–60.

Myla, S., Marella, S. T., Goud, A. S., Ahammad, S. H., Kumar, G. N. S., & Inthiyaz, S. (n.d.). *Design Decision Taking System For Student Career Selection For Accurate Academic System.* Academic Press.

Myneni, M. B., & Dandamudi, R. (2019). Harvesting railway passenger opinions on multi themes by using social graph clustering. *Journal of Rail Transport Planning & Management,* 10.

Nagageetha, M., Mamilla, S. K., & Hasane Ahammad, S. (2017). Performance analysis of feedback based error control coding algorithm for video transmission on wireless multimedia networks. *Journal of Advanced Research in Dynamical and Control Systems, 9*, 626–660.

NagaJyothi, G., & Sridevi, S. (2019). High speed and low area decision feed-back equalizer with novel memory less distributed arithmetic filter. *Multimedia Tools and Applications, 78*(23), 32679–32693.

Nagarajan, S. M., Muthukumaran, V., Beschi, I. S., & Magesh, S. (2021). Fine Tuning Smart Manufacturing Enterprise Systems: A Perspective of Internet of Things-Based Service-Oriented Architecture. In Handbook of Research on Innovations and Applications of AI, IoT, and Cognitive Technologies (pp. 89-103). IGI Global.

Nagarajan, S. M., Deverajan, G. G., Chatterjee, P., Alnumay, W., & Muthukumaran, V. (2022). Integration of IoT based routing process for food supply chain management in sustainable smart cities. *Sustainable Cities and Society, 76*, 103448. doi:10.1016/j.scs.2021.103448

Namisango, F., & Kang, K. (2019). Organization-public relationships on social media: The role of relationship strength, cohesion and symmetry. *Computers in Human Behavior, 101*, 22–29. doi:10.1016/j.chb.2019.06.014

Namugera, F., Wesonga, R., & Jehopio, P. (2019). Text mining and determinants of sentiments: Twitter social media usage by traditional media houses in Uganda. *Computational Social Networks*, *6*(1), 21.

Nan, Y., Lovell, N. H., Redmond, S. J., Wang, K., Delbaere, K., & van Schooten, K. S. (2020). Deep Learning for Activity Recognition in Older People Using a Pocket-Worn Smartphone. *Sensors (Basel)*, *20*(24), 7195. doi:10.339020247195 PMID:33334028

Narayana, V. V., Ahammad, S. H., Chandu, B. V., Rupesh, G., Naidu, G. A., & Gopal, G. P. (2019). Estimation of Quality and Intelligibility of a Speech Signal with varying forms of Additive Noise. *International Journal of Emerging Trends in Engineering Research*, *7*(11), 430–433. doi:10.30534/ijeter/2019/057112019

Nasrullah, S., & Khan, F. R. (2019). Examining the Impact of Social Media on the Academic Performances of Saudi Students - Case Study: Prince Sattam Bin Abdul Aziz University. *Humanities and Social Sciences Reviews.*, *7*(5), 851–861. doi:10.18510/hssr.2019.75111

Nazier, M. M., Khedr, A. E., & Haggag, M. (2013). Business Intelligence and its role to enhance Corporate Performance Management. *International Journal of Management & Information Technology*, *3*(3).

Nelson, R. A., Holzworth, D. P., Hammer, G. L., & Hayman, P. T. (2002). Infusing the use of seasonal climate forecasting into crop management practice in North East Australia using discussion support software. *Agricultural Systems*, *74*(3), 393–414. doi:10.1016/S0308-521X(02)00047-1

Neo, L. R. (2021). Linking Perceived Political Network Homogeneity with Political Social Media Use via Perceived Social Media News Credibility. *Journal of Information Technology & Politics*.

Neshenko, N., Bou-Harb, E., Crichigno, J., Kaddoum, G., & Ghani, N. (2019). Demystifying IoT security: An exhaustive survey on IoT vulnerabilities and a first empirical look on Internet-scale IoT exploitations. *IEEE Communications Surveys and Tutorials*, *21*(3), 2702–2733. doi:10.1109/COMST.2019.2910750

Ngai, K-L., S.S, K. M., Lam, E. S., Chin, S., Tao, S., & Eric, W. (2015). *Social media models, technologies, and applications: An academic review and Case Study.* Emerald Insight.

Nirmala, K., & Subramani, K. (2013). Content based image retrieval system using auto color correlogram. *Jisuanji Yingyong*, *6*, 67–73.

Nizzolino, S. (2020). Teacher Networking, Professional Development, and Motivation Within EU Platforms and the Erasmus Plus Program. In J. Zhao (Ed.), *Collaborative Convergence and Virtual Teamwork for Organizational Transformation* (pp. 195–218). doi:10.4018/978-1-7998-4891-2.ch010

Nizzolino, S., & Canals, A. (2021). Social Network Sites as Community Building Tools in Educational Networking. *International Journal of e-Collaboration*, *17*(4), 132–167. doi:10.4018/IJeC.2021100110

Nonaka, I., & Konno, N. (1998). The Concept of "Ba": Building a foundation for knowledge creation. *California Management Review*, *40*(3), 40–54. doi:10.2307/41165942

Nunes, M., & Correia, J. (2013). Improving trust using online credibility sources and social network quality in P2P marketplaces. *2013 8th Iberian Conference on Information Systems and Technologies (CISTI)*, 1-4.

O'Keefe, G., & Pearson, K. (2011). *The Impact of Social Media on Children, Adolescents, and Families.* American Academy of Pediatrics.

Obar, J. A., & Wildman, S. (2015). Social media definition and the governance challenge: An introduction to the special issue. *Telecommunications Policy*, 745–750.

Oberst, U., Chamarro, A., & Renau, V. (2016). Gender Stereotypes 2.0: Self-Representations of Adolescents on Facebook. *Media Education Research Journal, 24*(48), 81–89.

Odeh, A., Keshta, I., & Abdelfattah, E. (2021, January). Machine LearningTechniquesfor Detection of Website Phishing: A Review for Promises and Challenges. In *2021 IEEE 11th Annual Computing and Communication Workshop and Conference (CCWC)* (pp. 813-818). IEEE. 10.1109/CCWC51732.2021.9375997

Opdenakker, R. (2006). *Advantages and Disadvantages of Four Interview Techniques in Qualitative Research.* Retrieved on April 17, 2014, from: https://www.qualitative-research.net/index.php/fqs/article/view/175/391

Osmani, M. W., El Haddadeh, R., Hindi, N., & Weerakkody, V. (2020). The Role of Co-Innovation Platform and E-Collaboration ICTs in Facilitating Entrepreneurial Ventures. *International Journal of E-Entrepreneurship and Innovation, 10*(2), 62–75.

Osterwalder, A., Pigneur, Y., Bernarda, G., & Smith, A. (2014). *Value proposition design: How to create products and services customers want.* John Wiley & Sons.

Osterwalder, A., Pigneur, Y., Oliveira, M. A. Y., & Ferreira, J. J. P. (2011). Business model generation: A handbook for visionaries, game changers and challengers. *African Journal of Business Management, 5*(7), 22–30.

Othman, M., Hassan, H., Moawad, R., & Idrees, A. M. (2016). Using NLP Approach for Opinion Types Classifier. *Journal of Computers, 11*(5), 400–410.

Othman, M., Hassan, H., Moawad, R., & Idrees, A. M. (2018). A linguistic approach for opinionated documents summary. *Future Computing and Informatics Journal, 3*(2), 152158.

Ozbay, F. A., & Alatas, B. (2019). Fake news detection within online social media using supervised artificial intelligence algorithms. *Journal Pre-proof*, 21.

Pai, P., & Arnott, D. C. (2013). User adoption of social networking sites: Eliciting uses and gratifications through a means-end approach. *Computers in Human Behavior, 29*, 1039–1053.

Palanisamy, J. S. P. N., & Malmurugan, N. (2019). FPGA implementation of deep learning approach for efficient human gait action recognition system. International Journal of Innovations in Scientific and Engineering Research, 6(11).

Pandove, D., & Goel, S. (2015, February). A comprehensive study on clustering approaches for big data mining. In *2015 2nd international conference on electronics and communication systems (icecs)* (pp. 1333-1338). IEEE.

Papadimitriou, S. T., & Papadakis, S. (2021). E-Collaboration in Educational Organizations: Opportunities and Challenges in Virtual Learning Environments and Learning Spaces. *Collaborative Convergence and Virtual Teamwork for Organizational Transformation*, 120-146.

Parande, M., & Soma, S. (2015). Concealed Weapon Detection in a Human Body by Infrared Imaging. *International Journal of Science and Research, 4*, 182-188.

Parker, G. G., Van Alstyne, M. W., & Choudary, S. P. (2016). *Platform revolution: How networked markets are transforming the economy and how to make them work for you.* WW Norton & Company.

Park, J. H. (2014). The effects of personalization on user continuance in social networking sites. *Information Processing & Management, 50*(3), 462–475.

Park, N., Kee, K. F., & Valenzuela, S. (2009). Being immersed in social networking environment: Facebook groups, uses and gratifications, and social outcomes. *Cyberpsychology & Behavior, 12*(6), 729–733.

Park, S., Kim, H. T., Lee, S., Joo, H., & Kim, H. (2021). Survey on Anti-Drone Systems: Components, Designs, and Challenges. *IEEE Access: Practical Innovations, Open Solutions*, *9*, 42635–42659.

Pasi, G., De Grandis, M., & Viviani, M. (2020). Decision Making over Multiple Criteria to Assess News Credibility in Microblogging Sites. In *IEEE Conference on Fuzzy Systems*. Institute of Electrical and Electronics Engineers Inc.

Pathak, N., Deb, P. K., Mukherjee, A., & Misra, S. (2021). IoT-to-the-Rescue: A Survey of IoT Solutions for COVID-19-like Pandemics. *IEEE Internet of Things Journal.*

Pedersen, C. L., & Ritter, T. (2020). Use this framework to predict the success of your big data project. *Harvard Business Review.*

Pempek, T. A., Yermolayeva, Y. A., & Calvert, S. L. (2009). College students' social networking experiences on facebook. *Journal of Applied Developmental Psychology*, 227–238.

Pereira, R., Baranauskas, M. C. C., & da Silva, S. R. P. (2010). Social software building blocks: Revisiting the honeycomb framework. *2010 International Conference on Information Society*, 253-258. 10.1109/i-Society16502.2010.6018707

Peter, J., Valkenburg, P. M., & Schouten, A. P. (2005). Developing a model of adolescent friendship formation on the Internet. *Cyberpsychology & Behavior*, *8*(5), 423–430. doi:10.1089/cpb.2005.8.423 PMID:16232035

Pew Research Center. (2015). *Social networking fact sheet*. http://www. pewinternet.org/fact-sheets/social-networking-fact-sheet/

Pimmer, C., Chipps, J., Brysiewicz, P., Walters, F., Linxen, S., & Gröhbiel, U. (2017). Facebook for supervision? Research education shaped by the structural properties of a social media space. *Technology, Pedagogy and Education*, *26*(5), 517–528. doi:10.1080/1475939X.2016.1262788

Pintar, D., Humski, L., & Vranić, D. M. (2019). Analysis of Facebook Interaction as Basis for Synthetic Expanded Social Graph Generation. *IEEE Access*, *7*, 15.

Piyadasa, T. D. (2020). Concealed Weapon Detection Using Convolutional. *Neural Networks.*

Polamarasetty, V. K., & Reddem, M. R. (2018). Attendance System based on Face Recognition. *International Research Journal of Engineering and Technology.*, *5*(1), 4606–4610.

Pourkhani, Abdipour, Baher, & Moslehpour. (2019). The impact of social media in business growth and performance: A scientometrics analysis. *International Journal of Data and Network Science*, *3*, 223–244.

Prabu, S., & Kalaivani, M. (2019). An intelligent power adaptive model using machine learning techniques for wsn based smart health care devices. International Journal of Innovations in Scientific and Engineering Research, 6(12).

Pranav, Sreenivasarao, & Saheb. (2019). Augmenting the outcomes of Health-Care-Services with Big Data Analytics: A Hadoop based approach. *International Journal of Scientific & Technology Research*, *8*, 12.

Prenger, R., Poortman, C. L., & Handelzalts, A. (2019). The Effects of Networked Professional Learning Communities. *Journal of Teacher Education*, *70*(5), 441–452. doi:10.1177/0022487117753574

Prince, M. (2004). Does active learning work? A review of the research. *Journal of Engineering Education, Washington*, *93*, 223–232.

Qimei, C., Clifford, S. J., & Wells, W. D. (2002). Attitude toward the site II: New information. *Journal of Advertising Research*, *42*(2), 33–45. doi:10.2501/JAR-42-2-33-45

Qimei, C., & Wells, W. D. (1999). Attitude toward the site. *Journal of Advertising Research*, *39*(5), 27–37.

Quan, Xi., & Gang, M. Li. (2012). An Efficient Automatic Attendance System Using Fingerprint Reconstruction Technique. *International Journal of Computer Science and Information Security, 10*(1), 1–6.

Raacke, J., & Bonds-Raacke, J. (2008). MySpace and Facebook: Applying the uses and gratifications theory to exploring friend-networking sites. *Cyberpsychology & Behavior, 11*(2), 169–174.

Radcliffe, D., & Lam, A. (2018). *Social Media in the Middle East: The Story of 2017*. Academic Press.

Rafique & Khan. (2013). Exploring Static and Live Digital Forensics: Methods, Practices and Tools. *International Journal of Scientific & Engineering Research, 4*, 10.

Raghupathi, W., & Raghupathi, V. (2014). Big data analytics in healthcare: Promise and potential. *Health Information Science and Systems, 2*(1), 1–10. doi:10.1186/2047-2501-2-3 PMID:25825667

Raghuram, M., Akshay, K., & Chandrasekaran, K. (2016). Efficient user profiling in twitter social network using traditional classifiers. *International Journal of Intelligent Systems Technologies and Applications*, 399–411.

Raj Kumar, A., Kumar, G. N. S., Chithanoori, J. K., Mallik, K. S. K., Srinivas, P., & Ahammad, S. H. (2019). Design and Analysis of a Heavy Vehicle Chassis by using E-Glass Epoxy & S-2 Glass Materials. *International Journal of Recent Technology and Engineering, 7*(6), 903–905.

Rajput, Ganage, & Thakur. (2017). Review Paper On Hadoop And Map Reduce. *International Journal of Research in Engineering and Technology, 6*(9).

Ramaiah, V. S., Singh, B., Raju, A. R., Reddy, G. N., Saikumar, K., & Ratnayake, D. (2021, March). Teaching and Learning based 5G cognitive radio application for future application. In *2021 International Conference on Computational Intelligence and Knowledge Economy (ICCIKE)* (pp. 31-36). IEEE.

Ramane, D. V., Patil, S. S., & Shaligram, A. (2015). Detection of NPK Nutrients of Soil Using Fiber Optic Sensor. *International Journal of Research in Advent Technology, 1*(1), 66–70.

Rameshbhai, C. J., & Paulose, J. (2019). Opinion mining on newspaper headlines using SVM and NLP. *Iranian Journal of Electrical and Computer Engineering, 9*(3), 12.

Rap, S., & Blonder, R. (2017). Thou shall not try to speak in the Facebook language: Students' perspectives regarding using Facebook for chemistry learning. *Computers & Education, 114*, 69–78. doi:10.1016/j.compedu.2017.06.014

Rashid, A. (2016). LED Based Soil Spectroscopy. *Buletin Optik, 3*(1), 1–7.

Rasi, P., Hautakangas, M., & Väyrynen, S. (2015). Designing culturally inclusive affordance networks into the curriculum. *Teaching in Higher Education, 20*(2), 131–142. doi:10.1080/13562517.2014.957268

Raturi, G., Rani, P., Madan, S., & Dosanjh, S. (2019). ADoCW: An Automated method for Detection of Concealed Weapon. *Fifth International Conference on Image Information Processing (ICIIP)*, 181-186. doi:10.1109/CJMW.2008.4772439

Redecker, C. (2017). *European Framework for the Digital Competence of Educators: DigCompEdu*. doi:10.2760/159770

Reyaee, S., & Ahmed, A. (2015). Growth Pattern of Social Media Usage in Arab Gulf States: An Analytical Study. *Social Networking, 4*(2), 23.

Ritter, N., & Cooper, J. (2007). Segmentation and border identification of cells in images of peripheral blood smear slides. *30th Australasian Computer Science Conference Conference in Research and Practice in Information Technology*, 161-169.

Robertson, M., & Swan, J. (1998). Modes of Organizing in an Expert Consultancy: A Case Study of Knowledge, Power and Egos. *Organization, 5*(4), 543–564. doi:10.1177/135050849854006

Rosen, J., Polyzotis, N., Borkar, V., Bu, Y., Carey, M.J., Weimer, M.T., & Ramakrishnan, R. (2013). *Iterative mapreduce for large scale machine learning*. Academic Press.

Rosenberg, J. M., Reid, J. W., Dyer, E. B. J., Koehler, M., Fischer, C., & McKenna, T. J. (2020). Idle chatter or compelling conversation? The potential of the social media-based #NGSSchat network for supporting science education reform efforts. *Journal of Research in Science Teaching*, *57*(9), 1322–1355. doi:10.1002/tea.21660

Roseth, C. J., Johnson, D. W., & Johnson, R. T. (2008). Promoting early adolescents' achievement and peer relationships: The effects of cooperative, competitive, and individualistic goal structures. *Psychological Bulletin*, *134*(2), 223.

Rossi, E. (2002). *Uses & Gratifications/Dependency Theory*. Retrieved on June 15, 2014 from: http://zimmer.csufresno.edu/~johnca/spch100/7-4-uses.htm

Rothaermel, F. T., & Sugiyama, S. (2001). Virtual internet communities and commercial success: Individual and community-level theory grounded in the atypical case of TimeZone. com. *Journal of Management*, *27*(3), 297–312. doi:10.1177/014920630102700305

Roychoudhury, B., & Srivastava, D. K. (2020). Words are important: A textual content based identity resolution scheme across multiple online social networks. *Knowledge-Based Systems*, 17.

Rzheuskyi, A., Matsuik, H., Veretenikova, N., & Vaskiv, R. (2020). Selective Dissemination of Information – Technology of Information Support of Scientific Research. *Advances in Intelligent Systems and Computing.*, *871*, 235–245.

Sabit, H., Al-Anbuky, A., & GholamHosseini, H. (2011). Wireless Sensor Network Based Wildfire Hazard Prediction System Modeling. *Procedia Computer Science*, *5*, 106–114. doi:10.1016/j.procs.2011.07.016

Sadhasivam, J., Arun, M., Deepa, R., Muthukumaran, V., Kumar, R. L., & Kumar, R. P. (2021, July). Forex exchange using big data analytics. *Journal of Physics: Conference Series*, *1964*(4), 042060. doi:10.1088/1742-6596/1964/4/042060

Sadhasivam, J., Muthukumaran, V., Raja, J. T., Vinothkumar, V., Deepa, R., & Nivedita, V. (2021, July). Applying data mining technique to predict trends in air pollution in Mumbai. [). IOP Publishing.]. *Journal of Physics: Conference Series*, *1964*(4), 042055.

Saide, S., Inrajit, R. E., Trialih, R., Ramadhani, S., & Najamuddin, N. (2019). A theoretical and empirical validation of information technology and path-goal leadership on knowledge creation in university Leaders support and social media trend. *Journal of Science and Technology Policy Management.*, *10*(3), 551–568.

Saikumar, K., & Rajesh, V. (2020). A novel implementation heart diagnosis system based on random forest machine learning technique. *International Journal of Pharmaceutical Research.*, *12*, 3904–3916.

Saikumar, K., & Rajesh, V. (2020). Coronary blockage of artery for Heart diagnosis with DT Artificial Intelligence Algorithm. *International Journal of Research in Pharmaceutical Sciences.*, *11*(1), 471–479. doi:10.26452/ijrps.v11i1.1844

Saikumar, K., & Rajesh, V. (2020). Diagnosis of coronary blockage of artery using MRI/CTA images through adaptive random forest optimization. *Journal of Critical Reviews*, *7*(14), 591–600.

Salehi, M., & Hossain, E. (2020). Handover Rate and Sojourn Time Analysis in Mobile Drone-Assisted Cellular Networks. *IEEE Wireless Communications Letters*, *10*(2), 392–395.

Sallawar, N., & Yende, S. (2017). Automatic attendance system by using face recognition. *International Research Journal of Engineering and Technology.*, *4*(1), 2156–2159.

Salloum, S. A., Al-Emran, M., Abdel Monem, A., & Shaalan, K. (2017). A Survey of Text Mining in Social Media: Facebook and Twitter Perspectives. *Advances in Science, Technology and Engineering Systems Journal*, *2*(1), 7.

Santoro, Lampinen, Mathewson, Lillicrap, & Raposo. (2021). Symbolic Behaviour in Artificial Intelligence. *Deep Mind, 1*(1), 1-24.

Sarku, R., Van Slobbe, E., Termeer, K., Kranjac-Berisavljevic, G., & Dewulf, A. (2022). Usability of weather information services for decision-making in farming: Evidence from the Ada East District, Ghana. *Climate Services, 25*, 100275. doi:10.1016/j.cliser.2021.100275

Sarti, D., Torre, T., & Pirani, E. (2020). Information and Communication Technologies Usage for Professional Purposes, Work Changes and Job Satisfaction. Some Insights from Europe. *Lecture Notes in Information Systems and Organisation., 33*, 165–177.

Saunders, M., Lewis, P., & Thorhill, A. (2016). Research Methods for Business Students (7th ed.). Pearson Education Limited.

Savelyev, T., Zhuge, X., & Yang, B. (2010). Development of UWB Microwave Array Radar for Concealed Weapon Detection. *11th International Radar Symposium*, 1-4.

Sayed, M., Salem, R., & Khedr, A. E. (2019). A Survey of Arabic Text Classification Approaches. *International Journal of Computer Applications in Technology, 95*(3), 236251.

Schmid, R., & Petko, D. (2019). *Does the use of educational technology in personalized learning environments correlate with self-reported digital skills and beliefs of secondary-school students?* doi:10.1016/j.compedu.2019.03.006

Seghouani, N. B., Jipmo, C. N., & Quercini, G. (2019). Determining the interests of social media users: Two approaches. *Information Retrieval Journal., 22*, 129–158.

Selim, G. E., Hemdan, E. E. D., & Shehata, A., & El-Fishawy, N. (2021). An efficient machine learning model for malicious activities recognition in water-based industrial internet of things. *Security and Privacy, 4*(3), e154.

Senthilkumar, S. A., Rai, B. K., Meshram, A. A., Gunasekaran, A., & Chandrakumarmangalam, S. (2018). Big data in healthcare management: A review of literature. *American Journal of Theoretical and Applied Business, 4*(2), 57–69. doi:10.11648/j.ajtab.20180402.14

Sethi, P., & Sarangi, S. R. (2017). Internet of Things: Architectures, Protocols, and Applications. *Journal of Electrical and Computer Engineering, 2017*, 1–25. doi:10.1155/2017/9324035

Severin, W., & Tankard, J. (1997). *Communication theories: Origins, methods, and uses in the mass media*. Longman.

Shalini, A., & Jayasuruthi, L., & VinothKumar, V. (2018). Voice recognition robot control using android device. *Journal of Computational and Theoretical Nanoscience, 15*(6-7), 2197–2201.

Shang, W., Zhen, M. J., Hemmati, H., Adams, B., Hassan, A.E., & Martin, P. (2013). Assisting developers of Big Data Analytics Applications when deploying on Hadoop clouds. *Software Engineering (ICSE), 35th International Conference*.

Shang, W., Zhen, M. J., Hemmati, H., Adams, B., Hassan, A.E., & Martin, P. (2013). Assisting developers of Big Data Analytics Applications when deploying on Hadoop clouds. *Software Engineering (ICSE), 35th International Conference*. hipi.cs.virginia.edu/ 2014/

Sharma & Batra. (2014). Analysis of distance measures in content based image retrieval. *Global Journal of Computer Science and Technology*, 28-32.

Sharma, D., & Jinwala, D. (2015). Functional Encryption in IoT E-Health Care System. *Information Systems Security*, 345–363. doi:10.1007/978-3-319-26961-0_21

Sharma, G., Pandey, N., Hussain, I., & Kathri, S. K. (2018). Design of framework and analysis of Internet of things at data link layer. *2nd International Conference on Telecommunication and Networks, TEL-NET 2017,* 1–4. 10.1109/TEL-NET.2017.8343520

Sheen, D. M. (2006). Cylindrical millimeter-wave imaging technique and applications - art. no. 62110A. *Proceedings of SPIE - The International Society for Optical Engineering,* 6211.

Sheth, N., Newman, B. I., & Gross, B. L. (1991). Why we buy what we buy: A theory of consumption values. *Journal of Business Research, 22*(2), 159–170. doi:10.1016/0148-2963(91)90050-8

Shilo, S., Rossman, H., & Segal, E. (2020). Axes of a revolution: Challenges and promises of big data in healthcare. *Nature Medicine, 26*(1), 29–38. doi:10.103841591-019-0727-5 PMID:31932803

Shirky, C. (2011). The Political Power of Social Media: Technology, the Public Sphere, and Political Change. *Foreign Affairs, 90*(1), 28–41.

Shojafar, M., Cordeschi, N., & Baccarelli, E. (2016). Energy-efficient adaptive resource management for real-time vehicular cloud services. *IEEE Transactions on Cloud computing, 7*(1), 196-209.

Sideri, M., Kitsiou, A., Filippopoulou, A., Kalloniatis, C., & Gritzalis, S. (2019). E-Governance in educational settings Greek educational organizations leadership's perspectives towards social media usage for participatory decision-making. *Internet Research, 29*(4), 818–845.

Siegel, J., & Perdue, J. (2012). Cloud Services Measures for Global Use: The Service Measurement Index (SMI). *SRII Global Conference (SRII).* 10.1109/SRII.2012.51

Sierra Marín, S. D., Gomez-Vargas, D., Céspedes, N., Múnera, M., Roberti, F., Barria, P., Ramamoorthy, S., Becker, M., Carelli, R., & Cifuentes, C. A. (2021). Expectations and Perceptions of Healthcare Professionals for Robot Deployment in Hospital Environments During the COVID-19 Pandemic. *Frontiers in Robotics and AI, 8*(June), 1–15. doi:10.3389/frobt.2021.612746 PMID:34150856

Simon, C., Brexendorf, T. O., & Fassnacht, M. (2013). Creating online brand experience on Facebook. *Marketing Review St. Gallen, 30,* 10.

Sindhu, A., & Meera, S. (2015). A survey on detecting brain tumorinmri images using image processing techniques. *International Journal of Innovative Research in Computer and Communication Engineering, 3*(1), 16.

Singh, M., & Kim, S. (2017). *Blockchain based intelligent vehicle data sharing framework.* arXiv preprint arXiv:1708.09721.

Singh, C., & Preet Kaur, K. (2016). A fast and efficient image retrieval system based on color and texture features. *Journal of Visual Communication and Image Representation, 41,* 225–238. doi:10.1016/j.jvcir.2016.10.002

Singh, N., & Shaligram, A. (2014). NPK Measurement in Soil and Automatic Soil Fertilizer Dispensing Robot. *International Journal of Engineering Research & Technology (Ahmedabad), 3*(2), 635–637.

Siva Kumar, M., Inthiyaz, S., Venkata Krishna, P., Jyothsna Ravali, C., Veenamadhuri, J., Hanuman Reddy, Y., & Hasane Ahammad, S. (2019). Implementation of most appropriate leakage power techniques in vlsi circuits using nand and nor gates. *International Journal of Innovative Technology and Exploring Engineering, 8*(7), 797–801.

Sivarajah, U., Kamal, M. M., Irani, Z., & Weerakkody, V. (2017). Critical analysis of big data challenges and analytical methods. *Journal of Business Research, 70,* 263–286. doi:10.1016/j.jbusres.2016.08.001

Slavin, R. E. (2014). Cooperative Learning and Academic Achievement: Why Does Groupwork Work? *Anales de Psicología, 30*(3), 785–791.

Smith, S. D., & Caruso, J. B. (2010). *Research Study. ECAR study of undergraduate students and information technology* (Vol. 6). EDUCAUSE Center for Applied Research. https://www.educause.edu/Resources/ECARStudyofUndergraduateStuden/217333

Sommer, R., & Paxson, V. (2010, May). *Outside the closed world: On using machine learning for network intrusion detection. In 2010 IEEE symposium on security and privacy*. IEEE.

Somov, A., Baranov, A., Spirjakin, D., Spirjakin, A., Sleptsov, V., & Passerone, R. (2013). Deployment and evaluation of a wireless sensor network for methane leak detection. *Sensors and Actuators. A, Physical*, *202*, 217–225. doi:10.1016/j.sna.2012.11.047

Song, Y., Alatorre, G., Mandagere, N., & Singh, A. (2013). Storage Mining: Where IT Management Meets Big Data Analytics. *Big Data (BigData Congress), IEEE International Congress*.

Song, Y., Zhang, L., Chen, S., Ni, D., Li, B., Zhou, Y., . . . Wang, T. (2014, August). A deep learning based framework for accurate segmentation of cervical cytoplasm and nuclei. In *2014 36th Annual International Conference of the IEEE Engineering in Medicine and Biology Society* (pp. 2903-2906). IEEE.

Spasojevic, N., Yan, J., Rao, A., & Bhattacharyya, P. (2014). LASTA: Large scale topic assignment on multiple social networks. KDD, 1809–1818.

Sponcil, M., & Gitimu, P. (n.d.). Use of social media by college students: Relationship to communication and self-concept. *Journal of Technology Research*.

Srinivasa Reddy, K., Suneela, B., Inthiyaz, S., Kumar, G. N. S., & Mallikarjuna Reddy, A. (2019). Texture filtration module under stabilization via random forest optimization methodology. *International Journal of Advanced Trends in Computer Science and Engineering*, *8*(3), 458–469. doi:10.30534/ijatcse/2019/20832019

statista.com. (2019). *Active social media penetration in Middle East & North African countries in January 2018*. Retrieved from: https://www.statista.com/statistics/309668/active-social-media-penetration-in-arab-countries/

Steffens, M. S., Dunn, A. G., Wiley, K. E., & Leask, J. (2019). How organisations promoting vaccination respond to misinformation on social media: A qualitative investigation. *BMC Public Health*, *19*(1348), 1–12.

Stenbom, S., Hrastinski, S., & Cleveland-Innes, M. (2016). Emotional presence in a relationship of inquiry: The case of one-to-one online math coaching. *Online Learning Journal*, *20*(1). Advance online publication. doi:10.24059/olj.v20i1.563

Stieglitz, S., & Dang-Xuan, L. (2012). Social media and political communication: A social media analytics framework. *Social Network Analysis and Mining*, *3*(4), 15.

Suchitra, C., & Vandana, C. P. (2016). Internet of Things and Security Issues. *International Journal of Computer Science and Mobile Computing*, *5*(1), 133–139. https://www.ijcsmc.com/docs/papers/January2016/V5I1201636.pdf

Sujithra & Padmavathi, G. (2016, February). Internet of things - An Overview. *UGC Sponsored Two Day National Conference on Internet of Things*, 232–239.

Sultan, N., Khedr, A. E., Idrees, A. M., & Kholeif, S. (2017). Data Mining Approach for Detecting Key Performance Indicators. *Journal of Artificial Intelligence*, *10*(2), 59–65.

Sundar, P. P., Ranjith, D., Karthikeyan, T., Kumar, V. V., & Jeyakumar, B. (2020). Low power area efficient adaptive FIR filter for hearing aids using distributed arithmetic architecture. *International Journal of Speech Technology*, *23*(2), 287–296.

Sun, H., Yang, J., Shen, J., Liang, D., Ning-Zhong, L., & Zhou, H. (2020). TIB-Net: Drone Detection Network With Tiny Iterative Backbone. *IEEE Access: Practical Innovations, Open Solutions*, *8*, 130697–130707.

Sun, X., Ansari, N., & Fierro, R. (2019). Jointly optimized 3D drone mounted base station deployment and user association in drone assisted mobile access networks. *IEEE Transactions on Vehicular Technology*, *69*(2), 2195–2203.

Suyash, Dhabre, & Rahul. (2020). Automated Face Recognition Based Attendance System using LBP Face recognizer. *International Journal of Advance Scientific Research and Engineering Trends.*, *5*(1), 35–40.

Swamy, S. N., & Kota, S. R. (2020). An empirical study on system level aspects of Internet of Things (IoT). *IEEE Access: Practical Innovations, Open Solutions*, *8*, 188082–188134. doi:10.1109/ACCESS.2020.3029847

Swapna, B., Andal, C., Manivannan, S., Jayakrishna, N., & Samba Siva Rao, K. (2020). IoT based light intensity and temperature monitoring system for plants. *Materials Today: Proceedings*, *33*(1), 3409–3412. doi:10.1016/j.matpr.2020.05.269

Swapna, B., & Kamalahasan, M. (2020). Phase Measurement Analysis in Field Programmable Gate Array. *Test Engineering and Management*, *82*, 14225–14230.

Swapna, B., Kamalahsan, M., Sowmiya, S., Konda, S., & SaiZignasa, T. (2019). Design of smart garbage landfill monitoring system using Internet of Things. *IOP Conference Series. Materials Science and Engineering*, *561*(1), 012084. doi:10.1088/1757-899X/561/1/012084

Swapna, Divya, & Bharathi Devi, Sankari, & Pushpamitra. (2020). Soil Wetness Sensor Based AutomaticSprinkling Management System Using IC555. *Journal of Green Engineering*, *10*(1), 1835–1844.

Taha, B., & Shoufan, A. (2019). Machine learning-based drone detection and classification: State-of-the-art in research. *IEEE Access: Practical Innovations, Open Solutions*, *7*, 138669–138682.

Taher, B. H., Liu, H., Abedi, F., Lu, H., Yassin, A. A., & Mohammed, A. J. (2021). A Secure and Lightweight Three-Factor Remote User Authentication Protocol for Future IoT Applications. *Journal of Sensors*, *2021*, 1–18. Advance online publication. doi:10.1155/2021/8871204

Talwar, S., Dhir, A., Kaur, P., Zafar, N., & Alrasheedy, M. (2019). Why do people share fake news? Associations between the dark side of social media use and fake news sharing behavior. *Journal of Retailing and Consumer Services*, *51*, 72–82.

Tandoc, E. C. Jr. (2018). Tell Me Who Your Sources Are. *Journalism Practice*, *13*(2), 14.

Tang, T. A., Mhamdi, L., McLernon, D., Zaidi, S. A. R., & Ghogho, M. (2016, October). Deep learning approach for network intrusion detection in software defined networking. In 2016 international conference on wireless networks and mobile communications (WINCOM) (pp. 258-263). IEEE.

Tankard & James. (2000). *New media theory. Communication theories: origins, methods and uses in the mass media.* Reading. MA: Addison Wesley Longman.

Tan, W., Huang, P., Huang, Z., Qi, Y., & Wang, W. (2017). Three-Dimensional Microwave Imaging for Concealed Weapon Detection Using Range Stacking Technique. *International Journal of Antennas and Propagation*, *2017*, 1–11. doi:10.1155/2017/1480623

Tao, Y., & Grosky, W. I. (1999). Spatial color indexing: a novel approach for content-based image retrieval. *Proceedings IEEE International Conference on Multimedia Computing and Systems*, *1*, 530-535. 10.1109/MMCS.1999.779257

Tarek Kanan, O. S.-d., Kanan, T., Sadaqa, O., Aldajeh, A., Alshwabka, H., Al-Dolime, W., . . . Alia, M. A. (2019). A Review of Natural Language Processing and Machine Learning Tools Used to Analyze Arabic Social Media. *2019 IEEE Jordan International Joint Conference on Electrical Engineering and Information Technology (JEEIT).*

Taylor, K. (2020). How deep learning works for face recognition. *Hitechnectar.*, *1*(1), 1–10.

Team, I. (2019, September 13). *Biggest IoT Security Issues*. Intellectsoft Blog. https://www.intellectsoft.net/blog/biggest-iot-security-issues/

Teece, D. J. (2010). Business models, business strategy and innovation. *Long Range Planning*, *43*(2-3), 172–194. doi:10.1016/j.lrp.2009.07.003

Thenkalvi, B., & Murugavalli, S. (2014). Image retrieval using certain block-based difference of inverse probability and certain block based variation of local correlation coefficients integrated with wavelet moments. *Journal of Computational Science*, *10*(8), 1497–1507. doi:10.3844/jcssp.2014.1497.1507

Thomas, V. L., Chavez, M., Browne, E. N., & Minnis, A. M. (2020). Instagram as a tool for study engagement and community building among adolescents: A social media pilot study. *Digital Health*, *6*. Advance online publication. doi:10.1177/2055207620904548 PMID:32215216

Toet, A. (2003). Color Image Fusion for Concealed Weapon Detection. *Proceedings of SPIE - The International Society for Optical Engineering*, 5071.

Triantafyllou, A., Sarigiannidis, P., & Lagkas, T. D. (2018). Network protocols, schemes, and mechanisms for internet of things (IoT): Features, open challenges, and trends. *Wireless Communications and Mobile Computing*, *2018*, 1–24. Advance online publication. doi:10.1155/2018/5349894

Trust, T., Krutka, D. G., & Carpenter, J. P. (2016). "Together we are better": Professional learning networks for teachers. *Computers & Education*, *102*, 15–34. doi:10.1016/j.compedu.2016.06.007

Tseng & Aiello. (2014). Keynote. *2014 IEEE 7th International Conference on Service-Oriented Computing and Applications*. doi:10.1109/SOCA.2014.63

Tungkasthan, A., Intarasema, S., & Premchaiswadi, W. (2009). Spatial color indexing using acc algorithm. *7th International Conference on ICT and Knowledge Engineering*, 113-117.

Tykkyläinen, S., & Ritala, P. (2020). Business model innovation in social enterprises: An activity system perspective. *Journal of Business Research*.

Tyler, B. B. (2001). Complementarity of cooperative and technological competencies: A resource-based perspective. *Journal of Engineering and Technology Management*, *18*(1), 1–27. doi:10.1016/S0923-4748(00)00031-X

Urooj, U., Al-rimy, B. A. S., Zainal, A., Ghaleb, F. A., & Rassam, M. A. (2022). Ransomware Detection Using the Dynamic Analysis and Machine Learning: A Survey and Research Directions. *Applied Sciences (Basel, Switzerland)*, *12*(1), 172. doi:10.3390/app12010172

Van der Weken, D., Nachtegael, M., & Kerre, E. E. (2002). An overview of similarity measures for images. *IEEE International Conference on Acoustics, Speech and Signal Processing*, *2*, 3317-3320.

Vannoy, S. A., & Palvia, P. (2010). The social influence model of technology adoption. *Communications of the ACM*, *53*(6), 149–153.

Velliangiri, S., Karthikeyan, P., & Vinoth Kumar, V. (2021). Detection of distributed denial of service attack in cloud computing using the optimization-based deep networks. *Journal of Experimental & Theoretical Artificial Intelligence*, *33*(3), 405–424.

Venkataraman, S., Bodzsar, E., Roy, I., AuYoung, A., & Schreiber, R. S. (2013). Presto: Distributed machine learning and graph processing with sparse matrices. *Proceedings of the 8th ACM European Conference on Computer Systems*, 197-210.

Vijaykumar, G., Gantala, A., Gade, M. S. L., Anjaneyulu, P., & Ahammad, S. H. (2017). Microcontroller based heartbeat monitoring and display on PC. *Journal of Advanced Research in Dynamical and Control Systems*, *9*(4), 250–260.

Vinoth Kumar, V., Ramamoorthy, S., Dhilip Kumar, V., Prabu, M., & Balajee, J. M. (2021). Design and Evaluation of Wi-Fi Offloading Mechanism in Heterogeneous Network. International Journal of e-Collaboration, 17(1).

Vivian, R. J. (2012). *Students' Use of Personal Social Network Sites to Support their Learning Experience.* doi:10.13140/RG.2.1.2337.6484

Viviani, M., & Pasi, G. (2017). Credibility in social media: Opinions, news, and health, information—a survey. *Wiley Interdisciplinary Reviews. Data Mining and Knowledge Discovery*, 25.

Vuorikari, R., Punie, Y., Carretero Gomez, S., & Van Den Brande, G. (2016). DigComp 2.0: The Digital Competence Framework for Citizens. Update Phase 1: The Conceptual Reference Model. EUR 27948 EN. Publications Office of the European Union; JRC101254.

Vuorikari, R., Kampylis, P., Scimeca, S., & Punie, Y. (2015). *Scaling Up Teacher Networks Across and Within European Schools: The Case of eTwinning*. Springer. doi:10.1007/978-981-287-537-2_11

Vuran, M., & Silva, A. (2009). Communication through Soil in Wireless Underground Sensor Network: Theory and Practice. *Sensor Networks*, *1*(1), 25–36.

Waheeb Yaqub, O. K., Yaqub, W., Kakhidze, O., Brockman, M. L., Memon, N., & Patil, S. (2020Effects of Credibility Indicators on social Media News Sharing Intent. In *Conference on Human Factors in Computing Machinery-Proceedings* (p. 14). AMC.

Wallenius, J., Dyer, J. S., Fishburn, P. C., Steuer, R. E., Zionts, S., & Deb, K. (2008). Multiple criteria decision making, multiattribute utility theory: Recent accomplishments and what lies ahead. *Management Science*, *54*(7), 1336–1349. doi:10.1287/mnsc.1070.0838

Wanda, P., & Jie, H. J. (2019). URLDeep: Continuous Prediction of Malicious URL with Dynamic Deep Learning in Social Networks. *International Journal of Network Security*, *21*(6), 971–978.

Wang, D., Hu, P., Du, J., Zhou, P., Deng, T., & Hu, M. (2019). Routing and scheduling for hybrid truck-drone collaborative parcel delivery with independent and truck-carried drones. *IEEE Internet of Things Journal*, *6*(6), 10483–10495.

Wang, J., Crawl, D., Altintas, I., Tzoumas, K., & Markl, V. (2013). Comparison of distributed data- parallelization patterns for bigdata analysis: A bioinformatics case study. *Proceedings of the Fourth International Workshop on Data Intensive Computing in the Clouds.*

Wang, P., & Wang, Y. S. (2015). Malware behavioural detection and vaccine development by using a support vector model classifier. *Journal of Computer and System Sciences*, *81*(6), 1012–1026. doi:10.1016/j.jcss.2014.12.014

Wang, X.-W., Cao, Y.-M., & Park, C. (2019). The relationships among community experience, community commitment, brand attitude, and purchase intention in social media. *International Journal of Information Management*, *49*, 475–488.

Wang, Z. (2015). The applications of deep learning on traffic identification. *BlackHat USA*, *24*(11), 1–10.

Waruwu, B. K., Tandoc, E. C., Duffy, A., Kim, N., & Ling, R. (2020). Telling lies together? Sharing news as a form of social authentication. *New Media & Society*, 18.

Weaver, C. P., Lempert, R. J., Brown, C., Hall, J. A., Revell, D., & Sarewitz, D. (2013). Improving the contribution of climate model information to decision making: The value and demands of robust decision frameworks. *Wiley Interdisciplinary Reviews: Climate Change*, *4*(1), 39–60. doi:10.1002/wcc.202

Web, F. X. (2015). *What are data brokers–and what is your data worth?* Digital Marketing. https://www.webfx.com/blog/general/what-are-data-brokers-and-what-is-your-data-worth-infographic/

Weidlich, J., & Bastiaens, T. J. (2019). Designing sociable online learning environments and enhancing social presence: An affordance enrichment approach. *Computers & Education, 142*, 103622. Advance online publication. doi:10.1016/j.compedu.2019.103622

Wesley, R., dos Santos, R. M., dos Santos, W. R., Ramos, R. M., & Paraboni, I. (2020). Computational personality recognition from Facebook text: Psycholinguistic features, words and facets. *New Review of Hypermedia and Multimedia, 25*(4), 21.

West, R., & Turner, L. (2007). Introducing communication theory (4th ed.). New York: McGraw-Hill.

West, R., & Lynn, H. (2010). *Uses and Gratifications Theory. Introducing Communication Theory: Analysis and Application.* McGraw-Hill.

White, T. (2010). *Hadoop: the Definitive Guide* (2nd ed.). O'Reilly Media.

Widjaja, B. T., Sumintapura, I. W., & Yani, A. (2020). Exploring the triangular relationship among information and communication technology, business innovation and organizational performance. *Management Science Letters., 10*, 163–174.

WikiPedia. (2019). *Right to be forgotten.* https://en.wikipedia.org/wiki/Right_to_be_forgotten

Williams, T. D., & Vaidya, N. M. (2005). A compact, low-cost, passive MMW security scanner. *Proceedings of SPIE - The International Society for Optical Engineering*, 5789. 10.1117/12.603662

Wu, J., Sun, H., & Tan, Y. (2013). Social media research: A review. *Journal of Systems Science and Systems Engineering, 23*(3), 26.

Wu, J. J., Chen, Y. H., & Chung, Y. S. (2010). Trust factors influencing virtual community members: A study of transaction communities. *Journal of Business Research, 63*(9/10), 1025–1032. doi:10.1016/j.jbusres.2009.03.022

Xia, L., Luo, D., Zhang, C., & Wu, Z. (2019). A Survey of Topic Models in Text Classification. In *2019 2nd International Conference on Artificial Intelligence and Big Data.* IEEE.

Xiang, L. (2011). *Analysis on architecture of cloud computing based on Hadoop.* Computer Era.

Xiaochun, G., Yiwei, W., & Jingming, Z. (2017). *The Self-assessment in E-learning and Personalized Feedback Education.* doi:10.1145/3175536.3175571

Xu, Z., & Zhao, D. (2012). Research on Clustering Algorithm for Massive Data Based on Hadoop Platform. *Computer Science & Service System (CSSS), 2012 International.*

Xue, Z., Blum, R. S., Liu, Z., & Forsyth, D. S. (2004). Multisensor Concealed Weapon Detection by Using A Multiresolution Mosaic Approach. *IEEE 60th Vehicular Technology Conference, 7*, 4597-4601.

Xue, Z., & Blum, R. S. (2003). Concealed Weapon Detection Using Color Image Fusion. *Sixth International Conference of Information Fusion*, 622-627.

Xu, K., Wang, F., Wang, H., & Yang, B. (2020, February). Detecting Fake News Over Online Social Media via Domain. *Tsinghua Science and Technology, 25*(1).

Xu, M., Chen, H., & Varshney, P. K. (2011). An image fusion approach based on Markov random fields. *IEEE Transactions on Geoscience and Remote Sensing, 49*(12), 5116–5127. doi:10.1109/TGRS.2011.2158607

Xu, Z. Y., Dou, W. B., & Cao, Z. X. (2008). A New Algorithm for Millimeter-Wave Imaging Processing. *2008 China-Japan Joint Microwave Conference*, 337-339.

Yahav, I., Shehory, O., & Schwartz, D. (2015). Comments Mining With TF-IDF: The Inherent Bias and Its Removal. *International Transactions on Knowledge and Data Engineering, 14*(8), 14.

Yang, H.C., Dasdan,A., Hsiao, R.L., & Parker, D.S. (2007). MapReduce merge: simplified relational data processing on large clusters. *Proceedings of the 2007 ACM SIGMOD International Conference on Management of Data*, 1029–1040.

Yang, J., & Blum, R. S. (2006). A region-based image fusion method using the expectation maximization algorithm. *40th Annual Conference on Information Sciences and Systems*, 468-473. 10.1109/CISS.2006.286513

Yazedjian, A., & Kolkhorst, B. B. (2007). Implementing small-group activities in large lecture classes. *College Teaching*, *55*(4), 164–169.

Yeboah-Ofori, A., & Boachie, C. (2019, May). Malware Attack Predictive Analytics in a Cyber Supply Chain Context Using Machine Learning. In *2019 International Conference on Cyber Security and Internet of Things (ICSIoT)* (pp. 66-73). IEEE.

Yedida, R., Reddy, R., Vahi, R., & Jana, R. GV, A., & Kulkarni, D. (2018). *Employee attrition prediction.* arXiv preprint arXiv:1806.10480.

Ye, K., Piao, Y., Zhao, K., & Cui, X. (2021). A Heterogeneous Graph Enhanced LSTM Network for Hog Price Prediction Using Online Discussion. *Agriculture, 11*(4), 359. doi:10.3390/agriculture11040359

Ye, N., Zhu, Y., Wang, R., Malekian, R., & Qiao-min, L. (2014). An Efficient Authentication and Access Control Scheme for Perception Layer of Internet of Things. *Applied Mathematics & Information Sciences*, *8*(4), 1617–1624. doi:10.12785/amis/080416

Yiğit, İ. O., & Shourabizadeh, H. (2017, September). An approach for predicting employee churn by using data mining. In *2017 International Artificial Intelligence and Data Processing Symposium (IDAP)* (pp. 1-4). IEEE.

Yu, W., Xu, G., Chen, Z., & Moulema, P. (2013). A cloud computing-based architecture for cybersecurity situation awareness. *Communications and Network Security, IEEE Conference*, 488–492.

Zarrinkalam, F., Fani, H., Bagheri, E., Kahani, M., & Du, W. (2015). Semantics-enabled user interest detection from twitter. *WI-IAT, 1*, 469–476.

Zhang, J., Chen, R., Fan, X., Guo, Z., Lin, H., Li, J.Y., Lin, W., Zhou, J., & Zhou, L. (2012). Optimizing data shuffling in dataparallel computation by understanding user-defined functions. *Proceedings of the 7th Symposium on Networked Systems Design and Implementation (NSDI).*

Zhang, X., & Ghorbani, A. A. (2019, March). An overview of online fake news: Characterization, detection, and discussion. *Information Processing & Management*, 26.

Zhao, J. (2010). School knowledge management framework and strategies: The new perspective on teacher professional development. *Computers in Human Behavior*, *26*(2), 168–175. doi:10.1016/j.chb.2009.10.009

Zhao, J., Zhang, M., Zhou, Z., Chu, J., & Cao, F. (2016). Convolutional neural networks are used to detect and classify leukocytes automatically. *Medical & Biological Engineering & Computing*, *55*(8), 287–1301.

Zhao, Y., Hryniewicki, M. K., Cheng, F., Fu, B., & Zhu, X. (2018, September). Employee turnover prediction with machine learning: A reliable approach. In *Proceedings of SAI intelligent systems conference* (pp. 737-758). Springer.

Zhou, M., Duan, N., Liu, S., & Shum, H.-Y. (2020). Progress in Neural NLP: Modeling, Learning, and Reasoning. Elsevier Ltd.

Zulkernine, F., Martin, P., Ying, Z., Bauer, M., Gwadry-Sridhar, F., & Aboulnaga, A. (2013). Towards Cloud-Based Analytics-as-a-Service (CLAaaS) for Big Data Analytics in the Cloud. *Big Data (BigData Congress), IEEE International Congress*.

About the Contributors

Jingyuan Zhao is a research fellow at University of Toronto, Canada. She is a also professor at Beijing Union University (China). She obtained her PhD in Management Science and Engineering from University of Science and Technology of China (China) and completed a postdoctoral program in Management of Technology from University of Quebec at Montreal (Canada). Dr. Zhao's expertise is on management of technology innovation, technology strategy, regional innovation systems and global innovation networks, knowledge management, management information systems, and science and technology policy.

V. Vinoth Kumar is an Associate Professor in the Department of Computer Science and Engineering in MVJ College of Engineering, Bangalore, India. He is a highly qualified individual with around 8 years of rich expertise in teaching, entrepreneurship, and research and development with specialization in computer science engineering subjects. He has been a part of various seminars, paper presentations, research paper reviews, and conferences as a convener and a session chair, a guest editor in journals and has co-authored several books and papers in national, international journals and conferences. He is a professional society member for ISTE, IACIST and IAENG. He has published more than 15 articles in National and International journals, 10 articles in conference proceedings and one article in book chapter. He has filed Indian patent in IoT Applications. His Research interest includes Mobile Adhoc Networking and IoT.

* * *

Sampath Dakshina Murthy Achanta graduated with a B.Tech in Electronics and Communication Engineering in 2013 and an M.Tech in Digital Electronics and Communication Engineering in 2015. He is an Assistant Professor at Vignan's Institute of Information Technology's Department of Electronics and Communication Engineering (A). In Guntur, Andhra Pradesh, pursuing a Ph.D. in Electronics and Communication Engineering at Koneru Lakshmaiah Education Foundation Deemed University. Gait analysis in Pattern Recognition, Image & Video Processing, Fuzzy Logic, and Neural Networks are among his major interests. He has 60 articles published in international journals. He has four patents, has 2 books published, and has won 5 international and 5 national awards. He is a Life Member of the I.S.R.S., the IEI, the C.S.I., and the MIET. He is on the editorial boards of Scopus and SCI journals and serves as a reviewer for them. He has a combined teaching and research experience of 5 years.

Nagajyothi Aggala was born in 1982 at Visakhapatnam. She received her B.Tech (ECE) from Nagarjuna University and M.Tech(Radar & Microwave Engineering) from Andhra University College of

Engineering(A). She completed Ph.D in Radar Signal Processing from Andhra University College of Engineering (A). She has a teaching and research experience of 14 years. Presently working as an Associate Professor in the department of ECE, VIIT, Visakhapatnam. She has published papers in various National , International journals and conferences. Her areas of interest include Radars, Antennas and Nano technology.Completed project sanctioned under WoS-A. Completed ECRA –SERB New Delhi with sanction order no. ECR 2017-000256.

Badreya Al-Jenaibi is full Professor in Mass Communication at the United Arab Emirates University. She has a Ph.D. in International Communication and Public Relations (August 2008) from the University of North Dakota in the USA. Her MA (2004) was in Mass Communication is from the University of Northern Iowa in the USA. Her BA (2001) was in Mass Communication from the University of the United Arab Emirates. Her research interests include International Communication, Public relations, organizational communication, the uses and effects of mass media, new media, particularly the international level, as well as public relations and communication. Her doctoral dissertation was bout Press freedom in the Arab world. She is the author of The Scope and Impact of Workplace Diversity in the United Arab Emirates – An Initial Study, The Role of the Public and Employee Relations Department in Increasing Social Support in the Diverse Workplaces of the United Arab Emirates, Gender Issues in the Diversity and Practice of Public Relations in the UAE-Case study of P.R. male managers and female P.R. practitioners, The Use of Social Media in the United Arab Emirates, The Changing Representation of the Arab Woman in Middle East Advertising and Media, Public Relation Practitioners, Independency, and Teamwork in the UAE Organizations.

Radha Raman Chandan is currently working as Associate Professor & Head at the Department of Computer Science & Engineering, Shambhunath Institute of Engineering & Technology, Prayagraj, India. He completed his Ph.D. in Computer Science from the Department of Computer Science, Banaras Hindu University, and M.Tech in Information Technology from the Indian Institute of Information Technology, Allahabad, India. He has 12 years of teaching & training experience. He received an appreciation award under Students Project Grant Scheme by the Department of Science & Technology, Govt. of U.P. He has also received various reputed awards in his profession & academics. He has been invited as a key resource person in various National & International Workshops & Conferences. He has published various research papers in reputed National & International Journals along with he has published & granted national & International Patents to his credit. He is also a reviewer of some reputed journals (SCIE, Scopus indexed). His research interests include IoT, Artificial Intelligence, Machine Learning, Deep Learning, Cloud Computing, and Wireless Ad-Hoc & Sensor networks.

Abhay Chaturvedi received the B.E. degree in Electronics & Communication Engineering and M.E. degree in Microwave Communication& Radar Engineering from Dr. B.R. Ambedkar University, Agra, India in 2001 and 2004 respectively. He is pursuing Ph.D. from the Department of Electronics Engineering, Rajasthan Technical University, Kota, India. He is a member of the Institution of the Electronics and Telecommunication Engineers (IETE), Indian Society for Technical Education (ISTE), The Indian Science Congress Association (ISCA) and Systems Society of India (SSI). He is the author/co-author of forty plus research papers at International/National Journals/Book-chapters/conferences. His area of interest is RF Integrated circuits design for wireless communication systems. He has a total teaching experience of 18 years. Presently he is working as Associate Professor at the Department of Electronics and Communication Engineering, GLA University, Mathura, India.

Prasanna Ranjith Christodoss obtained Ph.D. in Computer Science from Bharathidasan University, Trichy, India in Dec 2017. Received Bachelor, Master of Science and Master of Philosophy in computer science in 1995,1997 & 2004 from Bharathidasan University, Trichy, India. Have more than 23 years of teaching experience at different Colleges and Universities in India, Libya, and Sultanate of Oman. Presently a Faculty & Research Chair, department of Information Technology at University of Technology and Applied Sciences, Shinas, Oman since October 2011. Organized various Workshops, Seminars and symposiums towards professional Research and development. Published many research papers in international journals and conferences of high repute. Hold ample experience in the field of Web Designing having widespread familiarity in ASP.Net (C#) & MS SQL Server. Awarded BEST FACULTY for 3 consecutive years. My research interests include Machine Learning, Deep Learning, Nature Inspired Algorithms, Soft Computing, Parallel Algorithms, Genetic Algorithms and Ad Hoc Networks.

Vijaya Lakshmi Dara received her B.tech in chemical engineering degree from Osmania University, Hyderabad and M.B.A, degree under Bangalore North University, Bengaluru, India and pursing Ph.D., degree in Participatory learning techniques from GITAM University, Bengaluru, India. Her current research interests include latest learning techniques in teaching, IOT, Technologies in Higher Education, Artificial intelligence.

Amira Idrees is a Professor in Data Science, Faculty of Computer and Information Technology, Future University in Egypt. Head of Information Systems Department, Faculty of Computer and Information Technology, Future University in Egypt. Head of University Requirements Unit, Future University in Egypt. Former Vice dean for Community Service and Environment Development and former Head of Scientific Departments, Faculty of Computers and Information, Fayoum University. Research interests include Data Science, Data Analytics, Knowledge Discovery, Text Mining, Opinion Mining, Sentimental Analysis, Cloud Computing, Software Engineering, and Data warehousing.

Rose Bindu Joseph is currently working as an Associate Professor in the Department of Mathematics at Christ Academy Institute for Advanced Studies, Bangalore. She received her Ph.D in Mathematics from VIT University, Vellore in the field of Interval Type-2 Fuzzy Theory. She has qualified NET for lectureship by CSIR-UGC. She holds a Master's degree and bachelor's degree in Mathematics from Mahatma Gandhi University, Kerala. She has more than 15 years of experience in academia and research. She has published more than 15 research papers in Scopus indexed journals and presented papers in many international conferences. Her research interests include fuzzy theory, machine learning, soft computing and artificial intelligence.

Satheesh Kumar received the MCA Post Graduate degree from Anna University and currently pursuing Ph.D in Visvesvaraya Technological University, Karnataka, India. Currently working as an Assistant Professor & Coordinator in Department of Computer Science, School of Applied Sciences, REVA University, Bangalore, India with 12 years of teaching experience and also a consultant advisor in Datalore Labs Pvt.Ltd, Blismos Solutions Pvt.Ltd, Bangalore. He is the author or coauthor of many papers in international refereed journals, Book chapters and many conference contributions. His research interests cover several aspects across Internet of Things (IoT), Network security, Data Mining and Data Security, ML, AI. He has given several invited talks at various institutions. He is the reviewer for many reputed journals.

Kayam Kumar is working as a research scholar in Koneru Lakshmaiah Education Foundation interested in medical imaging and deep learning.

Ajanthaa Lakkshmanan is currently an Research Scholar in the faculty of Engineering and Technology, Annamalai University, India. She had ten years of teaching experience in engineering colleges and university. Her area of interest is Image Processing, Machine Learning, IoT, Artificial Intelligence.

Ayeesha Nasreen M. is currently working as an Assistant Professor in RMD Engineering College, Kavaraipettai, Tamil Nadu. She received her Bachelor's Degree in Electronics and Communication Engineering in 2010 and her Master's degree in Applied Electronics from Anna University, Chennai in 2013. She is doing her part-time PhD under the guidance of Dr Selvi Ravindran, Assistant Professor in the Department of Information Science and Technology Anna University, Chennai. Her research interest includes Wireless Sensor Network, Wireless Body Area Network and Real-time prototype design for IoT applications.

Saradha M. received her B.Sc., degree and M.Sc., degree in Mathematics from Bharathidasan University, Tamil Nadu, India and the Ph.D., degree in Hybrid Fuzzy Fractional Differential Equations from Bharathidasan University, Tamil Nadu, India in 2017. Her current research interests include Fuzzy Sets, Cyber security Analysis, Technologies in Higher Education, Fractional Differential Equations and Error Analysis.

Gowtham Mamidisetti is currently working as Associate Professor in Department of Computer Science and Engineering, BRECW, Hyderabad. He received his Ph.D. in Computer Science and Engineering from Acharya Nagarjuna University, M.Tech in Computer Science from University College of Engineering, JNTU Kakinada and B.Tech in Computer Science and Engineering from JNTU Kakinada. He is GATE Rank holder in GATE 2010. He Published 18 papers in International Journals and presented 4 papers in International Conferences. He is having 2 Patents and 2 Book chapters on his name. He is having more than 8.5 years of teaching experience. He conducted and attended many workshops in different emerging technologies. His research area includes Cryptography, Information Security and Cloud Security.

Thillaiarasu Nadesan is currently working as an Associate Professor in the School of Computing and Information Technology, REVA University, Bengaluru, He has also served as an Assistant Professor at Galgotias University, Greater Noida from July 2019 to December 2020. He worked 7.3 Years as an Assistant Professor in the Department of Computer Science and Engineering, SNS College of Engineering, Coimbatore. Obtained his B.E., in Computer Science and Engineering from Selvam College of Technology in 2010 and received his M.E., in Software Engineering from Anna University Regional Centre, Coimbatore in 2012. He received his Ph.D., Degree from Anna University, Chennai in 2019, he has published more than 22 research papers in refereed, Springer, and IEEE Xplore conferences. he has organized several workshops, summer internships, and expert lectures for students. He has worked as a session chair, conference steering committee member, editorial board member, and reviewer in Springer

Journal and IEEE Conferences. he is an Editor board Member of editing books titled "Machine Learning Methods for Engineering Application Development" Bentham Science. He is also working as an editor for the title, "Cyber Security for Modern Engineering Operations Management: Towards Intelligent Industry", Design Principle, Modernization and Techniques in Artificial Intelligence for IoT: Advance Technologies, Developments, and Challenges" CRC Press Tylor and Francis, His area of interest includes Cloud Computing, Security, IoT, and Machine Learning.

Prabha Nair is working as professor in S. B. Jain Institute of Technology. She has 20 Scopus and UGC research papers.

Rajesh Natarajan completed his PhD in Computer Science from Bharathiar University, Master of Computer Application from Thiruvalluvar University and BSc Computer science from Madras University. He is currently working as a Lecturer at University of Technology and Applied Science, Shinas, Sultanate of Oman. His research interest includes data mining, machine learning, big data analytics, and Privacy preserving algorithms. He published articles in the National conferences, International Indexed journals, including SCI, WoS, Scopus.

Salvatore Nizzolino holds a Degree in Language Teaching and various Masters in the fields of Learning Design, Development of Teaching Programs and Knowledge Management. He serves as an Adjunct Professor for English at the Faculties of Information Engineering / Civil and Industrial Engineering and the Faculty of Economics at "Sapienza" University of Rome (Italy). He serves as a Lecturer and Tutor for the Advanced Course of European Project Management at the Faculty of Economics, "Sapienza" University of Rome. His research interests list Educational Networking, Lifelong Learning, Open Education, EU Projects, and Knowledge Management.

K. Raghu is working as Assistant Professor in Mahatma Gandhi Institute of Technology. Topics of interest are machine learning and WSN.

Saravanan S. received the B.E. degree in Electrical and Electronics Engineering from Arulmigu Kalasalingam College of Engineering, Virudhunagar District, affiliated to Madurai Kamaraj University, Madurai, Tamilnadu, India, in 1999 and M.E. degree in Power Systems Engineering from Thiagarajar College of Engineering, Madurai, affiliated to Anna University, Chennai, Tamil Nadu, India, in 2007. He is awarded doctoral degree from Kalasalingam University, Krishnankoil-626126, Tamilnadu, India, in 2015. He is presently working as a Professor in the Department of Electrical and Electronics Engineering, B V Raju Institute of Technology, Narsapur, Telangana India. Before to that, he worked as an Associate Professor in Department of Electrical and Electronics Engineering, Kalasalingam University, Tamil Nadu. He has published one text book titled Electrical Distribution System, NOVA publication, Delhi, India. He published more than 45 research articles in various renowned international journals/ conferences and five patents. His research interests include the power system load forecasting, Generation Expansion Planning, Applications of soft computing technics in power systems and grid connected renewable energy resources.

Venkatasubramanian S. received a B.E. degree in Electronics and Communication from Bharathidasan University and M.E. degree in Computer science from Regional Engineering College, Trichy.

He has 23 years of teaching experience. He is currently pursuing doctoral research in mobile Ad hoc networks. His areas of interest include mobile networks, Network Security, and Software Engineering. He has published 27 papers in international journals and 10 papers in international conferences. He has also registered three patents. At present, he is working as an Associate Professor in the Department of CSE at Saranathan College of Engineering, Trichy, India. He has also authored books on "Software Engineering" "Fundamentals of Mobile and Pervasive Computing" and "python programming". He is the receiver of the Global teacher award from AKS academy and Dr.Sarvepalli Radhakrishnan lifetime achiever national award.

K. Saikumar is working as professor in Koneru Lakshmaiah Education Foundation. Her research areas are image processing, WSN, machine learning.

D. Usha is currently working as Associate Professor in Department of Computer Science and Engineering, Dr. M.G.R. Educational and Research Institute, Chennai. She has 14 years of teaching experience. She completed her Doctorate of Philosophy in the year 2017 from Hindustan University. She has published more than 25 papers in various Conferences and Journals. Her area of research is Data Analytics and Data Mining. She is a member in IEEE and CSI. She is also playing the role of reviewer in International Journals.

Rajalakshmi V. has completed her Doctor of Philosophy from Vellore Institute of Technology, Vellore in the area of "Rural Women Empowerment through MGNREGS with reference to Vellore District in Tamil Nadu". She possesses four years of teaching cum research associate and one year of research experience. She is currently working as Assistant Professor of Commerce, REVA University, Bangalore. She has published more than 12 papers to her credit in Rural women empowerment, Sustainable development, Financial inclusion, Smart City, E-Learning and other in commerce and management areas at various scopus indexed journals, book chapters, national and international journals, seminars and conferences. She has honoured by Research award as best research article published in Scopus indexed journals of the year (2017-2018) from VIT University, Vellore and young research award from REVA University, Bangalore (2021) respectively. Her area of interest includes Research & Development in Women Empowerment, Human Resource Management and Finance.

Muthukumaran Venkatesan was born in Vellore, Tamilnadu, India, in 1988. He received the B.Sc. degree in Mathematics from the Thiruvalluvar University Serkkadu, Vellore, India, in 2009, and the M. Sc. degrees in Mathematics from the Thiruvalluvar University Serkkadu, Vellore, India, in 2012. The M. Phil. Mathematics from the Thiruvalluvar University Serkkadu, Vellore, India, in 2014 and Ph.D. degrees in Mathematics from the School of Advanced Sciences, Vellore Institute of Technology, Vellore in 2019. At present, he has a working Assistant Professor in the Department of Mathematics, REVA University Bangalore, India. His current research interests include Algebraic cryptography, Fuzzy Image Processing, Machine learning, and Data mining. His current research interests include Fuzzy Algebra, Fuzzy Image Processing, Data Mining, and Cryptography. Dr V. Muthukumaran is a Fellow of the International Association for Cryptologic Research (IACR), India; He is a Life Member of the IEEE. He has published more than 30 research articles in peer-reviewed international journals. He also presented 20 papers presented at national and international conferences.

Farah Yasser is a teacher assistant at Giza Institute for Managerial Sciences, Researcher at Faculty of Commerce and Business Administration Business Information Systems Department, Helwan University.

Index

E

F

G

H

I

K

L

M

N

O

Printed in the United States
by Baker & Taylor Publisher Services

Printed in the United States
by Baker & Taylor Publisher Services